BELL

Books by Robert V. Bruce

Lincoln and the Tools of War
1877: Year of Violence
Bell: Alexander Graham Bell and the Conquest of Solitude
The Launching of Modern American Science, 1846–1876

BELL

ALEXANDER GRAHAM BELL AND THE
CONQUEST OF SOLITUDE

ROBERT V. BRUCE

Cornell University Press

ITHACA AND LONDON

Cornell University Press edition first published 1990.
Cornell Paperbacks edition first published 1990.

A portion of this book has appeared in slightly different form in *American Heritage*.
Printed in the United States of America

Library of Congress Cataloging-in-Publication Data

Bruce, Robert V.
  Bell : Alexander Graham Bell and the conquest of solitude / Robert V. Bruce.
    p. cm.
  Reprint. Originally published: Boston : Little, Brown, 1973.
  Includes bibliographical references.
  ISBN 0-8014-9691-8 (pbk. : alk. paper).—ISBN 0-8014-2419-4 (alk. paper)
  1. Bell, Alexander Graham, 1847–1922. 2. Inventors—United States—Biography. I. Title.
TK140.B37B78 1990
621.385'092—dc20
[B]

♾The paper used in this publication meets the minimum requirements of the American National Standard for Permanence of Paper for Printed Library Materials Z39.48–1984.

To the Clan:
Alice and Frank,
Theo,
Marilyn and Wendy,
Debbie,
Connie and John,
Tess and Leah;
and to the memory of my father and mother,
Robert G. and Bernice I. Bruce,
who saw this book begun.

CONTENTS

Contents

PART THREE: AFTER THE TELEPHONE

Acknowledgments

First of all I am indebted to the late Gilbert H. Grosvenor, who consented in 1962 to entrust me with writing the long-postponed biography of his father-in-law, Alexander Graham Bell, and accordingly opened to me the indispensable collection of Bell's papers at the National Geographic Society in Washington. When I began work in earnest in 1965, I had the stimulus of dining with Dr. Grosvenor at Beinn Bhreagh and hearing his reminiscences of a quarter century's close relationship with Bell.

From the beginning I have been blessed with the cordial cooperation of Bell's descendants, both Grosvenors and Fairchilds, and in particular with the encouragement and understanding of Melville Bell Grosvenor, Bell's grandson and present literary executor. He and others of the family, especially Dr. Mabel H. Grosvenor, Mr. and Mrs. Joseph M. Jones, Mrs. Leonard Muller, and Mrs. Marston Bates, have contributed their own memories and suggestions on matters of style. But I have been given a completely free hand in what I examined, what I used, how I used it, and what I concluded from it.

I shall never think of my months in the Bell Room of the National Geographic Society without grateful memories of its longtime custodian and archivist, Mrs. Robert S. Cassilly — of her conviviality, her contagious enthusiasm for the Bell story, and her phenomenal knowledge of that vast archive. The Geographic family in general seems to have inherited the good-fellowship of its founders. Leonard Grant, Keilor Bentley, and Reed Dubois in particular were helpful and hospitable. In the latter stages of readying the book for press, I have depended on the unstinted help of Betty Tinley and Robert S. Patton.

In Georgetown, at the Alexander Graham Bell Association for the Deaf, I profited from the assistance and advice of its director, George Fellendorf, and its Volta Bureau Librarian, William E. Cwiklo (who besides being a fine librarian is, like Bell and me, a movie buff and therefore welcome company for an out-of-town scholar). Dr. Bernard S. Finn of the Smithsonian Institution read the first seventeen chapters and gave me expert advice and general encouragement. In New York, Lewis S. Gum of the American Telephone and Telegraph Company helped make my research in the Historical Collection enjoyable and rewarding. In Montreal, Robert H. Spencer, historian of Canadian Bell, was both guide and companion, and since then has been a sort of volunteer press agent for this book.

In England, the staffs of the British Museum and the Bath Public Library smoothed my way. In Edinburgh, those of the Edinburgh Public Library, the Scottish National Library, and the Scottish Record Office managed to come up with material that had escaped Bell's own searches; Jack Brewster, Property Secretary of the Scottish Life Assurance Company, took me through Bell's boyhood home, now owned by the Company, and even unearthed its original interior plan; and George Shepperson of the University of Edinburgh made me feel at home. So did everyone I met at Elgin, including Kenneth Wood and (no relation) James Wood, the staff of the Elgin Public Library, and the hotel manager who fortified me with Elgin lore and Scotch whisky on the house (in which dispensation the circumstance of my name may have entered).

Closer to home, I am grateful to my students and colleagues at Boston University. Rosario Tosiello did me a great service by writing an outstanding dissertation on the early history of the Bell System. My graduate assistant George Wise, with a grant from Boston University and the background of a master's degree in electrical physics, examined Bell's 210 laboratory notebooks page by page, painstakingly abstracted them for me, called my attention to many points of special interest in them, and formulated several persuasive insights into Bell's technical and scientific characteristics. Howard Gotlieb, Director of Special Collections of the Boston University Library, has for several years made a point of acquiring Bell manuscripts and other related material. Wilbert L. Pronovost, Professor of Communicative Disorders and Coordinator of the Deaf Education Section in the School of Education, in 1949 became Bell's successor (after a lapse of seventy years) as teacher of what Bell called "vocal physiology." He has read and commented on Chapter XXIX. Among my colleagues in the Department of History, David Hall read much of the manuscript and offered suggestions as to style. Saul Engelbourg, Arnold Offner, John Gagliardo, and others, not least our departmental secretary Ann Mechling, have given me good cheer as the book grew. Our genial chairman, Sidney Burrell, has shown such steadfast faith in its eventual completion that at last I came to believe it myself.

Acknowledgments

At Little, Brown, my editor, Llewellyn Howland III, has been acute yet gentle in his suggestions, patient with the delay of promised installments, and infallibly buoying to my spirits.

Through it all, my sister, Mrs. Wendell J. Greene, has shown patience and a talent for cryptography in retyping drafts, and with her husband, my uncle and aunt Mr. and Mrs. Frank Ritchie, my niece Mrs. John W. Fenner and her husband, and my niece Deborah Greene, has given me the same refuge and restoration that Alexander Graham Bell found in his own clan.

BELL

Prelude
Philadelphia, 1846

This is the story of how a young man came to make a great invention and what it then made of him.

It is also the story of how, troubled all his life by the timeless and universal human need to communicate, he met that need for himself, and meanwhile threw the weight of his ability and eminence against barriers that held millions apart from their fellowmen.

The wonder of his greatest creation is in its simplicity. All the science it required can be set forth in a single page. The mystery is in its long delay, for its two basic principles had been discovered many years before.

The first discovery was electromagnetism: the fact that an electric current generates a magnetic field about itself; the stronger the current, the stronger the field. So long as current is passing through a coil of wire, therefore, it magnetizes an iron core within the coil. The pull of such an electromagnet makes a telegraph sounder click when the circuit is closed by a telegraph key. It seems obvious — by hindsight — that if the current, instead of being made and broken by a key, were varied in strength with sufficient rapidity and subtlety, the electromagnet could vibrate a flexible iron diaphragm and create any kind of sound, including that of the human voice.

What could vary a current so quickly and subtly? Sound itself. Sound is a series of mechanical vibrations. These could act on the circuit through an element whose electrical resistance is varied by pressure or movement. The current in the circuit would thus be correspondingly varied, as if by a delicate valve.

The second discovery, induction, suggests another way of using sound to vary a current. Induction means that a *changing* magnetic field — chang-

3

ing in either strength or relative position — generates a current in a circuit. So if the sound of the voice vibrated an iron diaphragm in a magnetic field, the resulting changes in the field would induce a similarly varying current.

The trouble with such an electromagnetic transmitter is that it would have to generate the current changes from the limited power of the sound itself, whereas the variable-resistance type would modulate an existing current, which might be made as strong as desired. Nevertheless, the electromagnetic transmitter turned out to be a stepping-stone to the final achievement.

It follows that Hans Oersted, the Danish scientist who in 1820 awakened the scientific world to electromagnetism, knew enough basic science to have invented the variable-resistance telephone. Yet he did not invent it. In 1831 the English scientist Michael Faraday published his discovery of induction. The American scientist Joseph Henry discovered induction independently of Faraday and probably a few months before him, though Henry forfeited credit by delaying publication. Either of them could thereupon have invented the electromagnetic telephone. And yet forty-five years more were to pass before anyone invented either type. Why?

This is a recurring riddle of technological development. We are not likely ever to unravel it completely. Its answer lies deep in the workings of the human mind. But we cannot help trying. After all, the timing of technology has set the pace of modern history.

We may at least get nearer the heart of the problem by following the process through which some grand technological potentiality was at last realized. There could scarcely be a better example for study than the invention of the telephone, since the evolution of no other invention has been more fully and critically recorded. And so the first object of this book is to order and analyze that record. The second object is to understand the inventor by examining his life before and after the invention.

One element in the story is a man named Faber, who would indirectly inspire the inventor to his first attempt at mechanical reproduction of the human voice. Another is Joseph Henry, who at a crucial moment would rally the inventor to persist in applying electricity to the problem. A third is the city of Philadelphia, from which the news of success would ring out to the world. By one of history's numerous ironies, all three of these elements came together in 1846. And to them was added a blurred vision of the great goal itself. The incident shows how loud and close the knock of opportunity can come without awakening the strangely spellbound mind. And it thereby shows the greatness of first conceiving what in retrospect seems obvious.

Professor Joseph Henry of Princeton, the foremost American scientist of his day and a world leader in the study of electromagnetism and induction, visited Philadelphia in May 1846. While there he examined by invitation a mechanical "speaking figure" invented and exhibited for profit by a German named Joseph Faber. Henry expected to find a hoax, but there was no ventriloquist nor concealed speaking tube. Carved to look like a Turk sitting on a table, the wooden figure had a hinged jaw, an artificial tongue, and a rubber glottis or membrane with an opening through which air was forced. The artificial speech organs were worked by strings and levers controlled by keys like those of a piano. "I have seen the speaking figure of Mr. Wheatstone of London," wrote Henry to a friend, "but it cannot be compared with this, which, instead of uttering a few words, is capable of speaking whole sentences, composed of any words whatever."

But Henry saw more than entertainment in the device. "The keys," he wrote, "could be worked by means of electromagnetic magnets and with a little contrivance, not difficult to execute, words might be spoken at one end of the telegraph line, which had their origin at the other. . . . Thus if an image of the kind were placed in the pulpit of several churches . . . the same sermon might be delivered at the same minute to all. Or — but I will leave your own invention to make other applications."

To transmit speech by using mechanical power to regulate electrical impulses, which in turn would cause a sort of diaphragm to vibrate! The telephone was calling through a mist to the one man in 1846 who could have brought it forth if anyone could have. Yet Joseph Henry never did conceive the simple plan.

Indeed, not even the ultimate inventor of it had yet been conceived — at least not quite yet.

PART ONE

Before the Telephone

1

The Inheritance

He was born in Edinburgh, Scotland, on March 3, 1847, and christened Alexander Bell.

The time and place sounded the leading themes of his life. The telegraph had not yet reached Edinburgh, but it was on its way. "When the lines of communication are once fully established," remarked an Edinburgh newspaper on March 5, "a question asked in London will be received and answered by a person in Edinburgh or Inverness, in little more time than the words could be written on paper."

The editor naturally assumed that London would be asking and Edinburgh answering. With good warrant Edinburgh called itself the "Athens of the North" for its literature, its philosophy, and its science, in which last it surpassed London. Scientific notables spangled the University of Edinburgh faculty in the eighteenth and early nineteenth centuries, especially in medicine and geology, but also in chemistry and mathematics. They led in organizing the British Association for the Advancement of Science in 1831.

An extraordinary number of Edinburgh men had contributed to technology also. From men of that vicinity had come Nasmyth's steam hammer, Fairbairn's iron steamship, Ferguson's breech-loading rifle, and MacMillan's pedal bicycle. Edinburgh had nurtured Thomas Telford, the greatest civil engineer of his generation. The father, uncle, and grandfather of Edinburgh's Robert Louis Stevenson had all contributed to the technology of lighthouses and beacons.

And Edinburgh events suggested still another major element in the history of the newborn Bell, as well as of his ancestors and descendants: the power of the written and spoken word. On the day after the child was born, *Rob Roy* was presented at the Edinburgh Theatre-Royal. This dramatic version of Sir Walter Scott's novel, the most popular play in the

Scottish theater, had earlier been the setting for the Bell family's entrance on the stage of history, made by the child's grandfather.

Here is the cue for an essential scene in the drama of the child's life. For, much more than usual, his life work grew out of what his father and grandfather had done before him. And his character and ideas were deeply influenced by those two brilliant, charming, ambitious Scotsmen, both also named Alexander Bell. Biographies often begin with a pious sketch of family history. In this case, family history is a matter of neither pride nor preface. It is the start of the story proper.

The first Alexander Bell was born in St. Andrews on March 3, 1790, and became a shoemaker like generations of Bells before him. Not until after he married thirty-one-year-old Elizabeth Colville in 1814 did he put away his last and take to the stage.

His father-in-law may have set him an example of ambition and nerve. Andrew Colville, having studied medicine for a time, doctored his neighbors in Gauldry, Fifeshire, without complaint by those who survived. He was also skillful in wood turning and made good bagpipes. And he had the pluck to wage a long, futile legal battle on dubious grounds for succession to the title and estates of Lord Colville. The old man's mettle survived his defeat, and a visitor found Andrew Colville at ninety repairing the roof of his workshop.

The records of St. Andrews in 1817 referred to the first Alexander Bell as a comedian. He had a strong, regular face, square of jaw and broad of nose, perhaps too rugged to be called handsome but certainly not comic. A rough Scottish burr in his speech may have cast him. James Dibdin's *Annals of the Edinburgh Stage* makes much of the first performance of *Rob Roy* at Edinburgh in January 1818 and quotes a review that found the role of "Andrew Fairservice, spoken by Mr. Bell in the Scotch dialect" to be "very amusing." Bell played broad Scots characters and other parts for several years in Edinburgh. Meanwhile he and his wife pieced out their income by keeping a tavern. But by 1822 Bell's theatrical career had dwindled to that of a supernumerary in crowd scenes and to what he later called "the responsible and honorable situation of Prompter."

He may have been in the prompter's box late in 1821 for a play about the Abbé de l'Épée, eighteenth-century deviser and teacher of the manual sign language widely used by the deaf to this day. If so, the prompter would have heard the stage Abbé say, "The bare thought of prompting to the forgetfulness of nature — of calling forth the faculties of mind — gives strength, courage, and perseverance, to accomplish miracles."

The first Alexander Bell needed those qualities. His family grew by the birth of David Charles in 1817, Alexander Melville on March 1, 1819, and Elizabeth in 1822. Meanwhile his theater earnings fell off, and the tavern passed from his hands. Soon after 1822 Alexander Bell and his family quit Edinburgh for St. Andrews.

The retreat was no surrender. The prompter's box had been a student's stall. Bell had learned to speak English well. Instead of cobbling shoes, he began giving speech lessons to the children of St. Andrews's better sort.

Endowed with ruddy vigor, fine physical presence, a resonant and expressive voice, and the intelligence to use his gifts well, Bell won high praise from parents and from the St. Andrews Grammar School, which hired him as a boarding English master. His life onstage continued with occasional public recitations in the town hall. In 1826 Bell moved to the larger town of Dundee. By day he conducted classes in elocution; his evenings were spent tutoring in Dundee homes. More testimonials accumulated. The correction of stammering especially caught his interest.

Then his new prosperity, his self-esteem, his acceptance as a "gentleman," all went glimmering. His wife Elizabeth, to be sure, now enjoyed the luxuries of a carriage, servants, and a group portrait of herself and the children in the trappings of affluence. But in 1827, as time hustled her into middle age, she began a liaison with William Murray, rector of Dundee Academy. The lovers grew reckless. "The whole town of Dundee," said counsel later, "seems to have been aware of his wife's frailty, before the light flashed on the unfortunate husband."

The light flashed with the melodramatic quality that so often follows actors from stage to private life. Feeling ill just after Christmas 1829, Bell let his wife take a letter to a firm of Edinburgh publishers to whom he was offering a speech textbook. In Edinburgh the lovers met, took lodgings in a house of tarnished repute, and were seen together. Word came to the husband, and he confronted his homecoming wife in a fury. The panic-stricken woman hurried a letter by the guard of the Edinburgh tallyho coach to the returning Murray so that their stories might match, but the guard missed Murray's coach on the road, the letter fell into Bell's hands, and Bell read it. "I shall never betray you," it concluded. Mrs. Bell, in a state of collapse, asked the servant girl Elizabeth Baird to get her some poison, but the girl refused. So divorce rather than suicide ended the marriage in July 1831, after a legal contest that cost Bell more than eight hundred pounds.

During the long ordeal, Elizabeth Baird in the Kirk Session at Dundee proclaimed Bell the father of the illegitimate child she was carrying. Bell is not recorded as admitting paternity of Elizabeth Baird's baby. Moreover the provost of Dundee wrote in 1833 that "Mr. Bell's private character has been irreproachable."

Nonetheless Bell's name brought unsavory events to the mind of Dundee, and parents began taking their children out of his school. The affair distracted him from work, and court attendance devoured his time. Besides the out-of-pocket costs, his troubles reduced his income over three or four years by about eight hundred pounds. His earnings fell to forty pounds a year. His household furniture, "the only property he ever possessed," was sold off to pay his rent. On these grounds he sued William Murray for damages. But a legal technicality weakened Bell's case, and so in the final days of 1833 he accepted a settlement of only three hundred pounds. The money at least helped him to forsake not only Dundee but all of Scotland as well and to begin anew in London.

What the children thought can only be conjectured. A family friend testified in court that their father had been deeply attached to them. Their mother, now past fifty, went home to Gauldry, taking eleven-year-old Elizabeth with her. The fourteen-year-old Alexander Melville Bell, known affectionately as Sandy or more formally as Melville, went with his father to London.

Which one, if either, the sixteen-year-old David Charles Bell went with is not clear. The only clue to his doings is a letter he wrote his mother and sister in September 1837, three months after young Victoria's accession to the throne, when he was twenty. He had left London, perhaps after a visit, to take a job as teacher of English and elocution in a Durhamshire academy. "You will be anxious to know if I have given up the Stage," Davie wrote. "Most likely I will once again 'tempt the angry main.' Of my ultimate success either as a teacher or an actor I have no doubt. My fears, so far as the stage is concerned, arise from its uncertainty and bad name." As for his younger brother, "Melville's visit [with you in] Scotland has improved him very much; at least his good looks obtained him several compliments. He had much need of some fresh air after his long confinement."

Davie's father was by then well established in London. The handbill he issued there in 1834 laid out the course he would follow for the rest of his life:

Elocution and Impediments of Speech. Mr. Bell . . . attends Pupils for all departments of Public Speaking. Stammering, and other Impediments of Speech, completely and permanently removed. Schools attended on advantageous terms.

In London he had at last published his projected book, *The Practical Elo-*

cutionist, which used comma-like symbols to indicate word grouping and emphasis. As for his treatment of stammering, it included training the stammerer in breath management, teaching him how the vocal organs produced various sounds, and taking his "mental state" into account — in modern terms, recognizing the psychological basis of his difficulty. In 1836 Bell brought out a short volume called *Stammering, and other Impediments of Speech*, and in 1837 an edition of the New Testament, marked by his method. By 1838 a newspaper report of one of his public readings referred to him as "the celebrated Professor of Elocution."

Bell had become respected and comfortable again, even though not rich. And presently he took a new wife, the widow of a publisher. The second Mrs. Bell, a plain-featured homebody, gave her husband companionship, square meals, and no trouble to the time of her death some twenty years later.

By then both of "Professor" Bell's talented sons were off and away. Melville's illness, whatever it was, had not entirely yielded to Gauldry's fresh air. Nor had long hours as a London draper's assistant done him any good. So in 1838 his father sent the young man to St. John's, Newfoundland, where there lived a friend of the family.

In the New World, Melville found himself once more bound to commerce, clerking in a shipping house. His employer was "exceedingly kind," and even in the busy season young Bell managed to get up early two or three days a week for a row in the harbor — "the finest exercise in the world," he wrote home. His health revived completely. A watercolor portrait of him not long after (painted, to be sure, by his betrothed) shows fine dark eyes, luxuriant black hair, and a puckishly humorous face, more delicately molded than his father's square-cut visage. Bell pitched in to help get shorter hours for the clerks and shopmen of the town. He organized a class in the study of Shakespeare and directed amateur theatricals; and he treated faults of speech by the Bell method, winning grateful testimonials to his liberation of stammerers from their handicap. It was no wonder that when he left for home in 1842, his friends saw him off with a public dinner and a signed tribute to his "kind and gentlemanly traits."

The Newfoundland interlude started young Alexander Melville Bell on his lifelong career. It also gave him a lasting faith in the New World as a place for restoring lost health.

The twenty-three-year-old Melville rejoined his father in the spring of 1842 with a growing commitment to the field of speech. The drapers of London had to look elsewhere for help. By summer he was teaching speech two or three hours a day. He set about studying all the books he could find on the subject. And when he discovered that none of them covered the field completely, he began a quarter century of original investigation into the workings of the vocal organs.

Thus the tribe of elocutionary Bells increased. And a new round of posterity was under way. David Bell had married, settled in Dublin as a teacher of speech, and produced the first two of an eventual eleven children. Melville approved. "I intend to *get married* as soon as I can meet with a young lady to please me," he informed a friend in the summer of 1842.

An Edinburgh reunion with a shipboard friend in 1843 set Melville on his way to that goal. Through his friend he met an Englishwoman, Eliza Grace Symonds, a miniature-painter who lived with her widowed mother on Dundas Street in the New Town. "It was not exactly a case of love at first sight," wrote Melville to his grandchildren half a century later, "but it was a case of *struck* at first sight. I found the lady very pretty, slim, and delicate looking, and with the sweetest expression I think I ever saw. But she was deaf, and could only hear with the help of an ear-tube. My sympathy was deeply excited. But she was so cheerful under her affliction that sympathy soon turned to admiration."

Eliza Symonds met Melville Bell a day or two after her thirty-fourth birthday. The notice taken of her by this lively, good-looking, accomplished young man, almost ten years younger than herself, must always have seemed to her the best birthday gift of her life and probably the most unlooked-for. Her face did not conform to the soft, smooth Greco-Turkish oval of the 1840s' conventional ideal; the chin was too forthright, the nose too long and thin, the cheekbones too gauntly evident for that. Nature seemed to have cast her as a spinster.

Looking back from happier times Eliza once described her Dover childhood as "uncongenial." That understated it. The male Symondses could find fulfillment in doing their bit for the Empire. The naval career of Eliza's father, Samuel Symonds, spanned fifteen years of war with Napoleonic France. Beginning in 1798 as mate aboard the *Magnificent*, becoming surgeon on the *Osprey* in 1802, he served out the war in the latter capacity on ship after ship, dying in 1818 after a year of retirement. His self-reliant, strong-willed wife Mary raised three sons to imperial service: James, an officer in the Royal Navy; Charles and Edward, upper-level bureaucrats in New South Wales. Eliza, not only a girl, but also walled in by early deafness, had only intelligence and books to widen her world. But her cultivated mind, as well as her sweet temper, helped her win Alexander Melville Bell.

Melville Bell came to visit Edinburgh and stayed twenty-two years. For after outlining his hoped-for career, he had asked Eliza Symonds where she

Alexander Melville Bell, 1868   Eliza Grace Bell

thought he should settle. "There happened to be vacant rooms in the house where she had apartments," he recalled drily many years later, "and I took them." Six months later she accepted his proposal, and they were married in Edinburgh on July 19, 1844.

Few marriages are as long and happy as was that of Melville and Eliza Bell. Eliza firmly believed in the Bible and the tenets of her faith. Melville leaned toward skepticism. But each respected the other's views. Before their marriage Eliza had declared her dislike of smoking. "Yet," wrote Melville, "she came not only to tolerate the offensive thing, but to sit by the hour with book or needle in my smoke-laden study; and sometimes even to light my cigar, by taking the first puff." At the end of it all Melville wrote: "She was so kind, so gentle, so loving that during the fifty-two years of our companionship, I never saw a frown on her sweet face."

Early in 1845 Melville's first publication appeared, a pamphlet called *The Art of Reading*. An advertisement at the end announced confidently that "Mr. Bell undertakes the perfect and permanent eradication of Stammering, of however long standing, and however convulsive in its paroxysms."

Though Melville's earnings as a "Professor of Elocution" were at first small and seemed precarious, they continued unbroken and gradually increased. "We were not accustomed to luxuries," recalled Melville, "and therefore did not miss them." Eliza managed the household thriftily. A few months after their honeymoon the Bells were able to move to the elegant neighborhood of Charlotte Square, renting a second-story flat at 16 South Charlotte Street for twenty-two pounds, eight shillings a year.

At that address in 1845 was born their first child, Melville James. And as earlier noted, there also, on March 3, 1847, was born their second child, Alexander.

2

Aleck in Edinburgh

Charlotte Square still frames its green park with the classical eighteenth-century façades of Robert Adam, its magnificent north side most fully realizing the architect's grand design. Its west side has the domed, high-pillared front of St. George's Church, its east side a vista along George Street to the tall shaft of the Melville monument in St. Andrew Square. The south side, though harmoniously Georgian in style, presents no special feature. But it does have a historical landmark: the building in which Alexander Graham Bell was born.

At the moment of that event, another building in the square was up for sale as an "Excellent Self-contained House . . . with Coach-house and Three-stalled Stable, man-servant's apartment, laundry, and wash-house situated in the back green." The Bells had nothing so grand. They reached their flat through a plain doorway on Charlotte Street rather than the impressively arched and fanlighted entrance on the square proper. But a year or two later they moved to larger accommodations at 13 Hope Street on the southwest corner of the square, with a view of the square and St. George's Church. There in 1848 was born their third and last child, Edward Charles Bell.

Alexander in later life remembered little about the Hope Street home where he lived from his second to his seventh year. A miniature of him painted by his mother early in that period shows a serious little boy with light brown, almost blond hair, dark brown or black eyes, a high forehead, and a somewhat prominent nose. His earliest memory must have been of a family excursion, for it was not of Edinburgh but of a wheat field at Ferny Hill outside the city. The wheat stood taller than he did. He pushed his way into its golden mysteries and sat quietly in its heart for a long time, trying hard to hear the wheat grow. But there were only the sleepy sounds of a summer day. Presently it came to him that he was alone and could not tell which way to go. After searching a while he lay down and cried himself to sleep. He woke to his father's strong voice shouting "Aleck, Aleck!" Running toward it, he came out of the wheat to be caught up in his mother's arms.

It must have been in a darkened room at Hope Street that as a small boy, lying ill with scarlet fever in a big four-poster bed, Aleck saw a hooded woman standing silently at the foot. The eyes in her shadowy face seemed fixed on him. Young as he was, it occurred to him that the strange, dumb figure might be a figment of his illness; and with a mixture of curiosity and courage he crept up to it and touched it. It turned out to be his mother's cloak and cap hanging on the bed post. Nevertheless when he lay down again the eery shape still frightened him. And since it seemed to stare silently at him even after he took it down and laid it on the bed beside him, he dropped the flimsy shield of reason and yelled for help.

All his life Aleck remembered the day he found writing materials at the desk of a family guest and was seized with the desire to write a letter, though he did not yet know his alphabet. Some time later a servant girl laughed when he asked her to mail a scribbled "letter" in an envelope bearing one of the guest's postage stamps. Just then from another room his father's voice summoned him. Melville Bell by his early thirties had put on weight and may already have grown a beard; at any rate his small son looked upon him as something of a Jupiter Tonans. Until that moment Aleck had no consciousness of wrongdoing, but as he came before his father he put the letter uneasily behind his back. This drew the attention of his father, who laughed when he saw the document and asked Aleck where he had got the stamp.

Now Aleck knew his sin. He stammered that his mother had given it to him. "Ask your mother to come here," said the father; but Aleck instead ran upstairs and squeezed behind a large wardrobe in the guest room. From there he heard the hunt for him ranging all over the house. Through a crack in the wardrobe Aleck watched the posse look under the bed, rummage through the hanging clothes, even pull open the drawer at the bottom. He held his breath, sure that they must hear the beating of his heart. But they went away. Hours passed, dinnertime came and went, daylight faded. His father, now fearful that Aleck had somehow got out of the house, several times stood in the hall shouting in his powerful voice, "Aleck, come out and I will forgive you." At last a more thorough search revealed the fugitive, holding so tightly to his refuge that the wardrobe had to be moved out from the wall. He clearly expected retribution, and so down he was carried to his father's study, where the looked-for was administered. The Bell tradition had no place for anticlimax.

In that smoke-filled study Melville Bell had for several years been engaged in more constructive exercise, working and thinking night after night until two in the morning.

As an elocutionist and speech teacher he had the natural advantages of keen hearing and fine discrimination in pitch and tone, as well as the example and training of his father. Over the years, professionals and laymen alike remarked upon the purity, flexibility, and precision of the younger Bell's speech. He used it effectively in lecturing on elocution, with illustrative readings, in Edinburgh, Glasgow, and other Scottish cities and towns. But along with high praise in his press clippings went occasional telltale reproaches to the public for its meager turnout. Bell was said to read Dickens better than did Dickens himself, but the creator drew the crowds.

Perhaps that spurred Bell to be a creator too. In his study he pondered the workings of the human voice from two points of view: elocution and vocal production, especially phonetics. With a book published at Edinburgh in 1849, *A New Elucidation of the Principles of Speech and Elocution*, he began a series of contributions to both the art and the science of speech. Unlike most predecessors, he stressed the training of the ear, reasoning that one must recognize a tone in order to reproduce it. He insisted that intellectual and emotional understanding must precede mechanical skill in speech — that rhythm, for example, should be appropriate, not mere singsong. Speech training, he felt, was not for show but for the art and business of living. And his writings capped the elocutionary movement by ably summing up the best thought of others in the field.

Meanwhile he plunged into phonetics. In his 1849 book he suggested that it should be possible "to reconstruct our alphabet, and furnish it with invariable marks for every appreciable variety of vocal and articulate sound . . . with a natural analogy and consistency which would explain to the eye their organic relations." In short, "a really scientific alphabet could be easily constructed." The only difficulty lay in analyzing the physiological relationships of vocal sounds in fundamental and systematic terms.

But that goal had been the dream of elocutionists for nearly a century, and to reach it would cost Melville Bell fifteen more years of work and thought.

As the years went by he tabulated and compared all the oral sounds he heard, until existing combinations of letters fell short. Then he invented new symbols to help his memory. Gradually, as relations between the elements of speech came into focus, he adapted his symbols to suggest those relations. But a fully developed system still eluded him.

Out of his thought and experience came a shorthand system (which failed to disestablish Pitman) and a series of elocutionary pamphlets and readers, climaxed in 1860 by *The Standard Elocutionist*. In it Melville Bell, "assisted in compilation" by his brother David, summed up his elocutionary ideas briefly and followed with a long anthology of literary pieces arranged for elocution. The book made a lot of money for its publishers, running, it was said, to 168 printings in England alone by 1892; and in the United States about a quarter of a million copies were eventually sold. The Bells, however, somehow got little more than celebrity from it.

Nevertheless Bell's teaching gained from his research, while his lectures and pamphlets fattened his reputation and thus indirectly his bank account. His pupils included teachers, ministers, and the children of titled families. He became a regular lecturer at the University of Edinburgh. So in the fall of 1853, with some help from Mrs. Symonds, he was able to buy the house at 13 South Charlotte Street on the southwest corner of Charlotte Square, facing the side entrance of Aleck's birthplace. The new house was what Aleck ever after thought of as his Edinburgh home.

Bell's tenants occupied the first two floors, while the Bells themselves took the third and fourth, ten rooms in all. The third floor had the parlor and guest room at the front and Melville's study in the middle, looking out on narrow, close-built Rose Street. The nursery on the top floor also looked out on Rose Street to the gray side wall of the house opposite, garnished on the left by a mere sprig of Charlotte Square's green park. At the front of the top floor, however, Aleck's grandmother Mrs. Symonds settled down in a room with a dormer window overlooking Charlotte Square, a window slightly bowed and thereby offering an inquisitive child side glimpses of the Pentland Hills to the south and of the Firth of Forth and the Fife Hills beyond it to the north.

The three boys, Melly, Aleck, and Ted, like their father before them, got their early schooling at home. Eliza Bell taught them all that they would have learned in school and some things more. As to religious instruction, whatever trouble she may have had with the mischievous, quizzical Melly, she imbued Aleck deeply with her piety — "at least," as he later wrote, "until I reached years of discretion." The Bells attended church regularly, though Eliza could hear nothing of the service. The boys hunted up the hymns and Bible texts, and Eliza read along. The father used his shorthand to take down the sermon, which he would read to Eliza afterward. She would also ask the boys about what had been said. Aleck probably answered oftenest, since he was not only the most pious but also had a special ability to bypass his mother's ear tube and communicate in a low voice close to her forehead. She taught him the English double-hand manual

13 South Charlotte Street, Edinburgh, 1967   Aleck and Ted

alphabet so that he could spell out a general conversation to her without repeating it aloud.

Aleck observed the Sabbath more strictly in the dour Scottish tradition than did his father and brothers, who played as lightsomely as on any other day. Aleck's father, an enthusiastic amateur photographer in the pioneer days of the wet-plate process, once wanted to take Aleck's photograph on a Sunday. The small Sabbatarian admonished him that it was "a day for meditation," whereupon the father, laughing, gave his son leave to go and meditate. And Aleck did so, sitting by himself for a long time every Sunday, although in later years he could never remember what he had meditated upon.

His father may have helped to turn Aleck from the way of what Burns called the "unco guid." The elder Bell was one of the first to give public readings from the works of Charles Dickens, an act of daring that brought a church deputation to remonstrate. In answer, Bell changed his church and kept on reading Dickens.

Aleck never followed his mother's lead in drawing or painting, except for

21

an occasional notebook caricature; drawing became Ted's department. But his mother was also a good pianist, able to hear every note and shading by fastening her ear tube to her ear and resting the mouthpiece on the sounding board. She could play Scottish melodies so expressively that listeners could almost hear the words. She taught Melly and Ted to play well, but Aleck excelled them both. He learned to read music at sight with great facility. Music could flood his mind like wine, keeping him sleepless and intoxicated through the night and leaving him with a headache in the morning. Snatches of melody would ring in his mind for days. His mother came to call such seizures his "musical fever."

So evident was Aleck's talent that Mrs. Bell engaged a leading pianist, Monsieur Auguste Benoit Bertini, to give him lessons. Bertini felt that Aleck should make a career of music. In those days romanticism pervaded music as well as art and literature. Edinburgh's ruined abbey at Holyrood Palace had inspired Mendelssohn's Scotch Symphony. Chopin had played in Edinburgh the year after Aleck was born. The ideal of the romantic virtuoso still flourished in the person of Franz Liszt. Bertini reflected it, and the glow lit Aleck Bell's imagination. For some time he dreamed of such a role. Then Bertini died, and the boyhood dream in time passed away also.

Eliza Bell eventually reached the limit to which she could carry the boys' education. In 1857 Aleck and Ted, like Melly before them, entered James Maclaren's Hamilton Place Academy, an ordinary-looking stone building in an undistinguished street. Maclaren was no ordinary schoolmaster, however. In his half century as proprietor and headmaster he became one of Edinburgh's most popular and respected teachers. The Maclarens and Bells, moreover, were close friends, and so were their children. This extension of his horizon could not have been much of a wrench for Aleck.

Perhaps it prompted him to another extension. To be named Alexander Bell was no distinction in his family, so he decided to take a middle name. As he puzzled over possibilities, a young man from Canada, who had been a pupil of Melville Bell's for a couple of years, came to board with the family. Aleck probably took a fancy to the guest and certainly did to his name, which was Alexander Graham. On Aleck's eleventh birthday, March 3, 1858, wine was set on the table, and Alexander Melville Bell asked the company to fill their glasses. He then made a little speech about his second son's past and future and concluded by proposing a toast to the health of "Alexander Graham Bell."

The world of Alexander Graham Bell widened rapidly. Ben Herdman, a boy of Aleck's age who had been sent to Melville Bell for correction of stammering, enlarged Aleck's range northwest by a few minutes' walk to Bell's Mill and the stream called the Water of Leith.

In times unrecorded, as it twisted and turned in its struggle toward the sea, the Water of Leith had cut deep through the rising ground that later bore the New Town. When the city-builders occupied that ground, Thomas Telford's fine bridge carried the invaders over Leith Water's narrow valley and left a rustic enclave. Far below Telford's bridge the village of Dean kept about its business scarcely altered.

As late as 1790 the Water of Leith had powered the heart of the Scottish milling industry. But men and their hunger had since 1790 multiplied far beyond the power of Leith Water to satisfy. Bigger mills had grown up elsewhere, leaving only a half dozen in or near Dean Village in Aleck's time. Bell's Mill, which had its name from long-ago Bells unrelated to those of our concern, now belonged to John Herdman, Ben's father.

For Aleck and Ben, Bell's Mill had inexhaustible resources. Just downstream stood the Bell's Mill Bridge, once the principal route from Edinburgh to the north. Upstream from the mill, overhanging willow trees trailed their fronds in a still, deep pool. From that pool the millrace was led off along the foot of the steep left bank, below garden walls and lines of old thorn trees, leaving a green island in the stream.

Best of all were the mill works themselves, a trio of three-story stone buildings. Meal dust powdered the windows and the plain, solid walls, and floated mistily about the doorways. Inside, meal sacks stood about the dusty floor, a fine covert for hide-and-seek, and mill machinery whirled and shook to its own harsh music. Outside, behind the oldest building, the dark mill wheel turned unendingly under its burden of falling water, and its rumbling mingled with the rush of the millrace and roar of the caldron.

Back at Charlotte Square, Melville and Eliza Bell felt no concern at Aleck's ramblings. Inveterate hikers, they considered city life bad for children and took their own boys on country outings when they could. Melville Bell's growing prosperity permitted him in 1858 to do more. In the parish of Trinity, a pleasant area of fields and country homes a couple of miles northeast, he bought a two-story stucco house called Milton Cottage.

The cottage at Trinity had a large garden, enclosed by a high brick wall and embellished with winding gravel paths, stone urns, and latticework trellises. There was a brick tower two or three stories high, ready to serve as setting for the impromptu dramatics of childhood. From it may have been visible the Firth of Forth. Certainly there were alluring views of fields and roads roundabout for roaming. More unusual, and yet appropriate for the Bell family (perhaps it was Melville's doing), was an old pulpit in the garden, a standing invitation to boyish oratory.

23

The Bells spent at least two days a week at Milton Cottage during their remaining years in Edinburgh. At Charlotte Square the boys had to keep well dressed and well groomed. At Trinity they were paroled to racket about in old clothes and roll in the dirt if they chose. "Milton Cottage at Trinity was my real home in childhood," wrote Aleck in his middle age.

Melly, lively and whimsical, usually led in the boys' joint enterprises. Besides being younger, Aleck was more serious and Ted more mild-mannered. Their father, despite his bearded dignity and developing portliness, would sometimes join in their antics like the biggest brother of all. Other children came often to Trinity: the Maclarens, the three daughters of Eliza's brother James Symonds (who had married Melville's sister Elizabeth, retired from the navy, and settled in Edinburgh), and the Herdmans, two brothers and two sisters. There must have been days of shrill rantings from the garden pulpit and shrieking rides on a big old velocipede when Melville Bell needed his sense of humor and Eliza Bell her deafness.

Maclarens, Herdmans, and Symondses alike fondly remembered the Bell family's talent for entertaining. Melville and Eliza would play duets on flute and piano, Ted singing along in a sweet tenor voice. Aleck with flair and Melly with competence could substitute for their mother at the piano. Melly had extraordinary gifts for mimicry and sleight of hand, both of which supported his taste for practical jokes. In exercising his Bell inheritance, Melly found his element in the comic monologue. Ted had humor too, expressed through his talent for caricature. Aleck's specialty was an imaginary bee-chase, in which he would prowl and dart about almost like a dancer, perfectly imitating the buzz of the fugitive even to its muffled protests in his cupped hands. He showed special ingenuity at acting out charades and devising new games.

Yet Aleck seemed submerged by the ebullience of his father and older brother. His playmates later remembered him as thoughtful, restrained, even shy. His father photographed him at the age of eleven sitting before a vine-covered wall of Milton Cottage. Aleck's inward-slanting brows, apparently knitted in perplexity or suspicion, his slightly aquiline nose, his thin, straight upper lip, his unassertive chin, all suggest intensity and introspection. He has the look of a quick, alert, somewhat delicate boy.

And then again he may simply have had the sun in his eyes. A photograph may suggest mood and character but not certify them. Another photograph, taken at Trinity a few months later, shows Aleck as more robust than before, with stronger features, thicker and blacker hair, more vigor and self-assurance. Annie Herdman recalled Aleck Bell in his early teens as "tall and handsome, with long, black hair, which he had a trick of always throwing back." The photographic record, it is true, does not show any of the Bells as physically handsome or beautiful by conventional measures. But all of them, from every account, were decidedly attractive by force of personality and character.

Aleck often did find solitude congenial. "In boyhood," he wrote a few years later, "I have spent many happy hours lying among the heather on the Scottish hills — breathing in the scenery around me with a quiet delight that is even now pleasant for me to remember." A favorite spot was Corstorphine Hill, two or three miles west of Charlotte Square. There at the grassy viewpoint called "Rest-and-be-thankful" the sky was close, and below him circled the horizon. Detached and at peace, he could watch the world wag. Above him the birds made their mysterious way, and this more than all else stirred Aleck's envy and wonder.

Aleck was not content just to look and listen. He collected botanical specimens from Trinity, Arthur's Seat, Corstorphine Hill, the Water of Leith, even London and Dublin. At Melville Bell's urging, each was identified with proper Latin polysyllables. "That spoiled the whole thing for me," Aleck remembered later. He could work with great energy and endurance toward an object that caught his imagination, but the object had to be pinpointed. If botany required the dreary memorizing of nomenclature to no obvious purpose, then botany was not for him. He moved on to collecting birds' eggs. But experiments in planting one species' egg in another's nest led to a tragedy when a changeling pushed its smaller fellows from the nest. So he took up the study of animals.

Aleck and his brothers stocked Milton Cottage with dogs, cats, white mice, rabbits, guinea pigs, even frogs and toads. Aleck, more than his brothers, saw their pets as subjects of investigation. He had no heart to kill for science's sake; but when chance or nature supplied dead birds, rats, and mice, he explored their insides. On the top floor at Charlotte Square he had a "study" for his collections. Aleck and his friends started a "Society for the Promotion of Fine Arts among Boys," of which each member was an officer or a "professor," Aleck being "professor of anatomy." In one lecture before the "Society," he thrust a dissecting knife into the carcass of a small pig. Trapped air rushed out with the sound of a groan, whereupon the meeting adjourned out the door and down the stairs, led by "Professor" Bell. Thus we see the Edinburgh environment at work on the son of a university teacher.

And in a mid-nineteenth-century city of locomotives and telegraphs, of steamships and mills, the machine age gave him a technological cue. In his father's photographic darkroom at Milton Cottage Aleck and his brothers learned to coat the glass plates with collodion, put them through a silver nitrate bath, expose them properly, and finally develop them. It was at Bell's Mill, however, that he conceived his first invention.

One day when Aleck and Ben Herdman had been more than usually apparent about the mill, Ben's father called them into his office for a talking to. "Now, boys," he concluded, "why don't you do something useful?" When Aleck asked him what they might do, Herdman picked up a handful of grain and remarked casually, "If you could only take the husks off this wheat you would be of some help."

Through Ben Herdman Aleck got some samples and began to experiment. A hand-wielded nail brush seemed to work. How could its action be made automatic? Aleck thought of a machine containing rotating paddles, already used at the mill for some other purpose. If the interior were lined with brushes or some equivalent, perhaps the paddles might force wheat against it so that the husks would be brushed off.

To John Herdman's surprise, his son and Aleck Bell appeared before him with the cleaned sample and an explanation of Aleck's plan. Herdman had it tried. It worked, and for a while was put to use at the mill. "So far as I remember," wrote Aleck many years later, "Mr. Herdman's injunction to do something useful was my first incentive to invention, and the method of cleaning wheat the first fruit."

Annie Herdman later recalled that a mutual friend gave Aleck and Ben the run of a workroom, "where the boys amused themselves trying to invent things and experimenting in various ways." As to the wheat-cleaning invention, Annie wrote, "I do not think it ever came to anything." Hindsight may have shaped her memory of Aleck's interest in "the vibration of wires in a field fence, and transmission of sound by same." And yet, familiar with his mother's hearing tube, Aleck may well have played with such a notion.

Meanwhile his formal education had been proceeding. After a year at Hamilton Place Academy he and Ted together entered the Royal High School, where Melly had already become a student. In 1829 the High School had moved from the Old Town to a building on a shoulder of the Calton Hill. The most rigorously Grecian building in the city, its imposing center was joined to its two "temples" or lodges by lofty Doric colonnades. Behind were two acres of playground, and above that, the crest of Calton Hill itself.

"Of all places for a view," wrote Robert Louis Stevenson, "this Calton Hill is perhaps the best; since you can see the Castle, which you lose from

the Castle, and Arthur's Seat, which you cannot see from Arthur's Seat."
Come back to the hill, he urged his readers, "on some clear, dark, moonless
night, with a ring of frost in the air." There one might find perfect solitude, his only neighbor being the astronomer under the observatory dome,
silent in his search of endless space.

Here one morning in October 1858 came Alexander Graham Bell, aged
eleven, and Edward Charles Bell, aged ten, to sit with their new classmates
on the assembly hall's wooden benches and hear the sonorous injunction
laid annually upon the students:

The Rector and Masters of the High School feel it to be their duty to remind
the ingenuous Youth educated in this venerable Institution, that the moral well-
being of man is paramount even to his intellectual advancement. You are, accordingly, affectionately admonished to seek that fear of the Lord, which is the beginning of wisdom; and while you are not slothful in business, in the days of
your youth to remember your Creator, in the two great departments of duty —
love of God, and love to man.

Thereupon followed the disciplinary rules of the establishment.

At the High School Aleck kept much closer to Ted than to Melly, who
was in a higher class and had a different circle of friends. Aleck's mother
had impressed him with the wickedness of fisticuffs. Other boys found this
out. One slapped Aleck's face, was dutifully presented with the other cheek,
and slapped that too. Having run out of cheeks, Aleck offered the ag-
gressor a fist, and the pair went at it until a master stopped the match.
Thereafter Aleck had no more fights on his own account, though he occasionally picked up the gauntlet in Ted's behalf.

High School boys had other outlets for their energies. The Calton Hill
invited them to climb its slopes and dodge about its monuments. Saturday
mornings meant gymnastics. In those years also, almost every High School
boy came and went with a wooden bat or racket called a "clackan" hung
from his wrist, ready for a game of "shinty" or "hails." In 1860, during
Aleck's second year there, the High School got permission for its boys "to
play at cricket in Holyrood Park." This marked the beginning of modern
athletics at the school.

For compulsory learning Aleck felt no craving. Everyone had to take
Latin, English language and literature, history, geography, and elements of
science. Those in the Classical Department also took Greek; those in the
Modern Department might substitute for it one or more other courses.
Aleck's fate was classical, though he hated Latin and Greek. He took
mathematics and enjoyed the intellectual exercise, but was bored and hence

careless in working out the final answer once he knew the method. Natural history and chemistry had been introduced a dozen years before, but despite Aleck's extracurricular interests, he apparently did not take them.

A boy's first four years were customarily spent under one master, the remaining two in the Rector's class. Aleck's four years were with James Donaldson, a fine teacher·who later became rector of the High School and eventually president of St. Andrews University. Not even Donaldson could awaken Aleck's interest in formal studies. Aleck did well in reading or declaiming; he spelled fairly well; his bent for doggerel verse drew some notice. Otherwise he was academically undistinguished, or worse. Melly won first prize in his class for recitation in 1858, a special premium for reading and recitation in 1860, and second prize in French pronunciation in 1862. Neither Aleck nor Ted ever made the prize lists.

Thus the fifties drew to a close and the sixties commenced. Victoria's reign moved toward high noon, and reflections of its brilliance flashed from the arms of Highlanders in the Castle garrison. Aleck remembered only faintly the stir of the Crimean War, his mother's emotional accounts of Florence Nightingale, the distant shout of "Sebastopol is down." A sharper impression remained of the illuminations, the guns, the pipes, and the cheering when Edinburgh in 1858 welcomed home Colin Campbell's Highlanders from the relief of beleaguered Lucknow. That year, his first in the High School, Aleck saluted his sovereign in verse:

> Victoria, Queen Victoria,
> She rules a mighty band,
> Who'll stand by her for ever
> To guard their Native Land.

Of Melville Bell's three sons, the eldest seemed most promising. Melly finished the full six-year High School course. As to Aleck there is doubt. Many years later he wrote, after a frank account of his scholastic deficiencies, "I passed through the whole curriculum of the Royal High School, from the lowest to the highest class, and graduated, but by no means with honors, when I was about fourteen years of age." Perhaps he had come to think of the four-year course as "the whole curriculum." Perhaps he let the wish stand for the fact in his memory. Perhaps the supposed date of his first enrollment is wrong. The fact remains that in the fall of 1862, at the age of fifteen, Alexander Graham Bell left Edinburgh for London and a new way of life.

The call had come from his Grandfather Bell.

3

Pygmalion in Harrington Square

Grandfather Bell had once dreamed of riches. When Melville came back from Newfoundland in 1842, the father and son announced a course of "National and Simultaneous Eloquence" for five hundred pupils. But the mumbling masses did not rise to the promise of "harmonious modulation of the voice, perfect ease and gracefulness of delivery." The elder Bell had to fall back on correction of stammering and other impediments, not a matter for group instruction.

In the twenty years since Melville had moved to Edinburgh, Grandfather Bell had kept on teaching the English how to speak and had made a respectable though not a lavish living at it. He also made an occasional sally into the provinces for short speech courses and lectures; and in London he lectured on "National Education," on "Humbug," on "Parliament and the Social Order," as well as giving "Morning Shakesperian Readings." For a while he taught elocution and English literature in a school at Red Lion Square, and later in Cavendish College for the Instruction of Ladies in Languages, Arts, and Sciences on Wimpole Street. He even tried his own literary talent, in 1846 publishing *The Tongue*, 1232 lines of blank verse celebrating the glory and past power of elocution and mourning the decay of the art.

In 1847 Grandfather Bell published a five-act play called *The Bride*. On its first page the valet Allplace remarks: "How much I have improved the manners of this family. . . . Polishing a prosy lawyer into a tolerable baronet is a task to break a man's back. . . . I was taken into this family for the sake of example. The entire establishment, including Sir Cicero himself, was confoundedly vulgar." A printed copy surely reached David Bell's home in Dublin and in later years fell under the eye of his son Chichester's close friend, the intellectually omnivorous George Bernard Shaw. Grandfather Bell turned his plot into other channels, but Allplace may have been the embryo of Shaw's Henry Higgins, who in 1913 leaped to theatrical glory from the remark, "Well, sir, in three months I could pass that girl off as a duchess at an ambassador's garden party." Perhaps it was coincidence that Higgins played Pygmalion to Eliza's Galatea on Wimpole Street, a few doors from where Bell had instructed young ladies in language. Shavian scholars disagree as to the inspiration for Shaw's play. But if Shaw did take Bell's cue, there is irony in the preface to *Pygmalion*, which salutes "the illustrious Alexander Melville Bell" and leaves the original Alexander's memory to the custody of Dibdin's *Annals*.

Grandfather Bell's income must have come mostly from his private lessons. His fee for a three-months' course ranged from thirty to a hundred guineas, depending on the pupil's fiscal and vocal condition. Clergymen often enrolled. By 1849 he had an establishment of several rooms in Bond Street. Then, somehow, he suffered a financial reverse. He tried unsuccessfully for a position at King's College in London, even getting David and Melville to contribute testimonials. He gave up his Bond Street establishment and considered leaving London. But in 1850, aided by a small loan from Melville, he and his second wife rented an eleven-room house in Harrington Square on the Hampstead Road, a newly-built neighborhood near Regents Park.

The house, he wrote Melville, was "new, elegantly ornamented, and, certainly, *most* delightfully situated." Set in a row of identical houses separated by thin brick partitions, it had four stories with a narrow brick and stone façade. Above a small, twin-pillared portico and doorway, two floor-length windows opened upon grillwork balconies. The house faced a small park and had its own little garden in back. Rent and taxes came to a hundred pounds a year, but with his teaching fees and by renting out three or four rooms, Bell managed comfortably.

Soon after he turned seventy in March 1860, his second wife was laid to rest in Highgate Cemetery, leaving him alone in the house at Harrington Square. As the months went by he thought much about Aleck and his disappointing school performance. Probably Aleck seemed pushed into the family's background both by the older, more self-confident and outgoing Melly and by the younger, more delicate Ted. And Aleck's father, not realizing the sensitivity of Aleck's embattled ego, often put an edge of sarcastic humor on his guidance. Whatever may have been Grandfather Bell's analysis, the lonely old man saw comfort for himself and good for the boy in Aleck's coming to Harrington Square. Aleck needed more attention in his own right. Besides, his grandfather had done a good job of educating himself, his sons, and many others over the years.

Melville Bell consented. In July 1862 Aleck rounded out his four years in James Donaldson's class. Summer days went by and yielded to the smoky drizzle of an Edinburgh autumn. In October, bags were packed, Aleck was arrayed in his Charlotte Square best, and the Bells saw their fifteen-year-old son off to the metropolis.

Thus began what Aleck was to look back on, nearly half a century later, as "the turning point of my whole career."

Grandfather Bell began in the manner of Pygmalion by shaping the outward form. To the dismay of Aleck (and when they heard about it, to the chagrin of his parents), the boy's Edinburgh habiliments were at once pronounced unfit for public view. One of the best London tailors was summoned to deck out Aleck in conformity with current fashion plates. Grandfather Bell impressed on his new pupil the disgrace of stepping out of the front door unless dressed like a gentleman, complete with Eton jacket, kid gloves, and — most galling — a tall silk hat and a small cane.

His Charlotte Street clothes had been prison enough for a boy's high spirits. Now Aleck felt half-choked with tight clothes and self-consciousness. But then he had no occasion to cavort. Harrington Square had no young companions for him, no Maclarens, no Herdmans, no Melly, no Ted. For exercise and what passed for fresh air in London, his grandfather turned him out for an hour or so every day to saunter back and forth alone in Harrington Square Garden. That fenced-in triangle of greensward and young trees was pleasant enough on a good day, and across the Hampstead Road lay the larger park of Mornington Crescent. But both together fell far short of Corstorphine Hill in spectacle as well as elevation. The pacing youngster passed his hours there less in free reverie than in imagining the stifled amusement he furnished passersby.

Merely to spend a year in the company of an old man would have affected a boy in his middle teens. But Grandfather Bell's influence on Aleck was more positive than that. Many years later Aleck believed his grandfather wrong in divorcing Aleck's grandmother. If he thought that as a boy, it did not turn him from the old man. "We became companions and friends," Aleck later remembered. Even the daguerreotypes of his grandfather in old age suggest intellect, resolution, self-possession. In the calm, solid, regular face, with its aureole of thick white hair, was the look of a man who had grappled often with life and stood ready for another match. The companionship of such a man did much to steady and inspire the boy.

The grandfather set about perfecting Aleck's elocution and declamation. Together they read through several of Shakespeare's plays, and Aleck learned speeches from *Hamlet, Macbeth, Julius Caesar,* and others. His grandfather saw to it that he put aside light reading and applied himself

31

Alexander Bell, 1790–1865

to serious studies. Aleck at last tried the toughness of his mind and found exhilaration in the trial. His mind raced to catch up with his years, and he looked beyond to a university education.

Aleck must have absorbed something of his grandfather's outlook on men and affairs. Manuscripts of the grandfather's lectures reveal much about those ideas. Grandfather Bell believed, as he had reason to, that individual worth was not determined by class or heredity, that education would disclose ability as often in the humblest-born as in the highest. He scoffed at the notion that education might make the poor unruly. The poor did not need books to feel injustice. "The poor have eyes," he said. Education, on the contrary, would enable the poor to raise themselves by creating wealth rather than by despoiling others. Education might be a ladder out of even lower depths than poverty. "Our criminals, with few exceptions, are the neglected portion of the Community . . . our brethren, God's creatures," the products of "man's neglect, and not innate viciousness." Conversely, he held merit to be no more subject to bequest than was villainy, and so he felt that to reward achievements with hereditary titles was nonsense.

Grandfather Bell despised dogma and authoritarianism, whether in organized religion or in government. In a historical lecture he mocked the divine right of kings as claimed by that uncouth Scotsman James I; yet while he liked the Puritans' doggedness, he scorned their cant. He was no blind lover of America: "Everything is of huge growth in America . . . and their humbug is both huge and paramount." But he also believed that "the matured strength and sound intelligence of the government of the United States of America seem capable of preserving them against all assailants."

And Grandfather Bell cherished most of all the professional ideal in which Aleck was being raised: "Perhaps, in no higher respect has man been created in the image of his Maker, than in his adaptation for speech and the communication of his ideas. The Almighty fiat, 'Let there be light,' was not more wonderful in its results, than the Creator's endowing the clay, which he had taken from the ground, with the faculty of speech."

One piece of evidence suggesting that Aleck took his grandfather as a model fills fifty-six pages of a four-by-six-inch notebook, and bears the title "Play of Douglas [by] Alexr. Bell 1862." (Out of tact, Aleck did not use his self-chosen middle name that year.) Obviously inspired by his grandfather's *The Bride*, Aleck's "play" revolved about marriage and money. In its theme, its comic abandon, and its exuberant transcendence of probability, it calls to mind *The Matchmaker*, Thornton Wilder's modern re-creation of nineteenth-century farce. Through the play's whirl of puns, comic Irish laborers, mistaken identities, impersonations, and shamelessly compounded coincidences, there ran a serious assumption that money was a proper consideration in matrimony, that the groom owed his bride not only love but also security. Years later Aleck would have that feeling at a crucial stage of his own life.

The money theme may have reflected Aleck's greater financial independence in London. In Edinburgh he had had to ask for and account for everything he spent. Now his father sent him a regular allowance, and Grandfather Bell demanded no accounting. Thus, as Aleck later put it, "the spirit of independence arose in me." Indeed, as the year passed it was the old man who began to lean on the boy, for illness was coming upon Grandfather Bell by the spring of 1863. So Aleck tasted responsibility along with independence.

He turned sixteen in March 1863. Besides a necktie made by his mother, he got an extra half sovereign of pocket money from his father and a letter:

God bless you, my dear boy, and may you ever be as happy as I have tried to make you during your past life! . . . We miss you sadly when we assemble by the fireside at the cottage, but we are reconciled to your absence by the fact that you are good to grandpapa and have been a great comfort to him in his illness, and also that you are making good progress in your studies. You will have cause of thankfulness all your life that you had the benefit of such a training as my father has lovingly afforded you.

Melville Bell's prediction was borne out sooner than he expected. When he came down to London some months later to take Aleck home, he met a studious, thoughtful young gentleman of dignity and purpose. "From this time forth," Aleck later wrote, "my intimates were men rather than boys, and I came to be looked upon as older than I really was." The year with Grandfather Bell "converted me from a boy somewhat prematurely into a man."

4

The Young Teacher

Before leaving London, Melville Bell and his son called on Sir Charles Wheatstone, one of England's leading scientists. Wheatstone's fame rested especially on his work with electricity, including a respectable claim to priority in the telegraph. He had made contributions in other fields, from cryptography to the physiology of vision. But it was Wheatstone's work in the science of sound that brought Melville Bell to him.

In 1821, at the age of nineteen, Wheatstone had mystified Londoners at his father's music shop with his "enchanted lyre," in which tuned metal rods were sounded by vibrations coming from a distance through an unobtrusive solid conductor. Reporting this, *The Repository of Music* forecast the eventual broadcasting of operas from the King's Theatre throughout the city. "Perhaps," it added, "words of speech may be susceptible of the same means of propagation." A decade later, Wheatstone himself remarked that sending music in that fashion would be "of far less importance than the conveyance of the articulations of speech."

Wheatstone may have touched on this in talking with Professor Bell and his young son. They had come, however, to look at Wheatstone's improved model of Baron De Kempelen's eighteenth-century device for mechanical imitation of the human voice. Having seen the "speaking machine" of "Professor" Joseph Faber at the Egyptian Hall in London, Melville Bell had remembered or been told that Wheatstone had once made something similar. A quarter of a century had gone by since Wheatstone had made his "speaking machine." But he brought out his old device, and Aleck and his father "heard it pronounce, in a very mechanical manner, a few simple words and sentences." Sir Charles even lent Melville Bell his copy of De Kempelen's book on the subject, including plates and a full description of the apparatus.

Back home in Edinburgh, Aleck found himself being "treated as a boy again, after I had considered myself a man." His allowance was ended, and once more he had to ask for and account for every penny spent. Melly, two years older, likewise strained against the parental leash, having already made his debut as a public reader at Slateford the previous December. Perhaps sensing his sons' restlessness, Melville Bell challenged them to build a speaking machine of their own.

The project brought Aleck and Melly closer together than they had ever been before. Ted's bent was for art, but Melly shared Aleck's fascination with science and invention. And Aleck's recent stride toward maturity helped reconcile Melly to his company.

Aleck and Melly began by studying De Kempelen's book, then agreed upon a division of labor: Aleck to make the tongue and mouth of the apparatus, Melly to make the lungs, throat, and larynx. No available anatomical work told them all they needed to know about the larynx, and so with heavy hearts they decided to sacrifice their pet cat to science. They called upon a medical student, a friend of Melly's, to dispatch the cat painlessly. Instead, he took the cat into the Milton Cottage greenhouse and before the boys' eyes poured nitric acid down its throat. Only after it had raced around in agony for some time could he be persuaded to open an artery and end its suffering. Aleck and Melly renounced such expedients thereafter, and half a century did not erase Aleck's horror at the memory. The brothers settled for a lamb's larynx, given them by a butcher.

For lungs Melly thought of an organ bellows, but the impatient boys used their own lungs instead and simply blew into the "throat," a tin tube. To form a "larynx" at the other end, Melly experimented till he got a "musical sound," something like that of a tin horn, from two sheets of rubber meeting at an angle.

Taking impressions from a human skull, Aleck made gutta-percha replicas of the jaws, teeth, pharynx, and nasal cavities. His father guided him into substituting a purely functional resonance chamber for the nasal passages and restrained Aleck's theatrical yearning to give the machine a full cranium, a face, and even a wig. Aleck did provide it with soft rubber lips and rubber cheeks, which served a practical purpose as well as hiding the eerily disembodied grin of the gutta percha teeth.

Aleck's most ingenious arrangements were the soft palate of rubber stuffed with cotton, worked by a lever from outside, and the tongue, sliced crosswise into half a dozen carefully shaped wooden sections, each raised or lowered separately from below, each padded with cotton, and all covered with a single sheet of rubber. Thus the position not merely of the tongue as a whole but also each key segment of it could be reproduced as in life.

At last the brothers united their creations. Melly blew hopefully through the tin tube, Aleck manipulated the lips, palate, and tongue, and out came human-sounding gibberish in the quacking falsetto of a Punch-and-Judy show. By experiment they learned to approximate some consonants through adjusting the tongue sections. The easiest combinations of sounds — as every baby knows — came from opening and closing the lips while sustaining the basic vowel-sound "ah." The machine then cried out "Mama!" Aleck and Melly tasted triumph when a persistent demonstration of this feat on the common stairway at 13 South Charlotte Street brought a tenant down to see "what can be the matter with the baby."

That sort of show had been the boys' great goal. Melville Bell took a larger view of the project. From it his boys learned in the most lasting way possible how the organs of speech were made and how they worked to produce speech. Furthermore their father had kept them at it by praising, suggesting, questioning, even by offering a prize. Thus they learned that mistakes, wrong hypotheses, and spells of bafflement need not mean failure where there is the will to try again.

After the speaking-machine diversion, the boys began again to long for independence. Aleck remembered the grand eastward run of the view from Corstorphine Hill that led the eye so irresistibly over the New Town to Leith docks, the sea, and the wide world beyond. He decided to run away to a seafaring life. He even packed his clothes and fixed the hour for stowing away on an outbound ship. Then at the last minute he gave up the idea and instead scanned the newspapers for word of a position he might fill. Presently he came upon an ad for pupil-teachers in music and elocution at Weston House, a school "for the Board and Education of Young Gentlemen" at Elgin in Morayshire on the northern coast of Scotland.

Aleck applied as pianist and Melly as orator, giving the name of Professor Alexander Melville Bell as reference. James Skinner, the principal and classics master, had been one of Melville Bell's students. So the elder Bell learned at once of his sons' impatience to be on their own. A family conference resulted in Melly's being sent to the University of Edinburgh for a year and Aleck's being permitted to spend the same year, 1863–1864, as a teacher of both music and elocution at Elgin, in return for board, ten pounds, and instruction in Latin and Greek. Then Aleck would change places with Melly.

In August 1863, therefore, Mr. Alexander Graham Bell arrived in Elgin, a young gentleman of sixteen whose serious bearing and London-acquired *savoir faire* made him seem (fortunately for discipline) older than he was. Several of the pupils turned out to be older than he, but he kept the fact to himself.

37

Alexander Graham Bell at Elgin

In those first summer days there would have been James Skinner's welcome and discussion of duties. There would have been the introductions to and sizing up by the three other masters. There would have been Aleck's first stroll about the small town with its cathedral, its old stone church planted like an island in the middle of the main street, its concert hall, its town buildings and other landmarks. At the west end of town rose a green mound called Lady Hill, so abrupt and regular as to inspire legends of artificial origin, the raciest being that the Church had required erring girls to carry a basketful of earth there for every "bastard bairn" born in Elgin. It was a hundred feet high and six hundred yards about the base.

A train ride brought Elgin people to the summer resort of Covesea on the Firth of Moray, a haven of sparkling green sea, fine white sands, and fantastic rocks and caves with names like Sir Robert's Stable, the Gull's Castle, Hell's Hole, the Clucking Pow. In Aleck's time at Elgin he would slip away to scramble over the rocks of Covesea, explore its caves, and perhaps after tea join those who gathered on the warm grassy knoll of a high cliff in the light of the evening sun to watch the herring boats come out and stud the whole Firth with their sails. Aleck got into the usual scrapes of an active young man, rolling off a cliff, being trapped by the tide, getting stuck in a cave, but survived them all handily.

And so he settled into the small world of a boarding school, with its network of personal relationships, its rumors and sensations, its occasional dramas of confrontation and denouement, and its daily routine of classwork and sports.

Melly paid a visit to Elgin that fall for a public reading in the Concert Hall under the auspices of the Elgin Mechanics Institute. As "Lochinvar" he failed to stir the "large and fashionable audience"; but when he got to the humorous pieces, he brought down the house.

Aleck's young charges did him proud when Weston House held its annual public examination on the first day of summer in 1864. The crowded room was festive, with evergreens and flowers augmenting the colors of the wall maps. The *Elgin Courant* reporter heard Aleck's pupils read in measured tones with tasteful modulation, bringing out the sense of the author. In reciting it was the same — no mouthing or unnatural gesticulation, but the placing of emphasis on the really emphatic passages and words. The pronunciation was accurate to a remarkable degree, the very youngest boys being apparently thoroughly grounded in the proper sound of the vowels, and giving every one of them its full effect. Their teacher was identified simply as "the son of Mr. Melville Bell, Professor of Elocution in the University of Edinburgh."

It was a season of zestful enlargement for the older generation of Bells as well. In Dublin Uncle David worked up an invigorating rage at English oppression of Ireland. His denunciations of Sassenach misrule must have been thunderously impressive. At the age of eleven George Bernard Shaw attended a school where David Bell taught, and a generation later Shaw still remembered David Bell as "by far the most majestic and imposing looking man that ever lived on this or any other planet." The Fenian Brotherhood, dedicated to the forcible expulsion of the English, apparently agreed with Shaw. In October 1864 its New York headquarters notified members that David Bell would be sent over by the organization for an American lecture tour, the proceeds, less expenses, to go to the cause. The tour would "have a most stirring and beneficial effect," said the circular.

Whatever the effect of the tour (if it ever came off), one effect of the circular was to lodge David Bell in the Dublin House of Correction a year later, where he spent some time picking oakum. The treatment was not a cure. "I must bear it," he wrote Melville from prison, "still, however, looking forward to the proud watchwords — Ireland! Independence! No Saxon government, no base, bloody, and brutal whiggery, no Juggernaut of Palmerstonian policy shall depress my spirits." "Even as a boy," he added, "I looked to the '*stars* and *stripes*,' and now, with a life's experience, I believe that the *United States* is in many respects the best."

Early in 1864, while Aleck moved toward his first professional triumph at Elgin, his father in Edinburgh came to his greatest moment: a solution to the puzzle of a complete and universally applicable phonetic alphabet.

Others had been after it also. In 1854 Europe's leading phoneticians and philologists had met in London, agreed that a "universal alphabet" would require more knowledge of the production and relationships of vocal sounds than anyone yet had, and then adjourned. As far away as Washington, D.C., in the late fifties, Joseph Henry's diary recorded a Washington Scientific Club discussion of "the wave expression of the vowel sounds and the possibility of establishing a universal alphabet." Brilliant phoneticians like Alexander John Ellis of London persisted in the quest. But Alexander Melville Bell got to the goal first.

Instead of arranging the sounds and then analyzing the way they were produced, as the others did, Bell had placed his vocal organs in systematically varied positions and with his uncannily discriminating ear identified the precise sound made in each case. Year by year his catalogue grew, while his wastebasket filled again and again with discarded schemes of arrangement. At last, one evening early in 1864, the missing element fell into place: there were four, not three basic classes of vowels. The inspiration came with a flash, like a spark jumping a gap, lighting up the grand final structure, complete in all but details.

On its basis Melville Bell by April 1864 designed the sort of alphabet he had forecast in 1849. All consonants had a horseshoe curve standing for the tongue and facing up, down, right, or left, according to the part of the tongue employed. All vowels had a vertical line standing for the breath aperture. Bell identified a new class of sounds he called "glides," halfway between vowels and consonants, and gave them a set of symbols also. Modifying symbols, such as hooks or crossbars, were combined with the root symbols to specify certain vocal positions. And a few small symbols stood for actions like suction or trilling, and for tones and inflections. The various symbols, in short, told the reader how to dispose his vocal organs in order to produce the sound exactly, even if he had never heard it. They could even express a wheezing cough, a checked sneeze, braying, shuddering, growling, grunting, and so *ad nauseam* (which required only two symbols). They spelled sounds as the conventional alphabet spelled words. Bell called his system "Visible Speech."

Melville Bell saw his creation as the long-sought universal alphabet, as a

ALPHABETIC VOCABULARY OF TEST WORDS.—*Initial Vowels.*

(Visible Speech symbols, with language/dialect annotations:)

(Sc.) · (Sc.) · (F.) · (Sc.)

(Sc.) · (Pro.)

(Sc.) · (Prov.)

(Ga.) · (Am.) · (Ir.)

(Sc.) · (F.)

(Sc.) · (Ir.)

(Prov.) · (Prov.)

(Sc.) · (Sc.) · (Cock.)

(Sc.) · (Colloq.)

(Sc.) · (Port.) · (Ge.)

(Sc.) · (Sc.) · (F.)

(F.) · (Sc.)

(Sc.) · (Manx) · (F.)

(Cock.) · (Sc.)

(It.) · (It.) · (Sc.) · (Cock.)

---

* The accent is on the first syllable, unless otherwise expressed.

A page from Bell's *Visible Speech*

means of making phonetics and philology exact and therefore scientific, and as an instrument of teaching and communication open to an intoxicating multiplicity of uses. It deserved prompt, worldwide dissemination, he believed. To pay for special types, big printings, and teacher-training programs, nothing less than the backing of the British government seemed necessary.

Before seeking public and professional support, he grounded his three sons in the system. Aleck, just returned from Elgin, mastered it thoroughly in five summer weeks. In· Edinburgh and London, linguists dictated words to the father in a hodgepodge of languages, accents, and dialects and heard one or more of the sons, after coming from another room, reproduce them from Visible Speech slowly but faithfully. The brothers passed the test in American Indian, Arabic, Hindi, Persian, Urdu, and a dozen or more ancient and modern European languages. Though Melville Bell kept the workings of the system secret, pending government action, such demonstrations carried conviction; and long reports appeared in the Scottish and English press.

Grandfather Bell had been nettled in April when Melville pressed him for repayment of a loan, perhaps in order to print up a Visible Speech brochure. In remitting the money, Grandfather Bell wrote: "I *could* say much in reply to your *swaggering* note, but I forbear." By the end of June he understood. "I return your proof," he wrote. "*Send me a score of copies.*" If the system was as advertised, "your invention will certainly be esteemed as one of the wonders of this wonderful age." In August, after Melly left for Weston House, Melville Bell came with Aleck and Ted to use Harrington Square as his London base.

The greatest convert to Visible Speech during those summer weeks in London was Alexander J. Ellis. Bald, bearded, rotund, and jovial, he qualified at least by costume for a place on the long roll of eccentric English worthies. On most indoor occasions he wore a velvet skullcap, for which, as Bernard Shaw remembered, "he would apologize to public meetings in a very courtly manner"; outdoors he wore a white stovepipe hat with a wide black band around the bottom. Roly-poly as he was by nature, everything he wore still seemed a couple of sizes too big; but the slack was taken up by twenty-eight pockets, all crammed full, the outer coat with manuscripts, the inner with "articles for emergency," including string, knife-sharpener, scissors, and even a corkscrew, though he drank nothing but water. But for all his quaintness, Ellis's professional colleagues deeply respected his industry, learning, perspicacity, and integrity. After many years of effort and much expense, he had come closer than anyone before Bell to a "universal alphabet," though Ellis's ninety-four symbols, unlike Bell's, did not show by their written forms how their sounds were to be produced.

Ellis had never met Alexander Melville Bell but had long known and valued his writings. When Ellis read an account of Visible Speech in the *London Morning Star*, he wrote the editor at once. Visible Speech, as Bell's work, must be "a very ingenious and useful invention." But Ellis felt sure that Bell's thirty-four symbols, whatever they might be, could not express all the sounds of language.

Bell promptly invited Ellis to a private demonstration; and so on August 23 the round, rumpled figure of the great phonetician appeared at the door of 18 Harrington Square. Ellis himself told the story in a letter to the *Morning Star*. He gave the father "a most heterogeneous collection of sounds, such as Latin pronounced in the Etonian and Italian fashions, and according to a purposely rather eccentric theoretical fancy; various provincial and affected English and German utterances . . . Cockneyisms mixed up with Arabic sounds, and so forth," including some sounds not amenable to any known alphabet. Young Aleck came in, took the paper, and slowly "echoed my very words. Accent, tone, drawl, quantity, all were reproduced with remarkable fidelity, with an accuracy for which I was totally unprepared."

After that, demonstrations for other London savants, including Sir Charles Wheatstone, were anticlimactic. In September, Aleck and his father visited the prime minister, the redoubtable Lord Palmerston, to seek government support; but "Pam" kept his counsel. In retrospect the most significant event that day was an incidental remark in the *London Illustrated Times*: "We cannot pretend even to guess at the horizons opened up by such an alphabet in the training of the deaf, the dumb, and the blind." Melville Bell clipped the item, but could scarcely have guessed its portent for Aleck.

Much less could the elder Bell have foreseen how teaching speech to the deaf would lead Aleck to vindicate another offhand remark printed a year earlier. In the 1863 edition of his *New Elucidation of the Principles of Speech*, Melville Bell had alluded to the "wonderful mechanical adaptations of optical principles" that had grown out of a scientific approach to the faculty of sight. "Might not an analogous result attend the philosophical investigation of the faculty of speech," he asked, "and acoustic and articulative principles be developed, which would lead to mechanical inventions no less wonderful and useful than those in optics?"

Melville Bell, uneasy about Lord Palmerston's noncommittal attitude, petitioned Queen Victoria herself, offering to demonstrate Visible Speech to Her Majesty at Balmoral. But the Queen was not inclined to be thus amused. The petition went to the Foreign Office, which saw in it no relevance to foreign affairs and returned it to Bell with a reference to the Home Office. So began months of red tape and circumlocution. Melville Bell and his sons did not wait out the time in London but went home to Edinburgh in October 1864.

Aleck duly enrolled at the university, and his notes survive for November 1864 in Professor Blackie's Greek class and Professor Sellar's Latin class. Nothing more is known of his University of Edinburgh career, not even whether he finished out the term. Perhaps he did not, being involved in his father's demonstrations of Visible Speech in Edinburgh and Glasgow.

On April 23, 1865, Grandfather Bell died. Leaving Melly to represent the Bell clan professionally in Edinburgh, Melville Bell moved with his wife and two younger sons to the house on Harrington Square. He took advantage of his new London headquarters to keep interest alive in Visible Speech. The Ethnological Society with warm applause saw and heard him put Aleck and Ted through their phonetic paces in July. And during July and August a series of commentaries on Visible Speech ran in the *Phonetic Journal*. The *Journal* complained, like a growing number of Bell's professional colleagues, about the withholding of the system's details; but Bell would not be moved while government backing remained possible.

Late in that summer of 1865 Aleck went back to Weston House, this time as an assistant master rather than a pupil-teacher. Melly had done a fine job the preceding year, but had overworked himself and fallen ill. He was getting well again, Eliza Bell wrote Aleck in September. She cautioned Aleck to be careful of his own health. Melville Bell wrote also, handing down a table of half-joking commandments:

Don't lie down heated on the ground
Don't go crawling into caves
Don't go psalmsinging in a choir
Don't forget tomorrow in today
Don't bother your brains about ideography [philosophical speculation]
Don't neglect English *literature* & history
Don't forget what you know of languages
Don't fail to rank music last in your accomplishments
Don't let a week pass without writing to us
Don't hate London — for home's sake — if you can't love it
Don't gloom at my doubts, and
Don't doubt the constant affection of
              Your fond father

The Bells were to find their second son less heedful than before. Not that Aleck thought any the less of his parents. (During this very year he took to signing his name "A. G. Melville Bell.") But at eighteen he was emerging from adolescence.

Aleck was also making his own contribution to the understanding of speech production. His father had discovered that when certain vowels were whispered in a certain order, they seemed to form an ascending musical scale. He tried the sequence on Aleck, as having the best musical ear in the family. To Aleck the whispered vowels seemed to form a descending scale. For some time, father and son could not agree. Then Aleck perceived that each of the vowels was made up of two pitches, one rising in the sequence, the other falling.

At Elgin, as summer passed into fall, Aleck made that puzzling vocal phenomenon the subject of his first original scientific research. It began with a minimum of laboratory equipment. His roommate awoke one night to see Aleck grimacing into a mirror and snapping a finger against his throat. "What are you doing, Aleck?" the roommate called out with alarm. The dark-eyed intense young man explained that he was simply determining the respective pitches of his mouth and throat cavities in various vocal positions. If this did not enlighten the inquirer, at least it put him to sleep again.

Gloomy weather, chilly and wet, closed in on Elgin and held fast. Late in November 1865 a deluge flooded all the rivers in the district. The heather hills invited no rambles, and the wet gray town no evening strolls. Aleck spent much time in his third-story room at Weston House, theorizing and testing. He wrote a long report to his father. A hurricane of wind and rain was roaring down from the Firth of Moray, but Aleck's mood was sunny.

"I have experimented again," he began, "and I find the general results of my former trial correct — and I now *see the reason*." His line of argument was to compare the vocal cavities to bottles. When one blew a musical sound from an open bottle, the pitch was higher as the bottle was more nearly filled, or as the neck of the bottle was narrowed. So also with vocal sounds, as either the vocal cavity or the breath aperture was constricted. Furthermore, a section of the tongue formed a bottleneck between two cavities, as though they were two bottles, each with its own pitch. The two pitches could be distinguished by tapping successively against the throat and in front of the mouth.

More significant for the future, Aleck found a way of determining the precise pitch of certain vowels. He would sound a tuning fork or "pitchfork" before his open mouth, meanwhile moving his tongue through the positions of the vowel scale. One of the positions would make the fork sound loudly because of resonance. The pitch of the vowel made in that position would be precisely that of the fork. He could envision a variety of experiments along that line. "If I could only get a box of pitchforks!!!" he remarked.

The tuning-fork experiment was his first step on the road to inventing the telephone.

That winter Aleck got his "pitchforks" and analyzed the compound pitches of various vowels. When he came home to Harrington Square for the Christmas holidays there must have been much discussion, demonstration, and pitchfork whanging in Melville Bell's study. But at Weston House school duties took up a large part of his time, and much of the rest went to part-time teaching at an Elgin girls' school.

Meanwhile he locked horns with his father. Aleck wanted to set up for himself professionally in Glasgow, studying in his spare time for the degree examinations at the University of London (which in that period would grant a degree after examination, without requiring enrollment in one of its affiliated colleges). His father insisted that he should enroll at University College in London and study there for a year before going to Glasgow. He would overwork himself trying to teach and study at the same time, wrote his father, and besides at nineteen he was too young to start out on his own. "I think you should implicitly surrender yourself to Papa's judgement in this matter," wrote his mother.

Frustrated, perplexed, overworked, Aleck lay awake night after night. By day he was tired and depressed. His head ached, and a pain in his eyes interfered with his reading. "What I would not give if I could only get at you," his mother wrote anxiously. Keep off pickles, she warned him. Try a little beer, she suggested, and explain to Mr. Skinner that we approve it. Aleck should try cold water on his eyes. She feared there was some "faulty action of the liver." She sent a bottle of medicine for sleeplessness. He must give up all reading, for the sake of his eyes. She thought he might have had a bilious attack. "Its symptoms are lassitude, faintness, a lowering of the spirits, and a disposition to take a disheartening view of passing events. Under such circumstances one is incapable very often of forming a correct judgment. . . . In future when you feel a beginning of these symptoms, take a dose of Chlorodyne, and if not better next day, take another. A third will be seldom necessary."

So it went through the spring of 1866. As the weeks passed, Aleck's health improved. But neither argument, entreaty, Chlorodyne, resort to beer, nor renunciation of pickles seemed to cure his craving for independence.

Aleck found pleasanter distractions in Elgin. Anna Daun was a pupil in his class at Miss Gregory's school for young ladies. It would never have done for a gentleman teacher to betray a special feeling for a young lady pupil, certainly not in a small Scottish town during the reign of Queen Victoria. But a mutual friend managed to get Aleck a photograph of Anna. Forty years later, when both Aleck and Anna were grandparents, he confessed the incident of the photograph to her, adding that he had kept it all those years and had often thought of her.

He had a retreat that did him more good than Chlorodyne. Six miles southwest of the school, at the northern edge of a broad glen known as the Vale of St. Andrew, stood the remains of Pluscarden Priory, built six centuries earlier for a small order of monks, but now in the same state of romantic ruin that had inspired Mendelssohn in Holyrood Abbey. What was left stood quite roofless; ivy and tufts of wind-sown grass softened the ragged outlines of surviving walls; patches of turf carpeted the transept floors. Behind the abbey, pine-covered Heldon Hill lay like a long arm sheltering the ruins from the north wind, and southward in the Vale of St. Andrew stretched a fertile plain. There Aleck came to find peace on the fair spring days of his last year at Weston House.

In the solitude of Pluscarden, Alexander Graham Bell must have taken stock of his talents and wondered about his future. It would have to be bright indeed to outshine what his father was already doing. A letter from Ted that spring referred to "your wish to do something great, also the extreme poverty of thrilling ideas which you manifest in most of your prose and dramatic works."

Aleck by now seems to have known his literary limitations as well as Ted did. In his "wish to do something great" he began playing a stronger suit. In March 1866 he sent Alexander Ellis an account of his theory of vowel tones, as refined by his tuning-fork experiments over the winter. Ellis's reply set him off on a ten-year journey to something very great indeed.

At the time, Aleck read Ellis's letter with mixed feelings. "I find," wrote Ellis, "you are exactly repeating Helmholtz's experiments for determining the musical tones of the vowels." If this dashed Aleck's spirits, he could reflect that at nineteen he had independently worked out an idea that had engaged the mind of one of the world's half-dozen greatest scientists, an idea to which Aleck would have had clear title but for the accident of priority.

But that was not all. Ellis dropped another fruitful seed. Helmholtz had gone further, he told Aleck. By means of electromagnets the German scientist had kept several tuning forks at once in continuous vibration. He could adjust the loudness of each fork, and so had been able to synthesize vowel sounds. Ellis had not the time just then to explain in detail, but he gave a perfunctory description of the principle. "I should be very glad if you could repeat these very interesting experiments," Ellis added. "If you read German I shall be happy to lend you Helmholtz's book, which I can send by post." Unable to read German, Aleck resolved to talk with Ellis after returning to London.

Thus the spring of 1866 closed with dreams of glory. And there was even a taste of the real thing in the small world of Elgin. At Miss Gregory's school on the last day, Aleck's class fell self-consciously silent; Anna Daun rose shyly to make a presentation address; and Mr. Bell found himself for once tongue-tied with surprise and emotion, holding a new writing case in his hands. "This," he remembered many years later, "was the first present of the kind I had ever received and I was very proud of it."

5

Matters of Life and Death

While Alexander Graham Bell pondered life, love, and language at Elgin, the other elocutionary Bells had not marked time.

During the winter of 1865–1866, Melly invented a machine of some sort that was tested and pronounced perfect by two interested gentlemen, though nothing seems to have come of it. By the summer of 1866 he had settled in the family's former apartment at Edinburgh as his father's tenant. His spirits ran high as ever. He advertised the coming of a celebrated Russian prestidigitator, "the Great Loblinski," hired the Edinburgh Music Hall, appeared with false beard and accent, and carried off a successful performance. Meanwhile he found time to teach elocution and cure stammering and, on the evidence of testimonials, to do it well.

So did his father in London, giving the Italian-raised son of the poet Robert Browning lessons for several months in how to speak with a proper British accent. True, the elder Bell's clientele was embarrassingly small. But he taught at University College, gave occasional public lectures and readings, and so got along.

With his son Ted's help Melville Bell kept on demonstrating Visible Speech to popular audiences and learned bodies, still hoping for a government grant of two or three thousand pounds toward propagating it. His proposal bounced from the Foreign Office to the Home Office to the Exchequer to the Council on Education, all unavailing. Meanwhile his professional colleagues grew more impatient for an explanation of the alphabet.

Bell had explained it in confidence to Ellis and a few other scholars, who vouched publicly for its soundness. In private, by now, Ellis had tempered his initial enthusiasm a little. To the eminent American philologist Samuel Haldeman he wrote: "Bell . . . makes an a priori analysis which suits the special case with remarkable exactness. And he also enables a learner to train himself to read his combinations with singular accuracy." But though Bell had rare powers and his sons had special ability, even they necessarily fell short of absolute perfection. Most others would do less well. Nevertheless, Ellis wrote, "I think that the result shows a wonderful perfection in the *alphabet.*"

Americans had begun to hear about Visible Speech. In June 1866 Melville Bell had explained it to the visiting Haldeman, who later used it for American Indian languages. Bell sent Visible Speech brochures to other American scholars. And in July 1866 a Boston magazine of reprints copied a favorable English magazine article on it. The article asked how a man "who probably does not make any pretensions to learning" could have found what famous scholars had sought in vain, and answered that "he found it because he happened to take the way to the place where it was, while the learned men were misled by their learning to seek for it where it was not." The remark, as it happened, was prophetic of Aleck's future.

What Aleck did in the summer of 1866 left little trace. But we can be sure of one thing: he quizzed Alexander Ellis about Helmholtz's electric tuning-fork apparatus.

It had three main elements: a device to keep tuning forks in constant vibration by means of an intermittent current, a device to produce such a current, and a device to regulate the loudness of each fork.

To interrupt a current at precisely equal intervals, Helmholtz sent it through the arms of a tuning fork. At the end of each arm, a platinum wire barely touched the surface of a cupful of mercury, thus making and breaking the current as the arms vibrated. To keep the make-and-break fork vibrating continually, the intermittent current also passed through the coils of an electromagnet, placed so that its magnetic impulses reinforced the vibrations of the fork.

The same intermittent current activated eight other electromagnets, each having a permanently magnetized tuning fork between its poles. The forks were tuned so that the intermittent magnetic force coincided with each natural vibration of the lowest-tuned fork, every other vibration of the next fork, every third of the next, and so on.

Each successive fork was therefore less strongly reinforced in its natural vibration. But Helmholtz could adjust the loudness of each fork independently with a "resonator," a cardboard tube closed at both ends but with a small hole at the end next to the fork. Each resonator was just the proper length to amplify its fork's vibration by resonance, like an organ pipe. The volume of the sound could be reduced by sliding a lid part way across the hole.

With the lowest fork sounding loudly and the others more or less softly, Helmholtz produced a composite sound with an uncanny likeness to a vowel.

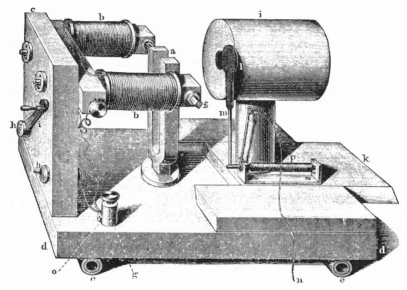

Helmholtz's tuning-fork sounder

Ellis and young Bell talked at length about Helmholtz's work, Ellis translating passages from the German. But the diagrams were complex, Aleck knew little about electricity, and Ellis himself did not fully grasp it all. So Aleck came away with only two basic notions, one of them mistaken. First, and wrongly, he thought that Helmholtz's device *transmitted* vowel sounds by electricity, rather than merely generating them. Second, and rightly, he saw how little he knew about electricity. And so he resolved to learn.

From his mistaken notion he jumped almost at once to the conclusion, which most experts would then have deemed equally false, that if vowels could be transmitted by electricity over wires, so could consonants, so indeed could speech, music, any sounds whatsoever. And he began telling his friends that someone, someday, would do just that. He did not see himself as the man. His determination to master the basic principles of electricity proceeded rather from Helmholtz's fruitful use of them in the science of speech, which was Aleck's first allegiance as his father's son.

The coming of September and school days brought Aleck to Somersetshire College, Bath. His new job, which his father had helped him get through London contacts, was only a hundred miles from London, as against Glasgow's four hundred. But in principle Aleck's desire for independence had apparently prevailed.

Bath ought to have pleased Aleck Bell, suggesting as it did a smaller, lighter, warmer, and — in its memories — more frivolous Edinburgh. The "dark satanic mills" of nineteenth-century industry had not spread down from the north to disfigure that green and pleasant land at the southern tip of the Cotswolds. Moreover the nearness, ease of working, and gleaming beauty of "Bath stone" had been discovered in time for the great rebuilding of the city in the late eighteenth century, when with its hot-spring baths and mineral waters it had become England's most fashionable resort. Though the whiteness of the fresh-cut stone had quickly mellowed to a warm creamy hue, the city seemed all of a piece, resting in its shallow bowl of hills like a pool of sunlight. As with Edinburgh, its topography offered fine viewpoints and prospects; but those of Bath circled politely around the city, rather than rearing up from its midst. And so a stroller, sighting down a street, could now and then catch a glimpse of rural hillside.

The architecture of Bath was that of Edinburgh's New Town, a little brighter, a little less massive, but with the same close harmony of the Georgian era, the same sweeping façades of Grecian columns in half relief exultantly repeated. As the New Town of Edinburgh had its eye-filling Georgian crescents and Royal Circus — the latter being a round park and concentric circle of classical buildings — so did Bath. Indeed, Bath's Royal Crescent and its Circus outshone anything Edinburgh could display in that category.

It was in one of the buildings of the Circus that Somersetshire College had been established a decade before. Though called a college, it prepared gentlemen's sons for the universities or the army. Like the older preparatory schools, it emphasized classics and mathematics; but having a smaller enrollment — less than a hundred boys — and a larger ratio of teachers to pupils, it claimed the advantage of more individual attention to students.

Taking rooms first in a plain building half a dozen blocks away, Bell moved in January to nearer and more elegant quarters in a lodging house on Bennett Street. Across the street in the magnificent Assembly Rooms, Regency rakes and their ladies had once gambled and gossiped. Now only memories hung about the place, like the incense from a snuffed candle.

The young teacher diverted himself with rowing on the River Avon, which meandered through the town past the Parade Gardens and the Abbey and under the shop-lined Pulteney Bridge, designed, like Charlotte Square, by the architect Robert Adam. In the winter he skated. He read an occasional novel, and now and then he saw a play at the Theatre Royal, including *Hamlet* with Charles Kean.

And Aleck worked and studied. He took French lessons himself and early in 1867 began teaching evening classes, four boys in Euclid, three in French. He also laid in a stock of glass bottles, acid, zinc, and copper to make electrical batteries. From his Bennett Street window he strung wires along a couple of houses and into the window of a friend in Mrs. Prankerd's boardinghouse, where the Somersetshire students lived and Aleck took his meals. With a couple of Wheatstone telegraphic instruments (which signaled by moving a needle) the two friends exchanged messages. In the process Aleck learned something about the workings of batteries, electromagnetism, and circuits, though not enough to shake his belief that Helmholtz had transmitted vowel sounds telegraphically.

All considered, his year at Bath should have been well remembered. Yet in later life it seemed dim in his mind, as though overlaid by some dark association. Almost certainly it was the tragedy of that spring that shadowed the memory of Bath for Aleck Bell.

Like Aleck, Ted Bell had dreamed and worried about his future. Early in 1866, when it looked as if he would succeed Aleck at Weston House, Ted wrote his brother for information about his probable duties. At seventeen Ted had grown rapidly, stretching to six feet four inches by the time Aleck left for Bath. He remarked once that he felt as if he had no backbone. Aleck had grown fast too, and Eliza Bell worried about both her younger sons' "outgrowing their strength." But Ted's weakness, it turned out, came from tuberculosis. He did not go to Elgin after all. That winter he was bedridden. Through January and February he could sit up only an hour or two at a time.

Writing Aleck on his twentieth birthday in March 1867, his mother showed concern about Aleck's own report of ill health. Mrs. Prankerd had been "kind and attentive," but Mrs. Bell nonetheless sent Aleck a little "mercurius" for bile and reminded him to take his Chlorodyne. Meanwhile Ted, quiet and uncomplaining, still kept to the drawing-room sofa all day and was too weak even to write Aleck his greetings. Aleck's health revived as spring came cold and wet to London and Bath. Yet when windows were opened to late April's warm air, Ted still lay coughing his life away.

On May 17, 1867, Aleck wrote in his diary: "Edward died this morning at ten minutes to four o'clock. He was only 18 years and 8 months old. He literally 'fell asleep' — he died without consciousness and without pain, while he was asleep. So may I die! AGB."

For everyone but Ted, life resumed. In Bath a "large and gay company" at the school athletic contests a few days later had to sprint more than once for the cricket clubhouse during intermittent showers. But spirits were not dampened, and the hurdle races, hundred-yard dashes, high jumps and all went soggily on. Every master gave a small prize, "Mr. Bell's prize" being for the older boys' hundred-yard race. A few days later the Somersetshire College eleven beat the Richmond Club in cricket.

In London, Aleck's father, giving up on government aid, had special types cast for the full revelation of Visible Speech. The book would be dedicated "To the Memory of Edward Charles Bell, one of the first proficients in 'Visible Speech,' whose ability in demonstrating the linguistic applications of the system excited the admiration of all who heard him; but whose life of highest promise was cut off in his nineteenth year."

At Edinburgh Melly announced his engagement to Miss Caroline Ottoway, whom he had met in London. Aleck wrote his mother gloomily, "I only wish I *were as fortunate as Melly is!!!* " School work filled most of Aleck's time, yet the days seemed to drag. He had been plagued by "nasty headaches" lately; "in fact the only idea I can form of this past week is *one immense headache*." Still, wondering whether his depression was mental or physical, Aleck leaned to the latter explanation and decided to take "plenty of Porter or Portwine every day, as a kind of medicinal course." And before he mailed the letter he was able to add: "I feel decidedly the benefit of a rest — Headache vanished — Health capital — but appetite ALARMING (owing to a small pull on the river before breakfast)."

At last, early in July 1867, Somersetshire College gave out its prizes for the annual examination, and young Mr. Bell could leave for London. Ted's death had ended Aleck's resistance to living with his father and mother, now left to themselves at Harrington Square.

While his parents vacationed in Scotland that July, Aleck settled himself in the house at Harrington Square and, having somehow encountered a seafaring native of Natal, set about transcribing Zulu clicks into Visible Speech. Duly credited to "the Author's Son (A. G. B.)," these made the final page of the Visible Speech book.

Even with the elder Bells away, the Zulu sailor was not Aleck's only acquaintance. James Murray, a pupil of Aleck's father in Edinburgh days, had turned up in London as a bank clerk after the death of his wife and child. Young Murray had found the Bell home at Harrington Square a tonic for his low spirits. That summer of 1867 Aleck Bell stood up as best man at Murray's second marriage. A few months later, Melville Bell brought Murray, an enthusiastic amateur dialectician, to the Philological Society. Murray joined it and eventually took over its faltering project

Alexander Graham Bell, about 1867

of a definitive English dictionary. By the time of his death some forty years later he had masterfully led the undertaking, like a philological Moses, to within sight of completion as the great *Oxford English Dictionary*. Among the authorities it cites is Melville Bell's 1863 *Principles of Speech*. Murray even considered using Visible Speech to show pronunciation, though in the end he used more conventional characters.

Then there was Adam Scott, five years Aleck's senior, orphaned as a small child and unhappily reared by a harsh, puritanical maiden aunt. Like Murray, young Scott had found a warm haven at Harrington Square, "the nearest approach to a home that I ever then knew." Fond of walking, he encouraged Aleck to join him. But Aleck, released from a school-day schedule, fell into his father's pattern of working, studying, or reading until two or three in the morning, then sleeping late. "We several times arranged to meet at half past six in the morning at King's Cross for a stroll before breakfast," remembered Adam Scott, "but he never turned up, and it always ended in my going on to Harrington Square, and finding him dead asleep in his bedroom." On Sundays, however, Adam and Aleck did take long walks.

Through his father, Aleck enjoyed the stimulating company of Alexander Ellis, Dr. Furnival, who was the secretary of the Philological Society, and Henry Sweet, the eccentric, quick-tempered, yet brilliant and somehow appealing phonetician, some of whose traits were incorporated in Shaw's Professor Henry Higgins. It may have been earlier, perhaps when Aleck's father had come to take him home from the year with his grandfather, that Aleck and his father dined with Prince Lucien Bonaparte, an occasion made memorable for the boy chiefly by the three waiters who would snatch away one course and set down another whenever he let his knife or fork rest on the plate — upwards of twenty times, he thought.

With these social and intellectual stimuli, Aleck must have found life in London less austere and lonely than in the year he had spent with his grandfather. And presently still other interests engaged his mind.

A stray Skye Terrier enlisted in the Bell household and was named "Trouve." By training the dog to growl on cue, and by manipulating his mouth and throat, Aleck learned to make him utter the sounds "ow, ah, ooh, ga, ma, ma." With a little prompting, visitors could hear them as "How are you, Grandmama?" Despite legends as to Trouve's repertoire, it went no further, and without Aleck's help he could only growl.

The Visible Speech system suggested a higher use for Aleck's teaching talents. Since 1864 the idea of teaching speech to deaf-mutes had grown in Melville Bell's mind from an incidental possibility to "one of the prominent utilities of the system," as he put it. This claim caught the eye of Bell's former pupil Susanna E. Hull, who now ran a private school for deaf children at South Kensington. In the spring of 1868 Miss Hull asked Melville Bell for help in following up the idea. Thus, on May 21, 1868, Alexander Graham Bell first tried his skill at teaching the deaf, his pupils being two "remarkably intelligent happy-looking little girls" named Lotty and Minna, aged six and eight.

On the blackboard he sketched the profile of a face, including "the insides of the mouth" (as he explained to the girls by finger spelling). Then he rubbed out all but the lower lip, the point, front, and back of the tongue, and the glottis. Those curved lines in their respective facings constituted the Visible Speech symbols for "back," "front," "point," "lip," and "voice." The girls immediately pointed out the parts of their mouths governed by each symbol. Before the lesson ended, Lotty and Minna had learned a dozen sounds.

Later they were joined by Kate and Nelly, "lovable little children, each about eight years of age." Nelly had been mistaught before. "Thus," wrote Aleck, "she not only has to learn but also to unlearn." But he persevered. By the end of the fifth lesson the children knew all the consonants and some of the vowels. Using these, Bell easily taught them many words. They looked forward to summer holidays and a chance to surprise their parents, and Miss Hull told him that after the fifth lesson Kate had gone about the house saying, "I love you, Mama. I love you, Mama."

In Kate's eagerness for home, Aleck may have seen some irony. His own home life at Harrington Square had its drawbacks. At twenty-one he chafed more than ever under his father's tutelage, which was often sharp and tactless. Even later, when Aleck was approaching twenty-three, Melville Bell could write to a friend that his son in many respects was "a perfect baby . . . and needs to be told when to wrap up in going out, when to change boots or wet clothes, etc. etc."

But Aleck's new interest in teaching the deaf may have helped divert him. So also may his studies. In June 1868 he passed entrance exams and matriculated as a student at the University of London. Fall classes at University College brought him not only another outside interest, but also more hope of early independence.

Within two years, however, events radically changed the Bell family's situation and ended Aleck Bell's dreams of earning a university degree.

6

Partings

In the summer of 1867 Alexander Melville Bell had at last published the "Inaugural Edition" of *Visible Speech: The Science of Universal Alphabetics*. Early in 1868 Alexander Ellis sent his American friend Haldeman a copy. "You will find Bell's book worth studying," wrote Ellis. "Don't go by his key words, as you will probably differ from him — I do — but go by his diagrams & descriptions." Professional opinion ran much like Ellis's. It accepted Visible Speech as the best system yet for recording vocal sounds. Ellis himself had publicly admitted in 1865 that until Bell's system came along, "alphabetics as a science . . . did not exist." But Ellis could not accept Visible Speech as perfect and final. And phoneticians echoed that opinion too.

The trouble was that by its very exactness and completeness, Visible Speech committed the user to Bell's theories and analyses. Phoneticians naturally resisted that. Furthermore, sound and simple as the basic concept was, its development became intricate, as in its twenty-nine symbols called "modifiers and tones" and its lengthy descriptions of fifty-two consonants, thirty-six vowels, and twelve diphthongs. To some phoneticians Bell's symbols seemed too outlandish and too numerous. And what experts found difficult, the public found impossible. In 1868 Bell brought out a sixteen-page pamphlet, *English Visible Speech for the Million*, intended especially for the teaching of illiterates and foreigners. But if "the Million" ever heard of Visible Speech, they did not choose to buy.

At this juncture David Bell, long an admirer of the United States, persuaded his brother to join him in an American lecture tour that summer and fall, one that might foster Visible Speech abroad and make up for some of the money spent on the book. Melville left Aleck to carry on in London.

Both brothers had pleasant memories of the New World, and the 1868 visit added to them. On August 2, even before the ship docked, Melville saw Staten Island as "a perfect picture of landscape loveliness." New York City captivated him. He considered Central Park "beyond comparison superior to any at home" and thought Wallack's Theater exceeded only by the Paris Opera House in "comfort and elegance."

Melville Bell responded to post–Civil War America just as his former pupil David Macrae had done a few months before. "Entering the States even from Canada," Macrae wrote in his book *The Americans at Home,* "is like pushing out from a sheltered creek into the current. Almost instantly you feel the catch of a swifter life. . . . The climate has something to do with all this. Even the passing traveller soon becomes conscious of the influence of that intensely clear vivifying atmosphere." More than climate inspired the Yankees, he went on to say. They rose to the vastness of the country, its political and social mobility, the consequent richness and openness of its opportunities. "The money-making instinct is next to universal. Young ladies speculate in stocks; children are commercial before they get out of their petticoats." Macrae heard of a minister who gave up his pulpit after patenting a new way of producing water gas.

"Everywhere, from the New England farmhouse to the Georgia plantation," Macrae wrote, "the fact that I was a stranger from Scotland seemed sufficient to secure me a kindly welcome." But Alexander Melville Bell got a far from kindly welcome from at least one American philologist. In the *North American Review* for July 1868 there appeared a long, sour review of *Visible Speech* by the Yale professor of Sanskrit, William D. Whitney. He gave the lie to Bell on several crucial points of his speech analysis and denied that Bell had advanced scientific understanding of the subject. The book, Whitney wrote, could be used only by trained phoneticians who knew how to make the sounds already. Some of his criticism carried the sting of truth, but much was unjust. Granting that Bell's sons had special talent and training, they had nevertheless produced sounds accurately from Visible Speech symbols without ever having heard them before. Nearer the truth was Whitney's more kindly conclusion that Bell's system, for all its faults, might stir up more interest in phonetics and advance it further than any predecessor had. Visible Speech obviously never became a people's alphabet, but it did become the direct ancestor of the present International Phonetic Alphabet, by way of Henry Sweet's adaptation of it into what he called "Broad Romic."

Fortunately, opinions from Yale seemed to carry little weight at Harvard. Whitney notwithstanding, President Thomas Hill of Harvard had learned Visible Speech from Bell's book alone. At a party given for him by Hill, Bell persuasively expounded his system to James Russell Lowell and other literary Brahmins, as well as a public-spirited Cambridge lawyer named Gardiner Greene Hubbard. The latter, Bell wrote home, "Aleck will remember as the head of an establishment for the Deaf & Dumb at Northampton. He is very anxious that I should visit his school." Because of his concern for the deaf, Hubbard was especially interested in an account Aleck had begun sending of his work at Miss Hull's school.

The Cambridge gathering turned out to be the salvation of the American tour. "Everybody at my house the other evening was delighted with your exposition," wrote Hill. And he had more to report. The Lowell Institute lectures had for thirty years been the most famous in the United States, and their trustee now had offered Melville Bell six evenings that fall.

Late in August, en route to Chicago, the Bell brothers paused in Ontario to see old Scottish friends now settled there. Alexander Graham, Aleck's chosen namesake, had taken a bride, "the very best wife he could have selected," Melville wrote Eliza. The Reverend Thomas Henderson and his family were "very comfortable in a small way" and had named their Ontario-born daughter "Eliza Bell." All of the Bells' emigrant friends looked "*as young* as when I saw them last. The climate, trying as it is in its extremes of heat & cold, evidently agrees well with our countrymen." And Eliza replied, "I shall not be at all surprised if you are smitten with a desire to settle there. It must be a glorious country!"

The travelers thought Chicago "a most wonderful place," its buildings "magnificent . . . grand in proportions and palatial in solidity and splendour." But the presidential campaign had begun, free oratory glutted the market, and so the brothers returned to Canada while the election ran its windy course.

From the Hendersons' home at Paris, Ontario, Melville wrote his wife significantly: "We went today to look at a beautiful cottage close by this . . . and we could not help agreeing with Mr. Henderson that if *James* [Eliza's brother in Edinburgh] were to become the possessor of such a place, he might 'live like a prince' on his pension in this country. . . . You and I would delight in the little mansion I speak of. Grass enough for a horse & a cow — room for pigs and poultry! only fancy!" As to lectures, "we feel," wrote Melville, "that we are doing good pioneer work . . . [and are] looking forward to a future trip *with* our beloved wives." "I doubt if James would have the nerve" to emigrate, Eliza replied (and so it turned out). "I think in his place *I* should prove the better man of the two, for I would not hesitate."

Through countryside aflame with fall colors Melville returned to Boston and his Lowell-endowed lectures that October. "The lectures have been quite a 'hit,' " Melville wrote Eliza afterward, "and I have no doubt of their leading to plenty of work when we return together." The *Boston Transcript* thought them "rich in instruction, of great popular interest, and given in a manner no less effective than it was unpretending." The six hundred dollars they fetched turned a net profit for the whole trip. Thanks to the Lowell Lectures, Bell came home to London that December satisfied with his transatlantic mission.

During the lectures Bell had described Aleck's success with deaf children. A Boston schoolteacher named Sarah Fuller, when asked later to take charge of the Boston School for Deaf Mutes, would remember what she had heard and would follow it up with momentous consequences for Aleck.

In his father's absence Mr. A. Graham Bell, though barely of age, had served admirably as the family's acting chief in London. Tall, dark-haired and dark-eyed, self-possessed, convincingly grave when he chose to be, provided by nature and training with a rich, clear voice free of adolescent catches, quavers, and twangs, he passed for at least twenty-eight. That helped. Yet only genuinely gifted teaching could have sustained his achievements.

September was slow and Aleck restless. Adam Scott, just back from Scotland, "fired Aleck with the desire to become a volunteer!!" Eliza wrote her husband. But business picked up in October and Aleck's spirits with it. Alexander Ellis sent him some advance sheets of a projected book and readily accepted two or three corrections as to Visible Speech. By November Aleck's mother thought that the responsibility had "much improved" him, and his father was "delighted to find that Aleck is justifying my hopes in him. . . . Not one man in ten thousand, of his years, could occupy his present position with credit! I am proud of him."

Already Aleck's reputation as a teacher of deaf children had begun to spread. Inquiries came in through family friends. When Miss Hull asked him to resume lessons at a lower rate, since the fees came out of her own pocket, Aleck happily left the terms to her. So in late August his weekly trips to Kensington began again. One day in October he came home elated over a sudden burst of progress by little Minna, who had been far behind the others until lately. He had made up a little story, and they had read it from his lips, Minna best of all.

In October Aleck began attending physiology and anatomy classes every weekday at the university. He joined the College Medical Society and after a few weeks watched surgical operations and visited hospital wards with the doctors. Despite this busy round, which probably helped confirm his night-owl habits and perhaps also increased what his mother called his "head-achey fits," Aleck now and then put business aside. He spent an August week in Dover, where his mother and grandmother were vacationing, and rambled over the hills and cliffs. He found it easier, however, to abandon himself to work than to play. A certain aloofness showed itself in this prematurely serious young man. "Rather sorry I went," he wrote his father about his Dover visit. "Was introduced to people I would not care to recognize if I went there again."

Nevertheless he had friends, and among them was Marie Eccleston, a buxom lady some years older than himself. Wherever and whenever they may have met, by the summer of 1868 Marie was writing Aleck long letters. Toward the end of September Eliza Bell wrote her husband that Marie might come to London soon, and if so, Aleck "may know his fate." Not until well into November, however, could Eliza report that Marie was on her way. "I am more and more convinced," wrote Eliza, "she would need to marry somebody with a purse and I doubt if she could sit down contentedly to home duties — without extra excitement." Considering that Aleck had yet "to *make* a purse," she added, "I think it would be better . . . if he selected somebody more like Carrie," Melly's wife.

Marie had not been at Harrington Square a day before Mrs. Bell warmed up to the idea of her marrying Aleck. "I suppose," Eliza wrote her husband, "it will cost him something to bring his courage up to the point. . . . I shall be sorry if he loves in vain — for he will not find many like Marie in strength of character." Having held her letter open for later bulletins, Eliza added: "She and Aleck . . . are off to Highgate." Then: "Our young folks got home *late* in a pouring rain. I have not been able to ascertain particulars — but A is *hopeful*." He had to make do with hope alone, however, for Marie went off after Christmas as a pupil-teacher in a German school.

So the only recruit to the Bell family that year was Edward Bell, born to Melly and Carrie at Edinburgh on August 8, 1868. He looked just as Melly had at that age, the older generation agreed, with eyes that presently lost their baby blue and turned dark and brilliant like his father's. "He has very *fine* eyes . . . and looks so hard at one sometimes," wrote his Grandmother Bell. Melly sent a picture of him, "a determined little fellow, with his large eyes and clenched fists." He certainly seemed determined to live, despite the obscure sickness that more and more evidently fastened on him. Carrie grew wan and tired with constant caring for him, and the cloak of jocularity wore thin about Melly's customary reference to the "little nuisance." At times he had seemed to be gaining, but early in 1870 he died.

Before the baby was two months old, Eliza Bell had begun to wish that

Melly himself were not so thin and pale. He put a brave face on matters as long as he could, both for others' sake and for his own. "Business has gone on wonderfully," he wrote his father in the fall of 1868; "as new expenses have arisen money has always come to meet them, sometimes just in the nick of time." He tinkered with his inventions in the old workroom at the top of 13 Charlotte Street and wrote enthusiastically of his new speaking machine, with "a very curious whispering and inflecting glottis."

Yet Melly's turn toward spiritualism that summer and fall may have been a response to intimations of mortality. Alexander Ellis, himself a convert to spiritualism, had once persuaded Melly's father to attend a séance. The elder Bell, by registering expectancy at the proper times, persuaded the table-rapper to spell out his sister's name as "Isabella" instead of Elizabeth. This test settled the question to his satisfaction, though he tactfully concealed the trick from Ellis. Melly nevertheless sent his father a book on spiritualism and urged him to "test the phenomena" for himself. And with Aleck, Melly made "a solemn compact that whichever of us should die first would endeavor to communicate with the other if it were possible to do so."

"Another day of fog and wintry cold," Melly wrote in an undated letter of this period. "I felt very coughified all night . . . I have a beastly headache. I have always a headache in thick fog. . . . I have not been up in my workshop for a long time. I don't know how it is, but though I have plenty of ideas, I have not got the energy to go up and work at them." "I don't know how it is" — perhaps he tried to conceal even from himself the suspicion that what had destroyed Ted was upon him too. He laid his increasingly longer and harder spells of sickness to bilious attacks or other causes, meanwhile struggling to keep up his teaching engagements. But at least once, probably in the spring of 1869, Aleck had to take over at Edinburgh while Melly recuperated in the country.

Aleck himself had not been happy as his father's assistant or junior partner in London during these last years of the sixties. The dream of self-reliance seemed to recede as fast as he approached it. Occasionally his father went off to Dublin or the provinces, leaving Aleck in charge again. Once he took Mrs. Bell with him in order "to give Aleck a sort of apprenticeship to housekeeping but . . . he left everything to the servants and they have made things fly. However he might be wiser in this respect on another occasion." Aleck at times studied into the early morning, whether as a university student or independently is not clear. His headaches recurred now and then, and his father thought him "far from strong."

Still Aleck did not let life slip from him as it was doing from Melly. Adam Scott remembered that "one winter [Aleck] got up at unheard-of hours to visit a cellar and take records of the barometer and thermometer. Why, I do not know." Scott recalled that Aleck had "tried his hand at an electric piano," though with little success. Playing the family piano, Aleck had been fascinated by the phenomenon of sympathetic vibration. He had noticed especially that by pressing the pedal that lifted the felt dampers from the piano wires and then singing into the piano, he could sound the wire that matched the pitch of his voice, the others remaining silent. He had noticed also that one piano would echo a chord struck on another nearby. (At 18 Harrington Square, thin walls separated the parlor from the apartments on either side, each of which had a piano.) His notion of an "electric piano" involved an electromagnet under each piano wire, all on a circuit with tuning forks arranged as in Helmholtz's apparatus. The vibratory current from a tuning fork would produce magnetic impulses of that frequency in all the electromagnets, but only the piano string tuned to the fork would sound, by the principle of resonance or sympathetic vibration. Aleck made no effort to build so elaborate a device. It remained only a concept, his vivid imagination doubtless playing out scenes of triumph like those of young Wheatstone with his "enchanted lyre."

The young pianist naturally thought in terms of chords, and so the faint foreshadowings of a momentous concept crept into his mind. Thus far they reached only to the assumption that if several tuning forks transmitted impulses simultaneously over a single wire, the whole jumble of vibrations would be felt electromagnetically by each piano wire, but the wire would respond only to that frequency in the mixture that it happened to be in tune with. The principle of sympathetic vibration would unscramble the mingled frequencies. Having been an amateur telegrapher since his Bath days, Aleck saw further that not only could a chord be transmitted over a single wire and unscrambled at the other end, but also a number of simultaneous Morse code messages sent in different pitches. His mind was hovering about the idea of a multiple telegraph, of a kind that would be called "harmonic."

But his mind did not alight just then. The demands such a project would have made upon time, money, and study were too formidable. Instead, in the spring of 1869 Aleck turned his mind back to the wheat-cleaning machine he had suggested to John Herdman, the miller, years before. From Edinburgh, where he was substituting professionally for the ailing Melly, he tantalized his parents with a promise of great news to come. His letters of this period do not survive, but from his mother's reply to the eventual announcement it seems that he had refined or elaborated on his boyhood notion with a view to patenting it and introducing it commercially through the Herdmans. Knowing the intensity of Aleck's enthusiasms, Mrs. Bell cautioned him not to be too excited and hopeful about it. "If the Herd-mans do bring anything out of yours," she wrote, "I hope they will be just, because it is not always the inventor who reaps the benefit. The man who possesses money to carry a scheme into effect, generally sweeps off the returns. So remember my maxim and don't build too high." From sub-sequent silence on the matter, we may guess her advice to have been sound.

Aleck demonstrated physical as well as intellectual resilience. Early that September he and his Dublin cousin Chichester Bell set out on a walking tour of Devonshire. They were eating "a solid breakfast before starting — and a tremendous tea-dinner in the evening," stoking their energies mean-while with splendid ripe blackberries along the roadside. They reached Bideford soaked in sweat, but then the air turned salty and "delightfully bracing . . . I felt as strong as a horse the moment I got near the sea." He and Chess walked twenty-three miles one day, twenty-seven another. Then Chess went south and Aleck took lodgings for a week at Ilfracombe. "This delightful watering-place," he wrote, "has quite charmed me, and I anticipate many a pleasant ramble among the coves and Elgin-like cliffs on the shore." One night he "astonished the neighborhood with a *velocipede* performance." The next day, with a novel and a telescope, he scrambled to a snug nook on the top of a two-hundred-foot cliff, whence he could see the dim outlines of Welsh mountains far across the water and with his telescope "could see all that went on on the parade or in the boats . . . even scan the faces of happy couples seated in romantic coves." Here was no pallid Victorian decline.

A year or so later Marie Eccleston wrote him: "Do you remember going to the 'Devil's Glen'? On our way home you took a race up a stony ascent while we walked up the hill. As you came back I thought 'if Aleck had but freedom amongst some hills for some months at a time every year, there would be no more pale cheeks or headaches.' "

Aleck's plans for marriage with Marie seemed to be in suspension, and as much by his will as hers. "I am glad to hear Aleck is 'deep in study,' " she wrote his mother at one point, "else I should have begun to think it was 'out of sight, out of mind' as far as I am concerned, or that I had *unwittingly* offended him." Aleck may have kept his distance from Marie in proportion to his distance from the independence he considered a pre-requisite to marriage.

By the summer of 1869 Aleck's father contemplated a move that would have given Aleck his independence by leaving the field to him in England. Though the world had not risen to the opportunities of Visible Speech as he had hoped it might, Melville Bell had gone on to develop a shorthand system along Visible Speech lines. The *Shorthand Writers' Journal* thought it clear, esthetically pleasing, correct in analyzing sounds, but not swift and flowing enough in the writing to displace Pitman's deeply entrenched system. The forecast was correct. Moreover, Melville Bell's lecturing drew no crowds. "A very good report," wrote his son Melly after one, "but the old story, a meagre audience. We hope you are not at all events out of pocket." At fifty, the intellectual and professional goal of twenty years now reached and passed, Melville Bell may have felt a slackening of optimism, a relaxation of ambition. Aleck's account of his Devonshire idyll started Eliza thinking about a Devonshire retirement for her husband and herself.

Her husband contemplated a more radical change. Whitney notwithstanding, the United States had been hospitable. In April 1869 a Boston elocutionist, Lewis B. Monroe, sent Bell a little manual of his own, which made a flattering acknowledgment of Bell's contributions to the field. Elocutionary studies had been unjustly ignored in Boston before Bell's visit, wrote Monroe to Bell, but a European luminary commanded more attention and respect than a native American. A couple of months later Thomas Henderson wrote from Ontario urging the Bells to settle in that place. Both Melville and Eliza liked the idea, and it struck them that the bracing climate might help Melly.

Melly hesitated. To emigrate would be to admit how seriously ill he was, and he resisted any such admission. Aleck, for his part, feared that Marie would not want to emigrate. By the fall of 1869 his parents were worrying about Aleck's health as well as Melly's and so did not want to leave him alone in London. That winter Melville Bell was invited to give twelve Lowell Lectures in Boston in the fall of 1870. Pending that trip, the Bells decided to leave open the question of emigration.

In February 1870 little Ted died. A few weeks later, as spring came to Edinburgh, Melly seemed at last to recognize the speed of his own decline. "My dear Aleck," he wrote,

I *do* wish you would give me a line now and then. . . . Is it that your scientific attainments are so much further advanced than mine that you can take no interest in the little I can say on such subjects? . . . As for your fighting shy of me, because I sometimes make fun of such of your ideas as strike me as ridiculous; it is absurd to give such a reason; and I suspect that this is the foundation of all I complain of.

Write soon like a good chap.

In retrospect Aleck must have recognized the despair and loneliness in that appeal. But at the time, far away in London and holding to hope as Melly himself had done for so long, Aleck and his parents awoke to the truth only upon an urgent call from Edinburgh. Aleck hurried north to take charge of Melly's professional engagements and help Carrie with household matters. He found his brother very near death from tuberculosis, though conscious and lucid.

With Melly's consent, Carrie took him away from the smoky dampness of Auld Reekie by sleeping train to London, Aleck remaining at Charlotte Street. At Harrington Square, Melly lay conscious and, by his father's account, unterrified for three days before he died on May 28, 1870. He was buried at Highgate Cemetery beside Ted and Grandfather Bell.

The news of Melly's death, though no surprise to Aleck, hit him hard. Even in his old age, the look on his face would impress his grandchildren with the depth of his feeling when the tragedy was mentioned. Aleck honored the "solemn compact" he had made with Melly the year before. "I well remember," he wrote several years later, "how often — in the still-ness of the night — I have had little seances all by myself in the half-hope, half-fear of receiving some communication . . . and honestly tried my best without any success whatever." The failure left him thoroughly skeptical about spiritualism.

It may have been after Aleck's return to London for Melly's funeral that his father, more worried than ever about Aleck's health and remembering his own restoration to well-being in Newfoundland thirty years before, asked his only living son to emigrate with him, Eliza, and Carrie to Canada. Aleck felt trapped. He remembered later having walked the London streets for a long time that night trying to see a way out. He was now his parents' only living child; he could not let them go alone. Yet in England, or better yet in Scotland, he saw ahead the life he loved, that of teaching, with the woman he loved or thought he loved, Marie Eccleston. The family reputation would give him a powerful start. Perhaps he would win the fame he craved. Besides, he felt strong ties of memory and sentiment with his native land.

Back at Harrington Square he found the light still burning in his father's study and entered the house thinking he would refuse or at least resist, seeing his whole future happiness at stake. His father and mother sat there silently, holding each other's hand and looking at him inquiringly. A sense of their loneliness undid his resolution, and somehow he found himself unable to keep back words of comfort, which they grasped at as consent.

A day or two later Aleck took the train back to Edinburgh to arrange the sale of his father's property and to wind up Melly's affairs. From Charlotte Street he wrote his father:

The dream that you know I have cherished for so long has *perished* with poor

Melly. It is gone and for *ever*. If you exult at this please have the heart not to let me know it. I do not wish to have it referred to again. Do not think me ungrateful because I have been unhappy at home for the last two years. I have *now* no other wish than to be near you, Mama, and Carrie, and I put myself unreservedly into your hands to do with me whatever you think for the best. I am, dear Papa

Your affectionate and *only* son, Aleck.

He may have hoped to be let off his promise, but his father took him at his word.

"The dream" that Aleck had "cherished for so long" may have been of independence, or of partnership with Melly, or of winning a university degree, or of marrying Marie Eccleston. In any case he wrote Marie a letter of farewell and enclosed two "cartes de visite" or small photographs of himself with more passionate expressions in shorthand.

The relief of decision and the excitement of preparing for the trip to America seemed to restore Aleck's buoyancy. A young Australian with a bad lisp had come to take lessons from Melly. In two weeks of concentrated work, Aleck managed to cure him. The young man wrote home: "I hardly ever now speak the old way though that habit of course had been of twenty-four years standing. Hurrah!" Aleck meanwhile paid bills and disposed of Melly's piano, conjuring apparatus, and household goods. One day he "collected all the things that were to be kept . . . locked myself in [the study] and tried to imagine myself in the Backwoods of Canada."

Marie Eccleston wrote him at last on July 2 in a tone of resignation perhaps a shade too cheerful for Aleck's self-esteem. She professed to have forgotten her shorthand and agreed with his father that the change would be good for Aleck's health. And she offered him advice that reflects something of young Aleck Bell's feelings and personality:

Don't grieve about your examinations, &c. — all the degrees in the world would not make up for ill-health. . . . Make a name for yourself away. Don't get absorbed in yourself — mix freely with your fellows — it is one of your great failings. . . . You see in all this arrangement your father is only *letting* out the love he has borne towards you your life long, tho' when sarcasm has made you feel thoroughly hurt, you half doubted he had it for you. As to your mother, she is so gentle you *experienced* hers. . . . How I shall miss you all! . . . I scarcely expect you will return, England would be too *slow* for you after America. Even your father caught the "go-ahead" infection the short time he was there.

Five years later Aleck would look back on his rejection by Marie as "the *one sore subject of my life*." Having by then found and won his true love,

68

he could assess the affair with less pain than before. "I had created an ideal which had no existence," he wrote. "Still I believe that I should have learned to love her had I not awakened to the conviction that she cared nothing for me — that she merely wished to give me sufficient encouragement to keep me — in case no other came forward." And he added: "Ever since Papa's unintended conduct in the matter of Marie Eccleston made me feel so bitterly . . . it has almost been impossible for me to approach him confidentially." That reference was not further explained, there or elsewhere.

Thus old ties were broken. The sailing was set for July 21, 1870. Having auctioned off the furnishings of 18 Harrington Square, the elder Bells took lodgings for a few days. Aleck returned to London, and on the last evening there Adam Scott came up from vacationing in the country to reminisce with them. Early the next morning Scott, the old family nurse, and the Skye Terrier Trouve constituted the farewell committee at Euston Station. The little dog seemed perplexed. He stood quietly, almost sadly, as Melville and Eliza Bell, the widowed Carrie, and Aleck, his young instructor in elocution, petted him in turn. Even when the emigrant party entered the carriage he made no attempt to follow them. As the train moved off (Scott later remembered), Trouve looked fixedly at the familiar faces in the carriage windows, "and his eyes followed till they disappeared to the view, then he stood staring at vacancy."

PART TWO

The Telephone

Alexander Graham Bell, 1876

7

Brantford and Boston

Just before leaving England, Aleck Bell had bought a French translation of Helmholtz's book *On the Sensations of Tone* and also a small notebook. Only the latter survives, headed "Thought Book of A. Graham Bell July 19th 1870." If he read the Helmholtz book on shipboard, he set down no thoughts about it in the notebook, only a nostalgic list of incidents in his past life and some telltale philosophizing: "A man's *own* judgement should be the final appeal in all *that relates to himself*. Many men . . . do this or that *because some one else* has thought it right."

The family's landing at Quebec on August 1, 1870, the shepherding of baggage, the transfer to a "Palace Steamer" for Montreal, the clear, blue sky and brisker life and novel sights of the New World, left no time for introspection and brooding. There followed a visit with the Hendersons in the town of Paris, Ontario, word of a likely property in the nearby town of Brantford, and on August 6 its purchase.

For $2600 the Bells got a comfortable house and outbuildings with ten and a half acres of land at Tutelo Heights, four miles west of town. The house, white with black trim, stood well back from the main road, screened by tall trees and thick shrubbery. Upstairs it had four bedrooms, downstairs a long parlor and a sitting room with French windows opening on a front porch with gingerbread trim, as well as a study, dining room, large kitchen, auxiliary kitchen, and bathroom. Jutting out from one side of the house was a glassed-in plant conservatory, and behind it a small workroom. That fall the lab equipment from Harrington Square arrived all smashed in its packing cases. But more was bought, and so the little workshop took its place in the history of technology.

Along with the main house went the stable, carriage shed, henhouse, pigsty, icehouse, well, and rainwater cisterns. In the orchard grew apple, pear, cherry, plum, and peach trees. The property ran back to a high bluff, around the base of which coiled the Grand River on its way to Lake Erie. From the tree-lined edge of the bluff could be seen the peninsula called "Eagle's Nest" and, several river twists and loops to the north, the spires and factory chimneys of Brantford, Ontario, population 13,000. Southward the heights curved sharply and sank down to the flat expanse of farm lands known as Bow Park, a peninsula of nearly a thousand acres, the property of the Honorable George Brown, born in Edinburgh, educated at the Royal High School, and now the influential editor of the *Toronto Globe* and a leading member of the Liberal party in the Canadian Parliament.

Among the birches at the edge of the bluff, Aleck, in the role of invalid, spread out rugs and pillows and dreamed away summer days with his books, including Helmholtz's work. In later years he liked to talk romantically of having come to Canada as a dying man. However near he came to the brink of the grave, the chill of a Canadian autumn presently chased him from the edge of the bluff. By early October he was jolting over "a most *awful* road" on his way to record Indian words in Visible Speech. A week later he and the others were "picking apples as fast as we can."

Meanwhile Melville Bell had been off lecturing in Canada and the United States. In October his Lowell Lectures on Shakespeare pleased the Bostonians generally and Sarah Fuller in particular. A year earlier, at the school board's invitation, Miss Fuller had organized a public day school for deaf children, the Boston School for Deaf Mutes (eventually renamed the Horace Mann School). Remembering Bell's earlier remarks about the use of Visible Speech for the deaf, Miss Fuller had looked up his books at the Boston Public Library, but had decided to wait for personal instruction when, as she confidently expected, he came back for an encore series.

Melville Bell told her that after working twenty years to develop the system he did not want to spend the rest of his days teaching it. But Aleck, obviously restive, presently wrote him that "I should not personally object to teaching Visible Speech in some well-known Institution if you would get an appointment — even if it was not renumerative." So in November 1870 Melville was able to write his son that a month's appointment to initiate Visible Speech teaching at the Boston School was in prospect, as well as similar engagements at such places as the Clarke School for the Deaf at Northampton. Boston's administrative mill ground slowly, but at last, late in March 1871, word came from the Boston School's original proponent and chief supporter, Dexter King: the $500 appropriation had been passed, and Aleck should come as soon as convenient.

Early in April 1871 an Edinburgh scientist, writing one in America, grew fretful at the handicap of distance. "Five minutes' conversation," he wrote, "is about as much as thirty pages of letter paper, and infinitely more intelligible. All the boasted civilization of the 19th century has not been able to give us anything even remotely suggesting an equivalent for a chat over a quiet pipe." Before that letter could reach America, there came a conjunction of the man and the place suited above all others to make the gift.

Boston met Alexander Graham Bell with a smile. Snow had vanished even from the woods and hills, and the countryside had gone freshly green. On his way from the station to his rooms on Beacon Hill, Bell heard the sounds of spring in the afternoon sunlight of April 5: the jubilation of birds, the rush of the northwest breeze through the trees on the Common, the shouts and laughter of children playing about the Frog Pond and along the Mall, the clatter and rumble of horse-drawn traffic in Boston's narrow streets. At 2 Bulfinch Place his landlady greeted him like an old friend.

Ascetically spare and erect, the Reverend Dexter King had the look of an Emerson: aquiline nose, strong chin, high cheekbones, and lean cheeks — a Yankee face softened by kindliness and white sideburns. In the morning he took Bell for a walking tour of the city. In accord with an old tradition, the governor had proclaimed the annual "fast day," set aside for "fasting, humiliation, and prayer." But the weather was fitter for exultation. People from the suburbs jammed the horsecars and converged on the city for religious services — and strolling. Under a warm sun the breeze blew cool and bracing. The two men may have paused in Louisburg Square — like a small London square transplanted — or among the formal flower beds and close-clipped lawns and splashing fountains of the Public Garden. But what struck Bell most forcibly was the evidence he saw of Boston's resources in learning: the brick and sandstone Public Library on Boylston Street, largest library in the nation and free to all; and standing out boldly on the flat, half-vacant, new-made land of the Back Bay, the Massachusetts Institute of Technology, an edifice with a Corinthian countenance but a thoroughly modern brain. That very evening at a meeting in the Institute building, Professor Edward C. Pickering was to give a paper on acoustical theory.

Before the sun had set, Bell began tapping Boston's reservoir of knowledge. He called on his father's friends and admirers, Professor and Mrs. Lewis Monroe, and almost at once their conversation turned to the science of sound. Monroe not only shed new light on the experiments of Helmholtz and John Tyndall but also gave his visitor a copy of Tyndall's new work on sound. In turn Bell fascinated the Monroes with his talk about the pitch of whispered vowels and his demonstrations of larynx-tapping.

Next day King took Bell to the "new" State House that Charles Bulfinch had built on Beacon Hill in the 1790s and that Oliver Wendell Holmes had more lately labeled the "hub of the solar system." Its dome, not yet gilded to the taste of the age, ruled Boston's skyline as St. Paul's did that of London. In the House chamber Bell heard a Negro member speaking on the Ku Klux conspiracy in the South and judged him to be "an educated gentleman" who read his speech "well but timidly." (That evening, at a play called *The Octoroon*, Bell noted wryly that "American Prejudice is too strong to permit of the young man *marrying* the Octoroon and so she dies.")

Most exciting that day was his discovery that the Institute of Technology had a complete set of Helmholtz's apparatus. "Prof. Monroe is going to *repeat Helmholtz' experiments with me* shortly," he wrote home elatedly. And when Saturday turned "intolerably *hot*," he retired to his room in shirtsleeves "to devour Tyndall's book" and also a bagful of oranges.

But the challenges of teaching were about to crowd acoustics to the back of his mind.

At the School for Deaf Mutes, he found some thirty boys and girls for whom Boston with all its bustling was the noiseless hub of a soundless universe. He liked them all at first sight. To the world they were uncannily mute, but only, Bell was sure, because the world was mute to them. Sarah Fuller, the principal, won him completely with her kind, firm, square-cut features, her obvious ability combined with modesty, her shining love for her young charges. "I never saw *Love*, *Goodness*, and *Firmness* so blended in one face before," Bell thought.

As he had done at Miss Hull's school in London, Bell drew a face on the board, had the children point to the corresponding parts of their own, rubbed out all but the elements conventionalized in Visible Speech symbols, and within half an hour had even the toddlers identifying four classes of symbols. The teachers were "thunderstruck," the children delighted. Miss Fuller had hoped for much but nowhere near so soon, she admitted. Then Bell plunged into individual assessment of and experiment with older pupils. The gaslight at Miss Fisher's boardinghouse shone that night on a young man happily lettering big symbols on cardboard for classes yet to come.

After only a day or two, Bell astonished the school's governing committee and guests with the results already achieved. The *Boston Journal* reported the exhibition and Bell proudly sent clippings to his friends and family. It had all been disarmingly easy. But in the weeks that followed, Bell faced the test of anticlimax.

By the beginning of May he had evaluated each pupil, instructed the teachers, and started systematic instruction, taking the most advanced group himself. A harder test came with the evening class of deaf men he began on his own account. On the third evening, grown overconfident with success, Bell pushed the men too rapidly. "*I made a mistake,*" he wrote home. The attempt was "a most *ignominious failure!*" They knew what to do but not how to do it, could bring forth only "the most lamentable *squeaks* imaginable," and went home with long faces. Bell resolved next time simply to write on the board whatever outlandish sounds they happened to make and thus help them find their own way. Only three came back next day, but for these the new way worked. This became a lifelong part of Bell's teaching technique: building confidence, one small triumph at a time.

Like his pupils, Bell grew in confidence with each success. He wrote his father proudly that he had been "obliged to decline no less than *three private pupils,*" two deaf girls and a stammerer. "I find myself making headway every day. New ideas are being constantly suggested by the defects arising in the school." He experimented with the use of a toy balloon to warn a deaf child of approaching danger in the street. If held tight against the body, he reasoned, it should vibrate in sympathy with street noises. But in practice it picked up too many harmless noises, whereas the child's feet could better distinguish the beat of hoofs and rumble of wheels.

At the suburban home of Miss Fuller and her sister Mrs. Jordan, he became "Uncle Allie" to the three young Jordan girls. From the Jordan home he took long hikes through the New England countryside in the warm sun of late spring. At last he was on his own again and flourishing. He decided to go on with his new life after the summer, to come back to Boston in September and give two or three months of private speech lessons. All he asked of his father was advice on what rates to charge. "I look to the establishment of a good profession here — perhaps ultimately to the foundation of a *Normal School for Deaf-Mute Teachers!*"

For a time he swung back from his euphoria. Progress in the Boston school began to seem much too slow for the effort it cost him and his pupils. Lewis Dudley, a trustee of the Clarke School for the Deaf in Northampton, wrote almost begging him to give a month to that school, yet Bell felt that he must go home to Brantford. Just then his father wrote something, perhaps disparaging his normal-school plan, that triggered an attack of nervous insomnia. Young Bell began to dread the coming public exhibition at Miss Fuller's school. He nerved himself and triumphed again. The superintendent of Boston schools thought the results "more than satisfactory; they are wonderful." The system "must speedily revolutionize the teaching in all articulating deaf-mute schools." Nevertheless it was a pen-

77

The Boston School for the Deaf, June 21, 1871. Top row: Dexter King, left; Alexander Graham Bell, right. Fourth row from top: Sarah Fuller, second from left; Mary True, far right

sive, homesick and rather hollow-cheeked young man of twenty-four who posed for a photograph on the first day of summer with Dexter King, Sarah Fuller, and the rest of the school on a flight of front steps in Pemberton Square.

His Boston visit ended, Bell discussed Visible Speech at Northampton with Lewis Dudley and the Clarke School's principal, Harriet Rogers. Dudley promised to send his deaf daughter Theresa to Bell's private classes in Boston that September, and he went with Bell to Hartford next day for a talk with the principal of the American Asylum for Deaf Mutes. Then it was home to Brantford.

Aleck found restoration in the bright days and cool nights of the Canadian summer, in the unhurried chores and pleasures of Tutelo Heights. That spring a big tree had toppled from the edge of the bluff, and in the shelf-like hollow it left, Melville Bell had put flooring and a wooden seat and had trained a wild vine around the little terrace. From there the view of the valley, the river, and Brantford was more sweeping than ever. Aleck called it his "dreaming place." One of Miss Fuller's young nieces wrote her "Uncle Allie" a child's news of Newton Lower Falls. "I am enjoying my holidays very much," he wrote back in August. "We have the loan of a nice boat and I often go for a row on the Grand River. What between fishing and shooting and swimming and boating and riding and writing my time passes very quickly. September will soon be here — and I am anxiously looking forward to seeing my little nieces again."

Refreshed in spirit, Aleck now looked back on his work of the past spring as the kind on which a full, proud life might be built. The children whose minds he had opened to the world had in turn opened his heart. A letter from Hartford asked his help, and Lewis Dudley promised Bell an engagement for March and April at the Clarke School.

So as he coursed the woods and river, Bell's thoughts kept returning to speech for the deaf. "I am hard at work on the Combination Exercises," he wrote Sarah Fuller in July. He fretted about a deaf boy he had encountered, "a smart little fellow, growing up with a few or no ideas because his parents *don't know what to do.*" Bell stirred up the boy's father and wrote Miss Fuller for the name of "some *picture book* suitable for the little fellow."

Late in July, while Bell found new health, Dexter King died in Boston. "He leaves a wife, but no children," said the newspaper report. Yet in a sense Dexter King left many children. His devotion to the cause of all children, especially the deaf, had strengthened Bell's own growing commitment. King's death ended a guarantee he had given Bell of pupils for September. But Bell did not waver. "I go to Boston in *faith*," he wrote Sarah Fuller in August; "please have the goodness to insert the enclosed advertisement at *once* in what you consider the most influential paper in Boston."

His course had been set. Thenceforward to the day of his own death, Alexander Graham Bell would proudly count himself above all else a teacher of the deaf.

8

Teacher of the Deaf

Only four pupils showed up when Bell came back to Boston in September 1871. But Theresa Dudley's hundred-dollar fee paid two-thirds of his living expenses. She was his first congenitally deaf pupil, a girl of seventeen who had never heard a human voice. Even after four years at the Clarke School, her speech was slurred and unpleasant, she had no voice control, some vowels and consonants were wrong. She could scarcely be understood.

Day after day that autumn Bell and Theresa worked two hours or more at a session. He sometimes wrote out his comments and instructions. In the first lesson, for example, "P" was one of the sounds. "The lips must be *blown apart*," he wrote down. "Hold this feather before your mouth. Now say 'P.' When you open the lips the feather scarcely moves — the air is not *trying to come out*. I *try to keep my lips shut* but the breath forces them open and blows away the feather. Bite this ivory plug. Now if you open the mouth the plug will fall out. Keep biting. Now sound 'P.' That is perfect."

A week or two passed. "When I push your chin up your head is moved back," he was telling Theresa now. "Try and move my head. You can push up my chin as much as you like; it is as solid as a rock. . . . Your voice will become better if the head is kept *firm* and the chin *in*. . . . That is beautifully done. I am very pleased." And another week went by: "You are progressing *most satisfactorily*. Today you have completely mastered *all the English sounds* — and you are rapidly getting command over your *voice*." Then came the challenge of inflections, of learning to make tones rise and fall properly. "You were doing it very nicely *slowly* when you laughed and *spoilt it!*"

"Does the voice in your throat *shake* while talking?" Theresa wrote once.

"The air is *always shaking* when I make voice. Put your hand on my throat and tell me what you feel."

"I feel your throat shaking."

"It is that *shaking* that we call 'voice.' You *feel* it, and I *hear* it."

On November 29 some of Boston's leading educators came to hear Theresa Dudley and other pupils demonstrate their progress. "I cannot describe to you the effect produced," Bell wrote home exultantly. Boston newspapers reported the demonstration with admiring wonder. "I believe," Bell wrote, "this experiment constitutes an epoch in the History of the Education of the Deaf and Dumb."

As a dutiful son he made much of Visible Speech in Theresa's triumph and otherwise. He gave a week's intensive training in it to a teacher sent from the Clarke School, he answered a couple of inquiries from schools for the deaf, and he wrote an article on it for the *American Annals of the Deaf and Dumb*. Appearing in January 1872, the article spread the news through the silent world of the deaf. Bell later mailed out 150 offprints. An address he made on Visible Speech in December 1871 to the American Social Science Association (which happened to have its headquarters next door to Miss Fuller's school in Pemberton Square) was later published in a popular magazine.

He drew the line, however, when his father urged him to be a wandering preacher of the Visible Speech gospel in a series of brief visits to deaf-mute institutions. Aleck still wanted to settle down with his own teacher-training school in Boston, especially after the Boston school superintendent publicly called on him to do so. He proposed to open it on March 1, 1872, "the birthday of the inventor of the system." This plan, he insisted to his father, would propagate the system more efficiently, would be easier on his health, and would put him "in such a position, that, if I still contemplated such a thing, I could be married" (although there were no prospects of that just then). The tug-of-war lasted well into the winter, even after Aleck's return to Brantford, punctuated by his sick headaches at moments of stress. His father's resistance delayed the plan past the time when Aleck could properly cancel spring commitments to Northampton and Hartford. But he firmly resolved to start his Boston school in the fall of 1872. That settled, his headaches let up.

Before Aleck left Boston in December 1871 there came two tokens of a future he did not yet foresee. One was a remark he made in his Social Science Association address: by using code numbers for Visible Speech symbols "a telegraphic dispatch may . . . be sent through any country without translation, and in the very words and sounds of the original message." The other was a two-week Boston engagement of "Professor

81

Faber" and his "Wonderful Talking Machine" at Horticultural Hall — the same machine seen by Joseph Henry in Philadelphia a quarter century before and by Aleck and his father in London in 1863, but not the same Faber. The inventor had died in the late 1860s and left the machine to his niece, whose husband was now exhibiting it under Faber's name.

Faber's machine had evidently lost its novelty or had been eclipsed by greater wonders of technology. At any rate, scarcely making enough to live on, the current "Faber" appealed to Joseph Henry for help at the Smithsonian Institution a year or so after this Boston appearance. Henry and other acoustical scientists could give him nothing more fattening than sympathy. The inventor's one triumph had reached a dead end. Yet it had not been entirely in vain, for it had once helped turn Aleck Bell's young mind to the mechanical reproduction of speech, and from that turn would eventually come something more profitable to the world.

At the moment, Bell had more pressing concerns than his long-past boyhood "speaking machine," among them his preparations to winter in the family home at Brantford, with Theresa along as a boarding pupil.

At Tutelo Heights during the winter of 1871–1872 Bell worked harder than ever with Theresa Dudley. "Every day's experience throws new light upon the difficult question — how to teach articulation," he wrote Sarah Fuller. "There are many new points I would like to discuss with you," he added, "but writing is to me a slow and tedious way of expressing myself. I long for one of our old confabulations!" He worked more and more like an experimental researcher, keeping alert to accidental hints. Playfully imitating Bell's lip movements, for example, Theresa produced what he called "nonsense-talking," and he seized upon it as a key to the problem of drawing proper inflections from her.

Early in March 1872 Bell brought Theresa back home to Northampton and began his own visit there. From the Clarke School buildings on Round Hill he looked out on a winter landscape. Bitter cold still gripped the Connecticut Valley below, and an icy wind swept through the hilltop grove of leafless chestnuts behind the school. But lengthening days warmed and softened the earth, and the Massachusetts countryside put forth a fresh green. In the sunlit spring afternoons Bell, rubber-booted for the boggy roads, took solitary rambles through the hills and valley, which he thought "most lovely." To give his mind a change he read up on geology and went rock hunting. "I am terribly afraid sometimes of having Visible Speech on the brain," he wrote home.

His enthusiasm and success inspired articles on Visible Speech in the nationally influential *Springfield Republican* and the widely read magazine

The Congregationalist. The Japanese commissioner of education turned up with two Amherst professors and looked into the application of Visible Speech to Japanese education, deaf-mute and otherwise. Bell gave the commissioner his last offprint.

But he was beginning to shift his emphasis. "I have been studying the subject of the Education of the Deaf and Dumb very deeply since I have been here," he wrote home in mid-April. "My feelings and sympathies are every day more and more aroused. . . . It makes my very heart ache to see the difficulties the little children have to contend with." He took up the study of sign language. Once he had regarded the education of the deaf as a way of applying Visible Speech. Now he began to see Visible Speech chiefly as a way of helping the deaf.

On April 8, 1872, the president of the Clarke School returned from a two-year tour of European articulation schools, and on that same day Bell showed him and the corporation what had been done and what was planned. "All went off splendidly," Bell wrote home. Thus auspiciously, Alexander Graham Bell and Gardiner Greene Hubbard at last met face to face.

Born in 1822, the son of a Massachusetts supreme court justice with American ancestry reaching back to 1635, Hubbard was named for his maternal grandfather, Gardiner Greene, who had brought a comfortable fortune with him from Ireland and eventually built it up to one of the greatest in Boston. Greene's famous estate included Pemberton Hill, later the site of Miss Fuller's school. Hubbard graduated from Dartmouth in 1841, studied law for a year, then entered a leading Boston law office. In October 1846 he married Gertrude McCurdy of New York City and settled down in Cambridge.

Hubbard, like Bell, had a concern for the welfare of others and acted on it with uncommon enthusiasm. Like Bell he also had a talent for communicating that enthusiasm. In a word, he was a promoter. As a commuter, he wanted quicker transit between Cambridge and Boston, and he did something about it in the mid-1850s by organizing the first street railway line outside the New York City area: the Cambridge Horse Railroad Company. In his race with a Boston rival to open first, he won in 1856 with hastily bought secondhand cars still bearing the painted Brooklyn destination "Greenwood Cemetery." He also got a water supply system for Cambridge and organized the Cambridge Gas Light Company.

As a lawyer in patent cases Gardiner Hubbard took a special interest in mechanical and electrical inventions. Telegraphy in particular fascinated him. As early as 1850, he wrote a long letter asking Joseph Henry's advice

Gardiner Greene Hubbard

about a Boston man's patent for a telegraphic recording device, which Hubbard contemplated backing. Nothing seems to have come of the scheme, but Hubbard's interest in telegraphy remained keen. He saw the importance of rapid communication, both local and long distance, in the fast-developing industrial society of mid-nineteenth-century America.

In 1847 Hubbard's only son died in infancy, and so did a daughter in 1867. Between them came Gertrude, Mabel, Roberta, and Grace. Mabel, born November 25, 1857, confronted her father with a harder problem in communication than he had encountered in telegraphy or transit; for when she was a bright little girl of five, scarlet fever left her totally and permanently deaf.

When after long suspense it became clear that Mabel's mind had not been damaged, Hubbard asked deaf-school officials if she could be taught to speak. They advised him to forget that idea and instead have her learn the standard sign language, devised by the Abbé de l'Épée a hundred years earlier. They knew, of course, that for three centuries deaf "mutes" had occasionally been taught to read lips and to speak. Thomas Braidwood's

Mabel Gardiner Hubbard, about 1863

school in Edinburgh, for example, had done that in the late eighteenth and
early nineteenth centuries. But the pioneer deaf-school founder in America, Thomas Gallaudet of Hartford, had preferred the French model, partly
because Braidwood refused to divulge his lucrative technique, partly because the "oral method" made heavier demands on the pupil's intelligence,
alertness, and perseverance. By Mabel Hubbard's time, all the New England states subsidized the education of their deaf children at the school
Gallaudet had organized in 1817, the "American Asylum at Hartford for
the Education and Instruction of the Deaf and Dumb." And the Hartford
school's prestige relegated the oral method to occasional private tutoring.

Gardiner Hubbard would not settle for sign language. He kept looking
for help until he heard of Samuel Gridley Howe's faith in the oral method,
which Howe had seen working in Germany twenty years before. At that
time, Howe had failed to get an articulation school started by the state of
Massachusetts. Now head of the Perkins Institution for the Blind in South
Boston and famous for his work with the blind and deaf girl Laura Bridgman, Howe was roused to a new try by Hubbard's visit. And what Howe

told Hubbard of the possibilities made Hubbard a redoubtable ally.

As chairman of the Massachusetts Board of State Charities, Howe swung the board's annual report like a saber against the American Asylum's methods, while Hubbard applied the weight of his business standing and high social caste. Their goal was a state-supported school using the oral method. Among their opponents, the Honorable Lewis J. Dudley of the Governor's Council, whose daughter Theresa had attended the Hartford school, did most to scotch their first effort in 1864. Howe fought all the harder, while Gardiner Hubbard and his wife Gertrude prepared a telling argument.

Finding that little Mabel could still speak a few intelligible words, the Hubbards insisted that she communicate orally. Her ability to speak survived, even grew. Then in a Maine village in the summer of 1865 they found Mary True, daughter of the town's minister, willing to join the family as governess. Miss True did much to enlarge Mabel's speaking vocabulary and general education. "She was my teacher for three years," Mabel said long afterward, "and my friend for all time."

Even before Mabel's deafness, her articulation had not been good, and Mary True could not now make it easily intelligible to strangers. But with effort and attention, even they could understand. And Mabel read, wrote, and thought in English, not in the sign language, which was remarkably expressive and more natural to the deaf, but far short of English in conveying abstract ideas. So Mabel scarcely thought herself apart from the speaking world in any respect. Interviewing her when she was nine, a Cambridge public school teacher reported her ahead of the average ten-year-old. "I am surprised," the interviewer added, "at the readiness with which she reads from the lips, as I have never talked with her before, and she understood me without difficulty."

Howe and Hubbard wore down the opposition. By 1867 they could point to Miss Harriet Rogers's new private school at Chelmsford, where the oral method had shown good results. In Northampton John Clarke offered to endow a deaf school handsomely. Most dramatic and impressive of all was the appearance of the eight-year-old Mabel before a legislative committee in January 1867. With unselfconscious ease, she confounded the doubters and converted no less an adversary than Lewis Dudley himself. The legislature chartered a school, and in October 1867, with Clarke's $50,000 endowment, it opened at Northampton as the Clarke Institution for Deaf Mutes, using the oral method exclusively. Harriet Rogers was principal, Gardiner Hubbard president, and before long Theresa Dudley became a pupil and her father a trustee.

Such was the train of events that brought the tall, thin, bearded Yankee and the black-haired young Scotsman together at Northampton in April

1872. Bell's impetuous enthusiasm and Hubbard's brisk assertiveness would now and then strike sparks in years to come. Fortunately for first impressions, no clash occurred this time.

On the education of the deaf, Bell and Hubbard thought alike. Three months before he met Hubbard, in fact, Bell had defended lipreading against the doubts of his own father, explaining that the context helped bridge the undeniable gaps in what could be seen of vocal action. As for telegraphy, Bell and Hubbard probably did not discover their common interest until months or years later. The Clarke School and the deaf in general were subjects too engrossing for small talk about sidelines. Moreover, after making a deep impression on the Clarke School, Bell soon moved on to the American Asylum in Hartford, that bastion of sign language, whose gates were by now ajar to articulation.

Bell arrived exuberantly in Hartford on May Day, 1872. Finding a rented piano still in his hotel room, he continued the rental of it. "I am *hungry* for music," he wrote home, "and you little realize the pleasure I feel in laying my hands upon the piano here." Nevertheless he forsook the keyboard that afternoon to watch the holiday games and stories at the American Asylum. During his two months' stint he carried on a demonstration class of ten congenital deaf-mutes and also gave vocal exercises and some Visible Speech instruction to all 250 pupils at once.

A correspondent for the *Silent World*, a magazine for the deaf, described the regular morning exercise. Standing up, the pupils stretched their arms back and forth to "open their lungs." Then

in a low tone and all together, they say what may sound like i i i i i i, as long as Mr. Bell wishes. He moves his hand, with thumb and forefinger close together, slowly from left to right, for this, and spreads out his fingers quickly when he wants them to stop. Then he begins again but with his thumb and forefinger wide apart, and such a roar comes up as makes the floor tremble, the windows rattle, and the hall resound again. With these simple motions . . . he has the whole two hundred and fifty voices, from deep bass to shrill treble, under sufficient control to make them roar in concert or die away softly. . . . The pupils like it. It is a new sensation to most of them. . . . People stop on the street to listen, and stare at the windows. The noise may be heard a quarter of a mile off.

Mark Twain may have heard it from his home on Forest Street but apparently did not investigate.

Bell's pupils all made striking gains in speech. In the year following, the two teachers he trained in Visible Speech found it to be the best way of teaching articulation, but doubted that the congenitally deaf, except for the most intelligent, could ever learn to speak readily. But by then Bell's own commitment to the deaf no longer depended on Visible Speech alone. At the graduating exercises he addressed the pupils in his newly acquired sign language, "a neat speech which all the pupils understood," reported the *Silent World* correspondent. The correspondent added: "He has won the friendship of everybody by his kindness, cheerfulness, and earnestness."

The plight of the deaf-mute possessed Bell's heart as well as his mind. "A poor boy died yesterday at the Asylum from dropsy of the lungs," he wrote home in May. "Nothing can show better how the too great use of signs tends to isolate deaf-mutes and constitute a class apart from hearing people than this boy's case. When he was dying, he did not *want to go home*. His friends, he said, could not *understand* him and he was happier in the Asylum." On a short visit to Northampton he saw the four-year-old deaf-mute daughter of a New York couple. Having no heart to correct her, the parents had let her become ungovernable. She broke things from sheer frustration, scratched her little brothers because they could not understand her, hugged them when they cried, then lay on the floor and cried herself because she could not understand her parents nor make them understand her. "The little thing gives one a most painful idea of what an uneducated deaf mute may be," wrote Bell.

"Quite well but just about worked out," Bell left for Brantford late in June 1872. At Tutelo Heights, he learned that Grandmother Symonds had died in Edinburgh at the age of eighty-four. But his mother had never looked so well as he found her. "I tell her that she has grown smaller, however," he wrote Sarah Fuller, "and she retorts that I have grown taller." His father still shook his head over the normal school plan as trading "a certainty for an uncertainty." But, wrote Aleck, "since he sees that I am determined to make the experiment, he will put no obstacles in the way and will assist me as much as possible. . . . My health is excellent."

Aleck's reputation also flourished. His article on Visible Speech came out that July in the magazine *Old and New*. In August, on an invitation arranged by Sarah Fuller, he met with six deaf-school principals in Flint, Michigan, at their annual convention, and read a paper on speech — a "mere motion of the air," he called it, a series of undulations, of which the frequency determined pitch, the amplitude determined loudness, and the shape determined quality or timbre.

Meanwhile the National Deaf-Mute College at Washington asked him to teach Visible Speech there for a while, perhaps even a year. In Washington, he speculated, he might persuade Congress to finance Visible Speech types and publications. But the president of the college saw no chance of that. And besides, "from all I hear," Bell wrote Miss Fuller, "Boston is the intellectual centre of the States, though Washington is the political centre." Boston it would be, come fall.

9

The Multiple Telegraph

No other section of Boston offered such architectural elegance and unity, so pleasing and coherent a street plan, as the South End. It was to Beacon Hill as Edinburgh's New Town was to its Old Town. But the herd instinct that sets fashion had begun to turn from it, one indication being the offering of rooms for rent. Late in September 1872, Bell took a two-room suite on the second floor of 35 West Newton Street, a tall, freestone-fronted house at the corner of Washington Street, facing on Blackstone Square's pleasant greenery.

A visitor would have alighted from the Washington Street horse car line and walked around the corner to the high front steps on West Newton Street. Up carpeted stairs from the entry hall he would have found a smaller hall from which a pair of sliding black walnut doors gave into Bell's octagonal reception room. Its big window overlooked Washington Street. He had furnished it with a marble-top parlor table, a suite of green furniture, and a black and green carpet. The adjoining room served as Bell's schoolroom by day. Its black walnut furniture was upholstered in black horsehair cloth. At night the sofa was opened out as his bed. Such was the first American home of which Alexander Graham Bell could call himself master.

During the school year a dozen or so pupils came to him, among them Theresa Dudley for two or three months, Susanna Hull from London for a month (somewhat to Bell's regret, since she had "very little ear" for speech), and Jeannie Lippitt, twenty-one-year-old daughter of a wealthy Rhode Island manufacturer, who like Mabel Hubbard had gone deaf from scarlet fever as a small child. Mabel Hubbard herself, back from two years in Germany with little improvement in her voice, for a while seemed likely to come, but she did not after all.

The pupil who stayed longest and needed most help was five-year-old George Sanders. Thomas Sanders, his father, came of an old Salem family.

George Sanders   Thomas Sanders

A lifelong fancier of horses, he had built up a prosperous horse farm in Vermont before coming back to Massachusetts to establish an even more profitable leather business in Haverhill. George, Sanders's first child, was born deaf. He was too young for the Boston Day School, but Sarah Fuller referred his father to Bell. So George came in September to live with his nurse at 35 West Newton Street. From the start Bell found him "a very teachable boy." "Little George progressing splendidly," Bell wrote home in November; "loving and lovable little fellow. . . . Expect great results with him if I can have him with me for two or three years."

On November 9, 1872, fire broke out downtown. The alarm came through Boston's electric fire alarm system, developed by the inventor Moses Farmer twenty years earlier as the first in the nation. But speedy notice did not save Boston's business district. The first raging outbreak carried brands high in the air to wooden mansard roofs out of hose range. Though the night was calm, the great blaze generated its own fire-filled windstorm.

Like half the male population of Boston, Bell walked the streets all night to watch the terrible spectacle, after having spent the day hiking in the country. He wrote an account of what he saw and sent it to George Brown's *Toronto Globe*. But the account does not survive, since the *Globe* already had all the stories it could use, and Bell kept no copy of his. Though he guessed he had walked forty miles altogether that day and night, he wrote home a few days later that he had been "sleeping like a top every night since." His health was "splendid," and he hoped to "bring home a little fat at Christmas."

About sixty acres of Boston's business district lay in smoking ruins. But as a sympathetic New Yorker observed, "the best treasures of Boston cannot be burnt up. Her grand capital of culture and character, science and skill, humanity and religion, is beyond the reach of the flame." What counted for Bell were Boston's "science and skill." Since the mid-forties, when it passed Philadelphia, Boston had been the nation's leading scientific center, and the opening of the Massachusetts Institute of Technology in 1865 helped it keep the lead during the seventies. It led also in making scientific apparatus, at least in physics (including electricity). And as chief city of New England, which was the cradle of American industry, Boston had plenty of skilled artisans, inventors, electricians, machinists, engineers — the technological elite of the nation.

Electrical inventors congregated at the Court Street workshop of Charles Williams, Jr., which made electrical instruments and apparatus to order. Moses Farmer often came there to have work done. Four years before, a countrified telegrapher named Thomas A. Edison, just three weeks older than Bell, had been fascinated by the improved fire alarm telegraphs being developed there. In a corner of the shop during his off-duty hours he tinkered with what had been his own dream for three years and would cost him seven years more of on-and-off labor. It was to be a multiple telegraph for the transmission of several messages at once over a single wire.

Emboldened by the cautious interest of some Boston capitalists, Edison quit his telegrapher's job to become a free-lance professional inventor, based at the Williams shop. His first patent, an electric legislative vote recorder, turned out to be just what the politicians did not want. His partners sold out control of his second, an improved stock ticker, for Edison had not yet learned the tricks of business in the Gilded Age. Troubles in perfecting his multiple telegraph ran him into debt. So in the spring of 1869 Edison left Boston for a fresh start in New York City.

Three years later his Scottish-born contemporary in Boston caught the multiple-telegraph fever.

The harmonic approach of sending different pitches simultaneously over the same wire and unscrambling them with tuned receivers — a different principle from Edison's — had occurred to Bell even before he left England. What set him on it again may have been an item of October 18, 1872, in the *Boston Transcript*, the paper in which he advertised his speech lessons. At the Western Union Telegraph Company's annual meeting, reported the *Transcript*, much was made of the company's having acquired the "duplex telegraph" of Joseph B. Stearns, a Boston customer of the Williams shop. The *Transcript* called the invention one of the most important in telegraphy since the industry began. The Stearns duplex system, unlike Bell's harmonic scheme, worked by means of a secondary parallel circuit at each end of the single main line. An ingenious device on each parallel circuit blotted out a message being sent from that end, but not one coming in. So a message could be sent in each direction simultaneously over the main line.

Other Boston men and events drew Bell toward science during October 1872. Besides reading Lewis Monroe's copy of John Tyndall's book on acoustics, Bell had also studied a Boston Public Library copy of Tyndall's book on electricity. So Bell made sure to attend the Lowell Lectures on light and heat which the great English popularizer of science gave in Boston that month. At the final lecture Bell heard Tyndall go on at length about the "undulatory theory" of light propagation. The idea of "undulatory" sound waves and electric currents would eventually be central to Bell's own greatest triumph.

And also in October 1872 Bell began attending free public lectures on zoology, geology, and experimental mechanics, part of the Lowell Institute lecture series (not the Lowell Lectures) at the Massachusetts Institute of Technology. The lectures on experimental mechanics were given by Charles R. Cross, assistant to MIT's Professor Edward C. Pickering, an acoustical physicist. Cross had just helped improve a device Pickering had contrived two years earlier for "the electrical transmission of sound." A vibrating tuning fork made and broke a battery-powered circuit, as in the Helmholtz device. The intermittent current passed through an electromagnet fixed near the tin-plate bottom of an open box. The intermittent force of the magnet vibrated the tin-plate sheet like a drumhead or diaphragm, thus making a loud tone of the same pitch as the tuning fork.

Pickering never thought of trying to send anything to his "tin-box receiver" but the simple tone of a tuning fork, nor could he have done so with what he had. Bell did not remember seeing the device in the fall of 1872, nor did Cross remember displaying it then. Its significance is that of a straw in the wind. As so often has happened in the history of technology, men's minds were converging on a single point.

To get there first would take a special set of talents, interests, temperament, and environment. Bell was born with the right talents and temperament. His upbringing gave him the special interests. Chance brought him to Boston, which perhaps more than any other city in the world provided the proper intellectual, technical, and economic environment for the invention of the telephone.

If a computer searching for the most likely inventor of the telephone had run through the punch cards of all the world's people and places at that time, one name at least would have been printed out. It would have been that of an enthusiastic, ambitious, intelligent, imaginative young man with unusually keen hearing and sense of pitch; a trained pianist; a man who knew the mechanics of speech and hearing and took a lively interest in telegraphy; one who lived in a society more eager for rapid communication than any before; one who had easy access to leading scientists, skilled technicians, an academic community, and a community of enterprising capitalists.

It was a tall order, but in the fall of 1872 Alexander Graham Bell, aged twenty-five, of Boston, Massachusetts, U.S.A., already filled it, except for the academic tie. And even that was soon to come.

The earliest surviving sign that Bell had taken up the pursuit of multiple telegraphy in earnest is a letter from his father, November 10, 1872, warning that he would waste his time trying to make "an instrument for transmitting vibrations" to help lip-readers distinguish between "P" and "B."

This did not stop young Bell. The lipreading aspect had been only an afterthought anyway. What he wanted, now that he was settled enough for sustained experiment, was to test his "harmonic telegraph" scheme. The first step was to duplicate Helmholtz's device for producing a sustained intermittent current of precise frequency, in order to see if it really would sound a fork of one pitch and leave others silent. To reach this simple goal would take much of his spare time over the next year.

Bell could not afford a patent lawyer, and besides he feared difficulty as a British subject in getting an American patent. So he relied on secrecy for protection. He had a table made with a cover that could be locked. Not trusting a hired technician, he made most of his own equipment, even winding the coils of his electromagnets. Now and then he showed his instruments to friends, but would divulge the principle to only a select few.

One of the few, Percival D. Richards, lived with his wife on the corresponding floor of the house next to Bell's, with a door giving access to Bell's hallway. Bell and the Richardses ate at the same table. And Richards was interested in electricity. So Bell ran wires to Richards's rooms and swapped messages as he had done with his Bath friend years before. Richards would drop in on Bell of an evening, watch his young friend's after-hours struggles with batteries and tuning forks and electromagnets, lend a hand now and then, and commiserate or cheer as results warranted.

Bell needed encouragement. He was not as dextrous at a workbench as at a piano. Readjust and tinker as he might, the current-interrupting fork would not keep vibrating but would quickly die away. "We are grieved but not surprised at your being unwell," wrote his mother in December. "You undertake too much. . . . We will talk of your invention, my dear boy, when you are at home [over Christmas]. . . . I want you not to think about it just now."

Christmastime at Brantford meant "one continued round of parties" for the homecomer. He returned to Boston in January 1873 refreshed and bearing some adjustable pole pieces hastily made by a Brantford smith. Then he

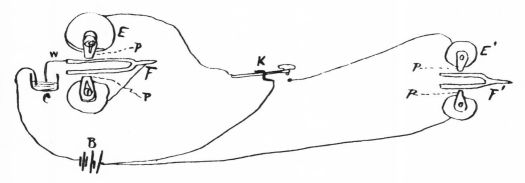

Tuning-fork rheotome with adjustable pole pieces, early 1873. The circuit on the left, including the wire w and the mercury cup c, maintains an intermittent current. When the key K is closed, part of the current passes through electromagnet E^1, causing the tuning-fork F^1 to vibrate between the adjustable pole pieces P

set about copying Helmholtz's arrangement more closely. The new pole pieces made better adjustments possible. He covered the mercury in the cup with a layer of alcohol to delay the oxidation which tended to cut off current entirely. Now the intermittent current lasted longer, though still not long enough for easy experiment.

All this meant night work. Teaching filled his days. Theresa Dudley helped now and then with Georgie Sanders, and in April a young lady, Abby Locke, came from Buffalo as a pupil and teaching assistant. But the strain showed. Two weeks beforehand he began fretting, with his usual "horrid debilitating headaches," about an address on Visible Speech for deaf-mutes he was to make early in June before the Massachusetts Medical Society. Afterward he wrote his father: "This has been a glorious day for Visible Speech. I did as you suggested and spoke entirely impromptu — and words came all right." More than a thousand medical men had given him "a most gratifying reception" and later "fairly besieged" him with questions about deaf patients. But his mood swung back as quickly to days of "dreary monotony" and more headaches, when he longed for home. Night after night he lay sleepless, hearing imaginary noises — perhaps phantom tuning forks.

So he gave in to another piece of his father's advice, sent when Aleck seemed tempted by a Boston School Committee appropriation to prolong his teaching another month. "Give up," appealed Melville Bell to his only living son. "I value your health above all things. . . . I prescribe a steamboat trip on the upper lakes." In the second week of July 1873 Aleck sent a farewell telegram to his neighbor Percival Richards and then headed north to Brantford.

10

Professor Bell and Miss Hubbard

At Tutelo Heights, Bell found a young Scottish friend, George Coats, of the thread-manufacturing family. Together Bell and his visitor set out by steamer through Georgian Bay to Lake Superior and the mining regions, then back through Lake Huron. From a frontier hotel, writing at the bar amid a noisy crowd of miners, Aleck in mid-course reported "fishing superb, scenery superb, fresh air — ad lib." He and George Coats both slept with loaded revolvers under their pillows. They planned to camp out in the hills for a week and wished it could be a month. Bell came back in early August "with renovated health and strength."

In the doorway to greet him stood that gentleman of peerlessly imposing mien, Uncle David Bell, come from Dublin for the summer with his wife Ellen and teen-aged son Charles. Later on, little Georgie Sanders arrived with his nurse to live nearby and take more lessons. Not until the end of September did Bell leave for a new home, Salem, and a new role, that of university professor.

Chartered in 1869, Boston University had entered the year 1872 with the largest single bequest given American higher education up to that time, $1,700,000 worth of downtown business property from the estate of the fish merchant Isaac Rich. It ended the year with much of that endowment consumed by the Great Fire and subsequent insurance company failures. Like Bell, however, the university had youth, resilience, and optimism. It had already shown pioneer spirit in admitting women, the first American university to do so from its beginning. In its brief, fire-blasted year of affluence, it had opened schools of law, medicine, and music. Fire and business depression notwithstanding, in 1873 it added a College of Liberal Arts. And as an experiment (which turned out in the end to be shortlived) it also sponsored a School of Oratory.

In those simpler days, universities often accepted ability and knowledge in lieu of academic degrees, as with Melville Bell at Edinburgh and London. Lewis Monroe, dean of the new School of Oratory, promptly offered Aleck Bell the professorship of "Vocal Physiology and Elocution," and before leaving for Brantford, Bell accepted. His university duties, he explained to his worried father, would occupy him no more than five hours a week, and would give him at least five dollars an hour and the use of a room at the university for his private pupils. He could commute by train from Salem, where Georgie Sanders's grandmother, Mrs. George Sanders, had offered him free rooms and board in return for instructing the boy.

Young Professor Bell arrived at Mrs. Sanders's Salem home in time for supper on October 1, 1873. Built in colonial days, the ample house with white clapboards and green shutters regarded Essex Street from behind a tidy little garden. About it the elms had turned golden with fall, and the maples flamed with crimson and orange. The back yard had a larger garden and a pear orchard, a big stable, and a barn with several carriages. Above the dormer windows of Bell's third-floor rooms, a railing enclosed the rooftop captain's walk from which, in the days of Salem's maritime glory, spyglasses had more than once been trained on the sails of stately East Indiamen, inbound with spices from Ceylon and Calcutta, Rangoon and Coromandel.

In the Sanders barn Bell stored his fine horsehair furniture. His telegraphic equipment, including the table with the locked cover, sojourned in the Sanders basement. With his lively temperament, fine dark eyes, and warm, resonant voice, Bell himself entered easily into the family circle. Despite his Old World upbringing, he found himself defending his university's forwardness in admitting women. "I come out strongly for the plaintiff in the action 'Woman vs. Man,'" he wrote, "because Mrs. Sanders *will* take the other side."

That fall and winter, having taught Georgie to read, Bell pasted small colored reproductions of dramatic and genre paintings in a blank book and carefully wrote out a little story to go with each — the Bear and the Baby, Jack and Jill, the Man with the Broken Leg, the Snowstorm, the Monkey's Ride, and so on — eighteen in all. He inked letters on the fingers of one of Georgie's gloves and taught him to communicate by finger spelling — the "manual alphabet." His night-owl habits now irrevocably fixed, Bell would come down late to breakfast after a long night of study, writing, or telegraphic tinkering and play "bear" with Georgie. In the late afternoon Georgie would stand at the window, waiting for his teacher to come home from Boston and spell out the excitement of the city streets, the hurdy-gurdy monkey, the little lost dog, the runaway horsecar, the caged parrot in the train. Bell could make the little boy hear in imagination the bustle

Three Happy Little Girls.

What are the names of these girls?

Mary, Kate, and Lizzie.
Which is Mary? The little girl
with the bread in her hands.
Which is Kate? Kate is the smallest.
She is throwing some crumbs to the dog.
Which is Lizzie? She is the tallest.
What is Lizzie doing?
She is holding Kate in her arms.
What is Mary doing? Feeding the chickens.
How many hens are there? How many
cocks are there? How many chickens are there?
How many pigeons are there?
What is the dog doing?
Which hen is the mother of the chickens?
How many leaves are there?
Where can you see any flowers? / There.
What color are the flowers? They are red.

Feb. 4th 1874

Pages from the Sanders Reader, 1873

and chatter and music of a world that would never speak to him in fact.

Dean Monroe chose his young professor to give the inaugural lecture at the new School of Oratory, and its success won him an invitation to repeat it before the whole university. Dealing with "Speech and the Instrument of Speech," he said much about the properties of sound — pitch, loudness, and timbre — and illustrated them by experiment and by diagrams from Helmholtz. He showed also a graph of undulatory sound waves and explained the workings of the ear and the vocal organs.

Then he settled down to his university duties at 18 Beacon Street: lecturing on Mondays, practical work on Wednesdays. His class opened with seven students and by the end of October had grown to twenty. That fall, shock waves from the Wall Street crash of September brought further ruin to the university's endowment. But the School of Oratory, largely self-supporting, stood apart from the wreckage.

Bell had other sources of support anyway. His work with Georgie gave him room and board. Abby Locke came back from Buffalo to take private pupils under his superintendence, dividing the fees with him. A stammerer and two deaf semi-mutes enrolled, children of the Rhode Island Aldriches and the Massachusetts Brookses and Hubbards — "just those calculated to do me good," Bell wrote home.

One of them was Mabel Hubbard, back from Europe with a thorough knowledge of the German language and other cultural baggage, but with little improvement in her speech.

Years later Mabel remembered having walked up Beacon Hill toward Boston University one day with her former teacher Mary True, lipreading Miss True's praise of Professor Alexander Graham Bell, and privately thinking that the man must be some kind of crank. Presently she and Miss True entered a dark green room one flight up, with a single window looking out on the Old Granary Burying Ground. When Bell came in, "I did not like him," Mabel remembered. "He was tall and dark, with jet black hair and eyes, but dressed badly and carelessly in an old-fashioned suit of black broadcloth, making his hair look shiny, and altogether, to one accustomed to the dainty neatness of Harvard students, he seemed hardly a gentleman."

Yet he had an engaging intensity of manner. And as Mabel had remarked to Miss True on the way, he might after all help improve her speech, a highly desirable object in case she should marry a rich man and have to keep up a position in society. (She was then about a month short of her sixteenth birthday.)

So the lessons began, mostly from Miss Locke, Bell himself taking charge only once every week or two. Mabel still doubted that a mere teacher could be a gentleman in the fullest sense. Yet as a mere teacher Bell proved irresistible — "so quick, so enthusiastic, so compelling, I had whether I would or no to follow all he said and tax my brains to respond as he desired." Before every lesson she hoped to find Professor Bell waiting for her, not Miss Locke. But it usually turned out to be Miss Locke.

Mabel had meant at first to make a brief trial of Professor Bell's skill. But by mid-November she was writing her mother in New York: "Mr. Bell said today my voice was naturally *sweet*. . . . He continues pleased with me. He said today he could make me do everything he chose. . . . I enjoy my lessons very much and am glad you want me to stay. Everyone says it would be such a pity to go away just as I am really trying to improve." So she settled down for the winter with her older cousin Mary Blatchford in Cambridge.

Just after her sixteenth birthday in November 1873, Mabel reported herself as "getting along very well with Prof. Bell, interest continues unabated. I went yesterday in a driving rain and was the only scholar who came." As Christmas neared, Professor Bell told her he had never had a pupil who improved so fast. After New Year's he repeated the compliment, "but," she wrote her mother, "I am much dissatisfied with myself. The days are slipping away so fast and hug them as I may they slip out and never come back."

The first week of February 1874 brought heavy snow and subzero cold. Mabel bundled herself up nevertheless and made her way to Beacon Street through the storm. "Both Miss Locke and Mr. Bell were surprised to see me, but," she wrote artlessly,

I did not want to lose a lesson when each cost so much, but Mr. Bell said he would not charge for the times I could not come. . . . Mr. Bell lost his train and when I was going out . . . took me back to the cars. We had a grand time running down hill through the deep snow and in face of snowflakes that were very nearly hailstones, they hurt my face so much. . . . I would have been almost sorry to get to the apothecary's, but I was quite out of breath, besides my waterproof and veil were flying about me and it was all I could do to hold them on. . . . If it is pleasanter tomorrow I will go, otherwise not, for Mr. Bell almost forbid me to go in and I'm afraid he will take it into his head that he must go back with me again.

The days kept slipping away. Spring came again to Boston, and on June 1 Mabel began a new notebook, in which her instructor wrote his comments. "I think nearly the *whole* peculiarity of your utterance arises from defective positions of the throat, and from the lack of *diaphragm* action," he informed her in a businesslike way, proceeding to an elaborate diagram and explanation of vocal action. "If you look into my mouth you will see that I have *two uvulas*. Now you may watch my soft palate while I move it up and down." All this, including the palatal calisthenics, displayed a properly clinical spirit, but something — was it the spell of a June day? — impelled him to add: "Your voice has a beautiful quality."

During that school year, Bell did his filial duty by Visible Speech. Perhaps he went beyond duty. A New York teacher, though committed to articulation for the deaf, claimed in print that the Visible Speech system demanded prior expertness in phonetics, and that Bell's innate teaching ability explained his success with it. Later, Bell himself conceded privately that except for his father's sensitivity in the matter he would have changed much in the system, including its name.

But in public Aleck betrayed no doubts. Instead he seized every chance to further the cause. In December 1873 he made a follow-up visit to the Clarke School, which led to his calling a convention of Visible Speech teachers at Worcester in January 1874. Sixty showed up to hear Bell give the principal paper and conduct the discussions, and some agreed to write for the "Visible Speech Pioneer," a manuscript periodical to be circulated among their institutions. Boston and Worcester papers ran long accounts. Dean Monroe was moved to require some training in the symbols for all School of Oratory students. The whole affair was "a glorious success," Bell wrote his father; "Saturday marked an Epoch in the History of Visible Speech and . . . the Education of the Deaf and Dumb." He announced another convention for June.

Bell worked hard to keep up momentum, addressing the School of Oratory on the subject, advocating a Braille version before the Perkins Institution trustees, getting the "Visible Speech Pioneer" off on schedule. But the most important consequence concerned not his father's future fame but his own and stemmed from a series of public lectures.

Now wearing the dignity of a professorship, he was invited in March to explain Visible Speech to the Thursday Evening Club, more "eminently scientific" than the better-known Saturday Club. Forty came, "the cream of Boston's intellectual men"; and as Bell wrote home, "of course it was a grand success." A month later, sponsored by MIT's Society of Arts and Sciences, he addressed an audience of four hundred, among them "the finest minds in Boston."

The MIT lecture, he rejoiced, "has at once placed me in a *new position* in Boston. It has brought me into contact with the *scientific minds* of the *city*." As a first dividend came "free access to the Institute of Technology

— and permission to experiment with Helmholtz' apparatus and with Scott and Koenig's 'Phonautograph' and Revolving Mirror Apparatus."

If Bell in later years ever cast up accounts as to Visible Speech, this indirect bonus alone would have balanced the books, and to spare.

11

Inventor and Scientist

Along with his other roles during the 1873–1874 school year — promoter, organizer, lecturer, tutor, professor, and occasional invalid — Bell played inventor in the fall and scientist in the spring.

More than once during his electrical experiments in November and December 1873 he was to grope his way into bypaths that could have brought him to the concept of the telephone. But he was not looking for that. So each time he would turn back toward what he considered his paramount goal: a harmonic multiple telegraph system.

As the need for secrecy more and more possessed him, he shifted his tinkerings from his Boston University room to his Salem study. Even there he carefully locked the cover of his table against "the hands and eyes of curious domestics." It was unlikely that "curious domestics" could have made anything of the tangle. Bell had trouble enough himself.

Not until mid-December did he manage to construct a Helmholtz-type current-interrupter or "rheotome" that would keep going for an appreciable time. Meanwhile he tried substitutes, like a steel strip or "reed," projecting like a springboard over a cup of mercury and having a platinum wire hanging down from the free end to make the contact. He could vibrate the reed mechanically by blowing down on it through a cone. And by using a screw clamp at the reed's fixed end, he could alter the length of the projecting portion and thereby change its pitch or natural rate of vibration — in short, tune it. This adjustability gave it a great advantage over a regular tuning fork.

Bell got the idea of substituting steel reeds for Helmholtz's tuning forks from a book in the Boston Public Library. *The Wonders of Electricity*, an English translation of a French book by J. Baile, had other interesting things to say. "Some years hence, for all we know," it remarked, "we may be able to transmit the vocal message itself with the very inflection,

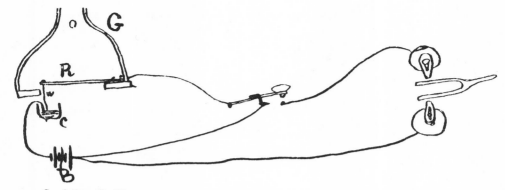

Steel "reed" (R) rheotome with cone mouthpiece (G), platinum contact wire (w), and mercury cup (c). Fall, 1873

tone and accent of the speaker." It suggested as a starting point the "acoustic telegraph." In this, a sound vibrated a metal plate, making and breaking an electrical circuit. The intermittent current, by activating an electromagnet, caused another metal plate to vibrate with the same pitch (like Pickering's "tin-box receiver"). As Baile pointed out, this device transmitted pitch only, not amplitude, and hence not speech.

Baile also described a device that must have recalled to Bell his old trick of singing into a piano and hearing the strings echo back. "A series of vibrating plates, answering to the strings of a harp, has been arranged," wrote Baile, "each of which vibrates when struck by a particular sound, and sends off electricity to create at the end of the line the same vibrations in a corresponding plate, or, in other words, to produce the same sound." In the following summer, a notion like this would help to inspire the basic concept of the telephone in Bell's mind.

But testing his harmonic telegraph concept came first now. In the fall of 1873 Bell connected two steel-reed rheotome transmitters of different pitches in parallel with each other and in series with two receivers consisting of steel reeds projecting over electromagnets. Each receiver was supposed to be tuned to a particular transmitter. Sure enough, the first transmitter sounded its corresponding receiver, while the other receiver, though subjected to the same current, remained silent, except for a tremor perceptible only to the touch. So far, so good. But when Bell tried the second transmitter, nothing happened. Neither receiver reed sounded, though he could feel both trembling silently.

Undismayed, Bell laid the trouble to improper tuning of the second receiver. So he took the balky receiver apart, remodeled its frame, shifted

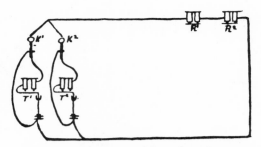

Harmonic telegraph, November 1873
(transmitters T^1, T^2; keys K^1, K^2; receiver reeds R^1, R^2)

the reed back and forth, and weighted it with drops of wax for finer tuning.

In the course of this tinkering, he tried pressing his ear against the uncooperative reed while both transmitters were sounding. What he heard, but did not recognize, was the first faint cry of the telephone receiver. For like the human voice it was a compound sound, a blend of both transmitter pitches. The pressure of Bell's ear had damped the reed's natural vibration and made it — like a diaphragm — the weak but faithful mimic of whatever frequencies bore upon it.

Since his goal just then was a harmonic receiver spurning all but its own frequency, the whisper in his ear was precisely what he did *not* want. He took note of it nevertheless, recognized it as sounding the pitch of whatever transmitter might be in operation, and saw a use in it. If a reed did not respond properly to a distant transmitter, he could "hear" the transmitter's pitch by damping the receiver reed with his ear. He could pluck the receiver reed with his finger to determine its own natural pitch. Then he could adjust the setting of the receiver until the two pitches coincided. Presently the practice of ear-pressing and reed-plucking became routine, automatic, unthinking.

By December 1873 Bell felt that the basic principle of his invention had been experimentally verified. Tension, work, an inordinate Thanksgiving dinner at the Saltonstalls', and two weeks of almost nightly partying brought a stubborn headache down upon him. Mabel Hubbard found him "miserable" after four days in bed. "He has his machine running beautifully," she told her mother, "but it will kill him if he is not careful." He gave up smoking for a while, then decided an occasional cigar would do more good than harm. Mrs. Sanders concocted an ointment of camphor and chloroform which seemed to help when rubbed on his neck. And withal he forsook his laboratory in favor of working out refinements on paper.

On paper Bell tried to design the circuits of a line with several stations. But as Baile's book said of another inventor's work on an "autograph" or facsimile telegraph, "the arrangements which would correct one inconvenience would often call another into existence."

A conventional telegraph system did not use a complete circuit of wire, only a single wire connected to the ground at each end, thus halving the amount needed. From one ground to the other, batteries along the line pumped a current, which the sending key interrupted in Morse code.

Bell's system could not easily do that. Each of his transmitters used an intermittent current of a special frequency. Several transmitters in series, interrupting a single current at different rates, would come near blocking it altogether, Bell assumed. So he sketched a line into which each transmitter would feed its own intermittent current, superimposing it on the others.

But if the main line were grounded at each end, a transmitter's current would flow from the transmitter's ground to each of the mainline grounds, dividing unequally in inverse proportion to the resistance in each direction. This would make the strength of a signal grossly different at different stations. So Bell sketched a system in which each transmitter had its ground, but the main line was grounded at only one end, thus reducing the discrepancies. The trouble now was that a transmitter could send only to stations between it and the grounded end of the main line.

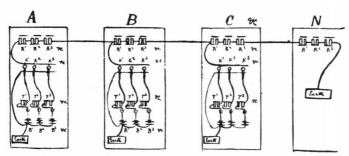

Harmonic telegraphic circuit of November 1873
(transmitters *T*, batteries *B*, keys *K*, receivers *R*)

Then Bell had an inspiration: he would put each transmitter on a separate local circuit, which would not be connected to the main line at all but instead would *induce* a current in it. Around an iron core would be coiled a length of the transmitter circuit and a length of the main line. A reduction of transmitter current would induce a main-line current in one direction, an increase would induce one in the other direction. The main line could be grounded at each end, like a regular telegraph line, and the alternating current would flow with the same strength through each receiver.

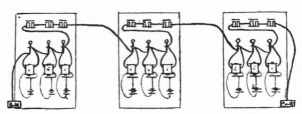

Harmonic telegraph circuit of December 1873
(induction coils *i*)

Now Bell saw a new difficulty. A receiver reed on the main line would feel an electromagnetic impulse both when the vibrating transmitter reed made a contact and when it broke one. So it would vibrate at twice the transmitter frequency. It would have to be tuned precisely an octave higher to respond most efficiently — a much harder thing to arrange than simply matching pitches.

Presently it occurred to Bell that whenever the main-line current reversed direction, the electromagnet of the receiver would reverse its polarity. So if the receiver reed were polarized by attaching it to one pole of a permanent magnet, the reed would be attracted by one impulse but repelled by the next. Thus the rate of its forced vibration would match that of the transmitter.

But now a fresh difficulty popped up. If the push and the pull were not evenly spaced, the receiver reed would be thrown off stride. So the make and the break of the transmitter current would have to be evenly spaced. And Bell could see no way to make sure that the platinum wire dipped into the mercury cup for exactly as long as it remained out of it.

After threading this maze of logic, Bell thus found himself back where he started: having to devise a new type of transmitter.

At this point he started down a path that he later followed to fortune. His inspiration had the originality and simplicity of genius. He would use a receiver as a transmitter. If a polarized receiver reed were mechanically vibrated, say by a stream of air, its motion would induce a fluctuating current in the coil of its electromagnet. Such a current would be *con-*

tinuous rather than intermittent. Sent through the coil of another receiver, it would presumably make that receiver's reed vibrate precisely as the first one was doing.

Such an arrangement would do away with both the transmitter batteries and the current interrupters with their finicking mercury cups. Bell saw this. He did not yet see that the device would reproduce in one reed the *amplitude* as well as the frequency of vibrations in another. This would be a long step — indeed, the crucial step — toward electrical transmission of complex sounds, including speech. But in any case, Bell took it for granted that a mere vibrating reed could not generate a current strong enough to travel over a long wire and make another reed vibrate audibly. In theory the receiver might make a sound. In practice, he was sure, no human ear would be keen enough to detect it. He was so sure that he did not even try what would have been the work of a few minutes, a mere matter of connecting two of the receivers he already had.

Instead, he went back to the idea of producing an intermittent battery-powered current with makes and breaks equally spaced. To that end, he sketched a cylinder with alternate strips of conducting and nonconducting

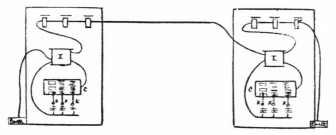

Make-and-break harmonic telegraph
(induction coils *I*, cylinders *C*, keys *K*)

material running lengthwise, the cylinder to be rotated against a contact point. This, combined with his transmitter induction-coil arrangement, seemed on paper to meet the requirements of an operating system.

Even in seeking patent protection, Bell ran into snags. In that period, an inventor could file a "caveat," stating his concept and his intention to develop it. This would establish priority and allow him a specified period in which to work out a patentable invention. Meanwhile the caveat would be kept secret from other inventors, unless one of them filed a conflicting caveat or patent application. But Bell was told that as a British subject he could not use the caveat procedure.

He could make a patent application. But while the law no longer required the deposit of working models, it permitted the Patent Office to require them by regulation, which the office still did. Bell felt that his own experimental contrivances were too crude and amateurish to submit. But he feared that if he hired an experienced electrician to help him, the idea might leak out. Besides, he could not afford that expense, along with a lawyer's fees and patent fees.

So in January 1874 he wrote the British Superintendent of Telegraphs, offering the invention to Her Majesty's Government. The answer from a British post office official struck him, he said later, as "almost a personal affront." "If you will submit your invention it will be considered," wrote the functionary haughtily, "on the understanding, however, that the department is not bound to secrecy in the matter, nor to indemnify you for any loss or expense you may incur in the furtherance of your object, and that in the event of your method of telegraphy appearing to be both original and useful, all questions of remuneration shall rest entirely with the postmaster-general."

Bell scorned the proposition.

Boston University classes, private pupils, Georgie Sanders's lessons, and especially the Visible Speech campaign kept Bell from telegraphy in January and February 1874. For his twenty-seventh birthday on March 3, Mrs. Sanders gave him a pair of gold sleeve-buttons and "A STUDY!! . . . a special room for me to carry on my experiments in." At once he set up his apparatus there and happily commenced testing his paper schemes. He verified "triumphantly," for example, a supposition ("the only part of the theory that I felt shaky about") that one battery could keep two tuning forks going at different frequencies.

In March he went to see Charles Cross demonstrate Helmholtz's experiments in a lecture at MIT and afterward talked with Cross about tuning forks and sympathetic vibrations. After Bell's own lecture at MIT in April on speech training for deaf-mutes, it was Cross who came up to offer him the use of the Institute's apparatus and laboratories. (Boston University and MIT already had an arrangement whereby the students of one could take courses at the other.) "I go to the Institute this afternoon to make *some experiments,*" Bell wrote home excitedly some days later.

These new experiments turned him away from telegraphy to acoustics.

Two devices especially interested him because they made speech "visible." One was the "phonautograph," invented by a Frenchman named Leon Scott and recently improved by one of Cross's students, Charles Morey. If one spoke or sang into the large end of a wooden cone, the sound vibrated a membrane diaphragm stretched over the small end. A thin wooden rod, hinged at one edge of the diaphragm, ran across the center of it and projected beyond the opposite edge. A bit of cork and glue joined the center of the diaphragm to the wooden rod, so that the vibrations were magnified at the rod's projecting free end. A bristle fastened to that end traced a curve on a piece of smoked glass being drawn past it. Each sound left a characteristic tracing.

The other device responded more delicately to sound vibrations but made no record of them. Called the "manometric flame," it was the work of another Frenchman, Rudolph Koenig. This also used a membrane diaphragm, stretched over a hole in a gas pipe. A speaking tube — "just like Mamma's tube," as Bell wrote home — carried the sound of the voice to the diaphragm. The vibrating diaphragm made the gas pressure fluctuate and thus varied the shape of a gas flame. To emphasize the shape of the flame, it was reflected from four mirrors on the circumference of a revolving wheel. In the whirling mirrors, the flame looked like a broad band of light with a characteristic pattern for each sound. One sound, Bell wrote, produced a sawtooth pattern, another a notched band, another "a most beautiful kind of lace-work pattern," and still another "a similar net-work of a blue light surmounted by *blobs* of crimson light at regular intervals."

"If we can find the definite shape due to each sound," Bell wrote home, "what an assistance in teaching the deaf and dumb!!" The teacher could show them what a sound looked like in the manometric flame and then help them experiment until they reproduced it. "In any future publication concerning Visible Speech, pictures of the vibrations due to each sound could be given, and thus the sounds be identified through all eternity."

His restless mind went further. He speculated, for example, that if a phonautograph tracing were cut through the wall of a cylindrical shell, and an air nozzle with a long, thin opening were held across the sinuous slit as the cylinder revolved, a "pencil of air" would be blown through which, in tracing the phonautograph line, might reproduce the original sound.

If Bell had thought of using a needle or stylus instead of a "pencil of air," and a groove instead of a slit, he might have anticipated Edison's phonograph by three years. Instead he devised clever adaptations of the phonautograph. He had thought a deaf-mute pupil might match practice flame patterns against phonautograph tracings of the proper sounds. But the flame patterns turned out to be different from the tracings. So he sought to make the phonautograph bristle's gyrations themselves directly visible to the pupil. For example, he arranged the bristle stylus so that it wagged back and forth along a slit of light. The revolving mirrors of

Koenig's apparatus turned the slit into a broad band, on which the black silhouette of the bristle made a characteristic pattern for each sound. Even more simply, he let the bristle tip vibrate just inside the edge of a ray of light, so that it drew a glowing pattern in the air. The free vibrations of the bristle likewise turned out to be different from the friction-dampened tracings. Nevertheless, Bell's ingenious ideas won him an invitation to describe them at Cross's own Society of Arts lecture on the phonautograph and manometric flame.

Bell still saw little if anything of Boston's ablest specialist in acoustical physics, Professor Edward Pickering of MIT. Pickering had difficulty in pronouncing the letter "R," and a friend urged him in May to see Bell for a cure. But apparently Pickering did not take the advice. Though Cross had demonstrated Pickering's tin-box receiver in January, Bell had not attended that lecture; nor did Cross mention the device to his friend from Boston University, so far as Bell could recall later. It had nothing to do, after all, with either multiple telegraphy or the visible manifestation of sound waves.

But Cross welcomed Bell as a full partner in "our acoustic experiments." In late April, writing about an idea for making a liquid-column gauge more sensitive to sound waves, Cross added, "If you are at leisure and care to try a few more experiments I should be glad to see you."

Since 1871 Bell had casually known Dr. Clarence J. Blake, a Boston ear specialist four years his senior. Aware, perhaps, that as a graduate student in Vienna Blake had made a special study of sound transmission, Bell quizzed him in the spring of 1874 about that subject and the mechanics of the human ear. Thus began a year of joint experiments that Blake recalled four decades later as "one of the joyous scientific experiences of a lifetime." Late that spring Blake and Bell began experimenting with temporal bones from the ears of two medical school cadavers, each bone including the eardrum and the small chain of bones of auriculation.

Along with all this, Bell began, but did not finish, writing an elaborate study of lipreading for his manuscript periodical, the "Visible Speech Pioneer." ("Do you think I could learn [lipreading] so late in the day?" wrote his mother wistfully. She never did.)

By late May, Bell was again "quite sick from overwork," Mabel Hubbard told her mother. Nevertheless he pushed on in the cause of Visible Speech. Trials in several institutions seemed to show its value for certain classes of the deaf: the semimute, the semideaf, and the more intelligent and determined of the congenitally deaf. Edward M. Gallaudet, son of Thomas Gallaudet, after visiting the Clarke School predicted success for the system. But one of Bell's Hartford teachers reported after two years' trial that the drill bored or discouraged many pupils, and that classroom use of the system ran up against wide differences in pupils' aptitudes and progress in it. "The labor of teaching," wrote the Hartford man, "is greater and more wearing than teaching by signs. No one who has not seen it can appreciate it. . . . Great patience and enthusiasm are necessary, . . . besides the ability to distinguish sounds accurately, and to translate them into the symbols of Visible Speech, and also a knowledge of vocal physiology." In short, not every teacher was an Alexander Graham Bell.

At the Second Convention of Articulation Teachers of the Deaf and Dumb, held in Worcester on June 13, 1874 — the last one for more than a decade — Bell was elected president. His final contribution to the proceedings was to exhibit and explain a phonautograph and a manometric flame apparatus borrowed from MIT.

On that same day in Washington a young scientist noted in his diary: "Went to the Smithsonian to see some novel experiments by a Mr. Gray, upon the telegraphing of sound. Very curious indeed!"

By 1874, at thirty-eight, Elisha Gray had painfully won his way to a place in life he had long dreamed of. As an Ohio farm boy, he had quit school and gone to work after the death of his father. Blacksmith work proving beyond his strength, he turned carpenter and boatbuilder. Meanwhile he read all he could about science, especially electricity and magnetism, and tinkered with homemade batteries and electromagnets. Encouraged by an Oberlin College professor, he worked his way through preparatory school and two years at Oberlin. But overwork nearly wrecked his health, and so he was thirty by the time he established himself as a professional electrician and inventor.

In the winter of 1866–1867, while working on telegraphic relays, Gray began experimenting with the transmission of tones. He connected a vibrating rheotome or circuit interrupter to a primary induction coil, and a "polarized relay" (much like Bell's receiving reeds) to a secondary coil. When a Morse key closed the primary circuit, the relay instrument vibrated with the frequency of the rheotome.

Though he overlooked the possibility of a multiple telegraph, Gray saw that melodies could be transmitted by an array of such instruments. But over the next half-dozen years he put off trying even that application, while he patented telegraphic repeaters, dial telegraphic instruments, a printing telegraph, an electric hotel annunciator, a telegraph station switch, and other devices. Meanwhile he moved to Chicago and formed a partnership, which in 1872 became the Western Electric Company, making electrical and telegraphic instruments. With Gray as superintendent the firm

Elisha Gray

grew in two years from less than fifty employees to more than a hundred. At that point Gray could afford to shift from management back to research.

In late January or early February 1874 he heard the refrain of the rheotome issuing from his bathroom, where he found his young nephew "taking shocks" to amuse the smaller children. With a vibrating rheotome in the circuit of a primary induction coil, the boy connected one end of the secondary coil to the zinc lining of the bathtub and held the other end in his hand. When the boy's free hand glided along the bathtub lining, it produced a whining sound in tune with the rheotome. Gray tried the effect and found that quick, hard rubbing made the noise even louder than that of the rheotome itself. When he varied the pitch of the rheotome, the noise followed suit.

Over the next few weeks, Gray made receiving instruments on that principle, such as a hand-cranked wheel against which the fingers or some equivalent "animal tissue" could be held. Then he set to work in greater earnest with an assistant. By late spring he was using a completely different

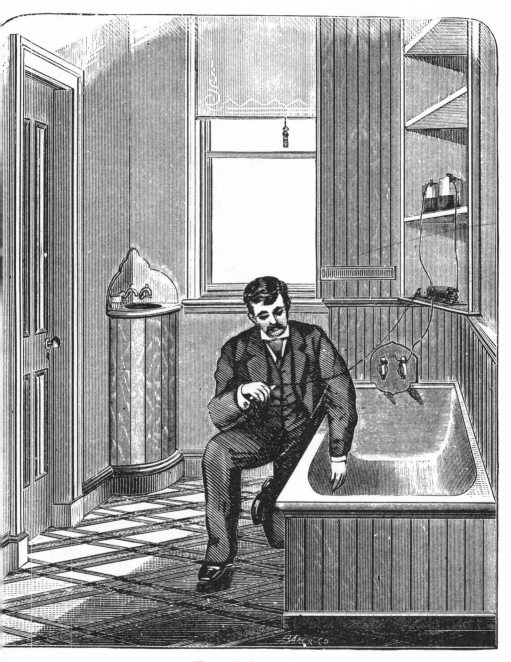

The musical bathtub

type of receiver, a metal diaphragm vibrated by an electromagnet, exactly like Pickering's tin-box receiver of several years before. In subsequent litigation it was dubbed "the wash-basin receiver." In May he made a transmitter with eight keys of different pitches, so that a simple tune could be sounded on either the "animal tissue" or diaphragm receivers.

The washbasin receiver

The great difference between Gray's instruments and Bell's in mid-1874 was that Gray's receivers would respond to any frequency, whereas Bell's were painstakingly constructed to respond selectively. Gray's receivers thus faced away from multiple telegraphy and toward the telephone. But his transmitters did not. They generated sounds; they did not pick them up from the air. All that Gray's system offered, therefore, was the transmission of simple tunes or chords, not speech or multiple messages. It was a sterile match of telephonic receiver with non-telephonic transmitter.

Nevertheless Gray showed his gadgets to Western Union Telegraph Company officials in May 1874 and to Joseph Henry and other scientists at the Smithsonian Institution in June. The young scientific diarist who thought the latter demonstration "novel" and "very curious indeed" may have referred to the "animal tissue" receiver, which had no future in either telegraphy or telephony. He could not have meant the other elements of Gray's device. For in August 1870 the diarist had attended the annual convention of the American Association for the Advancement of Science at

Troy, N.Y. On that occasion a member named Philip H. Van der Weyde had described "some further improvements in the method of transmitting audibly musical melodies by the electric telegraph wire," whereupon Professor Pickering had got up and described his tin-box receiver, which would make such transmissions even more audible. Pickering's remarks were "greeted with marked approbation," reported the *Troy Times*.

For a receiver, Van der Weyde had described an electromagnet mounted on a sounding box. Joseph Henry had reported years earlier that an electromagnet expands slightly when magnetized and contracts slightly when demagnetized. So an intermittent current makes it vibrate and produce a sound of the same frequency. Charles G. Page of Salem had discovered that effect and made "galvanic music" with it in 1837.

The transmitter described by Van der Weyde had been invented in 1860 by a German named Philipp Reis. Unlike the rheotomes of Bell and Gray, it presaged the telephone transmitter, for it picked up and transmitted the pitches of outside sounds, using a membrane diaphragm with a platinum

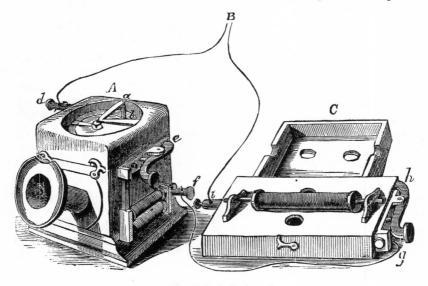

The Reis "telephone"

contact point in its center that made and broke a battery-powered current. But of course the intermittent current did not vary in strength to correspond with the volume of the circuit-breaking sound. And therefore, since Reis's instrument could not reproduce amplitudes or degrees of loudness, it could never transmit the subtle compound of many frequencies and amplitudes that constitutes speech. It could convey the rhythms and pitches of speech and thereby the illusion of intelligibility — but only the illusion. Reis did not choose to patent what he regarded as a purely scientific experiment. And indeed it had no conceivable commercial use then, nor would it ever.

Reis called his device a "telephone." He was not the first to use the word. As far back as 1796 another German had coined it, perhaps for the first time, from Greek roots meaning "far-speaking." Until Reis, the word referred to mechanical conductors of sound, such as speaking tubes. After Reis published an account of his "telephone" in 1861, it became well known among physicists and in lecture demonstrations. Eventually it appeared in several forms, including the one described by Van der Weyde. In all of them, a sound-vibrated membrane repeatedly interrupted a battery-powered current, and the intermittent current produced a sound in an electromagnet core, amplified by a sounding-box base.

On July 10, 1874, the *New York Times* ran an article on Gray's "music by telegraph," describing the keyboard arrangement but not the principle. With this device, which Gray called a "telephone," the *Times* reported, "anyone at the receiving end can distinctly hear, without the aid of electro-magnetism, the tune or air which is being played 500 or 1000 miles away." The receiving apparatus, said the article, "may be anything that is sonorous so long as it is in some degree a conductor of electricity," as for example, "a tin hoop, with foil paper heads stretched over it . . . a nickel five-cent piece, an old oyster can." A Western Union official forecast that "in time the operators will transmit the sound of their own voice over the wires."

The piece said nothing of multiple telegraphy. All the same, if the mercurial Professor Bell had read it he would have been unsettled, perhaps deterred from further work. He could have guessed that the "harmonic multiple telegraph" idea would (as in fact it did) soon occur to so ingenious an inventor as Elisha Gray. The statement that a transmitted tone could be received "without the aid of electro-magnetism" would have seemed to imply that Bell's receiver was both obvious and somehow outmoded. And the *Times* reported that Gray had applied for patents in the United States and Europe.

The *Boston Transcript* missed the story. But the *Boston Morning Advertiser* ran it verbatim on July 11, and the *Boston Morning Journal* on July 13. Bell saw neither. For on July 10 he had left Boston for a summer in Ontario. To the day of his death he never realized how near he had come to being shunted from the path to fame.

During the summer, news of Elisha Gray's gadget spread through the scientific community. In August, when Gray set out for Europe to get patents, he carried a letter of introduction from Joseph Henry to John Tyndall. As Gray passed through Boston, a patent lawyer there tried to get him an appointment with Pickering, who turned out to be away on vacation. It may have been that lawyer who sent Dr. Clarence Blake a clipping about Gray's musical telegraph. Blake wrote Gray at once about the phonautograph experiments by Bell and himself, offering to exchange information about their respective researches. But the letter did not reach Gray in Europe until late September.

In August, meanwhile, Professor Joseph Lovering of Harvard, who, though not a productive scientist, was an effective purveyor of other men's ideas, discussed Koenig's manometric flame before the annual convention of the American Association for the Advancement of Science, held at Hartford. And in the issue of the *Hartford Courant* that reported the closing session, the scientists could read the following item: "A curious and striking invention, called a 'telephone,' the effect of which is to telegraph musical sound, and even tunes, through any length of wire, has been made, it is said, by Mr. Elisha Gray, of Chicago. . . . Mr. Gray hopes one day to be able to transmit the sound of the human voice also by telegraph, . . . but towards this curious result nothing seems to have been done."

The *Hartford Courant* had no way of knowing that in Brantford, Ontario, less than a month before, Professor Alexander Graham Bell of Boston University, still unaware of Gray's "telephone," had at last seen how to do what Gray only dreamed about.

12

The Telephone Is Conceived

That summer of 1874 at Brantford, Bell found Mabel Hubbard more and more in his thoughts. But she was only sixteen, ten years his junior. Besides, he worried about his parents' reaction to her deafness. So he hid his feelings from her, from them, and for a while even from himself.

He thought about the deaf, too. A few days after coming home, he left for a week's visit in Belleville, Ontario, at the annual convention of American Instructors of the Deaf and Dumb. In an address there he referred to the difficulty lip-readers had in distinguishing between voiced and unvoiced consonants such as "B" and "P." "If some simple apparatus could be contrived to bring the vibrations of the speaker's voice to the hand of the lip-reader," he remarked, "one half of the ambiguities of lip-reading would disappear, and the awkwardness would be avoided of having the lip-reader place his hand upon the speaker's chest or throat."

Mostly he thought about acoustics, telegraphy, and certain related schemes indicated by a brief entry in his father's diary for July 26, 1874, five days after Aleck's return from Belleville: "New Motor (hopeful). Electric speech(?)."

The "new motor" scheme arose from the expansion of an electromagnet upon being magnetized. If a large number of such electromagnets were immersed in some confined fluid and subjected to an intermittent current, Bell speculated, the fluctuating pressure created by their expansion and contraction might be used to move a piston. As long as four years afterward, he was still playing with the idea, though the minuteness of changes in volume, the energy losses through hysteresis and friction, and other objecttions should have made it seem as impractical then as now. If Bell had been the sort to shy from the seemingly absurd, he would not have entertained such a notion. Neither would he have pursued the other notion of "electric speech."

Ever since his idea of the preceding fall about a vibrating, magnetized reed that would induce a fluctuating or undulatory current in an electromagnetic coil, he had been haunted by the ghost of a perception. Somehow, he felt, the secret of transmitting speech lurked in such a device. His thoughts reverted to it in Brantford, at the "dreaming place" on the bluff and in the little workshop back of the conservatory.

His first workshop project, however, was to construct what his family may have thought a rather grisly phonautograph from a dead man's ear

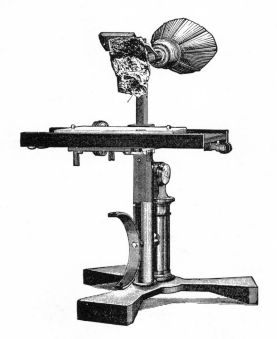

The ear phonautograph, 1874

procured for him by Clarence Blake. Making the membrane pliable with glycerine and water, he fastened a delicate stalk of hay to one of the bones, spoke through a speaking tube into the ear, and got tracings of complex waves on pieces of smoked glass drawn past the vibrating tip of the hay. Once again, as with previous phonautographs, it was borne upon him that the sum or resultant of complex sound vibrations could be conveyed through a single point and expressed as an irregular wavy line. In this case he was struck further by the way sound waves acting on a tiny membrane could move relatively heavy bones.

At the "dreaming place," watching the Grand River meander sinuously far below, he once more mulled over the idea of a multiple telegraph transmitter with air-vibrated polarized reeds that would induce an undulating or wave-like current. And considering the ear phonautograph, he was encouraged also to wonder if even mere sound waves might not be strong enough to generate an appreciable current. In the back of his mind, moreover, there lingered memories of the echoing piano strings and of the device mentioned in Baile's book: "a series of vibrating plates, answering to the strings of a harp, . . . each of which vibrates when struck by a particular sound, and sends off electricity to create at the end of the line the same vibrations in a corresponding plate."

Suddenly, perhaps on the day of his father's July 26 diary note, the jumble fused into a great insight: the fundamental principle of the telephone.

If one spoke or sang into Baile's "series of [tuned] vibrating plates" as into a piano, the plates or reeds would, if there were a fine enough gradation of pitches among them, echo the speech. And if, instead of Baile's or Reis's battery-powered intermittent currents, each vibrating polarized reed induced a continuous fluctuating current, then the amplitude of the current as well as the frequency would be properly proportioned for each. The resultant of all these currents could be transmitted through a single point, as in the case of the phonautograph stylus. At the other end of the line an electromagnet would transform that resultant current into pulses or undulations of magnetic force. The force, acting on another array of tuned reeds, would reproduce the original sound. And so speech would travel far with lightning speed.

The harp apparatus (never made), 1874

Instead of a separate permanent magnet to polarize each reed, Bell sketched a long iron bar shaped like a trough with its end walls knocked out, and with one side wall as north pole and the other as south. An array of reeds, like comb teeth, would extend up from each wall, taking on its polarity. Between the two rows, a single long bar with a coil around its length would serve as electromagnet.

It remained only a sketch. Bell lacked the skill and equipment to make such a thing there and then. Besides he still doubted that it could generate strong enough currents to work over a useful distance. And finally, the multiple telegraph took precedence in his mind. That required simple, steady frequencies. So instead of induced undulatory currents, he turned back to battery-powered intermittent currents. In his workshop he spent his time constructing a revolving make-and-break cylinder to produce such a current.

According to his testimony much later, Bell clearly remembered talking about the dead man's ear and the surprising power of its action, but felt less certain that he had talked about the "harp apparatus" to anyone in Brantford that summer. A letter he wrote home from Boston the following November 23 revealed it to his parents as if for the first time. For months, he wrote, "I have scarce dared to breathe [the idea] to anybody for fear of being thought insane," especially since "I was uncertain of the fundamental principle." But Moses Farmer, the electrical expert, had assured him of its soundness in electrical theory. "Please keep this paper," he concluded, "as a record of the conception of the idea in case any one else should at a future time discover that the vibrations of a permanent magnet will induce a vibrating current of electricity in the coils of an electromagnet."

The July date of the basic conception thus becomes a matter of historical rather than legal interest, because this letter of November 23, 1874, stating the concept clearly, itself preceded by a full year the earliest credible date of conception claimed by any rival.

More interesting is the puzzle of his failure to think in terms of a diaphragm transmitter and receiver sooner than he did. (The earliest independent evidence of a diaphragm in Bell's "electric speech" plan dates from late October 1874.) After all, he had been, and at the moment of his inspiration still was, working with phonautograph diaphragms as sound receivers. He knew that a diaphragm would be the form of receiver least confined to one particular frequency. He already had the habit of pressing an out-of-tune receiver reed against his ear to damp it down — turning it into a sort of diaphragm — so that he could hear it reproduce the transmitter's pitch. And both Pickering and Gray had resorted almost as a matter of course to diaphragms in their receiving devices (though Bell had not yet heard of their receivers).

The explanation probably lies in certain predispositions coming out of his reading and experience: Helmholtz's emphasis on speech sounds as composites built up from a number of pure tones; the trick of the echoing piano strings; Baile's suggestion of "vibrating plates." Gray probably had little or no knowledge of the first two items and so was not led astray by them. The scholarly Pickering probably had so broad a knowledge of acoustics that he could pick and choose among various approaches. Bell's formal knowledge fell between the two.

On the other hand, Bell would say in later years that if he had known more about electricity he would not have conceived the principle of the telephone. He may have meant that a practiced electrician would not have expected an electromagnetic armature to keep up with the speed and subtlety of sound vibrations, at least not well enough to translate them faithfully into electric current. That, at any rate, is as plausible an explanation as any for the myopia of the experts.

13

Allies and Adversaries

When Bell returned to Salem in September 1874, Georgie Sanders came running to meet him, full of health and spirit, no longer quiet and withdrawn. Bell's teaching prospered. Once more he gave the opening address at the Boston University School of Oratory. A newspaper ad brought half a dozen private pupils, and money came in from previous lessons. In a letter his mother urged him to bank it and "dissipate no part of it in mere experiments."

But he could not help himself. Dr. Blake admired his phonautograph tracings from Canada and suggested publishing their joint researches. Professor Cross invited him to talk over the summer's work "and perhaps plan for a little more." So Bell had the dead man's ear remounted and its membrane reglycerined.

Then events thrust acoustics into the wings and called telegraphy back to center stage.

One day Bell called at Gardiner Hubbard's home on Brattle Street in Cambridge. The Hubbard house embodied the elegance of genteel life in late nineteenth-century New England. Hedges and firs and fine red beeches shielded its five acres of grounds from the street. A crescent driveway led to the spacious frame house. Around the house lay broad lawns; the stable, gardener's cottage, and greenhouses; an old-fashioned formal flower garden full of periwinkle, foxglove, and larkspur, with straight gravel walks and at the end an old apple tree with a rustic seat; a summerhouse where japonicas bloomed in season; and a little pond, its banks lined with cobbles, set amid the green grace of great willows. Up the front veranda-posts of the main house climbed woodbine, and roses grew over frames outside the bays of the dining room and library.

On the mid-October day in 1874 when Alexander Graham Bell scuffed through leaves up the gravel drive to see Mabel Hubbard and her parents, outdoor color had passed from the flowers and been left to the sun and sky and lingering patches of foliage. But the solid interior ignored the seasons. Rich red velvet wallpaper gave warmth to the dining room, and crystal gas fixtures made rainbows of the daylight. Solid mahogany balustrades ran up and around the stairwell, which rose through three flights to the roof. In the big drawing room hung crimson damask curtains and heavy gilt valances. Enfolded by opulence, Bell readily consented to stay for tea. And afterward he played the piano, Mrs. Hubbard being a lover of music.

Suddenly Bell paused, turned around on the piano stool, and asked Hubbard if he knew that a piano would repeat a note sung into it. Then Bell demonstrated the trick with his pure, resonant voice. Did Mr. Hubbard know, he asked further, that a tuned instrument would also respond to a telegraphic impulse having the same frequency? Like a true Yankee businessman, Hubbard wanted to know what value there was in that. And Bell explained that on that principle one could send several messages at once over a single telegraph wire.

Now it was Bell's turn to be surprised. Full of excitement, Hubbard drew Bell aside for an explanation. Afterward he called his wife in to be enlightened also. "When she raised some objection to one point," Bell wrote home elatedly, "he answered it himself saying 'Don't you see there is only one *air* and so there need be but one *wire!*' "

Bell knew that Hubbard was in some way involved with telegraphy. What he probably did not know was that he had dangled the possibility of a great advance in multiple telegraphy before the one man in Boston, perhaps in the nation, most desperate to find just that. After his campaign for deaf-mute articulation schools in the mid-sixties, Hubbard had launched a new campaign for the public good — and his own advantage. The Western Union Telegraph Company, he charged, held back the power of the telegraph to broaden the nation's economy and enlarge the lives of its people. With cheap rates, telegrams could be sent like letters; and Hubbard insisted that the increase in business could easily be handled through improved technology without adding more lines. In a pamphlet of 1868, for example, he cited the Stearns duplex telegraph, Sir Charles Wheatstone's instrument for quick automatic transmission of specially prepared messages, and the new, cheaper, more efficient steel and copper wire. With increased volume, the lower rates would still be profitable, perhaps more so than before.

Hubbard did not propose government regulation of rates nor government ownership of lines. Instead he wanted Congress to charter a private corporation, the "United States Postal Telegraph Company," authorized to build lines along post roads and routes. The Post Office would receive and deliver telegrams at rates about half those then prevailing, and the new company would contract to send them. Among the named incorporators were to be Hubbard himself, his brother-in-law, and his Washington lawyer. Thus Hubbard would become a major figure in the nation's telegraph industry.

Hubbard had worked hard and selflessly for the deaf. Still, self-interest must have added zeal to his lobbying for "the Hubbard bill" through session after session of Congress. He shut up his Cambridge home and lived in Washington through the winter and spring of each session. His business and professional interests suffered. The Crédit Mobilier scandal of the early seventies, growing out of congressional subsidies to railroads, led some to suspect Hubbard's motives and methods. "I wish you were out of it all," wrote his wife from Paris in 1873. "How I hate to have you called a schemer & speculator. . . . Do you really think now that another year will finish it?"

Another year did not finish it, though the bill seemed to come closer to passage in 1874 than ever before. Hubbard began thinking of selling off his Cambridge land, perhaps house and all. His wife wrote that her father had "shown me again this morning all his books, and the amount you owe him seems larger every time I see it. . . . I am so sorry that your bank note clients are not satisfied with your proceedings."

In a Senate committee hearing that spring, President William Orton of the Western Union Telegraph Company, a burly man of formidable presence, kept Hubbard on the defensive. Orton prided himself on technological progressiveness, having lately acquired rights to the Stearns duplex. "But the day for new inventions has not passed," countered Hubbard desperately, "and these are not the only improvements that are to be made. . . . One wire, instead of being used for the transmission of only two messages at once, may be used for four or possibly eight messages. . . . The potentialities of the telegraph are boundless; no man dare say what the future will bring forth." The immediate future brought forth Edison's quadruplex telegraph, to which Western Union also promptly acquired rights. "The discovery," said Orton triumphantly, "may be called the solution of all difficulties in the future of telegraphic science."

So ran Gardiner Hubbard's affairs, when in October 1874 Alexander Graham Bell came to tea and offered to outdo Edison's quadruplex five or six times over.

For months, perhaps a year, Thomas Sanders had known about Bell's harmonic telegraph. When he heard that Bell had told Hubbard about it, Sanders grew concerned. He knew more than Bell about Hubbard's affairs,

William Orton

and the Crédit Mobilier exposures had made him wary of men in high places. He urged Bell to protect himself at once by patent.

"On inquiry," Bell wrote home, "I find I cannot secure a patent until actual models are sent to Washington." But he still lacked money to have models built by an electrician and to pay legal costs. So he consented to give Sanders a half share in the invention in return for the necessary funds.

On that same day, October 22, 1874, a letter from Clarence Blake jolted Bell. After inviting him to show some of his tracings at a meeting of the Boston Society of Medical Sciences, Blake added casually: "I have received a letter from the Royal Institute from Elisha Gray concerning his experiments in telegraphing vocal sounds, which you will be glad to see." Thus Bell first encountered the name of Elisha Gray.

Gray's letter actually had said nothing about telegraphing vocal sounds. Blake had read into it what he had heard earlier. Gray merely wrote that having experimented in London with Tyndall's help, he would be back soon and would like to swap acoustical information. But Bell, calling at Blake's office the next day, refused to look at Gray's letter or hear about

128

its contents, for fear he might later be accused of stealing Gray's ideas. Instead he revealed to Blake his own schemes for telegraphing vocal sounds, by now including membrane transmitters as well as the "harp apparatus." (Blake's offhand decision to save Bell's rough sketches eventually gave patent lawyers and historians their earliest dated record of the telephone conception.)

Blake immediately grasped the electrical principle of Bell's conception, and endorsed it as acoustically valid. He entered at once into a discussion of the proper type of transmitter membrane. Otherwise, convinced that Bell needed only to work out mechanical details and seeing the matter as outside his special field, Blake thereafter left Bell more and more to his own devices. The developing telephone had superseded the Bell-Blake acoustical partnership.

But while losing a partner in science, Bell gained another in business. After prudently searching the Patent Office for any possible anticipation of the harmonic telegraph idea, Hubbard offered Bell funds for experimenting in return for a share in patent rights. Disturbed by the looming up of Elisha Gray, anxious to join Mabel Hubbard's circle, flattered by the interest of a man whom he vaguely understood to be "the head of the telegraphic system of the States," the young inventor eagerly agreed, provided Thomas Sanders were willing. And Sanders, correctly valuing Hubbard's talents as organizer and promoter, did agree that the three should go equal shares, Bell putting up ideas, the others cash.

"I am tonight a happy man," wrote Bell to his parents at the end of October. "Success seems to meet me on every hand." Private pupils were "pouring in," thirteen in all; the Medical Society of Massachusetts had shown great interest in the ear experiments; and after taking the required oath of intention to become a United States citizen, Bell had put a first installment of his caveat in the hands of a patent lawyer. The enlistment of Hubbard encouraged "greater confidence in my own ideas," he wrote, and —in allusion to his boyhood invention— "I feel as if I may yet TAKE OFF SOME 'HUSKS'!!!" Meanwhile he was to give another lecture at MIT, this one on "the Education of the Deaf and Dumb," to promote the new articulation teachers' association and a future evening school for deaf adults. Lest his father feel displaced by multiple telegraphy, Bell added: "Should I be able to make any money out of the idea, we shall have Visible Speech put before the world in a more permanent form than at present." Somehow he forgot to proclaim (as on former occasions) the opening of a new epoch, though he might well have done so.

"Your wisest course," wrote his father characteristically, "would be to sell your plans to Messrs. Sanders and Hubbard. . . . You can't work out the scheme without neglecting your other business. . . . Take what you can get at once." Bell ignored the advice. Hubbard, on the other hand, pressed him to concentrate on multiple telegraphy, letting Hubbard and Sanders pay for a skilled assistant. Bell resisted this also. He shrank from taking money from his partners, as if it compromised his independence. And he honored the prior claims of his Boston University lectures, his special class of articulation teachers, his private pupils, and little George Sanders. So to Hubbard's annoyance, Bell worked on his invention mostly on Sundays, holidays, or late at night.

Nevertheless he worked on it. And as the rivalry of Elisha Gray took shape, he worked all the harder. On a trip to Washington, Hubbard had encountered William Orton, who spoke of Gray's tone-transmitting invention as "very curious" but apparently had no idea of its applicability to multiple telegraphy. All the same, Bell's lawyer, after talking with Gray's, warned Bell to hurry lest Gray come up with a multiple telegraph. After working "day and night," Bell sent off his caveat papers.

On that very day, November 14, 1874, the *Commonwealth,* a Boston periodical, reprinted without acknowledgment the *New York Times* article of July on Gray's invention. Bell's Boston University students called the item to his attention at his next class. The *Times* in July had said that Gray conceived his idea "about two months ago"; the *Commonwealth* in November reprinted that phrase also. But Hubbard promptly quenched Bell's elation at so long a head start by writing from Washington that Gray had applied for a patent on his "musical telegraph" as early as August.

Foreseeing litigation, Hubbard had already urged Bell to send him a dated and signed letter describing every experimental advance. Bell himself had just got Percival Richards to write down his recollections of the West Newton Street doings of early 1873. Now Hubbard's Washington lawyer Anthony Pollok persuaded Bell to withdraw his caveat, since Gray had already made a patent application for at least one element of the harmonic telegraph system. A caveat therefore would not protect Bell, but its existence might alert Gray to his progress. Bell's patent could be applied for soon; and then Gray's prior application could be overturned by proof of Bell's priority in conception.

Under such pressures, Bell finally consented to get expert help. He consulted Moses Farmer, "an electrician as celebrated here, as Wheatstone is at home," he wrote his parents; and Farmer assigned one of his skilled assistants, George Hamilton, to turn Bell's sketches into working apparatus. On Hubbard's advice, Bell also checked his multiple telegraph theories with Professor Lovering, who could see no flaws in them.

Bell even "ventured cautiously" to try his telephone theory on Farmer. "To my delight," Bell wrote his parents, "he said the theory was *all right* but that the difficulties of practically working the idea were such that it would take years to solve the problem practically. He advised me to publish the idea in the Philosophical Magazine after I had protected my telegraphic scheme." Professor Cross agreed with Farmer. But like Bell, Farmer and Cross thought the currents generated by the voice would be too feeble for practical use. So Bell put the telephone idea aside and concentrated on perfecting the multiple telegraph as Hubbard and Sanders wished.

"It is a neck and neck race between Mr. Gray and myself who shall complete our apparatus first," Bell wrote on November 23, 1874. "He has the advantage over me in being a practical electrician — but I have reason to believe that I am better acquainted with the phenomena of sound than he is. . . . The very opposition seems to nerve me to work." But, he added, "I feel that I shall be seriously ill should I fail in this now I am so thoroughly wrought up." Next evening he reported happily that Hamilton had finished a battery-operated transmitting and receiving instrument that worked well.

Two days later he made what seemed "a most extraordinary discovery." Grasping a vibrating receiver armature to damp it, he found that the tone continued. When he detached the armature altogether, he heard the sound coming directly from the iron core of the electromagnet. This could lead to a new type of receiver, he wrote excitedly on the spot to his father and Hubbard. The next day he found that a mere empty coil without a core also emitted a sound of the same pitch as the transmitter.

Elated, Bell invited Professor Cross and others to a demonstration. But Cross had to disabuse his young friend just as Alexander Ellis had done years before in the case of the double pitches of vowels. As Helmholtz had anticipated Bell's discovery then, so had Joseph Henry in the case of the vibrating electromagnet core, though not as to the empty coil. More than that, Cross told Bell of Philipp Reis's use of the effect in his "telephone" of 1861. If Bell had ever heard of the Reis "telephone" before, it had left no impression on his memory.

Other disappointments plagued Bell. Sparking oxidized the circuit-breaking contact. With Hamilton's help, Bell cured this by using a condenser. The transmitter pitch varied with circuit length and battery strength. This took longer to mend through new circuit and battery arrangements. Worst of all, Bell and Hamilton could not make Bell's ingenious induced-current circuit plans of the preceding winter actually work in practice. Nor did Bell's bar magnets on a revolving cylinder induce an "oscillating current" with the frequency he expected. So Bell decided to concentrate on direct-current circuits for the purpose of the patent, and to return later to induced currents.

Meanwhile he could not help coursing after variations on the phenomenon of the sounding coil. And all this at night while he taught by day. One December morning he arose feeling dizzy, unable to stand up for more than a few minutes without finding the room whirling about him. Two or three days in bed and a grudging reduction of telegraphic work to an hour a day helped to restore him.

By mid-December he was off on a new scent, having found that by applying the oscillating current from his revolving bar-magnet cylinder directly to his eardrums he could hear a sound, rising and falling with the speed of the cylinder. He also had a miniature organ made, using vox humana reeds as tuned transmitters, and transmitted tunes from the Sanders barn to the house through a resistance equal to that of a wire from Salem to Boston. On pressing his ear to a single receiver reed, Bell could even hear chords. The air-vibrated organ reeds, however, could not be tuned as easily or used as long as the electrically vibrated steel reeds.

Caught between teaching and his anxiety to get ahead of Gray, Bell thought of spending his Christmas holidays on telegraphic work in Boston. But at the last minute he went off to Brantford, arriving Christmas afternoon — with his hobby. "Al's experiments described," noted his father's diary next day. On the day after: "Long talk on multipl teleg and speech trans. Al sanguine." And two days later: "Talking half the night motor and telephone." Hurrying back to his Boston University classes after New Year's Day, he set to work again on his telegraph. But Visible Speech teachers were finding their labor intolerable, and so Bell developed schoolroom charts and looked into the cost of special type to reduce the work of writing out exercises. This further delayed telegraphic progress.

The new year brought Bell a new assistant. For two years Moses Farmer had, among other activities, been the official "electrician" of the United States Naval Torpedo Station at Newport, Rhode Island. Whether from physical or fiscal strain, he decided at the end of 1874 to close up his Boston operations and concentrate on his Newport work. Thus Farmer's assistant George Hamilton lost his chance at collateral immortality, and a still younger man instead become history's most famous listener.

Thomas A. Watson, son of a Salem livery stable foreman, had drifted from job to job for four years until July 1872, when at the age of eighteen he settled down at Charles Williams's shop on Court Street, Boston. A bright, quick boy, who had left school from restless ambition rather than incapacity and had later attended a commercial college while working, he took to his new job from the first day.

Thomas A. Watson, about 1872

Charles Williams

Charles Williams had been in business on his own for nearly twenty years, making a growing list of electrical devices in small quantities: mostly a variety of telegraphic and fire alarm apparatus, but also including call bells, gongs, hotel annunciators, school laboratory apparatus, batteries, and custom work for a motley procession of inventors. He and his twenty-five employees occupied the third floor and attic of the building. In the main shop, dust and soot had long since turned the whitewashed brick walls to a streaked and shadowy gray. Near the grimy front and back windows stood a dozen or more metalworking hand lathes and a couple of small steam-powered lathes. Under dusty ceiling beams, pulleys whirred and quivering leather belts raced along their endless courses. Wooden racks of steel, iron, and brass sheets and rods huddled in the center, and piles of rough castings lay about the floor. A small forge for annealing and tempering joined the steam engine in adding heat to the noise, dust, and movement of the place, but a partition shut all this off from the office and display room in one corner.

Dingy and cluttered as the place was, Watson in his later, more cultivated years recognized that he had unknowingly felt a sort of poetry in it all, a creative exultation as "I made stubborn metal do my will and take the shape necessary to enable it to do its allotted work." Even when given the job of making several hundred identical pieces, he put his mind to developing the most efficient sequence of motions and then making them almost automatic, an independent anticipation of scientific management. In like spirit he lay awake at night devising special tools to speed and improve his work. By 1874 he had a secure place, depression or not, as one of the shop's best men. He began also to study and think about scientific principles involved in his work and, for that matter, outside it. And so from early 1874 he came to be assigned almost exclusively to the custom work of inventors.

Some of the inventors were little more than crackpots whom Williams was ready to humor for pay. Others had genuine ability. Young Watson found especial profit, indeed "an important part of my electrical education," in talking with Moses Farmer, who occasionally came in to have parts made for an invention of his own or of someone consulting him. One of the latter, oblivious to shop protocol, came rushing out of the office directly to Watson one day in 1874 to have changes made in a couple of small instruments originally ordered through Farmer. This engaging subverter of discipline was, as Watson remembered him, "a tall, slender, quick-motioned young man with a pale face, black side-whiskers and drooping mustache, big nose and high, sloping forehead crowned with bushy jet-black hair. It was Alexander Graham Bell, a young professor in Boston University."

In January 1875, after Farmer and Hamilton left, Bell may have remembered young Watson and asked for him, or Watson may simply have got one more routine assignment to an inventor. In either case, Watson began working with Bell, not as Bell's employee but on assignment within the Williams shop. (He continued to have other assignments besides.) "No finer influence than Graham Bell ever came into my life," wrote Watson half a century later. Bell's table manners — he used a fork instead of a knife for conveying food to the mouth — his "punctilious courtesy to every one," his "expressive speech," his "clear, crisp articulation," were a "revelation." He talked of books which Watson in turn found absorbing, not only general literature but also the scientific works of Tyndall, Helmholtz, Huxley, and others.

Watson's first task in the new relationship was to make a brass cylinder, around which he saw Bell wrap a sheet of paper with evenly spaced square holes. Bell was making another unavailing try at getting the induced currents to come out right by means of his revolving-cylinder current-interrupter.

Then he put it aside and returned to the steel-spring rheotomes, of which Watson made several in a new design. These simplified the circuit problem by completely separating the local transmitter circuit from the main line. The transmitter reed had two contact points unconnected electrically, one making and breaking the local circuit that kept the reed itself vibrating, the other making and breaking the main-line current. Bell discounted his earlier fears that the current might be blocked by too many interruptions.

With this, Bell decided to apply for a patent. "Now comes the tug of war," he wrote. "There is no doubt that there is to be a struggle between Mr. Elisha Gray and myself."

He fretted over a letter he considered "dishonourable and blackmailing" from Joseph Adams, the lawyer who had handled the withdrawn caveat. Adams, who knew much about the invention and even had some of Bell's papers relating to it, wrote from Washington to ask if Bell had any further use for his services in the matter. If not, then Adams wanted to know if he were free to "act in the interest of other parties in matters involving the same subject." Bell wired Hubbard in Washington about the implied threat, and Hubbard evidently found, or at least believed he found, some way to keep Adams in line. The affair doubtless explains a sudden revival of Bell's earlier fears of chicanery or espionage.

He had planned to leave for Washington on February 12, but postponed the trip a week because of a hitch and a further inspiration.

His newly finished patent-model transmitters, brought home to Salem and anxiously tested, stopped vibrating after a few minutes. Watson came over to the Sanders house from his Salem home and worked with Bell until nearly midnight on the problem. Then Bell worked on alone until four in the morning. Next day he concluded that the trouble lay in the oxidation of the contacts. He and Watson struggled to find some way of fixing a platinum contact to the reed without spoiling· its temper. Bell finally thought of riveting a bit of platinum wire through a small hole in the reed. Transmitters so equipped ran through the night, while the older type all stopped. The problem was solved, he wrote his parents, but "I trust that you may never know the agony I endured all night and yesterday."

The inspiration that also delayed him was a device that would close an ordinary Morse receiver while the reed receiver was vibrating and open it when the reed stopped, thus recording messages as dots and dashes on a strip of paper, instead of requiring the operator to single out one pitch among several and take its signals by ear. The "vibratory circuit breaker," as Bell called it, was a long, light strip hanging from a pivot. When the receiver reed vibrated, it knocked the lower end of the strip to one side and kept hitting it before it could fall back again. While the strip remained in that position, a conductive crosspiece on it closed the local circuit, opening the circuit again when the reed stopped vibrating and allowed the strip to hang straight down. The battery-powered local circuit operated a Morse receiver.

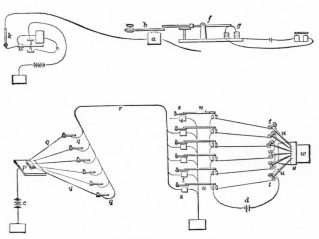

The autograph telegraph, 1875

Bell saw at once that this made the multiple telegraph adaptable to an "autograph telegraph" device. Instead of having each receiver stylus trace its broken line on a separate strip of moving paper, he could have as many as thirty styluses at once drawing closely spaced parallel broken lines on a single sheet. This could be made to come out like a picture or a document seen through a fine grid. If an original autograph message or a drawing were made with special paper or ink that would close a contact where written on and break contact where blank, it could be pulled along under contact points connected to thirty transmitters and thus be reproduced by the receivers — in theory. In existing autograph devices, a single receiver stylus had to traverse the sheet a number of times. "The new attachment," wrote Bell with characteristic optimism, "will render it possible to reproduce a long message almost instantaneously by a single turn of a cylinder."

After a last, triumphant test of the multiple telegraph at the Sanders house, Bell and Sanders took a night train for Washington on February 19 and arrived the following evening. The next day went to unpacking apparatus at Gardiner Hubbard's Washington house and setting it up for exhibition. "Now," wrote Bell, "comes the most important crisis of my life so far."

In the course of that hectic day, Bell at once suspected "some underhand work" when he was unable to get workable battery cells from the only electrician in town, and when the electrician's messenger "came right into the parlor and stared about, to see what kind of instruments I had got."

At that disconcerting moment Hubbard told him that President Orton of Western Union would be around in a half hour to see his instruments. "The Western Union," he wrote his parents later, "is probably the largest corporate body that has ever existed. It controls more miles of telegraph wire than there *are in the whole of Europe!*" Bell worked frantically to improvise the necessary additional cells from slop-basins, a sawed-up carbon, and substitute acids, and had the apparatus in working order "just half a minute" before Orton showed up.

The instruments had never worked better. Orton had told his longtime adversary Hubbard that he could only look in for a few minutes. But in the end he stayed at least an hour. And though he had to leave Washington that night, he urged Bell to give him another demonstration later in New York, using the company's actual long lines.

Elisha Gray had probably not suborned the capital's lone electrician. Nevertheless Bell had reason to worry. A month earlier, Gray had renewed earlier applications for patents on his "animal tissue" tone transmitter, about which Bell cared nothing, and also on a form of vibratory "musical telegraph," which came closer to Bell's conception. Gray's musical telegraph differed from Bell's device chiefly by using a single Reis-type electromagnet sounding-box receiver rather than a number of tuned reeds, and by inducing the composite intermittent current on a main line through a number of vibrating transmitters, which were started and stopped by the sending keys instead of vibrating continually. In using a single receiver, of course, it could not send multiple messages, except perhaps to operators with an unusual sense of pitch and power of concentration.

According to his assistant, Gray had conceived the idea of a harmonic multiple telegraph as early as the spring of 1874 (at least three years after Bell's original conception). But Gray's patent application renewals of January 19, 1875, did not include that feature. Not until February 23, three days after Bell's arrival, did Gray formally apply for a patent on the basic harmonic multiple telegraph system. This may have been coincidence. Then again, Gray's lawyer in Washington may have got wind of Bell's arrival and plans — perhaps on a tip from Joseph Adams — and moved to forestall him. If so, the Gray interest succeeded, for Bell did not file until February 25.

Bell's patent lawyers, Anthony Pollok and Marcellus Bailey, the partners in one of the capital's leading patent law firms, had at first been pessimistic. But Bell's recent inspirations of the vibratory circuit breaker and the autograph telegraph struck them as persuasive evidence of independent conception and development, since Gray had not yet reached that point. Also, one of the Patent Office examiners, though not in the electrical section, turned out to be a deaf-mute who knew Bell and could vouch for his character.

More than that, Zenas F. Wilber, the examiner who would actually handle Bell's patent applications, happened to be in Pollok's office one day when Bell came to call. "I had a long interview with him in which I explained everything," Bell wrote home, "and I can't help thinking that he must have been convinced of my independent conception of the whole thing."

On the advice of Pollok and Bailey, Bell applied for three patents: on the harmonic telegraph using direct current with a return wire instead of grounding; on the harmonic telegraph using induced currents on a single wire without grounding; and on the vibratory circuit breaker with its corollary of an autograph telegraph. "You need not fear," were Pollok's parting words, "we shall pull you through all right." The circuit breaker and autograph telegraph patent was indeed eventually granted on April 6. Bell rejoiced that it was the controlling one, without which he believed Gray could do little of commercial value. The other two were declared in interference with Gray's applications and also with that of one Paul La Cour of Copenhagen, Denmark, filed on March 3. Even so, since none of the interfering applications had been granted, the crux became not dates of filing but priority in conception. As to that, Bell with reason had high hopes of prevailing.

Meanwhile, on February 27, Bell, Sanders, and Hubbard put into writing their oral agreement to share equally in Bell's telegraphic inventions, including "any further improvements he may make in perfecting said inventions." Should the inventions prove valuable, a company was to be organized to control the patents, each of the partners to get one-third of the stock. This simple agreement began what eventually became the largest single business enterprise in the history of mankind.

On March 1 Bell called at the red brick Norman (or as some put it, "bastard Gothic") pile that housed the Smithsonian Institution, with its picturesque battlements looking out over the Mall to the Capitol in one direction and the Washington Monument in the other. He had a letter of introduction to the Smithsonian's director, Joseph Henry, but may not have needed it; some years earlier Henry's daughter had read and "thoroughly enjoyed" one of the elder Bell's books on elocution.

In his late seventies, Joseph Henry commanded greater prestige at home and abroad than any other American scientist then living, though his significant original researches had ended more than a generation earlier. His leonine head and dignified bearing befitted his past triumphs and present office. Patient, kindly, self-controlled, gently humorous, he could nevertheless plant a sting on sufficient provocation. He finally told one crackpot inventor, who had wasted two hours of his time with a plan to suspend a transatlantic cable from captive balloons, "that I thought he might claim the merit of originality because no two persons could possibly hit on such a chimerical scheme."

Since Henry had anticipated Bell's discovery of the sound from an electromagnet core, the young man told Henry of his other experiments, in order to find out which were new. Henry listened "with an unmoved countenance, but with evident interest" until Bell came to the sound from the empty coil. That effect had been described by a French scientist some years before, but Henry had apparently forgotten or overlooked the fact. He seemed to think it a new discovery. "Will you allow me, Mr. Bell," he exclaimed, "to repeat your experiments, and publish them to the world through the Smithsonian Institution, of course giving you the credit of the discoveries?" Bell replied that it would give him extreme pleasure and offered to show him the experiment at any time.

Henry prepared to go immediately to Hubbard's house, despite his age, a severe cold, and the rawness of the winter day, but Bell insisted on bringing his apparatus to the Smithsonian next day instead. After the demonstration, encouraged by Henry's evident interest, Bell nerved himself to broach his untested theory about how to transmit the human voice electrically by means of the "harp apparatus."

Henry may not have remembered his long-past speculations about the possibilities of Faber's talking machine, but he reacted now with the vision he had shown then. It was "the germ of a great invention," he told Bell.

As to Bell's notion of a membrane diaphragm transmitter, Henry showed less perspicacity than had Clarence Blake. The matter came up when Henry gave Bell his first sight of an actual Reis membrane transmitter. Henry thought a diaphragm would have a natural rate of vibration, which might unsuit it for speech transmission. But that cavil was far outweighed by his endorsement of the basic telephone principle.

As to the application of that principle, "what would you advise me to do," Bell asked (probably remembering Moses Farmer's comment), "publish it and let others work it out, or attempt to solve the problem myself?" Henry had bitter memories of the fame and financial security he himself had tossed away by letting Samuel Morse "work out" the commercial application of the telegraph. He urged Bell not to publish, but instead to perfect the great invention himself. And when Bell protested that he lacked the necessary electrical knowledge, Henry answered firmly and succinctly: "Get it!"

"I cannot tell you how much these two words have encouraged me," Bell wrote his parents a few days later.

I live too much in an atmosphere of discouragement for scientific pursuits. Good Mrs. Sanders is unfortunately one of the *cui bono* people, and is too much in the habit of looking at the dark side of things. Such a chimerical idea as telegraphing *vocal sounds* would indeed to *most minds* seem scarcely feasible enough to spend time in working over. I believe, however, that it is feasible, and that I have got the cue to the solution of the problem.

Good Mr. Hubbard was another of the *cui bono* people, though he could see somewhat further than Mrs. Sanders. In his view the harmonic multiple telegraph and its autograph offspring promised the quickest, surest, and largest profits and therefore deserved Bell's full attention. Talking by wire could come later. Bell himself felt the pull of ambitions that needed financing: a normal school for teachers of the deaf, a campaign to promote Visible Speech, and his "electro-motor" notion, "which I think as valuable an idea as this Telegraph." So he did not contest Hubbard's priorities.

At New York on his way home from Washington, Bell arranged for tests of the multiple telegraph on an actual line wire of Western Union. In mid-March he went back to New York for a long talk with President Orton and his chief "electrician" (electrical engineer) George Prescott on the theory of the invention. In a practical test his instruments "went like clock-work." "By a happy chance," he rejoiced, "they are much more perfect than I thought at first." Even with electromagnets much weaker than those normally used for such work, signals came through clearly over a two-hundred-mile circuit of line wire.

At Bell's suggestion Prescott sent out the instruments to be equipped with stronger electromagnets. That afternoon, when Bell returned for further trials, he was shaken by Orton's change of attitude.

The instruments had not come back yet, he was told, but after Orton finished some other business the two men sat down on a couch and began a cordial conversation. Orton leaned back casually, his feet on a chair, and drew from Bell a detailed history of his invention. Then Orton revealed that Elisha Gray had just called on him — an "ingenious workman," said Orton, with "good apparatus," compared to which Bell's seemed "crude." The Western Union was a great power, boasted Orton; its influence could make "the weaker party the stronger." Inventors were apt to overestimate the value of their work.

"By the bye," said Orton as Bell was about to go, "is Mr. Gardiner G. Hubbard interested in this matter with you?" "Yes," said Bell. "The Western Union," said Orton, "will never take up a scheme which will benefit Mr. Hubbard." Orton, "perfectly gentlemanly and polite in his manner," drove Bell to the hotel and promised him all the testing facilities Western Union could offer. But Hubbard had done too much to injure the company, Orton insisted, for it to aid any scheme in which he was interested.

Hubbard, who happened to be in New York that evening, volunteered to withdraw if he seemed to be hurting Bell's interests. Bell would not hear of it. So Hubbard advised him to put off his return to Boston one more day, get his instruments from Western Union, and take them across the street to Western Union's rival, the Atlantic and Pacific Telegraph Company.

Orton and Prescott had meanwhile been testing Bell's instruments with good success; and as Hubbard doubtless had hoped, the threat made them change their tune. His company, said Orton now, would not help *develop* any scheme for Hubbard's benefit, but mere sentiment would not keep them from buying a good thing if brought to them perfected. As for Gray, they would certainly back him if Bell went to their rivals, and they would not pledge future neutrality in any case. But at the moment they had no arrangement with Gray, nor did they contemplate making one.

In the upshot, then, Bell's bout with Orton added to both the hope and the urgency of his work with the harmonic and autograph telegraph systems.

14

The Telephone Is Born

Bell came back from New York in March 1875 "thoroughly worn out" by his race with Gray. His doctor heartily approved when he decided to cancel private classes. Now only his Boston University classes and Georgie Sanders's lessons competed with telegraphy.

That meant less income. Hubbard and Sanders paid only for equipment, supplies, and young Tom Watson's time, not for Bell's time and labor. By early June, Bell had borrowed $9.60 from Watson, who got $13.25 a week for his work at the Williams shop (including his work for Bell). But Bell's pride and his feelings toward Mabel Hubbard kept him from admitting his straits to Mabel's father, let alone asking for financial help.

Fortunately Dean Monroe of Boston University gave Bell vital support in June by paying him in advance for the next year's lectures. Without that aid, Bell said later, "I would not have been able to get along at all."

On March 11, Bell had "worked the instruments for the first time by an induced current." After his health revived in early April, he spent most of his workshop time trying to put multiple and autographic telegraphy into commercially acceptable working order. He felt bound to do so by his agreement with Hubbard and Sanders. And Hubbard saw to it that Bell did not forget that goal. A year or so earlier, in a published attack on the "Hubbard bill," William Orton had scoffed at existing facsimile telegraph systems as too slow and expensive. Now that Bell promised one ten times faster, Hubbard must have longed to make his adversary eat crow. More to the point, Hubbard saw the autograph telegraph as potentially the surest money-maker of all Bell's notions, besides which it was safely patented, beyond challenge from Elisha Gray.

Nevertheless, Bell was losing his zest for the chase. The autograph telegraph had become a matter of tinkering and fiddling to achieve more reliable multiple transmission. Signals meant for one receiver would break off halfway through or start beeping away on the wrong receivers. Then Bell would have to pluck the recalcitrant reed to determine its pitch, damp it against his ear to hear the transmitter pitch, and adjust the reed to make the two pitches coincide.

There was another frustration. Bell had talked of as many as thirty styluses drawing their broken parallel lines simultaneously. He had dismissed as a groundless theoretical bugaboo his earlier fear of interference among frequencies. Now, trying a mere half-dozen receivers, he found that problem a real one after all. And so he had to work out transmitter pitches that would interfere with each other as little as possible. It was not easy.

Late in May, Bell confessed to his parents that his inexperience in electricity was "a great drawback." But, he added bravely, "Morse conquered his electrical difficulties although he was only a painter, and I don't intend to give in either till all is completed."

Meanwhile he had to disregard other ideas that beckoned him away from the path of duty. He talked wistfully of them to Watson. Joseph Henry had replied with encouraging interest and respect to Bell's account of further experiments with the empty coil. Most seductive of all was Bell's idea for electrical transmission of speech. "In spite of my efforts to concentrate my thoughts upon multiple telegraphy," he recalled later, "my mind was full of it."

One evening at the Williams shop after a long day of unsympathetic vibrations, Bell roused his jaded young assistant by disclosing his scheme. "Watson," he said impressively, "if I can get a mechanism which will make a current of electricity vary in its intensity, as the air varies in density when a sound is passing through it, I can telegraph any sound, even the sound of speech." Then he went on to describe the cumbersome "harp apparatus," leaving the impression in Watson's mind that it would be about the size of a piano. Presumably Joseph Henry's doubts about a diaphragm transmitter and receiver still inhibited Bell's thinking.

On May 4, 1875, Bell wrote Hubbard about an idea he had conceived for overcoming the supposed practical objection to his "electric speech" scheme, to wit, "the feebleness of the induced currents" (which he still assumed without testing):

I have read somewhere that the resistance offered by a wire . . . is affected by the *tension of the wire*. If this is so, a *continuous current of electricity* passed through a vibrating wire should meet with a varying resistance, and hence a pulsatory action should be induced in the current . . . [corresponding] in *amplitude*, as well as in rate of movement, to the vibrations of the string. . . .

[Thus] the *timbre* of a sound [a quality essential to intelligible speech] could be transmitted . . . [and] the strength of the current can be increased *ad libitum* without destroying the *relative intensities of the vibrations*.

With this letter, Bell clearly established his priority in conceiving the final basic principle of the modern telephone: variable resistance.

Next day, he and Watson ran wires from Bell's Salem study to the neighboring music room of Bell's friend Manuel Fenollosa, a Spaniard who had come to America with a traveling band forty years earlier, married into one of Salem's leading families, and settled down as a music teacher. The strings of Fenollosa's piano turned out to be attached to a metal frame. Nevertheless Bell chose one as near the transmitter frequency as possible and listened while Watson, next door in the study, sent an intermittent current through it. Bell heard the faint whine of the transmitter, but could not be sure whether it came from the piano wire or directly through the air. Then he sent a steady current through the circuit and listened to a receiver in the study while someone plucked the piano wire. Again he could not be sure that what he heard faintly had been electrically transmitted.

But he cheerfully blamed the failure on the metal piano frame, not on the theory. When he had time, he planned to try passing current through a wire stretched across a membrane and hitched to its center. The membrane would vibrate the wire, varying its tension and thus also its resistance, making the current undulatory, or so he hoped.

The Fenollosa fiasco gave the autograph telegraph a reprieve. Bell returned to it dutifully. "I think I have at last mastered the Autograph," he began in his May 24 report to Hubbard, and added anticlimactically, "at least I see clearly one or two sources of difficulty . . . and I hope in a few days to have some good results to show." But on the same day he wrote his parents of what really filled his mind now: "I think that the transmission of the human voice is much more nearly at hand than I had supposed."

June 2, 1875, was unseasonably hot in Boston, hotter still in the attic of the Williams shop — and Bell did not like hot weather. It had been a frustrating day, piled on weeks of frustration — and Bell was an impatient and mercurial young man who longed to have done with niggling adjustments and get on to a grander quest. Among his tasks that day, as for weeks past, had been that of getting his transmitter pitches into as excruciating a dissonance as he could contrive, so that they would not excite each other's receivers — and he was a natural musician prone to headaches.

He had three transmitters and three correspondingly tuned receivers in his attic room, connected in series with a similar trio of receivers tended by Watson in the adjoining room. Bell pressed a key that sent one transmitter's

A replica of the attic laboratory at 109 Court Street, Boston, rebuilt from the original materials in consultation with Thomas Watson. (On exhibit at the New England Telephone Company general offices, Boston.)

tone through the circuit, and the proper receiver in each room sounded obediently. He sent another frequency through the circuit, and again the proper receivers sounded. He tried the third. But in Watson's room nothing happened.

It was a gallingly familiar experience. Now the pitch of the balky receiver would have to be checked and, if necessary, adjusted. But the first step in the routine was to make sure the free end of the reed was not simply set too close to the electromagnet pole just beneath. It might have touched the pole and frozen to it. If so, Watson might only need to pluck it free and perhaps minutely widen its separation by bending it. Bell called to Watson to try that expedient.

While Bell waited with the transmitters all off the circuit, he kept a weary and doubtless somewhat baleful eye on the twin of Watson's balky receiver. And then came the miracle.

Bell's reed vibrated. It vibrated by itself, visibly, with no transmitter current in the circuit.

With a wild surmise, Bell charged in upon the startled Watson to find out exactly what he had done. Keep plucking the reed, he told his assistant excitedly, and then he dashed back to see the marvel repeated.

Now, feverishly, Bell silenced the whining transmitters and reconnected the circuit so as to include only the three receivers in his room and the three in Watson's. Watson began plucking his receiver reeds one by one; and one by one, breaking into the expectant silence of Bell's room, the corresponding reeds not only vibrated but also sounded. It must have seemed to Bell the sweetest music he had ever heard.

He knew what had happened, and he knew what it meant. On the strength merely of some slight residual magnetism, the plucked reeds had induced the undulatory current he had postulated nearly a year before. And that preposterously feeble current, exciting the electromagnets in his room, had made their reeds vibrate in precisely the same way, vibrate vigorously enough to generate audible sounds. It dawned on him that he had grossly exaggerated the degree of vibration required to make a sound. An incredibly minute disturbance would do it.

While Watson in wonderment plucked at the reed, Bell pressed his ear against one of his own reeds not attuned to it, damping that reed's natural vibration. As he had dared to hope, he heard a faint sound, the pitch not of his reed but of Watson's. What was more, he heard its *timbre*. He heard faintly but faithfully just what Watson was hearing out of Bell's earshot. And as Bell knew at once, what he heard was no less complex and subtle than the sound of symphonies — and speech.

The telephone was born, and its wise father knew his child.

So the heat and the frustration, and for that matter the autograph telegraph, were all forgotten that sultry afternoon. The two young men, both in their twenties, set to work as if it were the morning of a bright new day. They plucked and listened, rearranged the circuit, and painstakingly reassured themselves by experiment that the sounds were indeed transmitted electrically and not by mechanical conduction. They sent a steady current through the circuit so as to excite the electromagnets and thus more strongly polarize the receiver reeds; and the sounds became even louder. They vibrated a permanent magnet over an electromagnet coil and got a loud sound from the receiver.

Having already heard the supposedly impossible, Bell now dismissed his doubts about diaphragms. Instead of wasting time on a "harp apparatus," he sketched a diaphragm instrument and asked Watson to make two of them. Watson studied the sketch on the midnight train home to Salem. To one pole of a U-shaped electromagnet, the end of a steel reed would be hinged, rather than clamped, thus freeing it from a predisposition to any particular frequency. The other end of the steel-reed armature would extend over the other pole of the magnet and would be attached to the center of a stretched membrane diaphragm.

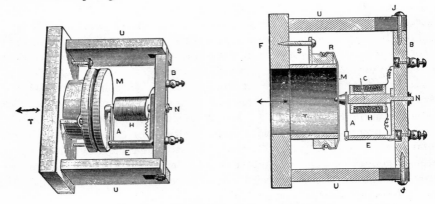

The first membrane diaphragm telephone, June 1875

A day or two later Watson brought in the first such instrument. But he had used delicate goldbeater's skin for the diaphragm and a heavier armature than Bell had intended. When Bell shouted into the device in the attic, Watson on the floor below heard a faint sound from a reed receiver. When they changed places, Bell could hear nothing. And before any conclusive tests could be made, the heavy armature first tore loose from the membrane and then, when more firmly fixed, ruptured it.

That mishap and Hubbard's impatience with any diversion were probably what put off a further trial for nearly a month. "When I was at your work shop the other day and saw your new arrangement," Hubbard wrote in mid-June, possibly referring to the diaphragm device, "I was almost convinced that it was of more value than the autography but further reflection brings me back to the autograph as the 'Ne plus ultra.'" Bell and Watson experimented with various arrangements and combinations of coils, batteries, and other elements to improve the strength of the newly discovered induced undulatory currents. But this, at least ostensibly, was directed as much toward improving the autograph telegraph as toward anything else. Bell saw the new undulatory currents as less subject to confusion or interference than intermittent currents, since they differed from each other in timbre or amplitude as well as in frequency.

By the end of the month, Bell could wait no longer. "At last," he wrote his parents, "a means has been found which will render possible the transmission . . . of the human voice." By the next afternoon, he told them, he would have ready "an instrument modeled after the human ear, by means of which I hope tomorrow (but I must confess with fear and partial distrust) to transmit a vocal sound. . . . I am like a man in a fog who is sure of his latitude and longitude. I know that I am close to the land for which I am bound and when the fog lifts I shall see it right before me."

On July 1 Bell and Watson tried the original transmitter with a heavier membrane and a lighter armature, connecting it as before with a reed receiver on another floor. When Bell sang into the transmitter, Watson heard the tune on the receiver. "Grand telegraphic discovery today," Bell wrote Sarah Fuller in a postscript to a note about Georgie Sanders. "Transmitted *vocal sounds* for the first time. . . . With some further modification I hope we may be enabled to distinguish . . . the 'timbre' of the sound. Should this be so, conversation *viva voce* by telegraph will be a *fait accompli*."

But so far Bell's device had done no more than Reis's. Hubbard was not impressed. He did not understand the new principle behind it. "I am very much afraid," he wrote Bell on July 2, "that Mr. Gray has anticipated you in your membrane attachment." A more impressive demonstration came soon after, when Watson at last brought in and connected a diaphragm receiver much like the transmitter.

Up in the attic that day, Bell sang and declaimed into the transmitter with all of his family's traditional sonority. There was a clatter on the stairs, and young Watson burst in. "I could hear your voice plainly," he reported excitedly, "I could almost make out what you said!" Bell rushed down and listened to the new diaphragm receiver as well as he could in the commotion of the shop, while Watson in the attic shouted into the transmitter. Bell heard nothing he could clearly distinguish as Watson's voice. The fault may have been in Bell's hearing or Watson's voice or both. Bell heard enough, however, to convince himself that the principle had been successfully demonstrated, and that only some tinkering and refinement were needed to make the invention fully usable. And indeed, identical instruments tried in a quieter place some years later succeeded in transmitting speech.

As it happened, Bell was to make no more improvements nor even experiments along those lines until after the patent had been granted eight months later. When the fog lifted in July 1875, he had indeed seen land dead ahead, but had not gone ashore.

Several circumstances explain the strange delay.

Gardiner Hubbard immediately brushed aside Bell's latest toy and hectored him back to work on the autograph telegraph. As July wore on, ill health and other distractions began to cut down the frequency of Bell's trips from Salem to his Boston workshop. By the end of July, the autograph telegraph had been largely perfected, at least on paper, but just then Watson fell ill. By the time Watson recovered, Bell was struggling through a personal crisis with the Hubbard family, followed by a convalescence of the emotions in Canada during most of September. Then came school days, his obligations to Boston University, the summons of his duty to advance Visible Speech, and a new motive for wanting to earn money. What spare time these calls left to him during that fall and winter went to working out and securing his patent.

15

Love and Locksmiths

The summer that determined Alexander Graham Bell's place in history also set the course of his private life thereafter. Each of those developments bore on the other. And so, to complete our understanding of the historic event, we must turn back to the personal crisis.

After Bell returned from Washington and New York in March 1875 he came more often than ever to the Hubbard home in Cambridge. "Our lessons continued irregularly," Mabel Hubbard recalled three or four years later, "but they were no longer confined to articulation, indeed they never had been. Alec used to give me information of the most miscellaneous kind and we often got into political discussions."

Nevertheless, until June, the month in which Watson's plucked reed signaled the birth of the telephone, Bell continued (convincingly, he believed) to play the role simply of teacher in his relationship with Mabel Hubbard. Then he learned that she was about to leave Cambridge for Nantucket, and that he might not see her again for months. So in the midst of his great technological triumph, he wrote Mrs. Hubbard to confess that he was "in deep trouble."

"I have discovered," he wrote, "that my interest in my dear pupil . . . has ripened into a far deeper feeling. . . . I have learned to love her." But he was not sure how Mabel's parents and, more important, how Mabel herself might feel about that. "I promise beforehand," he wrote too confidently, "to abide by your decision."

The Hubbards promptly decided that Mabel at seventeen was too young even to be told of Bell's feelings, that she should have more time to grow and look about and know herself before having to make so fateful a commitment. Until recently, when he told them he was twenty-eight, they had guessed him to be at least eight years older than that. Mrs. Hubbard asked Bell to hide his feelings for a year. Her husband next day regretted that the sentence of silence had not been two years, though he personally liked the young man very much.

At the Hubbards' that fine June evening Mabel's sister and a young cousin drew Mabel and her secret admirer into the moonlit garden and then mischievously ran off, his secret evidently being ill concealed. Mabel's hand rested on his arm. It seemed very hard for him to hide what he felt. The girls came back with flowers, and they all pulled petals for their fortunes. Aleck's came out "Love," and Mabel asked for his thoughts. He could not speak, but the girls laughed and said they knew and ran off to the house. Mabel ran after them. Flustered and breathless, Aleck sat at her feet on the veranda with the family.

"What," their poet neighbor Professor James Russell Lowell had asked the world, "is so rare as a day in June?" Aleck could have answered: a moonlit evening. Wherever summer was new and the moon risen, on Brattle Street and far beyond, other families sat together, and other lovers could show their love. But not Aleck Bell. On the Hubbard veranda Aleck bantered with the rest to hide his despair. How little there was about him, he thought, to attract such a girl. If he could only know what sort of man could win her, he could use his year of silence to make himself such a man. Before he could stop, he heard himself asking, "If you could choose a husband what should you wish him to be like?" Mabel passed off the question playfully, and the others laughed, and for a time he felt relieved.

That week he wrote his parents of what had come to him. Except for her deafness, he assured them, they would be proud to see him marry Mabel. She was beautiful, accomplished, well born, and more affectionate by nature than anyone he had ever known before. They were to tell no one, not even Carrie (who had married a neighbor) or Uncle David (who had settled in Brantford), with one exception: "I should like *Marie Eccleston* to know." Then, in a postscript, he told them of his "new and startling discovery" that "musical signals could be transmitted WITHOUT ANY BATTERY AT ALL!!" and of his hope next day "to transmit a vocal sound."

On the next day, July 1, Watson heard Bell's voice by wire. And Mabel left for Nantucket with her sister and her cousin Mary Blatchford.

That July, frustrations and deferred hopes stretched Bell's nerves taut. Lying awake at night, he played over and over in his mind the garden and veranda scenes of mischievous banter and blurted longings on that June evening at the Hubbards', and each time he felt deeper shame at losing his boasted self-control, at acting like a boy and not a man. Late in July he told the Hubbards that he could not conceal his feelings from Mabel except by avoiding her, that he *would* tell her, that he would go to Nantucket at once unless they forbade it. They persuaded him to await her return.

In Nantucket meanwhile Cousin Mary Blatchford, who disapproved of Aleck Bell, took it upon herself to tell Mabel what was what, so that she might make up her mind before going home. Mabel wrote her mother in a whirl of emotion:

I think I am old enough now to have a right to know if he spoke about it to you or Papa. I know I am not much of a woman yet, but I feel very very much what this is to have as it were, my whole future life in my hands. . . . Oh Mamma, it comes to me more and more that I am a woman such as I did not know before I was. I felt and feel so much of a child still. . . . Of course it cannot be, however clever and smart Mr. Bell may be; and however much honored I should be by being his wife I never never could love him or even like him thoroughly. . . . O it is such a grand thing to be a woman, a thinking, feeling, and acting woman. . . . But is it strange I don't feel at all as if I had won a man's love. Even if Mr. Bell does ask me, I shall not feel as if he did it through love. . . . You need not write about my accepting or declining this offer if it should be made. . . . I would do anything rather than that. . . . I feel so misty and befogged. . . . Help me please.

Mrs. Hubbard asked Aleck in and read him part of Mabel's letter. Next day Aleck appeared at the Hubbard home on his way to Nantucket, complaining that he was ill and that further delay and anxiety would unfit him for anything. The Hubbards' entreaties turned him back to Salem, but not for long. "The letter which was read to me yesterday," he wrote them later that day, "was not the production of a girl — but of a true noble-hearted woman — and she should be treated as such. I shall show my respect for her by going to Nantucket whether she will see me there or not. . . . I shall not ask permission from you now — but shall merely go."

On the way he called once more on the Hubbards. Earnest, agitated, impassioned, he overwhelmed their opposition. He would respect Mabel's wishes, he informed them. If she did not want to see him when he arrived, he would accept that. But to go seemed best to him, best for Mabel's peace of mind and for his own health; and he would do right as he saw it regardless of others' opinions. At the end of the interview, noted Aleck in his journal later, "both Mr. and Mrs. Hubbard said that they liked the way I acted about it."

At the Ocean House in Nantucket he passed another sleepless night. Next day a violent rainstorm deluged the island. Aleck spent all that evening on a long letter to Mabel. "I have loved you with a passionate attachment that you cannot understand, and that is to myself new and incomprehensible. I wished [in Cambridge] to tell you of my wish to make you my wife — if you would let me try to win your love." He told her of his promise to keep silent. "It did not occur to me to measure the breadth and depth of my affection so as to consider whether it was *possible* for me wholly to conceal it. My pride told me I could do it." When her letter had come from Nantucket expressing distrust of him (as her mother had reported without reading the passage to him), "I was so distressed as to be ill."

He reviewed what had passed between him and her parents. "It is for *you* to say whether you will see me or not," he told her. "You do not know — you cannot guess — how much I love you. . . . I want you to know me better before you dislike me. . . . Tell me frankly all that there is in me that you dislike and that I can alter. . . . I wish to amend my life for you."

Next day Mabel shrank from the interview after all, or so Mary Blatchford told him. To Miss Blatchford's visible surprise, Aleck readily departed. The letter, he rightly sensed, would serve the purpose, perhaps better than a face-to-face encounter.

From Salem he wrote a more temperate note to Mabel: "Now that I have done my best to show you what I am — I am contented — and can wait. . . . I shall not trouble you any more until the original year is out. And then if you still think of me as you do just now, I shall try to be happy in my *work*. If I may not be any nearer or dearer — believe me at all events, Your sincere *friend*, A. Graham Bell." And she answered with touching shyness and reserve: "Thank you very much for the honorable and generous way in which you treated me. Indeed you have both my respect and esteem. I shall be glad to see you in Cambridge and become better acquainted with you. . . . Gratefully your friend, Mabel G. Hubbard."

When Mrs. Hubbard got back from a visit to Mabel, the Hubbards had Aleck over for "a delightful and encouraging evening," as his journal reported. Mabel had feared, he was told, that her letter had not been written "warmly enough." Warned that the elder Bells might look coldly on the match, she had said she would make them like her.

Along with his other worries, Aleck had brooded over an allusion to Mabel in his mother's reply to his first full revelation of feeling. "You are of course the best judge," she had written, "but if she is a congenital deaf-mute, I should have great fears for your children." Infuriated especially by the "deaf-mute" remark, Aleck had written not a word to his parents since.

In mid-August he wrote them at length and resentfully. He had long been grieved that there was "so little confidence and sympathy" between him and his father and had hoped that the news would bring them nearer together. But only his mother had replied, he complained, and that perfunctorily. "Should this also be received with the same shameful neglect as my last," he concluded, "I feel that there is danger of a complete alienation of my affections from home." (He signed himself "Your loving son Aleck.")

His bewildered and stricken mother explained that she had written for his father as well, that the curtness of her comment had been in expectation of further details, that being deaf herself and yet happily married she could scarcely condemn Aleck's marrying a deaf girl. "We are the victims of your own excited imagination. . . . There are not so many of us left that we can afford to take up unreasonable offense against each other. . . . I can only excuse you by thinking that your mind must be unhinged by close and prolonged application to Telegraphic work."

Meanwhile Aleck had pushed his advantage with the Hubbards. "I feel myself still *hampered*," he informed them, "by promises that I should not have made. . . . I must be *free* to do whatever I think right and best — quite irrespective of your wishes — or those of other people." He still intended to put off the subject of marriage for a year or more, but only if his own judgment urged that course. "When I am sure of her affection I wish to be engaged to her. When I am in a position to offer a *home* I wish to marry her — whether it is in two years or two months! . . . If you do not like my conduct in the matter . . . you can deny me the house — and I can *wait*."

From Mrs. Hubbard came the glorious reply: "I give you back your promise, entirely, unreservedly. I believe your love to my Mabel to be unselfish and noble. I trust you perfectly. If you can win her love I shall feel happy in my darling's happiness." And she invited him to call.

The next day, August 26, 1875, he described at the time as "the happiest day of my life." That evening in the greenhouse he and Mabel at last talked freely and alone. She did not love him, she said, but she did not dislike him. It was enough. In the journal he had opened to chronicle his courtship, he wrote triumphantly: "Shall not record any more here. I feel that I have at last got to the end of all my troubles — and whatever happens I may now safely write: FINIS!"

Now he recognized how great a strain had been put on his mind and nerves by the events of the summer. (At one point during the crisis of the Nantucket visit, Mrs. Sanders had concluded that telegraphy had affected Aleck's brain; and she had been about to have him placed under restraint in Boston, when he explained everything to her.) Physically also he felt the need of rest and recruitment.

He wrote contritely to his father: "Please forgive me for my harsh allusions to you and do not think I do not love you. You do not know how much trouble I have passed through — and how ill I have been. . . . Do not think badly of me. I know you love me very dearly and I have longed to feel that I could talk to you as freely as I wish without fear of ridicule." And he packed his bags for Brantford.

Just before he left, a letter came from his father, crossing his own. "Come home and rest," Melville urged his son. "We . . . shall conclude that 'blood ill-tempered vexeth you.' . . . You will find no change in our affection and we still hope that we have left to us a 'good son.'" So, at last, early in September 1875, Aleck Bell set out for the cool tranquillity of Tutelo Heights.

16

Patent No. 174,465

Aleck's spirits rebounded the moment he climbed into his father's carriage at the Brantford depot on September 4, 1875. His father seemed determined to close the gap between them. "If he has not killed his fatted calf for me," Aleck wrote Mabel a few days later, "he has done everything else to make me happy and to show his affection for me."

At home all seemed flourishing: Carrie living not far away with her new husband; Uncle David with his wife and four youngest children planning to settle down nearby; Aleck's mother, spry and cheerful; Aleck's father, blessed variously with a successful year of teaching at a women's college in Kingston and with a phenomenal yield of grapes on the porch trellis. Aleck himself spent a few days on horseback in the countryside and came back as ebullient as he had ever been. He talked enthusiastically next day about his telephone and "electro-hydraulic" ideas. And to Mabel he wrote happily: "My father and I have come together as we have not done for years. . . . It is to me a new and delightful sensation."

He had not brought equipment for serious experimentation, but now and then he put some ideas on paper, including a method for eliminating the troublesome sparking of intermittent-current contacts without having to use a condenser. He had a couple of undulatory-current instruments with him, which he exhibited, but did not demonstrate, at the local telegraph office. If a person sang or spoke into one instrument, reported the *Brantford Expositor* on faith, "not only the words would be heard" at the other, but also "the tones of the voice." The *Toronto Globe* copied that remarkable story. Without having quite heard his latest invention talk, Bell had no doubt that it could; and he had a knack for communicating his convictions to others.

He drew on that knack hoping to get free of an increasingly oppressive dilemma: the conflicting claims about to be made upon his time by Boston

University, Georgie Sanders, Gardiner Hubbard, Visible Speech, and the telephone quest. If he could somehow get money without sacrificing time, he might at least return the advance given him by Dean Monroe of Boston University and thus eliminate that teaching obligation. His agreement with Hubbard and Sanders covered only United States rights to his inventions, leaving foreign rights as a salable asset. So, giving up on the unresponsive Herdmans, Bell looked about for more accessible customers.

Bell's first thought was of the Canadian financier Sir Hugh Allan and his Montreal Telegraph Company. For an introduction, he called on the Bells' friend and neighbor the Honorable George Brown, proprietor of the *Toronto Globe* and a politician who had the questionable distinction of having been Prime Minister of Upper Canada for a single day. Brown and Allan being at political odds, Brown did not incline toward putting Allan onto a good thing. Instead he and his brother Gordon offered to go shares in foreign rights themselves, provided that Bell would in confidence furnish convincing particulars, and provided also that Bell would promise to take no steps in the matter of a United States patent that might prejudice patent rights abroad. Bell promised.

When Bell returned to Salem early in October 1875, he sent George Brown a description of his intermittent-current multiple telegraph, on which United States patents had already been granted or applied for, and made a cautious reference to the undulatory-current concept, which he now regarded as his most original and important. No hint of the latter, he assured Brown, had appeared in press notices of his work. In fact, he wrote, "I have only spoken of the matter to two or three scientific friends who have been assisting me with their advice."

In mid-August, before leaving for Canada, Bell had urged on Hubbard the transcendent importance of the undulatory-current discovery and the wisdom of filing at once for an American caveat or patent. Hubbard cautioned against filing a caveat, which amounted to an admission that the idea had not yet been perfected and which might tip off rivals as to what was up. Perfect the invention first, he advised, and meanwhile establish priority of conception by writing him dated notes and by consulting scientific men. After all, Hubbard pointed out, it was priority of conception and not date of filing that would decide a patent controversy.

Bell brought back the beginnings of an American patent specification early in October and worked on it at intervals through the month. "My present invention," it said, "consists in the employment of a vibratory or undulatory current of electricity in place of a merely intermittent one, and of a method and apparatus for producing electrical undulations upon the line wire." In specifying the method, Bell dwelt on electromagnetic induction by a vibrating armature. The surprise, the wonder, the promise of the June 2 breakthrough had come near crowding the variable-resistance idea from his mind. But not quite. Though in a perfunctory and somewhat obscure way, the vibrating piano-wire notion of May appeared unmistakably in the penciled foolscap draft as an alternative "method of producing an undulatory current by the vibration of a body through which a continuous current of elect. is passed." One use of the undulatory current, however produced, was set forth as that of "transmitting vocal utterance telegraphically."

Now, however, it was Bell who hung back from filing. To file in America might jeopardize an English application, and he had just promised George Brown to avoid that.

Bell waited in vain for the written agreement and financial support George Brown and his brother had promised to send on receipt of Bell's confidential letter. The time arrived to honor his obligation to Boston University, but no money to bail him out of it. So his university classes resumed, while in the Williams shop that fall and winter no Bell work came Tom Watson's way.

Bell resolved to take his dilemma by the horns. As soon as he finished drafting his patent specifications, he would suspend all inventive work and throw himself into teaching. His past proselytizing for Visible Speech among institutions for the deaf had created a demand from that quarter for trained teachers. So he would organize a teacher-training course, would meanwhile canvass for private pupils, and after a while would shift much of the burden of the latter onto his student teachers. This would provide practice for them and income for himself, while freeing some of his time for a return to invention. It was a bold and ingenious scheme — and a feasible one, if Gardiner Hubbard could be pacified meanwhile.

Bell went at it with energy and even a sort of exhilaration. "I feel as if I am a new man," he wrote his father in mid-October, "younger and more hopeful than I have ever been — and more on the look-out for No. 1!" Over the next three or four weeks he wrote hundreds of short notes announcing his coming normal course, placed ads, sent out printed cards, delivered half a dozen lectures on Visible Speech at normal schools and other institutions, and made arrangements with a deaf printer for the future publishing of Visible Speech texts and exercise books. Withal he spent two or three hours a day teaching Georgie Sanders, who was at the moment his only source of income. "Telegraphy dormant! George Brown unheard from!" he wrote his parents a few days before the mid-November opening of the normal class, and then added triumphantly: "Professional visitors are beginning to *pour* in. . . . Every prospect of a *large* class." And so it turned out.

But Gardiner Hubbard would not hold still. "I have been sorry to see how little interest you seem to take in telegraph matters," he wrote Bell at the end of October. To Hubbard, Bell's frenzy of lecturing seemed to "confirm the tendency of your mind to undertake every new thing that interests you & accomplish nothing of any value to any one." "Your whole course since you returned," he concluded, "has been a very great disappointment to me, & a sore trial."

Somehow Bell managed for a time not only to mollify Gardiner Hubbard but also to court Mabel. "You must not think that my visits to Cambridge are hindering me in my work," he wrote her; "they are having just the *opposite* effect." Nevertheless, harried and hard-pressed, he wrote her an emotional plea not to encourage him out of pity, but to let him know at once if she could not sincerely return his love. Then he had to soothe away her hurt at being suspected of thoughtless trifling.

Gardiner Hubbard, whose patience had narrow bounds, brought matters to a climax and resolution just before Thanksgiving. He demanded that Bell choose between teaching and Visible Speech on the one hand and telegraphy and Mabel on the other, offering to furnish Bell's living expenses in the latter case.

Bell exploded. Precisely because he loved Hubbard's daughter and prized Hubbard's respect, he would not accept any special favors. The existing agreement was perfectly fair. As for Visible Speech, it had been the life work of Aleck's father; to further it would be Aleck's life work, whatever Hubbard or anyone else thought of it. He had become a teacher because he was needed, and he was still needed. He had the power to free deaf and stammering children from what amounted to life imprisonment for no crime; he would not withhold it, at least not until he had qualified others to take his place. Fortunately for his hope of marriage, which to his mind required the means of proper support, his work was at last bringing him a respectable income — currently at the rate of four or five thousand dollars a year, he estimated. He was sorry if Hubbard did not like his profession, but he did not propose to change it. And if Mabel loved him, she would marry him anyway.

As in the crisis of the past summer, Bell had grasped the nettle and remained unstung. He wrote Hubbard in a more temperate vein, not abating any of his views, but apologizing for any seeming disrespect and gratefully acknowledging Hubbard's "sincere interest in my welfare independently of any pecuniary interest . . . in my inventions." "Please bear with me for a little longer," he concluded.

Perhaps as much as Aleck's firm stand and conciliatory letter, Mabel's reaction helped bring her father around. She resented her father's presuming to make Aleck choose between her hand and his profession. And her mother, seeing her distress, told her not to go on that way, but either to give Aleck up or consent to be engaged to him at once. So when Aleck came to call on her eighteenth birthday, November 25, 1875, which was also Thanksgiving Day, Mabel told him that she loved him better than anyone but her mother, and if that much satisfied him, she would be engaged to him on the spot.

Mabel took Aleck by surprise at what had seemed to him a time of utter despair. Conscience compelled him to remind her of her youth and how little she had seen of other men. But she told him that she knew she would never find anyone else to love as well. And so the commitment was made. For the rest of their lives, Thanksgiving Day would have a special meaning for Aleck and Mabel.

"I am afraid to go to sleep lest I should find it all a dream," Aleck wrote Mabel that night, "so I shall lie awake and think of you." And he sent the good news to his parents. "My heart is too full to allow me to write much to you tonight," he told them, "so I scribble off only these few lines that you may know of my happiness." But he managed to send a few scraps of other news also. "I am quite well and getting stout! Chest all right — cough gone — business satisfactory — telegraph promising — George Brown not heard from." Cheerfully yielding to his future wife's preference in spelling, he signed himself "your happy son, Alec."

Alec's father wrote Mabel to express his "great gratification" at the news, along with his congratulations, best wishes, and a fatherly assessment of her prospects. "Alec," he wrote (acceding to the new orthography), "is a good fellow and, I have no doubt, will make an excellent husband. He is hot-headed but warm-hearted — sentimental, dreamy, and self-absorbed, but sincere and unselfish. He is ambitious, to a fault, and is apt to let enthusiasm run away with judgement. . . . With love you will have no difficulty in harmonizing. . . . I have told you all the faults I know in him, and this catalogue is wonderfully short." The catalogue was also to prove remarkably accurate, given the imminent gratifying of Alec's ambition by world-wide fame.

Bell made some concessions to his prospective father-in-law. In early December he let Hubbard borrow the undulatory-current specifications to show their patent lawyer Anthony Pollok in Washington. He also promised Hubbard that he would work on telegraphy four days a week from nine to two. But he would not let Hubbard file a patent application until after one more appeal to George Brown over the Christmas holidays. And Bell's promise to work regular hours at invention soon went down before heavy professional demands on his time. He had not at first acquired enough private pupils to satisfy the practice-teaching needs of his normal class. So he had started free evening classes at Boston University for adult deaf-mutes, recruited by addresses before local deaf-mute societies. Then private pupils had begun to show up after all.

For years, moreover, Bell had done his best work late at night, when all was still and he could focus his thoughts. But Mabel had been pressing him to conform with the diurnal majority. "I have been doing violence to my own instincts in the hope of working a reformation," he pleaded that winter, "but the result has been that I have been unable to do any serious thinking at all . . . and in the meantime telegraphy and everything else is at a stand-still."

So December passed with no progress toward a patent. Alec reached Brantford on Christmas Day, just in time for "a regular merry old-fashioned Christmas dinner . . . with Turkey and Goose — and Plum-pudding all in a blaze — and Holly — regular English holly." He showed photographs of the Hubbard family, whereupon his cousin Charles, one of David Bell's sons, stared for five minutes at a photo of Mabel's sister Roberta and then "remarked — in the most matter-of-fact way in the world — that he had often wished to see Boston — and that he thought he would ask a holiday from his Bank and run down to pay me a visit soon!" (Charlie eventually married Roberta.)

Two days later, having returned to his Toronto home, Charlie Bell wired Alec that the Honorable George Brown was still in town and wanted to see Alec immediately. After a long conference in Toronto, Alec formally agreed to give the Browns a half interest in all his foreign patents, in return for which they were to pay the costs of taking out and defending those patents and were to pay Alec fifty dollars a month until the patents were obtained, up to a maximum of six months. The money was not for Alec's support — he felt prosperous enough now to do without such help — but for the rent of rooms away from the Williams shop, where strangers (spies for Elisha Gray, Alec suspected) were rumored to have been taking notes on his apparatus.

In small matters and large, the new year 1876 — the Centennial Year of the American Republic — showed signs of becoming a turning point in Alec Bell's life.

For one thing, he grew perceptibly stouter. His clothes got too tight, and he changed his collar size, not for the first time. "If things go on at this rate," he complained to his parents, "I shall have to have my measure taken again soon." For another, he turned over the education of Georgie Sanders to one of his student teachers and moved from Salem in mid-January to a couple of rooms at Exeter Place in Boston, where rent was only sixteen dollars a month and board five dollars a week in the restaurant downstairs. He made one room his laboratory, taking up the carpet and bringing in a battered old kitchen table. "I want a table upon which I can hammer and saw and carve to my heart's content," he wrote Mabel, "and a floor upon which I may spill acids without fear of damages."

Meanwhile, in early January, after his return from Canada and before his move to Exeter Place, he tried in what spare time his classes left him to make final revisions in his new American patent specifications, and to prepare copies of those and his previous American patent and pending applications for George Brown to file in England. That fortnight was a crucial time in Bell's career. It was a hectic and muddled time, too, of which certain obscurities will probably never be completely dispelled.

On January 6 Bell wrote his father from Salem that he was "hard at work" on the specifications for Brown. On January 7 he wrote Mabel that he had just sat down with the intention of working on those specifications. But his pen had somehow commenced a letter to her instead. "My afternoon at the University has been a very hard one," he wrote, "and I am so tired out and nervous that I feel it would be madness in me to goad my mind to any serious work to-night." It was in this letter that he pleaded to be allowed a return to his old habit of nocturnal thought and study: "I have made various commencements — at different times during the past three months — upon the [undulatory-current] specification that I feel to be so important — but it is no further advanced than it was in October!"

The feeling haunted him, as he recalled afterwards, that something vital was missing from that all-important specification, that he had left a "hole" in it. Later he testified that the missing element had been the concept of using a variable resistance to create the undulatory current. Yet, as we have seen, that concept had been included, though somewhat obscurely, in an earlier draft still in existence that January. Had the variable-resistance clause been dropped from later drafts? If so, why? The dubious results of the Fenollosa piano experiment? Sheer forgetfulness, arising perhaps from his absorption in the June triumph of *induced* undulatory currents? If the latter, then why would not a comparison of drafts have revealed the oversight? Had the earlier draft been mislaid or perhaps left in Canada? Or was the "hole" in fact not an overlooking of the entire variable-resistance principle, but rather the lack of some more convincing application than the vibrating piano wire? Whatever the reason, the fact that the George Brown specification did not mention even the principle of variable resistance suggests that the whole idea had also been left out of the American specification draft as it stood early in January 1876.

On January 12, Mabel did not feel well. Her father was in Washington, her mother and sister had an evening engagement. So Alec stayed at the big Hubbard house in Cambridge "to see that burglars did not enter." While Mabel rested upstairs and Alec waited for Mrs. Hubbard and Roberta to get home, he busied himself in Hubbard's book-lined library copying the undulatory-current specification to send Hubbard the next day. At about midnight, Mabel leaned over the mahogany balustrade and called down the stairwell for Alec to keep his promise and stop his night work. Many years later she recalled the scene vividly. "He came running up all eagerness and begged that I leave him alone just that one night, for he was at last on the track of that 'hole' in his telephone specification." She gave in. (Later she told him she was painting his portrait to hang in his Exeter Place room; on delivery it turned out to be a painting of a great white owl.)

Either Bell had remembered the variable-resistance clause that night or had been moved to reinsert it by the inspiration of a new technique. The inspiration came from the spark arrester he had conceived at Brantford in September and for which he was now about to write a patent specification. To prevent sparking because of self-induction during the breaks in an intermittent-current transmitter, he proposed to bypass the contacts with a high-resistance shunt, consisting of a cup of slightly acidulated water into which two leads were to be dipped. One lead would remain fixed. The other's depth of immersion could be adjusted by hand. Dunking it deeper would increase its area of contact with the liquid and thus decrease the resistance of the shunt; raising it would increase the resistance. So the shunt resistance could be precisely varied until a small amount of current — not enough to affect the receiver — would flow through it instead of sparking when the main-line contact was broken.

Apparently it was on January 12 that this resistance-varying device fused in Bell's mind with the membrane-vibrated armature of his magneto transmitter. That evening in Hubbard's library, Bell accordingly inserted a vital clause:

Electrical undulations may also be caused by alternately increasing and diminishing the resistance of the circuit. . . . For instance let mercury [Bell wanted to avoid electrolytic action] or some other liquid form part of a voltaic circuit. Then the more deeply the conducting wire is immersed in the liquid the less resistance does the liquid offer to the passage of the current. Hence the vibration of the conducting wire in a liquid included in the circuit occasions undulations in the current.

Bell sent the specification, hole now plugged, to Hubbard next morning. On January 15 Hubbard wrote from Washington that their attorney Anthony Pollok had gone over it, was "very much pleased with it and says he does not think it will require any alteration." Hubbard warned that "there is now much excitement in this branch of Telegraphy and you in conversation may unwittingly communicate information, which will set others on the look out. Mr. Pollok gave me sufficient reasons to think we should lose no time in the application." One good reason undoubtedly was that Elisha Gray had just arrived in Washington to hurry along his telegraphic patents.

Meanwhile George Brown had wired Bell to meet him with the specifications in New York on January 25. The intervening days were busy. Along with his teaching duties, Bell began his spark-arrester specification on January 16, moved to Exeter Place on January 17, finished the spark-arrester specification and began copying his old harmonic and autograph telegraph specifications for Brown on January 19, swore to and returned on January 20 a copy of the telephone specification that Hubbard had lately sent back unaltered from Washington, and left for New York on January 21, arriving next day.

In New York on January 25, George Brown met Bell and Hubbard, seemed "much pleased" with the specifications, and said he would go over them in England with the eminent scientist Sir William Thomson (later to become Lord Kelvin). Somehow Bell and Hubbard failed to notice that Brown's copy of the undulatory-current specification did not include the variable-resistance principle. Presumably Brown's copy of that specification had been drawn up before the night of January 12 and by an oversight, in the flurry of Bell's activities, had not been revised.

A postscript was added to the instructions accompanying the specifications: "N.B. — It is understood that Mr. Bell will not perfect his applications in the American patent office until he hears from Mr. Brown, that he may do so without interfering with European patents."

George Brown sailed for England the next day, having sealed the bargain with a first payment of fifty dollars. It turned out to be a last payment, also. At the end of February, almost a month after the Honorable George arrived in England, his brother got around to notifying Bell that the whole deal was off. Bell's "invention" could not be made to pay in England, the Browns believed. When George Brown came home, he surprised his brother by reporting that while he had consulted an electrician about the harmonic multiple telegraph, he had never considered the telephone idea worth mentioning.

Meanwhile, back in Boston, Bell had his spark-arrester specification notarized and sent off by the beginning of February. His other specifications were in the Patent Office at Washington, but not to be formally filed until Brown was heard from. Hubbard had been pleased with them and now reported that "the people in the Patent Office" credited Bell with knowing more about electricity than all the other inventors put together. Bell proudly sent his parents his only copy so that they could "see the specification that has cost me so many sleepless nights." "I really think I have struck the solution of a great problem," he added.

By the second week in February he should have got a cablegram from Brown that the English patents had been applied for. It failed to come. But Bell seemed too busy to fret. "I rush from one thing to another and before I know it the day has gone!" he wrote his parents on February 12. "University — Visible Speech — Telegraphy — Mabel — visiting — etc., etc. — usurp every moment of my time — and I cannot manage all that I lay myself out to do. . . . There is a great deal of hard work before me — and now that I am well again I am ready to fall to with a vim!" He would have to stay up late that night, he complained, to answer a letter from the Patent Office inquiring about certain points in the application he had filed almost a year before for a patent on the harmonic multiple telegraph.

Apparently he never got around to that letter, or else it did no good, for in the Patent Office on February 16, Examiner Zenas F. Wilber handed Bell's attorneys Pollok and Bailey an official notice declaring Bell's harmonic multiple telegraph application of February 25, 1875, to be in interference with Elisha Gray's application of February 23, 1875, and Paul La Cour's of March 3, 1875. The question of priority in conception would now have to be determined by formal hearings, a prolonged and costly undertaking.

Probably this hitch occasioned the emotional exchange of letters between Mabel and Alec on February 16 and 17. The letters themselves do not quite specify the issue. The best the historian can do is to quote salient passages verbatim.

Mabel, in womanly fashion, did not date her letter. "All this which you told me tonight has worried me so much," she wrote Alec. "This slowness and procrastination which is I think your great fault, is mine also, and I don't know what we are going to do if this goes on." She proceeded to admonish him at some length on that score and to urge reform for both of them.

And then she wrote:

I love you so much dear Alec I cannot bear that anyone should write to you and with too much justice, as my father has done. [Presumably he had telegraphed the bad news that day.] And I cannot bear that procrastination should rob you of the fruits of hours of hard study and of the great abilities God has given you. I know that ill health has the last week or two prevented you from working as hard as usual. But it seems to me when the thing to be done was so very important, when if it failed you lost the reward of past toil and suffering, you might *perhaps* have put even that aside for the time.

"Procrastination is indeed my besetting sin," Alec replied on February 17.

Help me, Mabel, to conquer it. I will not disguise from you dear what a blow this misfortune has been to me — for I had looked to this new patent to avoid all conflict with Gray — & to place the control of the new system of Telegraphy entirely in my own hands. I feel it more deeply than anything that has ever before happened to me — except one thing! and you know what that is. . . . Just received a kind note from your father. If any telegram comes for me today please send it to the University tomorrow morning.

All this seems like an extravagant reaction to the news that what for a year had been a preliminary suspension of a pending patent had now become a formal interference — especially since Bell had already been granted the vibratory circuit breaker and autograph telegraph patent that he considered a key element in the system. The self-reproach also seems extreme, since the points of information about which Examiner Wilber had inquired on February 7 were mere matters of definition and elaboration.

In retrospect, the intensity of feeling would seem more appropriate to certain remarkable events of February 14, which jeopardized the new telephone patent. Yet this alternate hypothesis presents still greater difficulties. The "thing to be done" which Alec had not done in that case was simply to authorize the filing of the patent application — certainly not a matter of "working as hard as usual," and not so much a case of "procrastination" as of scruple. Furthermore, official notice of what had happened in the Patent Office on February 14 did not reach Pollok and Bailey until February 19 (though they may possibly have had some earlier unofficial word).

The affair of February 14 involved the doings of Elisha Gray. Since mid-January, Gray had been in Washington preparing and filing "a number of cases relating to telephony [that is, tone transmission] that had been accumulating through a number of months" (as he testified later). Gardiner Hubbard surely knew about Gray's presence and activities in general. He may well have suspected further that Gray had clues to what Bell was now about to patent. After all, Joseph Adams a year earlier had threatened in scarcely veiled terms to sell Gray his inside knowledge of Bell's work; others, for all Hubbard knew, might actually have done so. Since then, Bell had told several Boston scientists precisely what he had in mind; and Gray had contacts in the scientific community. In September the widely read *Toronto Globe* had reported Bell's progress toward electrical transmission of speech. Someone who knew of Gray's related interests might well have mentioned the item or sent him a clipping.

Whatever the immediate reasons, Gardiner Hubbard gave up on George Brown toward the middle of February. Notwithstanding Bell's commitment to delay, Hubbard went ahead without consulting Bell and formally filed the undulatory-current and telephone application at the Patent Office on the morning of February 14. Later that day, Elisha Gray's attorney William D. Baldwin entered Gray's caveat for a speaking telephone on the liquid variable-resistance principle.

On many other occasions in the history of science and technology, two or more men have arrived independently at the same conception at nearly the same time. The harmonic multiple telegraph is an example. But the almost exact simultaneity of Bell's telephone patent application and Gray's caveat has given this episode a special fascination. "Such a coincidence has hardly happened before," Bell wrote his parents when he found out about it.

Was it entirely coincidence? If Gray had prevailed in the end, Bell and his partners, along with fanciers of the underdog, would have suspected chicanery. After all, Gray did not put his concept on paper nor even mention it to anyone until he had spent nearly a month in Washington making frequent visits to the Patent Office, and until Bell's notarized specifications had for several days been the admiration of at least some of "the people in the Patent Office." The evident candor and fairness of Gray's early reactions to his telephone disappointment rule out suspicions of wrongdoing on his part. Nevertheless, it is easier to believe that a conception already forming in Gray's mind was precipitated by rumors of what Bell was about to patent, than to believe that chance alone brought Gray to inspiration and action at that precise moment.

In later testimony, Gray gave a plausible account of how his conception had evolved. As early as the winter of 1874–1875, he had noticed the ability of a resonant-box magnet receiver — the same kind as Pickering's tin-box receiver of 1870 — to emit a variety of electrically generated sounds, such as the peculiar noise of a sparking intermittent-current transmitter or the groan-like tones of a make-and-break cylinder rotating at an irregular speed. If only some way could be found to pick up and transmit vocal sounds electrically, he speculated, such a diaphragm receiver could probably reproduce them also. But he pursued the thought no farther.

Not until late in the fall of 1875 did Gray, even by his own account, conceive the idea of a liquid variable-resistance transmitter. The stimulus came from his first sight of the "lovers' telegraph" or "thread telegraph" being vended as a toy in Milwaukee in November or December. This novelty (which Bell did not hear of until the following September) consisted of two tin cans, each with one end open, and a taut string connecting the centers of the closed ends. A sound uttered into one can would travel by mechanical conduction along the string and come out of the other can.

According to Gray, the gadget showed him that if the vibrations caused by vocal sounds at a single point on a diaphragm were reproduced at a single point on another diaphragm, the second diaphragm would reproduce the original sounds. He saw a way to reproduce those vibrations electrically, based on the well-known fact that two conductors, separated by a poorly conducting fluid, would transmit current of a strength varying inversely as their separation. In particular, he testified, he remembered a "water rheostat," made by his Western Electric Company several years earlier, "which would produce a rise and fall in the tension of an electric current by sliding a metal rod, by hand, in or out of the liquid." Gray, in other words, arrived at a liquid variable-resistance device in just the way Bell had. The difference between the two devices was that Gray's worked by varying the distance between the two leads, whereas Bell's worked by the varying area of liquid contact with one lead.

"I claim as my invention," Gray wrote in his caveat, "the art of transmitting vocal sounds or conversations telegraphically, through an electric-circuit." The accompanying sketch showed a man bending over to talk through a mouthpiece into a horizontal diaphragm. As the diaphragm vibrated up and down, a thin metal rod extending down from the center of the diaphragm into a vessel of water would (the caveat explained) approach and recede from another lead immersed just below but not touching the rod. This would vary the current in the circuit in proportion to the voice vibrations, thus reproducing the voice in a simple electromagnetic diaphragm receiver. Or so Gray reasoned, though he had not yet made any such instrument.

This was a brilliant insight, all the more remarkable if, as he claimed, Gray's only prompting had been the "lovers' telegraph." Yet Gray showed some confusion in going on to say: "I contemplate, however, the use of a series of diaphragms in a common vocalizing chamber, each diaphragm carrying an independent rod, and responding to a vibration of different rapidity and intensity, in which case contact points mounted on other diaphragms may be employed." The idea of a diaphragm that would respond

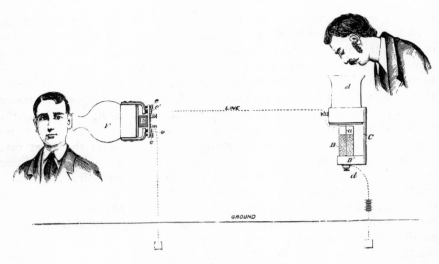

Diagram for Elisha Gray's caveat

to a vibration of particular frequency runs counter to both Bell's telephone and the "lovers' telegraph," which used a single diaphragm for the whole range of the voice. And to speak of "contact points" rather than armatures makes no sense at all.

In coming up with his telephone idea, Gray broke away more sharply from his current line of thought and research than had Bell. This fact adds probability to what was certainly possible as early as September 1875: that some general hints, if not details, of Bell's new goal had reached Gray and vibrated in his mind like the sympathetic response of a tuned reed. This hypothesis does not require Gray to have spied on Bell or stolen his ideas; Bell himself, despite patent contests with Gray, rejected any such suspicions after he came to know Gray better.

There is firm evidence, however, that some knowledge of the undulatory-current principle would have made it relatively easy for an ingenious and knowledgeable man independently to invent such a device as Gray's. A few months later, in an offhand way and knowing little if any more about the principle than Gray might have picked up in his pre-caveat frequenting of the Patent Office, Edward Pickering did just that. Pickering, it will be recalled, had anticipated Gray's tin-box receiver while teaching physics at MIT in 1870. Now director of the Harvard College Observatory, Pickering attended a lecture on the electromagnetic telephone given by Bell early in May 1877. While listening, Pickering let his thoughts run to the problem of making a telephonic relay. By then, of course, he knew the fundamental principle of the undulatory current and had probably heard Bell mention the variable-resistance transmitter principle without specifying the means of applying it.

Immediately after the lecture, Pickering wrote Bell a note freely offering the idea that had occurred to him:

The problem is to utilize a local current, so that in a given circuit it shall be proportional to the current induced by the magnets. I think this may be done by attaching to the plate of the receiving telephone a fine wire dipping in water and nearly touching a wire connected with a second telephone and battery. . . . Now as [the first wire] vibrates the interval will alter, and with it the total resistance, and consequently the current. . . . [Thus] a feeble exciting current may regulate a powerful local battery. . . . This instrument might also be used as a telephone, but would probably not give as good results as yours. As a relay it would alter the *timbre* since the law of variation is unlike that of the original current.

In any case, hypotheses aside and accepting without question Gray's own undocumented assertion of the date of his conception, Bell had clear priority at every point except the liquid-transmitter embodiment of the variable-resistance principle (which embodiment turned out to have no commercial value). Even on the latter point, moreover, Bell had notarized his conception before Gray spoke or wrote a word about his.

Gray may, as he claimed, have worked out the idea in his mind before coming to Washington, but his chief assistant knew nothing of it until the second week of February 1876, when Gray gave him a penciled sketch to guide a draftsman in Baldwin's law office. Not until February 14 did Baldwin, in Gray's presence, dictate the caveat specification to his clerk, after which it was transcribed into longhand, corrected, and recopied for filing. And since records show that the Patent Office fee for the caveat was paid separately in order to meet the two o'clock fee deadline for that filing date, the caveat itself could not have been filed until after that hour.

It was not, of course, priority of conception that was at stake at that point. In a fair hearing, Bell's claim to priority of conception, and hence to patent rights, could not have been (and in fact was not) successfully refuted, even on Gray's own showing. The time of filing nevertheless meant a great deal practically to Bell and his partners. However strong their case in law and justice, the suspension of the patent during long-drawn-out hearings would have made it difficult if not impossible for them to get backing for the fight or to establish a profitable enterprise meanwhile. Western Union, once persuaded of the telephone's potential, would thus have found it easy to force the Bell interests into selling out their contested claim cheaply (they came close to that as it was), after which the telephone might or might not have been allowed to compete with the telegraph, depending on Western Union's calculations of profit. Analogous situations suggest that the day of the telephone would have been long delayed.

Events did not take that course, however. On February 19 Examiner Wilber handed Pollok and Bailey official notice that the patent application had been suspended because of interference with "a pending caveat." On February 24, Pollok and Bailey wrote Acting Commissioner of Patents Ellis Spear that upon inquiring the date of the interfering caveat (as was Bailey's custom in all such interferences), they had found Bell's application to have been filed early on the same day; and so they requested that the office records be checked further to see which filing had been first. Wilber contended that the time of day did not matter. Nevertheless, when the records clearly showed Bell's application to have been filed before Gray's caveat, Commissioner Spear overruled Wilber, who therefore on February 25 notified Bell and Gray separately that the interference had been dissolved.

Meanwhile Wilber had discovered what seemed to him another interference, this time with a caveat filed by Gray on January 27 for the use of a continuous but varying current. Hubbard had already arranged for Bell to leave for Washington on February 29 in connection with the harmonic multiple telegraph interferences. In view of this new interference, Hubbard wired Bell to come as soon as he could. So Bell arrived in Washington on the morning of Saturday, February 26.

Bell settled the latest question easily. In accordance with regulations, Wilber could divulge no more about the Gray caveat of January 27 than the point at issue: the use of sudden changes in a current without actual interruption of it. Bell was able to point out a passage covering that type of current in his own application of February 1875, a passage that Wilber had to confess he had overlooked in that connection. Furthermore Bell explained the difference between an abruptly changing current, which he called "pulsatory," and the "undulatory" or gradually changing current upon which his new application rested. As the law provided, Wilber thereupon allowed Bell to amend his application for the purpose of clarifying his terms. (The amendments, of course, had no bearing on the variable-resistance interference, which had already been settled in Bell's favor.)

That cleared away the last remaining obstacle in the way of the new patent. "If I succeed in securing that Patent without interference from the others," Bell wrote his father on February 29, "*the whole thing is mine — and I am sure of fame, fortune, and success if I can only persevere in perfecting my apparatus.*"

Though Wilber had not required a complete model of the autograph telegraph in 1875, Bell on this occasion was prepared with models for all but the telephone transmitters. However, Wilber as before did not ask Bell to submit models. This was probably because of the patent's emphasis on the undulatory current as its central figure, and its almost perfunctory description of the telephone as one of several special applications. (Our modern awareness of the tail's mighty dog-wagging makes the patent seem misshapen, unless we regard it with the eyes of Bell and Wilber at the time.) Bell could scarcely deposit a specimen of the undulatory current his patent went on about at such length; and the instruments he described were all so simple as apparently to need, in Wilber's view, no physical examples.

Wilber even relaxed to the point of answering Bell's question as to what had been at issue in Gray's other caveat; though since that interference had already been dissolved, the revelation was unnecessary and therefore — technically, at least — improper. He did so simply by telling Bell what part of his patent had been threatened by the February 14 caveat: to wit, the liquid variable-resistance clause. Bell's curiosity was natural, and so was Wilber's obliging response. The caveat interference seemed to be a dead issue; and besides, so far as Wilber then realized, nothing of huge importance was at stake. Only later, under the microscope of litigation, would this casual indiscretion appear gross.

While Bell waited in Washington for the final granting of his patent, he had spasms of distress. Pollok and Bailey cast him down by advising him not to bother even filing his spark-arrester application; it had been fully anticipated by earlier patents. He found the multiple-telegraph interferences in a "curiously muddled condition." Not only were Gray and Paul La Cour of Copenhagen in conflict with him, but also "another whose name was withheld but I have discovered it to be Mr. Edison of New York, who has evidently been employed by the Western Union Telegraph Company to try to defeat Gray and myself."

After seeing early in 1875 what Bell and Gray were up to, President Orton that summer had indeed quietly hired the brilliant young professional inventor Thomas Edison to leapfrog the other two. Perhaps Orton even speculated about a "speaking telegraph," for he provided Edison with a German report on Philipp Reis's intermittent-current "telephone" of 1860. Since then Edison, despite his handicap of partial deafness, had been tinkering with the Reis approach and other aspects of harmonic telegraphy. He had even tried unsuccessfully to make a "speaking telegraph." Like Bell's and Gray's devices, some of Edison's used the varying resistance of a lead moving in water; and the caveat Edison filed on January 14, 1876, used electromagnetism to make a diaphragm emit a sound. But in using both those techniques, as Edison conceded later, he had only the harmonic telegraph, not a speaking telephone, in mind.

"You can hardly understand the state of uncertainty and suspense in which I am now," Bell wrote his father just after his talk with Wilber. Mabel wrote to console him on the loss of the spark arrester. "But my dear Alec," she added, "I really don't see what the other failures and disappointments are. . . . So far as I can see from your letter it is Mr. Gray more than you who is to be disheartened by the way the patent stands now."

As usual, Bell's spirits were more pulsatory than undulatory. Included in his letter to his father was a zestful account of his stay at Anthony Pollok's mansion, a local showplace, and of being introduced to "some of the elite of Washington." Pollok and he had called on Joseph Henry that day, and Pollok was soon to give a party in his honor. "So you see I am having a gay and happy time."

Just about then, Elisha Gray left Washington for Philadelphia to prepare exhibits for the coming Centennial Exhibition and to show his multiple telegraphy apparatus to prospective investors. In Philadelphia also lived Dr. Samuel S. White, a dentist who had become a millionaire in the manufacture of porcelain teeth, and who had been Gray's chief backer in all of his inventions since 1874. On March 3 Gray's patent attorney William D. Baldwin wrote White about the telephone caveat. "We could still have an interference by Gray's coming down to-morrow and promptly filing an application for a patent," Baldwin wrote, ". . . but my judgment is against it, as Gray made the invention, as I understand it, while here, after Bell's application was sworn to."

Later that day Baldwin talked with Bell in Pollok's office about the harmonic telegraph interferences, though not about the telephone patent. Reporting this interview to White, Baldwin wrote that Bell "has been very much annoyed by spies set upon him, probably by the Western Union, and . . . thinks they are trying to play him and Gray off against each other." Baldwin assured Bell that Gray had nothing to do with the annoyance. As for the telegraph interferences, Bell seemed receptive to the idea of joining forces with Gray against Western Union. All in all, Baldwin found Bell "intelligent, gentlemanly, and well-disposed."

Bell had reason to be amiable. On that day, his twenty-ninth birthday, the Patent Office examiners formally approved the telephone patent. Elisha Gray did not choose to contest it further. Two days later Bell left triumphantly for Boston. On the next day, March 7, 1876, his great prize, United States Patent No. 174,465, was officially issued.

The news of March 3 had delighted Mabel Hubbard. Her father being

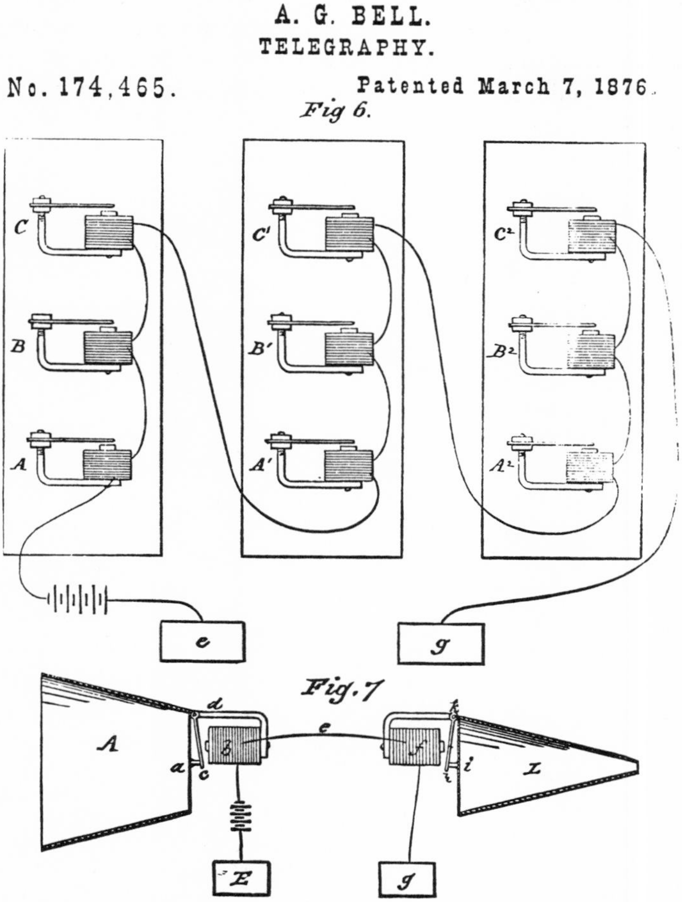

A. G. BELL.
TELEGRAPHY.

No. 174,465.

Patented March 7, 1876.

Fig 6.

Fig. 7

A page from Bell's patent on "Improvement in Telegraphy."
Figure 6 shows the tuned reeds of a harmonic multiple telegraph.
Figure 7 is a magneto-electric telephone.

a man of affairs in the Gilded Age, she knew enough to be filled with wonder that mere right should have overmatched the "colossal power" of Western Union and of William Orton, "almost the most powerful man in this country, and willing to spare no expense, honest or dishonest, to conquer you. Just now too when [Secretary of War] Belknap's iniquity, coming after all those other stories and scandals, makes us feel as if there were no justice in such a sink of corruption as Washington."

After savoring the news for two days, she wrote, "I only begin gradually to comprehend what a triumph your success is." In that comprehension she still had a long way to go. United States Patent No. 174,465, the basic telephone patent, would turn out eventually to be one of the most valuable patents ever issued.

17

The Telephone Speaks

By the time of his trip to Washington, Bell had finally carried out his plans of the preceding fall. So high did he now stand professionally that in mid-February 1876, at the age of twenty-eight, he was sounded out about succeeding the late Samuel Gridley Howe as head of the Perkins Institution. By then he had also been able to terminate some of his private lessons or delegate them to his student teachers, and still scrape by on fees from the latter. So at last he and Thomas Watson could resume the autograph telegraph experiments they had suspended six months before.

This time Bell and Watson stalked their quarry not in Williams's shop but in the privacy and relative quiet of the new Exeter Place rooms. Nevertheless Bell needed time to warm up to the work again. For a week he repeated and elaborated on some previous experiments with reed receivers, tested various combinations of armatures, cells, and circuits, tried a new design of autograph stylus, and adapted the manometric flame to measuring the electromagnetic effects of undulatory currents. The telephone meanwhile waited in the wings for its first speaking role. Then came the summons to Washington and a twelve-day hiatus in experiments.

On his return to Boston, Bell took a different tack. He needed no urging to work toward electrical transmission of speech. On the contrary, Gardiner Hubbard had often been hard put to restrain him from it. Furthermore, Bell's search for the "hole" in his specification and his dramatic midnight plugging of it insured that whatever else happened he would sooner or later try a liquid variable-resistance transmitter. By indicating the subject of Elisha Gray's caveat, Examiner Wilber had revealed to Bell not that Gray had successfully tried such a transmitter but rather, from the very nature of a caveat, that Gray had *not* tried it. The hint did let Bell know that an able rival also considered the possibility worth exploring. But the effect of that influence, if any, could only have been to hasten Bell's investigation by a few days, or at most a few weeks.

Bell's Exeter Place laboratory in March, 1877, painted by William A. Rogers from his sketch at the time

Resuming experiments on March 8, 1876, the day after his return to Boston, Bell did not make variable resistance the first order of business. He began as he had left off, in more or less random experiments with a reed receiver. Using undulatory currents induced by the vibrations of a tuning fork, he tried various arrangements of both the receiver and the tuning-fork transmitter. The receiver responded when the fork vibrated near successive electromagnets of different resistances, but not when the electromagnet battery was cut out of the circuit or the fork was vibrated near the wire alone.

Then, as if merely trying another permutation, Bell took the electromagnet out of the circuit (leaving the battery in) and substituted a dish of water, one lead being draped over the edge into the water, and the other lead connected to the handle of the tuning fork, the arms of which he vibrated and then partially immersed at an angle. From the reed receiver he heard "a faint sound" — the work of a current made undulatory not by induction, but by the fluctuating resistance of the battery-powered circuit, as the fork vibrated in the water and thus varied the area of its contact.

178

Faint as it was, the sound stirred Bell's interest. He added some acid to the water, increasing its conductivity, and the sound became "much louder." He improved the liquid contact of the stationary lead by connecting it to a ribbon of brass in the water, and the sound got "very loud." Finally, and inexplicably, he substituted a handbell for the tuning fork and immersed the vibrating bell straight down in the water, getting no sound at all.

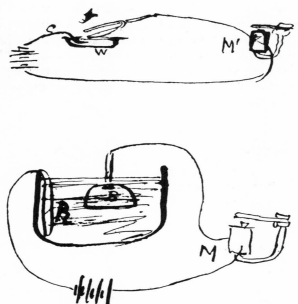

Sketches from Bell's notebook, March 8, 1876

Presumably he tried the bell simply because it happened to be near at hand. The puzzle is why he thought it might produce perceptible undulations in the current. True, the lateral vibration of the bell metal would vary its separation from the other lead and therefore vary the resistance also. But the two were so far apart in the water that the proportion of variation was negligible. And since the bell was held straight down in the water, the area of its contact could not have varied at all.

Whatever his reasoning at that moment, observation led him to conclude that the vibrating lead's area of liquid contact should be small and that of the stationary lead should be large. That view comported with his earlier spark-arrester concept and with his later theoretical analysis on March 20, both of which proposed to vary resistance by varying the area of contact.

On that basis, the less the vibrating lead was immersed, the greater would be the proportion of change with vibration and thus the more pronounced the effect.

So it was probably both correct observation and sound theory that led him that day to sketch a horizontal membrane diaphragm from the center of which a mere needle dipped its tip into a dish of water, the other lead lying at the bottom. With contact area rather than separation distance in mind, he raised no objection when Watson next day, March 9, set up the arrangement with the stationary lead hanging at the side instead. That afternoon Bell sang into a hole at the top of a box that had the diaphragm as its bottom. The varying pitch came through on the reed receiver. "When Mr. Watson talked into the box," Bell noted, "an indistinct mumbling was heard at [the receiver]. I could hear a confused muttering sound like speech but could not make out the sense."

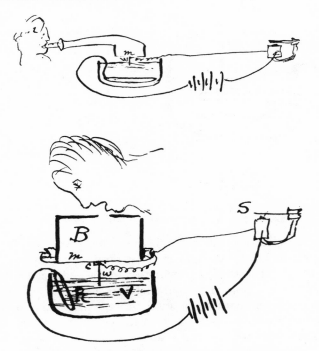

Sketches from Bell's notebook, March 8 and 9, 1876

On the morning of Friday, March 10, 1876, Bell had Watson make a similar device, with a brass pipe extending into the water as the stationary lead and a platinum needle dipping down from the diaphragm. For the box, however, Watson substituted a chamber with a speaking-tube mouthpiece. That afternoon in the Exeter Place rooms, Bell and Watson got around to trying the new model. Its dish or bowl contained water with a little sulphuric acid. By the bureau in Bell's bedroom, Watson waited with the reed receiver pressed against his ear; in the laboratory at the far end of an entry hall, with two intervening doors closed, Bell leaned over the transmitter.

In his laboratory notebook two days later, Bell described what ensued:

I then shouted into M [the mouthpiece] the following sentence: "Mr. Watson — Come here — I want to see you." To my delight he came and declared that he had heard and understood what I said. I asked him to repeat the words. He answered "You said — 'Mr. Watson — come here — I want to see you.'" We then changed places and I listened at S [the reed receiver] while Mr. Watson read a few passages from a book into the mouth piece M. It was certainly the case that articulate sounds proceeded from S. The effect was loud but indistinct and muffled. If I had read beforehand the passage given by Mr. Watson I should have recognized every word. As it was I could not make out the sense — but an occasional word here and there was quite distinct. I made out "to" and "out" and "further"; and finally the sentence "Mr. Bell do you understand what I say? Do — you — un — der — stand — what — I — say" came quite clearly and intelligibly. No sound was audible when the armature S was removed.

So ran Bell's own account of the first intelligible communication by telephone.

Its details agree with those given in a letter he wrote his father on the very day of the event "to announce a great failure [George Brown's desertion] and a great success." The elation Bell and Watson felt that day comes through more clearly in the less formal account. "This is a great day with me," Bell exulted. "I feel that I have at last struck the solution of a great problem — and the day is coming when telegraph wires will be laid on to houses just like water or gas — and friends converse with each other without leaving home."

Rich reward had come at last to Bell's rare combination of qualities: the ability — call it intuition or genius — to conceive the incredible goal, the stubborn faith to keep grasping at straws, the luck to find the magic needle in the haystack, and the wit to recognize it.

Since the summons to Watson has become one of history's best remembered utterances, it deserves some further comment, if only to satisfy antiquarians.

In a memorandum book that evening or the next day, Watson set down his version of it, to wit: "Mr. Watson I want you," the words "come here" then being inserted after "Watson" with a carat. Evidently Watson either missed or forgot the words "to see," which Bell included in both his lab report and his letter.

The question of what Bell said bears on a further uncertainty. In Watson's autobiography, published fifty years later, the famous sentence is described as Bell's instinctive call for help after spilling some acid on his clothes. If so, it was fortunate for the innocence of future schoolchildren that his historic outcry was so genteel. But neither Bell's two contemporary accounts nor Watson's sketchy notebook entry nor Watson's testimony on the witness stand half a dozen years later mention such an occurrence. And Bell's own version of the first telephone transmission casts further doubt on the spilled-acid story. The mishap, if it occurred, may have been a matter for assistance but scarcely for consultation. Watson probably confused that occasion with another, more satisfyingly dramatic.

Then again, Bell may have omitted the matter of the spilled acid as being undignified.

It should also be said that far from becoming an instant byword, like Stanley's opening remark to Livingstone, Bell's call to Watson did not even go on public record until Watson's court testimony in August 1882.

The two young men let the late winter dusk settle over Boston while they tested and tinkered with this eloquent materialization of Bell's dream. "How do you do?" the little steel reed asked Watson, and then it began singing "God Save the Queen." Bell tried adding more battery cells, but with stronger current the brass and platinum leads in dilute acid generated gas bubbles, which drowned out his voice with their fizzing. Two cells of battery seemed to be the best compromise.

A black deposit had to be scraped from the wire frequently, and water kept splashing up and softening the membrane. So Bell detached the new membrane transmitter and worked off some of his excitement by more experimenting with the tuning fork in the water. In that way he confirmed that the vibrating element should be kept to a minimum of immersion, but found that with a stronger and hence more highly conductive acid solution the depth of the stationary lead made little difference.

Thinking over the muffled quality of some transmissions, he suspected that the receiver reed had to be pressed too tightly against the ear. So he sketched an electromagnetic receiver with its steel reed attached to a membrane in a closed chamber, from which a hearing tube would bring the sound to the ear. But next day, when Watson finished the hearing-tube receiver without the membrane, no sound could be heard. Much better re-

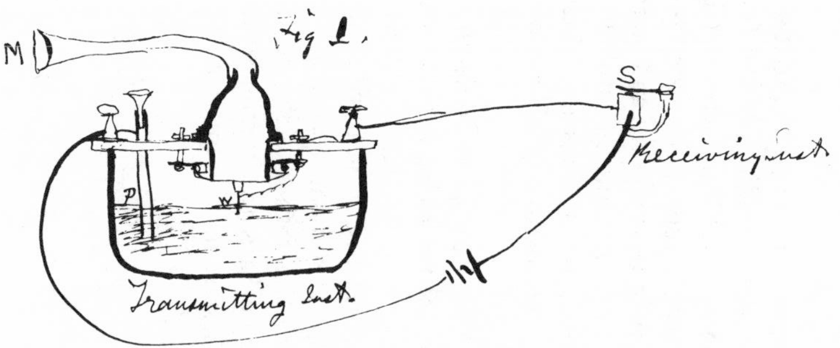

Fig. I.

M

S

Receiving Instrument

Transmitting Instrument

1. The improved instrument shown in Fig. I. I was constructed this morning and tried this evening. P is a brass pipe and W the platinum wire M the mouth piece – and S the armature. The Receiving Instrument.

Mr. Watson was stationed in one room with the receiving instrument. He pressed one ear closely against S and closed his other ear with his hand. The Transmitting instrument was placed in another room and the doors of both rooms were closed.

I then shouted into M the following sentence: "Mr. Watson – Come here – I want to see you." To my delight he came and declared that he had heard and understood what I said.

I asked him to repeat the words – He answered "You said 'Mr. Watson – come here – I want to see you.'" We then changed places and I listened at S while Mr. Watson read a few passages from a book into the mouth piece M. It was certainly the case that articulate sounds proceeded from S. The effect was loud but indistinct and muffled.

If I had read beforehand the passage given by Mr. Watson I should have recognized every word. As it was I could not make out the sense – but an occasional word here and there was quite distinct.

I made out "to" and "out" and "further"! and finally the sentence "Mr. Bell Do you understand what I say? Do – you – un – der – stand – what – I – say" came quite clearly and intelligibly. No sound was audible when the armature S was removed.

Bell's notebook entry for March 10, 1876

sults came on the following day upon removing the chamber and tube and adding the membrane.

On Monday, March 13, Gardiner Hubbard and Dean Lewis Monroe climbed the stairs to verify Bell's astounding report of talking by telegraph. Bell had now made both transmitter leads out of platinum to minimize the noisy generation of gas, and had incorporated a wooden shield to protect the membrane from splashing. Even so, using the reed receiver, neither visitor could hear anything for some time. "Indeed," Bell wrote in his lab notebook, "both seemed at first rather sceptical, and I presume thought that the imagination had a good deal to do with the sounds." But Bell and Watson accurately transcribed from the receiver what Monroe, his "full rich tones" quite recognizable, dictated into the transmitter. Hubbard learned not to press the reed too hard against his ear, after which he heard "articulate sounds," though too indistinct to understand. Monroe also heard them.

Acting on Hubbard's shrewd suggestion, Bell on March 14 tested the acoustical properties of the transmitter. Heard directly through a tube let into the transmitter chamber, Bell's voice sounded to Watson as muffled and indistinct as it had through the reed receiver. So Bell sketched a simpler form of mouthpiece, flaring outward from the diaphragm like an inverted cone or wide-mouthed funnel. This type became the best known of Bell's early variable-resistance transmitters, and some historians later assumed it to have been the historic one used on March 10. But in fact Watson did not get around to making that model for some time, its first recorded test being on March 23.

On March 14 Bell also tried different arrangements of the variable-resistance transmitter, using a simple, continuous tone-sounding device for a test signal in order to free Watson for other work. Spiral leads were tried, double points in series, different resistances, different liquids. Bell turned up no startling improvements. Next day he decided to stop tinkering and start theorizing, and on March 20 (having made no experiments meanwhile) he wrote out an elaborate analysis of his previous observations and their meaning, including graphs of calculated current fluctuations with various assumed combinations of voltages and resistances.

His father came down in late March, after Alec had spent some time arranging lecture engagements for him; and as usual the elder Bell pleased his Boston audiences mightily. He pleased his son, too. "I feel as if I had never known him till now," Alec wrote Mabel. "It is a new pleasure to me to be able to talk freely and fully to him. . . . I feel somehow as if it was your love that has brought us together." And after his father left in April, the Exeter Place rooms seemed empty to Alec. Perhaps the young man's recent triumphs in career and courtship had at last relieved him from the fear of domination so long underlying his admiration for his brilliant and impressive father.

Upon the variable-resistance transmitter, drawing power from a readily expansible source, the future telephone industry would rest. Bell saw its great advantage clearly enough. Yet by the end of April he had drifted away from it and back to the magneto transmitter, which depended on the puny power of the sound waves themselves to induce its current. Several factors combined to send him off course.

Apparently he lacked confidence in his quantitative theorizing as an approach to the variable-resistance transmitter (as well he might have, since he did not allow for the bulk of the liquid and so assumed for it a grossly exaggerated resistance). Certainly he failed to continue or repeat that approach or even to check its conclusions experimentally.

During the next month, moreover, ten or twelve more days of experiments with sundry variable-resistance arrangements yielded no conclusive or even much improved results. On March 23 he sketched a thin slab of acid-moistened sponge sandwiched between two platinum plates, one fixed and the other vibrated by a membrane diaphragm with a point contact. Bell might have pursued this line to the pressure-sensitive solid conductor that at last solved the problem. But that would have taken much time, money, and patience, and he had none of those. Not even the initial notion was actually tested. Instead he veered off that same day to a scheme of vibrating battery plates — also not tested — and thence to fruitless experiments with platinum foil.

On April 1 a new inspiration crowded out earlier ideas: the vibration of an imperfect solid conductor such as carbon or animal tissue in a good liquid conductor such as mercury. The next day he got good results by vibrating plumbago (graphite) from a lead pencil in a cup of mercury. At two the next morning he gave up trying to sleep and wrote a letter to Mabel: "I try to stop thinking but it's of no use — I cannot get the reins of my mind! There is a picture before my eyes — a moving picture — a little lead-pencil vibrating in mercury! . . . I am only disheartened at the immensity of the horizons opened out to me. . . . I feel like the first mariner in an unknown sea — uncertain which way to go."

His mercury-plumbago devices might eventually have brought him nearer a feasible variable-resistance transmitter. He kept on with them at intervals for another three weeks. But his perfectly sensible expedient of using pure tones for testing lent itself subtly to pressures from his father and Hubbard to concentrate on the autograph telegraph. Thus tested, his new devices confirmed his belief that undulatory current would accommodate simultaneous musical notes that would have interfered with each other in an intermittent current — and the chief use of that property was in the autograph telegraph. "Stick to the autographic till you bring it into practical use," wrote his father in mid-April; until then "I would suspend all other experiments." Hubbard may have put the elder Bell up to that. At any rate, Hubbard concurred a few days later. "If you would take Mr. Williams' man [Watson] as I proposed and work with him or let him work steady on *one thing* until you had perfected it you would soon make it a success," wrote Hubbard. "While you are flying from one thing to another you may accidentally accomplish something but you probably will never perfect anything."

Once again Hubbard brought his daughter into the struggle to keep Alec on what seemed to be the right track. "If you could make one good invention in the telegraph," he wrote Alec, "you would secure an annual income as much as the Professorship and then you could settle that on your wife and teach Visible Speech and experiment in telegraphy with an easy and undisturbed conscience." Hubbard even persuaded Mabel early in May to tell Alec she would not marry him until he finished the autograph telegraph. This, Alec wrote Mabel that night, "almost broke my heart. . . . I want to marry you, darling, because I love you . . . and I wish to feel that you would marry me for the same reason." He had no intention of giving up on the autograph, he wrote, but her words had made him wish he had never heard of telegraphy.

He had been working on the autograph telegraph anyway, and he kept on with it. That involved time-consuming experiments with conductive writing fluids or embossing or the depositing of metal on lead pencil marks, none of which helped advance telephony in the slightest.

His professional work, though not so demanding as before, kept breaking into his train of experimentation. New applicants for the class had to be interviewed; lessons had to be given; exams had to be graded.

He had scarcely finished grading exam papers in early April when word came that a long-pending multiple-telegraph interference with Gray was about to be decided in Washington. So he had to spend much time that month writing an elaborate statement of every stage in the development of his multiple telegraph from its first conception to March 1875. Meanwhile he worked up a Visible Speech exhibit for the coming Centennial Exhibition at Philadelphia that summer. By the end of April he was laid up again with his old complaint of headache and had to decline a number of requests to write articles for encyclopedias and journals. "I am worn out with work and anxiety," he wrote home.

Either he or the Perkins Institution had apparently not chosen to follow up the feeler of February, but in late April he gave some thought to the offer of a professorship at the deaf-mute college in Washington, D.C. Hubbard uged him to reject it. "If you were tied up to rules and hours," he wrote Alec, "it would be very irksome and hard. . . . You are not like other men and you must therefore make allowance for your peculiarities. . . . If you could work as other men do you would accomplish much more. . . . But you must overcome these habits by your own will, and not by rules imposed by a college." Hubbard knew his future son-in-law.

The wonder, considering the foregoing catalogue of duties and distractions, is not that Bell failed to perfect the variable-resistance transmitter at once, but rather that he managed to give it as much attention as he did. And the more wonder, since the magneto transmitter, so much simpler and more convenient, had meanwhile begun to talk.

Bell backed into his revived emphasis on the magneto transmitter. With his father watching and listening on March 27, he tried a couple of unrewarding experiments with the liquid variable-resistance transmitter. His father noticed that the receiver reed sounded loudest when its reed did not touch its electromagnet pole. Alec's attention being thus called to the receiver, he tried plucking it as a magneto transmitter, using a diaphragm receiver. Then he tried a magneto diaphragm instrument at each end of the line, as in his patent drawing. Finally he tried a small, flat-coiled clock spring as the transmitter membrane armature. And the arrangement transmitted the first intelligible word by that type of instrument. The word was "Papa."

For a time thereafter, Bell experimented with both magneto and variable-resistance transmitters. But it was the former type he used for speech, and the latter for autographic transmission.

After only two days of experiments in May with the magneto transmitter, he recorded no more experiments of any kind until the last day of June. But those two days in May confirmed Bell's predilection for the magneto instrument. On May 5, at Hubbard's house, a few sentences came through a membrane transmitter and receiver "perfectly." Two or three times Bell distinctly heard the self-consciously awed exclamation "What hath God wrought," even to the "th" sound in "hath." On May 22 at Exeter Place with a double-pole magneto transmitter and a simple reed receiver, Bell found to his delight that "consonants & vowels were equally intelligible" (consonants having been elusive till then). And although his ear was pressed tightly against the damped reed, he heard with utmost distinctness the pregnant question: "Mr. Bell, are you going to the Centennial?"

18

Philadelphia, 1876

The question put to Bell on May 22 was not the earliest coupling of the Centennial and the telephone. On March 22, 1876, only twelve days after Bell's first intelligible speech transmission, the *New York Times* had run an editorial, "The Telephone." This was inspired not by Bell's work, of which the writer evidently knew nothing, but by that of one "Prof. Reuss" (presumably Philipp Reis), the reputed inventor of a device "intended to convey sounds from one place to another over the ordinary telegraph-wires." "By means of this remarkable instrument," the writer suggested, "a man can have the Italian opera, the Federal Congress, and his favorite preacher laid on [like gas] in his own house." Perhaps, he added, there was "a sinister purpose . . . of the enemies of the Republic" to cut down on Centennial Exhibition receipts by letting people hear the music and speeches without going to Philadelphia.

Half-mocking, half-eager, the *Times* editorial foreshadowed public reaction to the real thing when it presently came along. Incredible, people would seem to say on first thought; and on second thought, high time. For Bell, a great deal depended on the length of the interval between first and second thoughts. The precise speed of public acceptance would matter little to mankind at large. Once demonstrated, the telephone was bound to make its way in fairly short order against whatever skepticism it might encounter (unless Western Union suppressed it for a time). But to the individual fortunes of Alexander Graham Bell, Gardiner Hubbard, and Thomas Sanders, the immediacy of that public acceptance was crucial. If it were delayed a few months, perhaps only a few weeks, discouragement and inability to raise capital might induce or force the pioneers to sell out cheaply to some such late-coming giant as Western Union. Even Gardiner Hubbard, with a personal stake in the matter, seemed slow to appreciate the possibilities of the telephone as compared to the autograph telegraph. How much slower, then, might be the general public? The challenge now before Bell remained one of communication, but communication this time in a cultural and commercial sense, not a technological one.

Bell rose to the challenge. The Hubbards' Cambridge connections helped in rounding up five Harvard professors of science, along with Pickering of MIT, for a demonstration of Bell's magneto diaphragm telephones on May 2. Bell risked the trial with remarkable faith in his newborn and still rather croupy brainchild; and it did him proud. Hamlet's soliloquy went into one clock-spring armature and "articulate sounds" came out of the other, perhaps requiring the cue of familiarity, but nevertheless suggestive enough to impress the savants.

In the Boston Athenaeum on May 10, before one of the nation's oldest and most respected learned societies, the American Academy of Arts and Sciences, Bell unveiled his telephonic theory, both variable-resistance and electromagnetic, and gave some account of its practice. That he felt diffident before what he later described as a "dignified assemblage of grey heads" is suggested by the learned references (most of them hastily copied from Kuhn's *Encyclopoedia der Physik*) with which he armored his paper. They included a reference to Reis's work. It seems likely that caution also held him back from trying to demonstrate speech transmission on the spot. Attempts at that, his paper confessed, had thus far been only occasionally intelligible. Bell showmanship could not be totally repressed, however. As Bell began talking, he sent a telegraphic signal to Mabel's twenty-six-year-old cousin William Hubbard, who waited at a "telegraphic organ" (using intermittent, not undulatory current) in Bell's Boston University room a few doors up Beacon Street; and to the astonishment of the audience, a box on the table emitted the jubilant sonorities of "Old Hundred." At the close of the paper, the audience applauded heartily, something that another Hubbard cousin, the biologist Samuel Scudder, remarked that he had not known to happen before in his dozen years of membership. Next day, MIT asked Bell to repeat the talk later that month before its Society of Arts. "There is nothing like a bold front after all," Bell wrote his parents. "I feel myself borne up on a rising tide."

The MIT lecture before a large audience on May 25 was "another glorious success," concluding with what was probably the first public demonstration of telephonic speech. A clock-spring membrane telephone in the hall communicated with another in an adjoining house. "Vowels are faithfully reproduced; consonants are unrecognizable," the *Boston Transcript* reported; "occasionally, however, a sentence would come out with startling distinctness, consonants as well as vowels being audible." "My name is sure now to be well-known to all scientific men," Bell wrote his mother; and he envisioned a professorship in physics at Boston University, along with money and fame from lecturing on "Telephony."

But his present duties to teaching and Visible Speech could no longer be scanted. Bell's father had complained weeks before that Alec "is forever poking over experiments and puts off necessary work in the old style." By June, Alec's own conscience had chimed in. Haunted by exam papers that should long since have been graded, he wrote Mabel that "the more I examine my life and character . . . the more am I frightened for your sake — I do not see there the kind of man that should marry at all. . . . However if love and affection can make amends for bad defects of character, I promise you that." So the public heard no more of telephony for a while.

What happened in mid-June must have eased Bell's conscience as to Visible Speech. A month after having participated in opening the Centennial Exhibition at Philadelphia on May 10, the junketing Emperor Pedro II of Brazil visited Boston; and Bell wrote him urging that he visit the Boston School for the Deaf to see the Visible Speech method in use. Dom Pedro (as he was popularly known) was a portly, full-bearded gentleman of democratic manners and lively mind, who not only looked like Santa Claus, but also shared that legendary figure's love of children. To the delight of Bell, Sarah Fuller, and the children, the emperor accepted the invitation. On June 14, His Majesty observed the class's work with interest, shook Bell warmly by the hand, and asked that a couple of copies of the elder Bell's book be sent him to take back to Brazil.

Somehow the great Centennial Exhibition at Philadelphia kept intruding itself into Bell's crowded ken. Few Americans could have been unaware of it that hundredth spring of independence, what with Centennial sermons and articles and symposia and even dance parties (one of which Alec and Mabel attended). For Bell, his meeting with the Exhibition's most imposingly titled foreign tourist gave extra emphasis to the national jubilee. Bell's old Boston friend Percival Richards had been appointed agent for the committee on the Massachusetts education and science exhibit, and Bell's intended father-in-law, Gardiner Hubbard, was one of the three committee members. Naturally, the education of the deaf found a place in the Massachusetts exhibit, along with a display of Visible Speech books and charts. More incongruously, space was set aside for the eventual display of Bell's telephone and multiple telegraph, since the deadline for their regular entry in the electrical section of the Exhibition had passed before Bell deemed the telephone ready for showing.

A couple of days after Bell's meeting with Dom Pedro, Hubbard telegraphed Bell to hurry down to Philadelphia, where the great scientist Sir

William Thomson and other experts were soon to judge the Exhibition's electrical entries. Without Bell's enthusiastic presence, Hubbard feared, the honors in multiple telegraphy might all go to Elisha Gray, who was already on the scene and well prepared.

The call came at a bad time for Bell. He still had not graded his examination papers nor completed his speech course, and time was running out. Now it was Mabel who urged the claims of telephony, and Alec who resisted. Protesting all the way, having "not the remotest intention of leaving Boston," he let her take him to the depot. There in the June twilight, seeing "how pale and anxious she was about it, I could not resist her and here I am," he wrote his mother from New York the next morning. "What I am going to do in Philadelphia . . . I cannot tell."

Gardiner Hubbard met him in New York, however, and took him in charge. At Philadelphia Hubbard installed him in the same hotel as that of three judges in the electrical department of the Exhibition. Before the night was over, Bell had met two of them and talked at length about his projected exhibit. One judge, Professor James C. Watson of the University of Michigan, later recalled that Bell "stated to me in detail the character of his inventions. . . . While sanguine as to practical results from his multiple telegraph, his great invention [he said] was the speaking telephone, which he believed he had discovered, and in respect to which there was no rival claimant." Next day, because of a light remark by Hubbard about having to "smuggle" the telephone equipment into the exhibition building, Bell threatened on grounds of conscience to give up the whole business and go back to Boston; but Hubbard mollified him by producing the proper permit and a proof of the official catalogue with Bell's exhibits duly listed as "Visible Speech" and "Telegraphic and Telephonic Apparatus."

So Bell's usual nervous headache passed away, and he began to enjoy himself. "I really wish you could be here to see the Exhibition," he wrote Mabel. "It is so prodigious and so wonderful that it absolutely staggers one. . . . Just think of having the products of all nations condensed into a few acres of buildings." The Chinese were making "a splendid show. . . . How presumptuous of us to think ourselves so far superior to them. If I had only a little money to spare, I would buy some of the Chinese works of art for you." In the Brazilian Department he marveled — like a proper Victorian — at "a table of solid silver weighing 4002 pounds." But it was science and, better yet, scientists that entranced him. Rudolph Koenig, the inventor of the manometric capsule, was there in person with "a splendid exhibit of tuning-forks and scientific apparatus." They had a long talk on scientific subjects, Koenig speaking French and Bell English.

Best of all, on June 21 Bell found Sir William Thomson, just as the latter

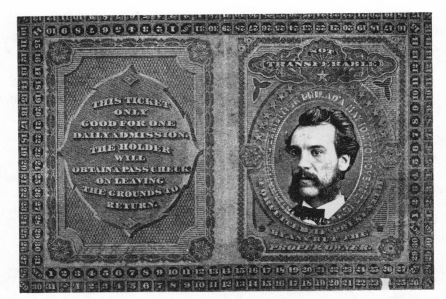

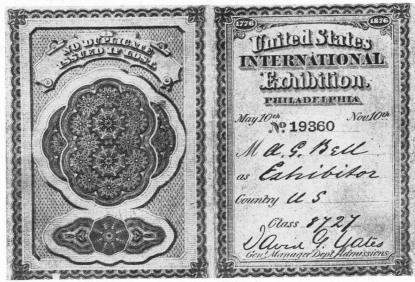

was examining Elisha Gray's apparatus. "I verily believe too," wrote Bell, "that it was Elisha Gray himself who stood beside him when I handed him my card!" Sir William turned out to be "a splendid, genial, good-hearted and wise-headed looking man. . . . What was my delight, when he addressed me, to hear a good broad *Scotch* accent tinging his utterance!!" Thomson wanted to see the Bell telephone that day. But he lost track of time in looking at some Western Union instruments. So he apologized, remarked that he and other judges were coming with Dom Pedro on Sunday, June 25, to look at Gray's instruments, and asked Bell to submit his for judging at the same time.

Bell accepted the suggestion, though with private misgivings. Gray had already enjoyed the advantage of time and ample facilities in preparing for the occasion, whereas Bell's equipment was still arriving piecemeal, some of it damaged in transit. But Bell knew "one thing I can stand my own upon, and that is *theory*. I can *talk* and *explain* and Sir William will understand." Gardiner Hubbard promptly conferred with Professor George F. Barker of the University of Pennsylvania, who was in charge of arrangements for the Sunday exhibition and had invited a number of leading American scientists to attend. Bell set to work nursing his battered equipment back into shape. He confessed some pessimism about it in a letter to his parents: "my only chance consists in having my apparatus for the transmission of *vocal sounds* a success." A wire to Boston summoned Willie Hubbard, whose "telephonic organ" rendition of "Old Hundred" had so effectively charmed the Boston savants. Exhausted by effort, excitement, and the baking heat of that summer weekend, Gardiner Hubbard left Philadelphia for home by the same Saturday midnight train that brought his young nephew Willie to the Centennial City. And so the stage was set and the cast assembled for the dramatic scene that would always be one of Bell's favorite reminiscences.

Sunday, June 25, 1876, dawned hot in Philadelphia and grew hotter as the midsummer sun beat down on Fairmount Park and its motley exhibition buildings. Darkness still covered another great encampment far to the west on the Little Big Horn River, that of some five thousand Indians who on that same day would fall upon and destroy General George Custer and much of his command. Bell's destiny was kinder than Custer's. It awaited him in the gargantuan shed known as the Main Building, reputedly the largest man-made shelter in the world. The Massachusetts education section occupied six rooms, comprising the whole east gallery except for a central space given over to an exhibit of organs.

There would be peace that day in the Centennial grounds, if not on the

Little Big Horn, for Sabbatarianism closed the grounds to the general public. That was why a Sunday had been chosen for judging exhibits so dependent on the accurate perception of sounds. Up to the Main Building that quiet morning rolled two or three open carriages, bearing the emperor and empress with their party of Brazilian ladies-in-waiting and officers. Professor Barker led his party from their assembly point in a nearby hotel. When Bell joined them at Gray's exhibit, prominently displayed under the aegis of Western Union near one of the main aisles, the group amounted to some fifty people. Among them, in addition to Sir William Thomson, were such notable scientists as Rudolph Koenig, the Canadian geologist T. Sterry Hunt, the astronomers James C. Watson of Ann Arbor and Henry Draper of New York, the scientist brothers John and Joseph LeConte of Berkeley, and the geodesist Julius Hilgard of the United States Coast Survey.

The sun poured hotly through the multi-paned glass walls of the gigantic structure while Bell listened to what seemed an interminable disquisition by Gray on the development and objects of his work. Bell noted with some satisfaction that Professor Barker had to help out Gray occasionally in matters of theory. Attempts at multiplex transmission of Morse code by Gray's method struck Bell as failures, and yet the novelty of Gray's transmission of musical tones seemed to impress the listeners. When the presentation finally ended, Dom Pedro, who had already recognized Bell in the crowd, walked over to him, shook hands, and thanked him for the Visible Speech books.

In later years, Bell liked to make this encounter the suspenseful climax of

his dramatic reminiscence. Gray's time in the sun had indeed run much longer than originally planned. As Bell told it later, only Dom Pedro's personal interest kept the perspiring judges from abandoning the schedule, which called now for trudging to the end of the building and up a flight of stairs for Bell's exhibit. That may have been so. But in a letter to his parents two days after the event, Bell simply described the emperor's greeting and added, "Sir William and Dom Pedro then came to see my apparatus." In any case, Sir William, Professor Watson, and others had by this time (thanks to Bell's explanations) grasped the significance of Bell's work firmly enough to guarantee it a serious hearing within another day or two — a hearing that Willie Hubbard, as it turned out, was competent to conduct even in Bell's absence.

Whether drawn by duty or Dom Pedro, the party proceeded to the East Gallery. At its south end, Bell had placed some chairs around a small table. The latter held a set of two current-interrupters with correspondingly tuned receivers, each receiver being equipped with one of Bell's latest patented vibratory circuit breakers for operating Morse sounders; this arrangement was to demonstrate harmonic multiple telegraphy. A Koenig manometric capsule was fitted with an electromagnetically vibrated diaphragm so as to show, by the dancing flame and revolving mirror, how an undulatory electric current might be visually observed. And there waited also what later came to be called the "Centennial iron-box receiver," a device Bell had hit upon that spring. He had been considering the possibility of an electromagnet shaped like a three-pronged fork, the center prong carrying the coil and forming one pole, and the two outside prongs being of the opposite polarity and bridged by the steel reed. Then he noticed a tubular magnet in the Williams shop and recognized its adaptability to this approach. The resulting instrument consisted of a hollow iron cylinder, three inches long and closed at one end, with an iron rod inside running up from the closed end and wrapped in a coil of insulated copper wire. A current through the wire would turn the device into an odd-shaped electromagnet, the rim of the little cylinder at its open end constituting one pole and the free end of the rod the other pole. Thus when a sheet-iron lid was fastened on the open end of the cylinder, and the inside rod adjusted so as to come close to the lid without touching it, the lid would respond as a diaphragm to the electromagnetic fluctuations induced by the undulatory current.

Wires ran along the railing of the gallery, past the Hook and Hastings organ exhibit, to Bell's transmitters, some hundred yards or so away at the northeast corner of the great Main Building. These consisted of a variable-resistance transmitter with a slender rod projecting down from a membrane into a small cup of conductive liquid; and two electromagnetic trans-

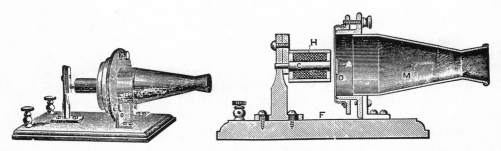

The Centennial single-pole transmitter

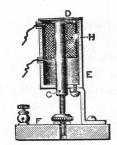

The Centennial iron-box receiver

mitters with small pieces of thin sheet iron glued to their membrane diaphragms, one of the transmitters having a single straight-cored electromagnet, the other having two, side by side. Here also, presumably, the telephonic organ awaited Willie Hubbard's touch.

Bell explained the principles of his harmonic telegraph, how much cheaper and simpler it was than Gray's, and how the vibratory circuit breaker relieved the telegrapher from having to pick out one Morse code message from others simply by its different pitch. Dom Pedro and Sir William successfully transmitted signals singly and concurrently. "I then explained the 'Undulatory Theory' and offered to test the transmission of the human voice," Bell wrote his parents a couple of days later. "I stated however that this was 'an invention in embryo.' I trusted they would recognize firstly that the pitch of the voice was audible and secondly that there was an effect of articulation." He withdrew to the far end of the gallery and began singing into one of the membrane telephones, leaving Willie Hubbard in charge at the receiving end.

Sir William, sitting at the table with the lid of the iron-box receiver pressed against his ear, was transfixed by a disembodied song. Then, eerily attenuated but with startling distinctness, there followed the words, "Do you understand what I say?" He clapped the little cylinder to his ear again. "Yes! Do you understand what I say!" he shouted. "Where is Mr. Bell? I must see Mr. Bell!" Willie Hubbard started off along the gallery, but Sir William ran past him and came on the young miracle worker still shouting "Do you understand what I say?" into the cone of the membrane instrument. "I heard the words 'what I say,'" Thomson reported breathlessly to Bell, asked him to sing again and recite something, and hurried back for more of the marvel.

Now it was the emperor's turn. Although Sir William's reaction must have prepared him for something remarkable, Dom Pedro started from his chair and cried out, "I hear, I hear!" He reapplied the little cylinder and repeated the words "to be or not to be." Bell, still declaiming Hamlet's soliloquy, presently heard a pounding noise and saw Dom Pedro rushing toward him "at a very un-emperor-like-gait," followed by a retinue of curious onlookers. Back at the receiver end, the soliloquy came by bits and pieces to the ears of successive listeners as they jostled for a chance to confirm what seemed to be happening. Among them was Elisha Gray. "I listened intently for some moments," he later testified, "hearing a very faint, ghostly, ringing sort of a sound; but, finally, I thought I caught the words, 'Aye, there's the rub.'" It was all too appropriate an introduction for Gray to telephonic speech, the first (as he freely and publicly admitted afterward) he had ever heard. "I turned to the audience, repeating these words," he said later, "and they cheered."

Bell had scored what (with some pardonable lack of originality) he reported to his parents as "a glorious success." Now he could and must go home that night in order to give his final exam on schedule next day, and he told Sir William as much when the latter asked for further demonstrations and a chance to bring Lady Thomson to them. It was arranged for Willie Hubbard to do the honors on Monday evening after the Exhibition closed for the day.

Later that Sunday, Elisha Gray came to Bell's hotel room for a long, friendly talk about the harmonic multiple telegraph. The telephone was not at issue; Gray laid no claim to that. "We have explained away all matters in dispute," Bell wrote two days later, "and have decided that it may be advantageous to both of us to unite our [multiple telegraph] interests so as to control the Western Union Telegraph Company, if those associated with us can be brought to a mutual understanding." The alternative, Bell and Gray agreed, would be protracted lawsuits, "and the ultimate result will be that the Western Union can step in and buy up whatever part they choose." The multiple telegraph interests were not joined, as it turned out. Nevertheless, that conversation with Gray, wrote Bell a few months later, "satisfied me that he was an honorable man & an independent inventor — & that my suspicions were unfounded."

Before the eventful day was over, Professor Hunt wrote Bell a congratulatory note. Sir William Thomson, he reported, declared Bell's telephone to be "the most wonderful thing he has seen in America. You speak of it as an embryo invention, but to him it seems already complete; and he declares that before long, friends will whisper their secrets over the electric wire. Your undulating current he declares a great and happy conception." Sir William, wrote Hunt, was already spreading the word among his scientific friends and compatriots.

Bell took the train to Boston that night. The next evening in Philadelphia, Willie Hubbard met and mastered the unexpected problem of having to move the telephones to the judges' pavilion (at Professor Watson's request) for the demonstration to Sir William and Lady Thomson. Even stricter precautions were taken against mechanical transmission of sound, the transmitters being placed in a different building from that of the receiver. As before, the liquid transmitter was not tried, but both the single-pole and double-pole membrane transmitters performed with indisputable success. Willie fired off a triumphant telegram to Bell immediately afterward: "Sir William is entirely satisfied with the experiments. Never was more successful. A large number of sentences understood."

Sometime after June 25 and before leaving Philadelphia about July 10, Elisha Gray had his assistant make a model of the liquid variable-resistance transmitter described in the Gray caveat — the first such model constructed — and tried it with Gray's own crude "sounding-box receiver." But it failed utterly. Perhaps this happened between the Sunday demonstration and that of Monday evening, which Gray also attended. At any rate, in talking with Professors Barker and Watson after the latter occasion, Gray apparently suggested that if the Bell telephone wires had been suspended lightly enough the results might have come from mechanical transmission. Watson later recalled that Gray had asserted the impossibility of any other explanation; but Gray denied having said that. Whatever he may have thought or said about Bell's achievement, Gray felt certain at this time that he had no grounds, legal or moral, for claiming it himself; and so he did no more in telephony during the Centennial year.

19

The Word Travels

In the rush of preparing the Centennial exhibit, Bell and Hubbard had evidently neglected to brief or cultivate the press properly. Only one press notice appeared before September, a tepid comment on July 8 in the *Boston Transcript* (to which the telephone was not entirely new anyway), and that item may have been inserted at the instance of Bell or Hubbard. The *Scientific American* of the same date reported that "Dom Pedro has again visited the Exposition, and has made a minute survey of its contents. The marvelous work of the Walter press . . . is said to have astonished him more than all else." No mention was made of Bell's exhibit. This imperial judgment, if accurately reported, had probably been handed down before June 25, since the weekly *Scientific American* made no effort to report late news. Several thick volumes descriptive of the Centennial Exhibition appeared in subsequent months, none mentioning the telephone. Like the *Scientific American* item, their texts had probably been prepared before Bell's triumph.

But the captivated scientists who had seen and heard Bell's telephone insured, of course, that it would not be forgotten. "I understood it at once," recalled Professor Joseph LeConte, "and on my return to Berkeley gave the students and faculty of the University a lecture explaining it." Boston, the hub of the American scientific world, knew all about it. Sir William clearly intended to be its herald in Europe. Bell doubtless understood all this. He seemed not at all concerned about press silence in the weeks following.

On the contrary, he showed signs of being a little dizzy with success. He offered to transmit speech from Boston to Sir William in Philadelphia if the use of telegraph lines could be had; fortunately, the lines turned out to be too busy for what would surely have been a failure. He did transmit — or thought he transmitted — speech over telegraph lines 120 miles from Boston to New Hampshire and back to the same building. This induced a wild dream of transmitting without a battery (that is, with residual magnetism) over the entire Atlantic cable. Thus far the idea of developing an all-metal transmitter diaphragm had not occurred to him; he had previously, in a casual way, tried using the iron-box receiver as a transmitter, but with little success. On July 11, however, he greatly improved the performance of the membrane transmitter by enlarging its dime-sized metal armature almost to the full size of the membrane. Only a slight and obvious further step remained.

Next day Sir William Thomson, en route through Boston, sought out Bell; and that night they tried another long-distance transmission over telegraph lines. Sir William suggested that the transmission might have been over the short part of the circuit rather than the long lines (which was probably the case). He also thought that in the harmonic multiple telegraph, Bell, Paul La Cour of Denmark, Cromwell Varley of England, and Elisha Gray were all on the same track; and he urged Bell to unite with Gray. Perhaps a bit crestfallen, Bell took comfort in the short-line performance, and he gave Sir William duplicates of the membrane transmitter and iron-box receiver to carry back to England. Then Bell spent three days on a device to generate electricity, through leads brushing the middle and end of a cylindrical magnet revolving on its long axis. The federal government indulged this fancy on August 29, 1876, with Patent No. 181,553, to which neither Bell nor the rest of the world seem to have paid any further attention. The experiments may have helped to work off whatever chagrin he felt at Sir William's deflation of extravagances. On the other hand the interlude tended to divert him from the track of the metal diaphragm.

"We have all of us our hands full with Alec," Mrs. Hubbard wrote her husband at about this point, rather as if she were rehearsing for the stock role of mother-in-law.

He has not yet sent in his report to the Bureau of Awards & I cant make him *do* it. He says that his brain won't work & he cannot make it. Then he has had applications from two Lecture Bureaus . . . for lectures on Acoustics & Electricity. He wants to lecture because he enjoys it, & as a matter of dollars & cents. Then he is crazy at the idea of Mabel's going away next winter & wants to be married. Then he would give up V. S. [Visible Speech] or the Telegraph; he says he cannot & I believe he ought not to try to carry them on together. Which shall it be? Which will pay immediately? Then he ought to go to Portland to see . . . about starting a [public school for the deaf]. . . . Then he wants to stop at Toronto & see George Brown, & he must be at Brantford on Tuesday — & more than all he wants to talk with you. . . . Poor May will have a busy life if she attempts to keep him up to present duty. . . . As to . . . lecturing, I . . .

begged him as he valued his own & Mabel's happiness to stay at home & mind his proper business.

And Alec unknowingly confirmed the analysis by writing Mabel that "those experiments of the past few days have quite unsettled me." But he suggested a sense of priorities: "there is a sort of telephonic undercurrent going on [in my mind] all the while."

Mabel was in Nantucket; the *Boston Globe* had done its duty to the coming age with a full account of the Boston experiments observed by Sir William; and telephonic inspiration seemed to flag. Bell had little reason to linger in Boston. On July 20 the temperature reached ninety degrees for the twenty-seventh day in a row; and Bell, who detested hot weather, reported that "my headache has taken root in my left eye and is flourishing!" He packed up exam papers, telephones, electrical supplies, and personal belongings; and though he got off the train in the town of Paris, Ontario, at one o'clock on the clear, cool morning of July 24, he lost no time in hiking ten miles to his parents' Brantford home, Tutelo Heights.

The Bell clan was there in strength, recently augmented by visitors from Australia, Mrs. Bell's brother Edward Symonds and his three daughters. Bell missed Mabel nonetheless. "Separation from you renders me as nervous and miserable as can well be imagined," he wrote almost immediately. She wrote him, too, with wise advice to focus on one thing at a time, and to make telephony the first object, all the better eventually to support and extend Visible Speech. "Try the lines between Brantford and Paris," she suggested, "and do your utmost to induce some one to take up your foreign patents and to allow you to go on working."

Whether Mabel's advice influenced or merely coincided with Bell's own views, he conformed to it. His Uncle Edward, shortly to leave for Europe, promised to find out what had become of the Herdmans, of old Edinburgh acquaintance, and whether they would take up the telephone patents abroad. Meanwhile Bell set about trying the telephone over considerable distances.

On August 3 Bell drove with his equipment in a hired buggy to the quiet village of Mount Pleasant and set up a telephone receiver in the general store, which served also as local office of the Dominion Telegraph Company. Some of the townspeople gathered in the store that evening to hear snatches of the prearranged transmission from the Brantford office, five miles away, over the telegraph line. With the iron-box receiver pressed to his ear, Bell recognized the rich voice of his Uncle David declaiming into the membrane transmitter at Brantford, and he was able to understand occasional words. The human voice had been transmitted intelligibly over a distance of several miles, farther than ever before by any medium, if the dubious results of early June are discounted. Others took their turn, those at Brantford singing or speaking, those at Mount Pleasant listening. Now Bell knew that "my undulatory current can be used upon telegraph lines."

On the next evening at Tutelo Heights, Chinese lanterns shone from the trees for a champagne supper in Uncle Edward's honor. Bell needed neither champagne nor lanterns to feel aglow; for to the assembled gentry, listening one by one, he and two of his Symonds cousins successfully transmitted three-part songs through a specially constructed triple mouthpiece, the wire being run between an outbuilding and the veranda. To Bell this meant that "with the undulatory current a single transmitting instrument will suffice for *any number of simultaneous messages* — while with the intermittent current there must be a distinct instrument for each message sent. . . . The more I think of it the more I see that the undulatory current is the thing." In this, of course, he had multiple telegraphy in mind. Bell's vision was evidently not yet sharply focused on the telephone as a means of vocal communication exclusively.

The guests, including two members of the Canadian parliament and other dignitaries, may have been more impressed by the program of entertainment transmitted from Brantford, four miles away. Bell had bought up most of the iron wire (generally used for supporting stove pipe) that he could find in Brantford and had run it from the house along wooden fences for a quarter mile to the nearest point on the Brantford–Mount Pleasant telegraph line. From the Brantford telegraph office to the veranda at Tutelo Heights came a Shakespearean recitation by David Bell and songs by one of his daughters and a local singer. One might suppose that a sense of magic suffused that moonlit summer night. If it did, it failed to touch the *Brantford Expositor*'s correspondent, whose account began with the guest of honor and proceeded to list sixteen of the most prominent guests before commenting on the telephone program. The item was headed simply "An Evening on Tutelah Heights." The *Toronto Globe*, owned and edited by the Honorable George Brown, reported the event in a column of trivia on its third page a week later, citing the *Expositor;* the telephone demonstration, it remarked, "afforded much pleasure and information to those present."

After some dutiful work at amending a patent specification, reading exam papers, and other overdue matters, Bell arranged a still more impressive demonstration for August 10. By courtesy of the Dominion Telegraph Company, Bell connected his iron-box receiver to the line at Paris, Ontario, while the triple-mouthpiece transmitter was connected at Brantford, eight miles away. As Bell listened at the receiver amid the crowd of excited Paris townspeople jamming the telegraph office that evening, he at first heard "perfectly deafening noises . . . explosive sounds like the discharge of distant artillery . . . mixed up with a continuous crackling noise of an indescribable character." Through all this, he could hear "vocal sounds in a far-away sort of manner." But he had provided for a quick change from low-resistance to high-resistance coils on both the transmitter and receiver electromagnets. He signaled for this by telegraph. When the change was made, voices came through clear and strong (though only occasionally were complete sentences intelligible), and the crackling interference became less annoying. He recognized his Uncle David's voice, as expected. Others followed. Then, after a pause, came what sounded like the voice of his father, who had not been expected there. At Bell's telegraphed inquiry, the answer came that he had guessed right; his father had indeed walked to town and joined the group. By the time the fascinated listeners at Paris and the exhilarated performers at Brantford were content to stop, three hours instead of the allotted one had passed.

In late August, Bell came back to Boston to confront the dregs of summer heat, ungraded exam papers, and unanswered letters. But presently the heat wave broke, and Mabel pitched in to help with the letters. More important, Hubbard and Sanders, at Bell's urging, persuaded young Tom Watson to forsake his steady job with Williams and risk full time commitment to telephone work, in return for his Williams pay of three dollars a day and the added inducements of free room and board and a one-tenth interest in all of Bell's patents, including the telephone. Any patents taken out by Watson were to be the property of the "Bell Patent Association" (as the group was later referred to); and since these eventually numbered about sixty, Watson felt in the end that he had given as good value as he got. Watson had been commuting from Salem, but a vacancy occurred in the other half of the Exeter Place boardinghouse attic that housed Bell's lab and rooms, and so Watson moved in. On September 11 he and Bell resumed their experiments. On that day Bell reported himself "almost in despair at the condition of professional matters," but to Mabel he seemed "brighter and better than I have seen him for some time." His "arrangements with Mr. Watson," she added, were "already a great relief and help to him."

Under pressure from Hubbard, Bell and Watson passed the remainder of September mostly in unproductive experiments with armatures for intermittent-current transmission. On October 1, Bell put these firmly aside and concentrated on the telephone. A review of his laboratory notebook reminded him of the July 11 experiment with a large metal disk on the transmitter membrane. He could not recall nor imagine why it had been relinquished, so he tried it again and found it, as before, a great improvement. Between two rooms of the Exeter Place house on Friday, October 6, 1876, Bell and Watson held history's first telephone conversation, using two membrane transmitters alternately as transmitters and receivers, one of them having a large disk of thin steel glued to the membrane. "The utterance was perfectly distinct," Bell wrote his parents on Sunday. "If we can only keep it always so our fortunes are made. The success (pecuniarily) of Telegraphy [that is, telephony] is no longer an uncertainty. I *know* that my fortune is in my own hands. I *know* that complete and perfect success is close at hand."

On Monday, October 9, an improved instrument was used on each end of the line, and conversation became even easier. By permission of the Walworth Manufacturing Company, Bell took one instrument to the company's Kilby Street office that evening, while Watson went with the other to the company's East Cambridge factory, which was linked with the office by a two-mile private telegraph line. Watson in Cambridge at first heard nothing and had to cajole an incredulous watchman into letting him search the factory for the trouble, which turned out to be a telegraph relay. Once that was cut out, Watson listened again and heard Bell still shouting hoarsely and impatiently at the Boston end of the line. For some time, each recorded what he said and heard; and the lengthy notes later appeared in parallel columns of the *Boston Advertiser* to attest the almost perfect intelligibility of transmission. The two men chatted long after their note-taking ceased, reveling in their unprecedented triumph over separation by distance. Bell next day called the occasion "the proudest day of my life, as marking the successful completion of Telephony" — though he conceded that "much doubtless yet remains to be done in perfecting details of apparatus."

On Wednesday evening, Bell reported the Boston-Cambridge trial to the American Academy of Arts and Sciences and let the members try the instruments themselves; once again he won a round of applause. Mabel had "seldom seen him so pleased or excited about anything" as about the Boston-Cambridge experiment and the Academy's reaction. His elation spread to other things. On Friday, October 13, he opened at Boston University with what he thought was "the best lecture I ever gave"; on Saturday he lectured on Visible Speech with great success at the Framingham normal school. A few days later, on October 19 or 20, he finally eliminated the transmitter membrane altogether and tried a thin sheet of steel as both armature and diaphragm. "The articulation was heard . . . much more distinctly than we had heard before." Watson promptly made two wooden-box transmitters, each with a six-inch-square steel sheet over its open end, close to a single electromagnet pole. He also made a hole in the top of each box, through which to talk and listen.

At this heady moment, events combined to discourage experiments for two or three weeks. Watson spent a week in Philadelphia and soon afterwards came down with typhoid fever. Mabel prepared to leave for a transcontinental tour with her father, who had been appointed chairman of a special federal commission to investigate railway mail transportation; Alec therefore spent nearly every remaining evening with her. And what little time he could spare from his daily teaching duties went to drawing up specifications for a British patent.

By the second week in November, Mabel was gone, and Watson was back on the job, waking Bell every morning at seven and keeping him awake until he had washed and was safe from relapse. The Cambridge Observatory had a telegraph line for time signals during the day to various places in Boston, and the astronomer William A. Rogers readily cooperated with Bell and Watson in using the line for telephone trials during the evening. Bell ran a wire from the line's Boston terminal at the electrical firm of Stearns and George to his Exeter Place room, where friends gathered to marvel at his invention's growing effectiveness. He and Watson began systematically to vary each of the telephone's principal parts — electromagnet, coils, resistances, diaphragm, mouthpiece — while keeping the others constant, in an effort to determine its best form. For these tests Bell was stationed evenings at the Observatory (which was not far from the Hubbard house), and Watson at Exeter Place. By mid-November, Professor Rogers was able to tap casually on his instrument in Cambridge whenever he felt talkative, and Bell in Boston, if he chose to hear the signal, would chat with him by the hour as easily as if through a speaking tube. "The telephone reminds me of a child only it grows much more rapidly," wrote Bell after one such session on November 14. "What is before it in the future, no man can tell — but I see new possibilities before it — and new uses."

As November wore on, the Boston-Cambridge line tested Bell's determination as much as it did the telephone's performance. Everything that could go wrong on that line seemed to do so. Bell and Watson had to cope with leakage from the underwater cable and between wires, with short circuits between crossing wires, with interruptions by opened keys or by reversals of magnets or batteries, with unearthly and inexplicable caterwaulings in the background (and sometimes in the foreground) like a feline debate adjourned from Kilkenny to Hell. But the indefatigable pair tracked down and subdued every defect, and Bell philosophically congratulated himself on "gaining experience as to the kind of difficulties to be overcome in bringing Telephony into general use."

In the course of these trials, Bell found that the speaking hole in the top of the wooden box was a detriment, not a help, and that much better results came from simply speaking under a hand cupped against the steel diaphragm and listening with one ear pressed against the diaphragm. Later he fastened one end of a speaking tube at that point. He learned by trial and error what proportion the resistance of the electromagnet coil should have to that of the total circuit. And he discovered what would be one of his last commercially significant contributions to telephony: that instead of sending a battery current through the electromagnet coil to magnetize an iron core, a permanently magnetized core would serve just as well to provide the magnetic field needed for the diaphragm-armature to generate an undulatory current in the coil.

In this last, Bell was returning to his earliest breakthrough in telephony, the mysteriously twanging reed of June 2, 1875. The weak undulatory current that had unexpectedly set that reed going, it will be recalled, had been induced by Watson's plucked reed, vibrating in the residual magnetic field of its own electromagnet, even though no battery current was flowing to excite the latter. Now and then thereafter, Bell had cut out the battery current from a circuit and heard, in full awareness of its cause, telephonic transmission on the same basis. He had gone on to try permanent magnets in the coils, but had given them up when electromagnetically magnetized cores seemed more effective. And in early July 1877 he had envisioned talking across the Atlantic on the strength of residual magnetism. In the chastened, even somber mood of November 12, however, Bell was startled when, with the battery cut out, residual magnetism sent a Moody and Sankey hymn across the Charles. The improvements lately made in other elements of the telephone, he now perceived, might make permanent magnets worth trying again. Sometime in December, Watson, who made a practice of reading Public Library books on electricity, came across a description of a so-called "quick-acting" permanent magnet, the designation of which caught his attention; it was simply a horseshoe magnet of laminated steel with soft-iron cores fastened to the ends to carry the coils. Watson made one that worked so well that by early January batteries were omitted entirely. Not until variable-resistance telephones began to displace the magneto type in commercial use some two years later were batteries reintroduced into telephony.

As for the variable-resistance method, Bell remained aware of it as an alternative. In mid-November, reflecting on certain inherent, though not

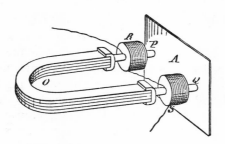

The "quick-acting" compound
magnet, about January 1, 1877

serious, distortions of sounds by magneto phones, Bell turned his mind to
an analysis of the action of a liquid variable-resistance transmitter. But he
concluded that the distortions by that method would be no less. While not
specifying other objections to it, he must have been influenced by the
growing practical effectiveness of the simpler, more durable, and less easily
disarranged magneto phones. At any rate, he made no recorded or recol-
lected experiments with variable resistance.

The great and eventually decisive advantage of variable-resistance transmission — the fact that it permitted an increase in the strength of the telephonic signal as desired — must have nagged at Bell's mind for a time. In
the latter part of December, he and Watson developed a miniature wet-cell
battery "about three times the size of a thimble" that seemed as effective
over a long line as the old type. This suggested to Bell that the undulatory
current might be obtained in any desired strength by vibrating "the battery
itself," by which he presumably meant one or both of the electrodes. But
the usual Christmas visit to Brantford, followed by patent business and
other matters, interrupted (permanently, as it turned out) a line of investigation that might well have led him to a practical variable-resistance transmitter.

On November 27, Bell and others in Boston held easy conversation over
the Eastern Railroad telegraph wire with Watson in Salem, sixteen miles
away. "Professor Bell," reported the *Boston Post*, "doubts not that he will
ultimately be able to chat pleasantly with friends in Europe while sitting
comfortably in his Boston home." That he meant it is suggested by the
experiment of December 3, in which Watson stationed himself on the same
railroad wire, but this time 143 miles away at North Conway, New Hampshire (thereby conceiving what turned out to be a lifelong passion for
mountains). "We could hear each other," Watson remembered many years
later, "but the telegraph line was in such bad shape with its high resistance
and rusty joints that the talking was unsatisfactory to both of us." That un-

derstated it. "A roaring rushing sound like wind mingled with the crashing of branches and all the noises of a storm utterly prevented us from hearing the faintest trace of Mr. Watson's voice," Bell wrote Mabel three days afterward. "It seemed as if a cyclone had been imported express by telegraph for the occasion." But at last, after a mortifying hour of tinkering, Watson could be heard faintly through the telephonic storm singing "The Last Rose of Summer." After another half hour, Watson and Bell managed to hold a ragged conversation. Years would have to pass and much be done before conversation over such long distances would be practicable. But in these trials the permanent magnet arrangement gave as good results as the main-line current and thus hastened the adoption of permanent magnets in magneto phones.

On the morning after the Boston-Salem tests, four Boston newspapers carried full accounts. Journalists now knew what at least two lecture bureaus had guessed early in July — that the telephone would be prime news once its existence was known and its authenticity established. The press silence of July and August, in retrospect so unaccountable, had at length been decisively broken in September — and by developments in two different quarters at that.

The Brantford trials of August awakened the *Scientific American* to the telephone. That journal, which reached inventors, manufacturers, businessmen, and newspaper editors, as well as many others who had merely an onlooker's interest in technology, on September 9 ran an article about the trials (citing the *Toronto Globe* account) and described the magneto phones. In its supplement of November 25, it reprinted the Boston-Cambridge conversation of October 9.

Even more effective was the fanfare sounded in August by Sir William Thomson's opening address, on science in America, to the annual meeting of the British Association for the Advancement of Science. After praising Elisha Gray's "spendidly worked-out electric telephone," by which he meant multiple telegraph, Thomson described the feats of articulation he had heard performed by Bell's invention, which he referred to not as a "telephone" but as "the greatest by far of all the marvels of the electric telegraph." Thomson may not have helped to clarify nomenclature, but his remarks, so authoritative in their source, so solemn in their setting, so categorical in their praise, scotched for those who read them any suspicion that Bell was either a hoaxer or a crackpot. The English journal *Nature*, leading organ of the British scientific community as a whole, published the address on September 14. Some weeks later, having just received Bell's May 10 American Academy paper, Alexander Ellis wrote the elder Bell from

England; what Sir William Thomson had said about the telephone, Ellis thought, "was enough to show the great scientific value of the discovery independently of its practical results." In "mere acoustics," Ellis pointed out, the telephone constituted "a most beautiful confirmation of Helmholtz's vowel theory." By the end of September, the *Nature* report had been picked up and the passage on Bell reprinted by the *Boston Advertiser* and other American publications. Late in December, a report over Sir William's signature accompanied a Centennial Exhibition award for the Bell telephone, which Sir William saw as being "of transcendent scientific interest." He and other judges, he reported, had been "astonished and delighted," and he expected Bell eventually to extend the telephone's range to "hundreds of miles."

By mid-December, Bell had heard from a professional lecturer that the mention of telephony would draw crowds. "In Salem," he wrote Mabel, "his audience *applauded* and *cheered* at the mention of my name."

The response of the American press and public to Sir William's endorsement encouraged Bell to look for material rewards, which were his self-imposed prerequisite for marriage. "The telephone is mixed up in a most curious way in my thoughts with you," he wrote Mabel in November. "Even in its present condition I think the instruments can be made a commercial success — so I give you fair warning that it won't be very long before I claim a certain promise — oh! I forgot! It was I that made the promise and not you! However I shall claim it all the same!" But something else also made him long for an early profit from his invention. "I want to get enough," he wrote Mabel,

to take off the hardships of life and leave me free to follow out the ideas that interest me most. Of one thing I become more sure every day — that my interest in the Deaf is to be a life-long thing with me. I see so much to be done — and so few to do it — so few *qualified* to do it. I shall never leave this work — and you must settle down to the conviction that whatever successes I may meet with in life — pecuniarily or otherwise — your husband will always be known as a "teacher of deaf-mutes" — or interested in them.

And he went on to a plan of organizing day schools for deaf children in cities large enough to warrant them. Parents were to be mobilized in the cause of giving their deaf children schooling without depriving them of the great advantages of home and family life. Bell would train the teachers. He was already trying to organize such schools in Chicago, St. Louis, and Marquette, Michigan.

Elisha Gray tried to console himself with the notion that he had lost little by Bell's telephone priority. "As to Bell's talking telegraph," Gray wrote his attorney on November 1, "it only creates interest in scientific circles, and, as a scientific toy, it is beautiful; but we can already do more with a wire in a given time than by talking, so that its commercial value will be limited so far at least as it relates to the telegraphic service." There were already others besides Bell who would have disagreed. After widespread press coverage of the October 9 trials, Stearns and George, the partners whose electrical firm in Pearl Street served as Boston terminus for the Cambridge Observatory wire and who had done some work for Bell, had been approached by would-be telephone customers who believed them to be dealers or agents for Bell. By early November, one company had ordered ten instruments from them to be delivered as soon as available.

"I often feel like giving up in despair as I note the gathering pile of letters upon my table," Bell wrote his mother at the end of November. "Letters from friends — letters from strangers — business letters — telephonic letters from all parts of the country — offers to buy telephones — offers for the right to use them in certain parts of the country (I like such letters as these — they show the dawn of a brighter day) — letters from graduates now at work with Visible Speech requesting advice."

Among Bell's new correspondents was John Ponton, born in Edinburgh five years before Bell and now editor of the *Titusville* (Pennsylvania) *Morning Herald.* In his editorial capacity, Ponton had seen the September press accounts of the telephone and had promptly written Bell to propose a Titusville-Boston test. By the end of October, Ponton was applying for the right to introduce the telephone commercially into the Pennsylvania oil regions around Titusville. His view of its prospects there delighted Bell, and also his suggestion of routing telephone calls through a central exchange, rather than putting up private lines each of which would connect only two points. Ponton's plan, Bell wrote Mabel, "is exactly what has always struck me as the most feasible method of bringing in an immediate return. When people can order everything they want from the stores without leaving home and chat comfortably with each other by telegraph over some bit of gossip, every person will desire to put money in our pockets by having telephones." Bell dutifully referred Ponton to Gardiner Hubbard in all his proposals, even including that of a long-distance trial. But he cherished Ponton's portrait of him as a prospective "millionaire."

Bell looked forward impatiently to foreign profits also. His agreement with Hubbard and Sanders covered only American rights to the telephone, and so he needed other backers for overseas enterprise. Hearing nothing from or about the Herdmans, Bell drafted an agreement with the British actor and impresario Dion Boucicault, then playing in Boston; but Boucicault backed out before signing, doubtless to his later regret. Bell had already drawn up specifications for a British patent and employed a solicitor, who filed them on December 9. Meanwhile, pending backers, Bell began saving up for patent fees in other nations.

That December, Watson continued to rout Bell out of bed as early as he could, and Bell assured Mabel that he was giving mornings and evenings to improving the telephone instruments. But he also tried to keep up with his students, do some reading in the physiology of hearing, encourage the new day schools in Chicago and Marquette, design a "plaiting" machine for Mrs. Hubbard's Christmas present, answer letters, and correspond with Mabel. At the Centennial Exhibition, the Japanese commissioner had become interested in Visible Speech as a possible phonetic alphabet for the Japanese language as well as an aid to teaching speech to deaf Japanese; and the Japanese government now sent a student, Shūji Isawa, to Boston University for study with Bell. So crowded and various a schedule did not allow for radical advances in telephony. But Watson was on the job; and it seems to have been in the latter part of December that, following Bell's experimental line, he came across the "quick-acting" permanent magnet.

Bell took his usual Christmas vacation at Tutelo Heights, where he showed up with the remnant of a cold and was promptly doctored with homeopathic pills and mustard plasters. "On Saturday," he wrote Mabel, "I was a man, full six feet high with whiskers & moustache of the most unmistakable kind, and within twenty-four hours I have dwindled down into a little boy once more!" With fame beginning to touch him, he could afford to accept a little mothering with good humor. On December 28, back in Cambridge, he talked to Watson in Boston without difficulty over eight miles of wire, using only permanent magnets.

On the last day of 1876, Bell arrived in Washington to give testimony in the harmonic telegraph patent interference hearing. He got his first taste of what would become familiar in later years, and he rather liked it. In the office of his patent attorney, Anthony Pollok, he sat in a comfortable armchair and testified confidently, while a stenographer for each side recorded his words. Gray and his lawyer, William D. Baldwin, sat listening, and Pollok stood in the middle of the room with his feet apart, his hands in his pockets, and an expression of contented mastery on his face. Pollok's smugness had good grounds. Bell proved capable of responding to one of Pollok's leading questions from ten in the morning until four in the afternoon without a break. Western Union soon gave up the harmonic telegraph claims it had based on Edison's work. This narrowed the contest to one between Bell and Gray.

Bell and his counsel began to suspect that Western Union had not actually quit the harmonic telegraph fight, that it had instead dropped Edison and come to some understanding with Gray. But Bell now felt that "even if they have, it does not matter, as my latest inventions with which there is no *interference* render it a matter of slight importance how the matter goes." At last his harmonic multiple telegraphy was assuming the place in Bell's mind, and probably by now in Hubbard's also, that history would assign it so far as Bell was concerned: that of a stage in the evolution of the telephone.

Bell, Gray, La Cour, Edison, and other experimenters with harmonic telegraphy faced inherent difficulties that they did not fully understand. Bell had realized early that merely superimposing signals by interrupting the main-line current at different frequencies would soon reduce the current below effectiveness. He tried to counter this by linking his transmitters to the main line by induction only. Gray's gambit was to give each transmitter its own battery so that each added signal would increase, not reduce, the main current. After much trial and error, Bell and Gray also recognized that they had overestimated the very slight power needed to start and maintain resonant vibration and consequently to generate spurious signals by overtones; and they discovered the value of a purely sinusoidal current in minimizing the generation of such signals.

Despite such expedients, the tuning of receivers still required more delicacy and attention than was commercially practicable. Bell, Edison, and others furthermore did not understand the effect of capacitance in actual lines, which distorted and attenuated alternating current signals; and so they found difficulties in practice, which had not been present in laboratory lines using artificial resistances as the supposed equivalent of distances. Finally, the use of mechanical resonators, such as reeds or tuning forks, limited the precision of discrimination between signals and exposed the receivers to disarrangement or false response by such mechanical disturbances as air currents or jarring. Even after electrical resonant circuits replaced mechanical resonators in the nineties, twenty years of work, including the development of the electric wave filter, the vacuum tube modulator, and the vacuum tube oscillator, remained before the harmonic telegraph would become the modern, commercially important, frequency multiplex telegraph.

In downgrading multiple telegraphy, Bell did not have to foresee the magnitude of the work that remained to be done on it. News from Boston that first week of 1877 was enough to give telephony first place. Watson wrote that the new telephone instruments worked splendidly, better than any before, and he sent them forward to Washington. Then he constructed

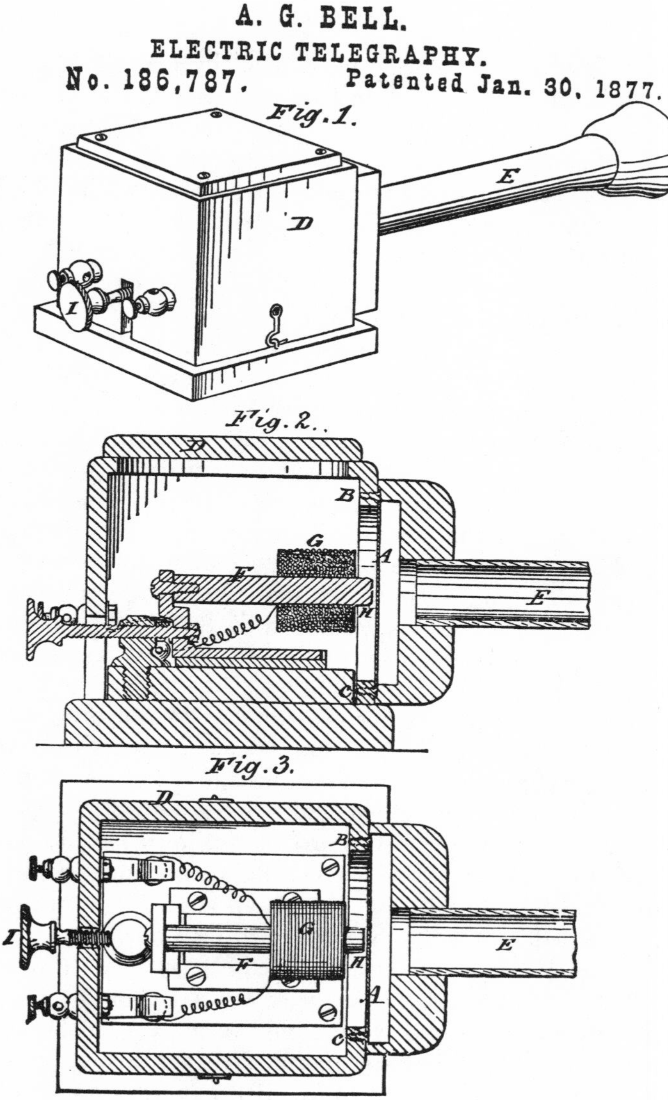

Fig.1.

Fig.2.

Fig.3.

A page from the second basic telephone patent

the model Bell had asked him to make just before leaving, one "using compound permanent magnets in place of batteries." On January 5 he telegraphed Bell, "New instruments exceeded greatest expectations. Secure patent without delay." Bell's attorneys Pollok and Bailey also urged Bell to draw up a new patent, incorporating telephonic advances since the first one and nailing down whatever seemed to have been left loose then. They had been uneasy at the brevity of Bell's description of the telephone in the first patent.

The urgency of the proposed new patent must also have been heightened by information from Percival D. Richards shortly after Christmas that a friend of his, Professor Amos E. Dolbear of Tufts College, had independently thought of using permanent magnets instead of electromagnets and intended soon to test his idea experimentally.

So Bell got to work on specifications with the help of Pollok and Bailey. "However much you may deplore the habit of night work," he wrote Mabel, "still it is the only way for me to accomplish any important thing." He finished the specifications on January 13, they were filed on January 15, and the patent — No. 186,787 — was issued on January 30, 1877. Covering the "box" phone as transmitter and receiver, the patent included the metallic diaphragm-armature and the permanent magnet with coil, including the "quick-acting" compound horseshoe form. Pollok and Bailey saw to it that the claims included, so far as consistent with the first patent, the general principles of telephony as thus far developed by Bell. They did their work well, and so Patent No. 186,787 became one of the two fundamental telephone patents.

In still another way, Bell improved his time in Washington. Hubbard's influence got him a government line, and he talked successfully from the War Department to an electrician's office. Joseph Henry had not seen Bell himself demonstrate the telephone in Philadelphia, but as a judge of electrical exhibits he had later seen and heard the realization of his own long-past dream of speech transmission and of his more recent advice to the young Boston University professor to go ahead and get the electrical knowledge he needed. At the Smithsonian Institution on January 13, Henry was delighted with the new instruments, called his daughters in to see them, and invited Bell to show them that night at a meeting of the Washington Philosophical Society. Bell did so with complete success, to the fascination of Washington's foremost scientists. Joseph Henry paid public tribute to "the value and astonishing character of Mr. Bell's discovery and invention."

Notwithstanding the pile of unanswered letters awaiting him, Bell was in high spirits when he picked up Mabel in New York on January 15 and escorted her back to Boston.

20

Circuits and Connections

Less than a week after leaving Washington, Bell wrote Gardiner Hubbard from Boston that "by increasing the resistance of the coils, by converting the cover into a sounding-box, and by supporting the magnet upon rubber," he and Watson had "increased the loudness of the sounds so much that they are audible all over my experimental room." Mabel had never seen him "so bright and encouraged about the future." One regret haunted him. "I only wish, darling," he wrote her, "that you could hear my instruments."

On January 20, the telephone transmitted what Bell later called "the first foreign language" spoken over it. Shūji Isawa, Bell's Japanese student at Boston University, brought in two compatriots studying at Harvard, Kentarō Kaneko and Jūtarō Komura, and found that Bell's undulatory current accommodated Japanese without difficulty. (Brantford tradition holds that William Johnson, an Indian chief who lived nearby, had discoursed in Mohawk over the telephone in the summer of 1876, but Bell either did not count Mohawk as "foreign" or — more likely — tradition dated the chief's talk a year too early.)

For a time, Bell and Watson went on with their systematic variation and testing of telephone elements. Increasing the thickness of the diaphragm, they found, brought greater clarity but less volume. Increasing the diameter, however, brought greater volume. So Bell, logically enough, had Watson make a diaphragm that was both thick and large, one of boiler plate a quarter inch thick and two feet square; fortunately for future users, it turned out to be no improvement. Increasing the resistance of the coils also seemed to make transmission louder. "I believe," Bell wrote on January 21, "we can make the vocal sounds of almost any loudness we desire by increasing the resistance of the coils . . . sufficiently."

This belief must have tended to confirm Bell's policy of sticking with the magneto phone rather than exploring variable-resistance transmission. The path ahead thus seemed one of plodding, of trial-and-error refinements rather than exciting breakthroughs. As early as January, probably reflecting Bell's mood, Mabel Hubbard had referred to the telephone as "very nearly done now." Certainly Bell's telephone research became sporadic after mid-February, and by Bell's own admission, such as went forward owed as much to Watson as to Bell.

The letup in Bell's research can be explained also by the demands of other activities. Mabel occupied some of his time and more of his thoughts. His Boston University class took up several hours a week. More than that, "letters — letters — letters — continue to pour in!" he complained. Early in March eighty-three letters piled up in four days, and in mid-April he mentioned a stack of sixty waiting to be answered. And most time-consuming of all, there was the lecture circuit.

Mabel's mother had disliked the idea of Bell as a popular lecturer from the moment it was first seriously broached in July 1876. She thought it beneath his dignity as a scientist and educator. Her husband shared her misgivings. By early 1877, moreover, the telephone scarcely needed an elaborate publicity campaign to attract attention; newspapers throughout the nation were copying reports of its triumphs. But aside from Bell's own taste for the work, lecturing seemed his readiest source of money, badly wanted for patent expenses abroad, as a release from nontelephonic work, and as a condition for marrying Mabel.

Early in February he borrowed two hundred dollars from his father toward foreign patents, but a misunderstanding about it led to a sharp exchange and the huffy return of the check. ("Muddle, muddle, muddle!" wrote his father; "when will you learn wisdom and common sense?") Applications for telephones or for the right to supply them at such places as Detroit, Akron, and Syracuse came in upon Bell in growing numbers; but he referred them all, presently by a printed postal card, to Gardiner Hubbard in Washington, who withheld action pending refinement of the instruments and formulation of a commercial policy. One question may have been whether to rent telephones or sell them outright; another was of how much control to surrender for capital and in what form to do so. "I do wish your father would decide at once what to do with the invention," Bell wrote Mabel on February 13; "applications for telephones continue to pour in, and we could be making money *now* if we chose."

By the end of January the telephone was reliable, audible at a greater distance from the receiver than before, and hence adaptable to popular entertainment. At the office of the Boston Rubber Shoe Company on the last day of the month, Bell gave a demonstration, using a private wire to the home of the company's president Elisha Converse in Malden, about six miles north, where Watson was stationed. The listeners in Boston, comprising businessmen, Midwestern railroad officials, and electricians, smiled (said the *Boston Transcript*) with "mingled pleasure and surprise" at the intelligibility of Watson's words and the ease of conversation with him. More suggestive of lecture possibilities, they listened as a group "with rapt attention" to the voice of a "fair cantatrice" in Malden singing "The Last Rose of Summer" with "a distinctness equal to that attainable in the more distant parts of a large concert room."

The clincher came on the evening of February 12 at Lyceum Hall in Salem, where Bell, without compensation, gave a telephone demonstration and lecture as part of a series sponsored by the Essex Institute. The Institute sold all its tickets in advance, standees packed the aisles and doorways, and a large crowd was turned away at the door. Mabel came to watch. The stage remained empty for twenty minutes after the scheduled time of beginning, and she was distressed to feel the pounding of canes and the stamping of feet. There had been a brief mixup in connecting the Atlantic & Pacific Telegraph Company line, lent for the occasion, with the little room eighteen miles away in Exeter Place, Boston, where Watson and a few others waited. But at last Bell appeared and placed a brown box on a table.

There the box sat portentously while Bell gave an account of his researches. Displaying the same prudence as in his American Academy debut nearly a year before, Bell began the demonstration with the dependable tone of an intermittent current from Boston, "a noise very similar to a horn," wrote the *Boston Globe* reporter. This simple fanfare brought forth a great burst of applause. "Auld Lang Syne" and "Yankee Doodle" followed on the "telephonic organ." Then Bell explained how he had learned to transmit vocal sounds and "paid a graceful tribute to Mr. Watson," a Salem man. From the box came Watson's voice shouting "Hoy! Hoy!" (the salutation that Bell had adopted from the start and insisted upon all his life thereafter, notwithstanding the public's infatuation with "Hello").

"As I placed my mouth to the instrument," Bell wrote next day, "it seemed as if an electric thrill went through the audience, and that they recognized for the first time what was meant by the telephone." The sound of Watson's songs, remarks, and news report of an engineers' strike just begun on the Boston & Maine Railroad could be heard throughout the small hall, though understood only by those nearest the receiver. Several prominent Salem men spoke to Watson, the Reverend E. S. Atwood perhaps achieving a historic telephone first by asking Watson if it was raining at

The Salem lecture, February 12, 1877

his end of the line. (It was not.) Watson recognized the voice, or at least the accent, of Shūji Isawa speaking from Salem. The *Globe* reporter pronounced the affair "an unqualified success."

People crowded up onto the stage afterward; and not even the ruse of turning down the gas and pretending to remove the telephone discouraged a score of diehards. Their persistence was rewarded by a chance to hear Bell dictate the *Globe* reporter's enthusiastic account to Watson and a shorthand stenographer in Boston. The report appeared next day under the proud heading: "SENT BY TELEPHONE. *The First Newspaper Despatch Sent by a Human Voice Over the Wires.*"

Interest in the Salem lecture reached far beyond the *Globe*'s readership. Bell pasted stories from a dozen newspapers and journals into his scrapbook. The *Springfield* (Massachusetts) *Republican*, which had a national circulation, let its imagination slip the leash. Political campaigns might be transformed, it suggested; if the disputed presidential election were awarded to neither Tilden nor Hayes by March 4, a half-dozen speakers by telephone might be enough for the new election. Good music might be popularized by telephone broadcasting. Sermons might be carried by wire from the preacher's study to a whole parish. Newspapers might be supplemented by broadcasting of news commentaries. "Infinite are the uses to which the new invention could be put." The *New York Daily Graphic* carried an illustrated account of the Salem lecture, along with a full explanation of the telephone principle. At the end of March, *Leslie's Illustrated Weekly* and the *Scientific American* did likewise. In the latter, the story monopolized the whole first page and spilled over to others; "phenomenal capabilities" were seen in the new device. The *Athenaeum* of London picked up the story in March; and April in Paris brought an article in *La Nature* on "Le Télégraphe Parlant: Téléphone de M. A. Graham Bell." Just turned thirty, Bell was becoming internationally famous.

"Pecuniarily it must have been a good thing for the Essex Institute," Bell commented thoughtfully the day after the lecture. He suggested to the secretary of the Essex Institute that he might repeat it, since so many had been turned away; and thus prompted, the mayor and prominent citizens of Salem invited him to do so. This time the proceeds would be his. Five hundred tickets at fifty cents each would amply reward him for one night's work, he wrote Mabel; two or three such evenings would free him from teaching during the spring; and by summer, surely, the telephone would be earning money. Invitations to lecture had come from two other New England organizations, and he would ask two hundred dollars for each.

Not only the lecture but also the triumph was repeated at Salem on February 23. The account of the *Providence Star* reporter, under the heading "Salem Witchcraft," described Bell as "a tall, well-formed gentleman in graceful evening suit, with jet-black hair, side-whiskers and moustache, light complexion, forehead high and slightly retreating, nose aggressive and black eyes that could look through a water commissioner." His "scientifically beautiful utterance" was "of itself a pleasure worth going far for." This time the program from Boston was augmented by a male vocal trio and a cornet player. Again the little hall was filled with some five hundred people, who gave the demonstration "deafening applause." Bell cleared $149, the first money he had ever earned directly from the telephone. The first thing he bought with it, for $85, was a little silver model of a telephone for Mabel; forty-five years later she called it "perhaps the most historically interesting thing I have."

The silver telephone model, dated February 23, 1877

If Bell had not just then taken the plunge into lecturing for pay, an item in the *Chicago Tribune* of February 16, 1877, might well have pushed him into it:

The real inventor of the telephone — Mr. Elisha Gray, of Chicago — . . . concerns himself not at all about the spurious claims of Professor Bell. . . . Mr. Gray's claims . . . are officially approved in the Patent Office at Washington, and they have already brought in large returns in money as well as in reputation to the inventor. Talking by telegraph and other sport of that description Mr. Gray has not paid much attention to as yet.

(This, of course, used the term "telephone" for Gray's intermittent-current transmission of tones, but the public was unlikely to have understood that.) Then, on February 21, Gray wrote Bell for permission to demonstrate Bell's telephone in a public lecture, promising to give full credit for it to Bell, though Gray remarked ruefully that "I was unfortunate in being an hour or two behind you" in filing a caveat on the idea. "There is no evidence," he added, "that either knew that the other was working in this direction."

Bell hotly replied by telegram that Gray had the requested permission, provided he refuted the *Tribune*'s "libel" both in his lecture and in the *Tribune* itself. Gray answered that he had not seen the "libel," had never said a word against Bell in the public prints, and should not be held responsible for everything the press said. "I have always," he remarked, "defended you when I have heard disparaging remarks made about you . . . [and] am always willing to correct any wrong done you." This brought a more temperate, even apologetic letter from Bell, granting that Gray could not control the press and thanking him for his honorable policy, which Bell had himself endeavored to follow toward Gray. The use of the word "telephone," Bell pointed out, had created confusion. He knew nothing of Gray's caveat, and not even for sure that it was for a speaking telephone, only that "it had something to do with the vibration of a wire in water, and therefore conflicted with my patent."

Gray, mollified, replied from Chicago on March 5 that he had subsequently seen and deplored the *Tribune* item and had done Bell full justice in his lecture. Then he added a genial paragraph which, when at last his telephone disappointment festered into grievance, he would deeply regret having written:

Of course you have had no means of knowing what I had done in the matter of transmitting vocal sounds. When, however, you see the specification, you will see that the fundamental principles are contained therein. I do not, however, claim even the credit of inventing it, as I do not believe a mere description of an idea that has never been *reduced* to *practice* — in the *strict sense* of that phrase — should be dignified with the name invention.

In March, while Gray followed the lecture circuit, Bell struggled to catch up with schoolwork and other business. But in April and May, he delivered three public lectures in New York City, three in Boston, and one each in Providence, Lowell, New Haven, Springfield, and Lawrence. For at least half of them, being in larger halls than the one in Salem, he put up two or three additional wooden-box receivers above or behind the audience. The programs were elaborations of those at Salem. Bell would recount the work of others in the field, giving scrupulous credit to Reis, La Cour, and Gray for their work, but also pointing out the crucial distinction between their transmission of tones and his of speech. He would then describe his own researches, explain the workings of his invention, and forecast its busy future as a link to friends, family, police, firemen, shops, and business offices, all through a central exchange. His first lecture of the spring, at Providence, inflicted an excessively long theoretical discussion on the audience to cover up an hour's delay in getting adequate transmission over the snow-covered wires from Boston, but later ones (except the last) escaped this embarrassment. By May, he had also prepared lantern slides to illustrate technical points.

After Bell's formal talk would come the demonstration. Now and then, despite Mrs. Hubbard's allusions to Barnum, Bell tried more elaborate en-

Prof. Bell

My Dear Sir.

I have just rec.d yours of the 2nd inst. and I freely forgive you for any feeling your telegram had aroused. I found the article I suppose you referred to in the personal Column of the "Tribune" and am free to say it does you injustice.

I gave you full credit for the talking feature of the telephone, as you may have seen in the associated press dispatch that was sent to all the papers in the country— in my lecture in McCormick Hall Feb. 27th. There were four different—

tertainment than at Salem: the Boston Cadet Band, a Brown University quartet, a couple of Italian male opera singers. But the professional singers could not bring themselves to press their lips to a mouthpiece and extrude bel canto like spaghetti. In the end, the most reliable and effective turns proved to be the old telephonic organ, a cornet playing some universally familiar piece like "The Last Rose of Summer," and a song from Tom Watson. Watson's operatic pretensions were nil, but he had learned how to enunciate clearly from association with Bell, had had more practice in shooting the

*papers represented at the lecture
but only one—the tribune—alluded
to my mention of you— ~~except~~ except
the "press" dispatch. I described
your apparatus at length by
diagram.*

*Of course you have
had no means of knowing what I
had done in the matter of transmitting
vocal sounds. When however you
see the specification you will
see that the fundamental principles
are contained therein. I do not
however claim even the credit of
inventing it, as I do not believe
a mere description of an idea that
has never been reduced to practice—
in the strict sense of that phrase—
should be dignified with the name
invention.*

*Yours very truly,
Elisha Gray*

rapids of the undulatory current than anyone but its discoverer, and could therefore launch into a Moody and Sankey hymn like "Hold the Fort" with little purity of tone but enough power to shiver the snow from Boston to Providence and bring down the house at the receiving end. (In order to forestall eviction from Exeter Place for excessive noise one night, he later recalled whimsically, he invented the world's first phone booth, made of blankets and a barrel hoop. It worked.)

Press reports remind us, as we need reminding, of how unprecedented in human experience was so totally disembodied a voice as that of the absent Mr. Watson. Edison's phonograph was yet to come. Speaking tubes and tin-can telephones were of a different order from this lightning translation of a human voice from one city to another and its reconstitution in a little box. "As Prof. Bell says," reported the *Providence Press*, "the publication of his discovery has spoiled one of the best spiritualistic opportunities ever known, and it is indeed difficult, hearing the sounds out of the mysterious box, to wholly resist the notion that the powers of darkness are somehow in league with it." The *Providence Star* went further, testifying on one occasion that "the sensation felt in talking through eighteen miles . . . leaves the spiritual seance away back in primeval darkness." In Manchester, New Hampshire, the first keening from the little box produced an "uncanny" sensation; "had the hall been darkened," the *Manchester Union* remarked, "we really believe some would have left unceremoniously." Nor was this mere up-country innocence. The *Boston Advertiser* thought that "the weirdness and novelty were something never before felt in Boston," and even in sophisticated New York City, the *New York Herald* called the effect "weird and almost supernatural."

But, as the *Manchester Union* put it, "this feeling soon wore off," and then the audiences paid a somewhat more critical attention. Luck was generally with Bell; indeed, he racked his brain in a vain effort to fathom why the telephone had suddenly worked so well at the first Providence lecture after the ominous opening delay. When transmission was poor, Bell told the audience in advance what they should expect to hear; and of course, most of them heard it. When all else failed, Watson had the knack of somehow making the audience catch one particular phrase: "Do you understand what I say?" — which, it will be remembered, also happened to be the first intelligible sentence he had transmitted to Bell on the historic afternoon of March 10, 1876, and the first sentence Sir William Thomson had heard at the Centennial trial. The press did not always praise the quality of transmission, on occasion likening it to "some one a mile away being smothered" or talking with "his mouth full and his head in a barrel," but it assured the reader that the sounds got more intelligible with some practice in listening. The facility of Bell and Watson in understanding each other was evidence of this.

Men of influence and celebrity attended. The Boston lecture of May 4 had to compete with a benefit reading by Oliver Wendell Holmes and other literary leaders; yet the audience had "Boston written all over its face," one newspaper reported, and included such notables as Charles Francis Adams, Jr. Despite oppressive heat and poor advance publicity, the New York lecture of May 17 drew the presidents of Columbia, New York

University, and Stevens Institute, Orton of Western Union and Eckert of the Atlantic & Pacific Telegraph Company, a deputation sent by the British government to study American telegraphic advances, several leading scientists, and Cyrus W. Field, the promoter of the Atlantic cable over which Bell so confidently looked forward to talking. Several of these and some other prominent men had been in the group of about fifty for whom Bell had given a successful demonstration at a New York hotel a few days before, transmission being from the A&P Telegraph office in Brooklyn.

A similar private demonstration in Providence on March 11 and subsequent public lectures there involved Bell and his telephone with a group of scientists in that city. Eli W. Blake, Jr., grandnephew of Eli Whitney and professor of physics at Brown University, had first seen Bell and the telephone at Exeter Place early in February. John Peirce, professor emeritus at Brown, a shy and unassuming bachelor, attended the March 11 demonstration. Their mutual friend William F. Channing, son of the principal founder of Unitarianism, William Ellery Channing, had earned an M.D. from the University of Pennsylvania and an independent income from collaboration with Moses Farmer in developing the electric fire alarm system.

For several months Blake, Peirce, and Channing worked at telephone experimentation for the excitement of the chase, assuring Bell that anything they bagged was his to use without obligation or even acknowledgment. For all their good intentions, they embarrassed Bell by coming up independently with several minor refinements, most of which he had already arrived at — in some cases months before. Bell publicly praised some interesting (though not especially significant) uses of the telephone as an instrument of scientific research, devised by Blake. John Peirce made the most original and valuable contribution: a mouthpiece cut like a very shallow funnel into the cover of a transmitter, with a small opening into an air chamber just deep enough for the diaphragm to vibrate freely. This much improved the clarity of transmission, and it became standard. For this Bell gave Peirce full public credit, though Peirce had steadfastly protested that he wanted nothing but his own private satisfaction at having made a small contribution to a great advance.

Channing turned out to be less generously self-effacing than Blake and Peirce. In May 1877 he designed what came to be called a "hand telephone" or "butter-stamp" telephone, shaped something like the receiver of an early twentieth-century phone and small enough to be held in the hand. Bell had independently designed a similar one at about the same time. But Bell's first model used a narrow horseshoe magnet, whereas Channing's used a straight bar magnet. The latter was adopted for commercial telephones during the year or two remaining before variable-resistance phones displaced the magneto type. Although Bell had used straight bar magnets in other telephone models long before and had already ordered one tried in a second "hand telephone," Channing began to brood and then complain about what he considered Bell's failure to credit him with that feature. Bell was troubled by Channing's complaint but could see no justice in it. In 1901, Channing's obituary notice in the University of Pennsylvania *Alumni Register* gave him sole credit for making the telephone "commercially practicable," but the claim seems otherwise to have been buried with the claimant.

Channing, who like Bell had a notable father to outdo and who had already tasted the financial rewards of successful invention, may have been especially prone to envy the fame that was gathering around Bell that spring. Bell reported proudly to his father and mother that he had been elected to fill one of the vacancies, opened only by death, in the American Academy of Arts and Sciences. This, he confessed, "has been for the last two years the summit of my ambition." And his father replied, "You have fairly won the honour by rendering yourself one of the foremost men in the United States." Mabel, who had already refined the spelling of Alec's nickname to her taste, perceived that it was now or never for his public appellative. "Why do you let them speak of you as *A.* Graham," she wrote him just after the first Providence lecture. "I perfectly hate it when I think how handsome the full name is." He informed her at once that, as she commanded, he would now and henceforth style himself "Alexander Graham Bell" for the autograph hunters who were writing him and in all other public circumstances besides.

The public lectures undoubtedly hastened Bell's celebrity. But his prime motive in them had been to raise money. Even before the Providence debut, Hubbard had not been encouraging about that. Aside from the doubtful propriety of public shows, Hubbard told Bell frankly, "you have no more capacity to make a *fair* contract than you have to talk Hebrew or Chinese intelligently without ever having studied the language." Mrs. Hubbard added her own dash of cold water. In New York City, she wrote Bell at the very outset, Elisha Gray's lectures of early April had satiated curiosity, and so Bell would probably not draw an audience. And quite aside from these doubts, Bell faced a dilemma. "I often feel as if I shall go mad with the feverish anxiety of my unsettled life," he wrote Mabel. "I do so *long* to have a home of my own, with you to share it." But for that he needed money, and lecturing still seemed the quickest way to get it.

At the second Salem lecture Bell had met and been much taken with

Frederic Gower, the young editor of the *Providence Press*. It was Gower who proposed and arranged the first Providence lecture, which drew an audience of two or three thousand despite a spring snowstorm. "I like him exceedingly," wrote Bell of Gower, "both as a friend and as a business man." When a dual lecture — Bell in New Haven, Gower in Hartford — lost money under other management because of poor advertising (as Bell saw it), Bell made Gower the manager of his remaining engagements, except for New York. But the three-part Boston series, which Gower had expected would net at least $1200, maybe $2000 or $3000, cleared only $150 over expenses, apparently in part because of losses in other cities where Gower similarly lectured at the other end of the line from Bell. Already, it appeared, it was as much the inventor as the invention that drew crowds. By late May, even with Bell in person, the attendance was disappointing at the first lecture in New York, and Hubbard increased it at the second chiefly by giving away several hundred tickets.

"Alec is blue and bright by turns," Mabel noted in mid-May. The lecture platform was his element, but the dwindling financial returns depressed him. He saw marriage receding. Headaches occasionally plagued him, and then a nervous rash over much of his body. The crusher may well have been the fiasco at Lawrence on May 28, where Bell, assisted by Gower, was to lecture and introduce Watson's voice from Boston. Watson believed that telegraph operators along the line, having discovered that they could get some faint intimations of the program by connecting their highest resistance relays, overdid it and choked Watson's transmission off entirely. Whatever the cause, the consequence was a dead silence.

Thus ended Bell's telephone lectures, not with a bang and in fact, so far as concerned Watson, not even with a whimper.

Though the profits from lecturing fell short of Bell's hopes, Thomas Watson wrote long afterward that they made the difference between following Mrs. Hubbard's recommendation of selling telephones outright for an immediate return and Hubbard's plan of leasing them. No contemporary correspondence bears this out. But Watson's published recollection is decided and unequivocal; and its view of Mrs. Hubbard's motive, namely her impatience to see Alec in a financial position to marry Mabel, agrees with contemporary evidence. It may be that Alec's platform earnings, or the prospect of them, did soften Mrs. Hubbard's opposition in private family counsels to the plan of leasing, and that the decision was close enough to have turned on that.

Hubbard's own view is not hard to understand. He had been an attorney for the McKay Shoe Machinery Company (nucleus of the later

United Shoe Machinery Corporation), which had done very well by leasing its machines for a royalty on every pair of shoes made by them. Chauncey Smith, the telephone company's first counsel, had also been an attorney for the McKay Company, and he supported the leasing policy. So did John Ponton of Titusville, pointing out, as did Smith, that if telephones were sold outright, it would be difficult to prevent unauthorized manufacture of so simple a contrivance. Another result, perhaps not so clearly foreseen. at the start, was in the standardizing of equipment, which in turn facilitated both efficient repair and universal interconnection. Leasing also encouraged the rise of a single dominant corporation in the telephone field, as in shoe machinery.

Newspapers hailed the inauguration of the world's first regular telephone line on April 4, 1877, connecting Charles Williams's Boston shop with his Somerville home. "I went into his office this afternoon," Bell wrote Mabel that day, "and found him *talking to his wife by telephone.* He seemed as delighted as could be. The articulation was simply *perfect*, and they had no difficulty in understanding one another. The first Telephone line has now been erected *and the Telephone is in practical use!*" The widely reported success of the Williams line stirred up public interest in the telephone as a utility rather than a mere curiosity. Inquiries and orders poured in, and within a month the leasing plan became practice.

Williams, being the manufacturer of the telephones, did not pay for them and so could hardly be counted as the telephone's first regular customer. That place in history fell to a young friend of Williams, Roswell C. Downer, a junior member of the Boston banking firm of Stone and Downer. On or about May 1, two rented telephones were installed on a telegraph line connecting the firm's State Street office with the Downer home in Somerville. Roswell Downer and his younger brother Frank heartily endorsed the telephone and were so quoted in a circular put out by Hubbard in the latter part of May. Their eventual payment, however, was not to be the first income from a telephone customer, since James Emery, Jr., on May 30 paid twenty dollars in cash to Charles Williams, as a year's advance rental for a telephone connection installed early that month between his house and that of his brother Freeman in Charlestown. Not knowing what to do with the money, Williams carried it around in his pocket until he could consult Gardiner Hubbard and at his direction write it down in the Williams cash book under the date of June 8, 1877, as "To A. G. Bell & Co., telephone account from J. Emery, Jr., $20.00."

As Williams's uncertainty suggests, the spring of 1877 passed without a more formal organization of the telephone enterprise than the Patent Association of Bell, Sanders, Hubbard, and Watson. This meant a chronic shortage of money, especially if returns were to be through rentals rather than sales. Thomas Sanders had recently prospered in his leather business and so could put up money for modest current expenses. But to Anthony Pollok, in explanation of a delay in remitting legal fees, Gardiner Hubbard wrote early in July that "my money matters are . . . entirely deranged by the adverse circumstances of the last few years, and with a very large amount of property I am more in need of money than if I owned nothing, for I do owe some debts and have large sums to pay for taxes and have no income from my property."

At some period during the late fall or winter of 1876–1877, Hubbard seems to have offered Western Union all rights to the telephone for one hundred thousand dollars. As with Mrs. Hubbard's supposed attitude on the leasing question, Bell's private correspondence is strangely silent on this matter, though it was crucial to his own fortunes and could not have been carried through legally without his consent. Nevertheless, other concerned parties later testified to the offer and its rejection by President Orton of Western Union. Watson vividly recalled his own disappointment at losing the chance to realize ten thousand dollars for his interest in the telephone patent. Several factors may have entered into Orton's rejection of what seems in retrospect to have been one of the great financial bargains of the century. Possibly Orton's disinclination to do Gardiner Hubbard any favors had some influence. More likely, Western Union saw the telephone as either a passing fad or a disturbing influence rather than an opportunity, and felt that in any event it could be matched by other inventors or laid low by money-draining patent suits.

The great corporation already had Thomas Edison working on improvement of the telephone, and to good effect despite Edison's own deafness. Following up the idea of variable resistance, Edison spent the fall and winter of 1876–1877 — the probable period of Hubbard's negotiations with Western Union — trying various solid materials whose electrical resistance varied with pressure. A rheostat he used in his quadruplex telegraph experiments had introduced him to this property in graphite or carbon granules. On January 20, 1877, he "succeeded in conveying over wires many articulated sentences," using points sticking into a dish of loose carbon granules. That spring he progressed to a graphite disk, writing Peter A. Dowd of Western Union on May 14 that while the Edison telephone was not yet ready for introduction, it was better than Bell's. "You need have no alarm about Bell's monopoly," he assured Dowd. Edison went on that summer to have his staff try some two thousand chemicals as variable-resistance materials. Tinkering and experiment continued laboriously through the fall and winter. At last Edison arrived at a button molded from lampblack and fastened to a small metal diaphragm. In February 1878 he applied for a patent on this carbon transmitter. Edison had a habit of announcing success before he had quite achieved it, and some such false dawn months earlier had probably made Orton readier to reject Hubbard's offer.

During the early months of 1877, while Hubbard talked with applicants for agencies in particular localities, he also explored two broader propositions. One involved Elisha Converse of the Boston Rubber Company, who had for some years been associated with Hubbard in a Nova Scotia coal company. Converse had been impressed by the demonstration connecting his home and office in January. Furthermore, he wrote Hubbard a week later, "I like Mr. Bell very much, he is in every way a gentleman." Also involved was Governor Henry Howard of Rhode Island. Howard may have first heard about Bell from a predecessor in the governor's chair, Henry Lippitt, whose deaf daughter Jeannie had profited from Bell's teaching some years earlier. Howard told Bell that he suffered from heart disease, but that "after seeing your deaf-mute school and your telephone, *I want to live*, if it is only to see what is coming out of it all!" He had all the wealth he cared for, but wanted his name connected with something great like telephony. Hubbard approved of Howard's plan to join with Converse and others in a company to control both United States and foreign patents, "a company controlling several millions of dollars," as Bell paraphrased Howard's remarks. Two or three days later Howard seems to have talked a good deal smaller. "The commercial results of your inventions are as yet somewhat problematical," he wrote Bell, suggesting that immediate working capital be ten thousand dollars and adding, "I cannot afford to give much of my time to it, unless it is likely to prove an affair of magnitude." Whatever caused the shift in tone — perhaps a colder reception than expected from prospective investors — it presaged the eventual abandonment of the scheme, of which no more was heard after February.

The other major scheme originated with John Ponton. On March 8, 1877, Hubbard agreed to give Ponton a nine percent interest in the telephone patents, to be made up out of the shares of the other associates, if Ponton would raise ten thousand dollars in four months; later, by this agreement, a new company would be organized, Ponton and his backers to receive a specified additional interest in return for further capital. This agreement also fell through, because of a bank failure in Titusville and the feeling of those whom Ponton approached for funds that he was asking too much for too little in a bad time for business. Early in May, Ponton made another proposal, but Hubbard rejected it; and so Ponton settled for the consolation of the Titusville telephone agency. Hubbard had by then grown confident that capital could be raised on better terms than Ponton could offer. Orders were coming in, lines were being projected, agencies were being negotiated. "We shall probably keep the general control in our own hands," Hubbard wrote Ponton, "interesting parties in different cities to introduce the patent in their locality."

During June, therefore, Hubbard, Sanders, Bell, and Watson prepared for the formal creation of a telephone company all their own. On July 9, 1877, the Bell Telephone Company was born as a "voluntary association," unincorporated and therefore without any stated capitalization. Its birth certificate was a "Declaration of Trust" assigning all of Bell's telegraphic and telephonic patents, past and future, to Gardiner Hubbard as trustee. The patent associates were to get approximately the same proportion of the company's five thousand shares as their respective patent interests and would have proportional voting rights in electing the board of managers, under the authority and supervision of which the trustee would manage the company's affairs.

Early in May, perhaps encouraged by what at that point promised to be a goodly nest egg from lecturing, Gardiner and Gertrude Hubbard came over to the idea of a June wedding for Alec and Mabel. Then, dashed by the falling off of receipts, Alec himself hesitated at setting so early a date. Once again help came from Providence. As Bell told it jubilantly to Sarah Fuller on June 25, a Providence cotton and cotton goods broker named William H. Reynolds

tempted me to negotiate with him for the sale of a part interest in my English patent on the telephone and he has kept me [traveling?] backwards and forwards between Boston and Providence for the last week or ten days arranging matters — but at last the matter has been settled, and I have sold him a portion of my patent for five thousand dollars cash. The result is that I shall leave Boston on the eleventh of July *with my wife!* We shall spend the summer and autumn abroad, returning in October. The wedding is to be very quiet and very few people are to be asked but I *must* have my dear friend Miss Fuller.

Formally engaged for a year and a half, Alec and Mabel had already learned much about each other's ways and views. Mabel being settled in faith, Alec composed a careful report to her on the state of his religious thought in January 1876. It was then pretty much what it would be for the rest of his life.

"I cannot believe in the inherent wickedness of man," he told her. "The world seems very beautiful to me, and there seems to me to be more good about mankind per se than bad. . . . Concerning Death & Immortality,

Salvation, Faith and all the other points of theoretical religion, I know absolutely nothing & can frame no beliefs whatever." Men should be judged not by their religious beliefs but by their lives. He asked that much and no more for himself. He was glad for Mabel's sake that she had religious faith, and he did not want his agnosticism to weaken it. Neither did he want her to take him for what he was not. Two days later, in response to her further questions, he wrote that he hoped for life after death but could not accept it as certain. "My religious beliefs, or rather non-beliefs," he added, "are a source of great grief to my poor mother who prays constantly for her 'misguided son.'" Later still, he denied that science had blunted his love of nature. On the contrary, "I catch glimpses of the harmonies of nature, of how one part fits into another like the wheel-work of a complicated machine."

Bell's views were characteristic of most scientists of that time and since. As for Mabel, through a long, happy marriage she would hold to her beliefs and respect his, as he had hoped.

In politics Alec took a lighter tone. He cheerfully escorted Mabel, her sister, and her mother to see a grand torchlight procession in Boston for Hayes and Wheeler, the Republican candidates, during the campaign of 1876; and he thought the spectacle, as seen from the MIT building on Boylston Street, "magnificent," with its twenty thousand torchbearers, its banners and devices, its pyrotechnic wagons erupting crimson and blue fireballs. But he felt detached enough to tease Mabel with "Democratic heresies" when first reports had Tilden elected — prudently adding, however, that "I may as well . . . confess that I . . . would rather have had Hayes win than not." Hayes, of course, did win, and went on to become the first President with a telephone.

Somewhere between politics and religion, perhaps, stood the mother-in-law question. If Bell had been a man of smaller mind, Mrs. Hubbard might have ruffled him with her discerning eye for his foibles; but she saw the good in him too, and Bell appreciated that. The surest evidence of her basic feeling toward him, after all, was her happiness at his approaching marriage to Mabel. "You have no idea," Mabel wrote him, "how much she has helped you with Papa. I am afraid you would have never got along together but for her helping you with her knowledge of you both." Alec had no reservations: "I think I love your mother about as much as it is possible for me to love any one in this world, she is so sweet and good, and as unselfish as anyone can be." In some thirty years of life remaining to Gertrude Hubbard, nothing in her son-in-law's words or acts ever belied this profession. As for her husband, not withstanding his freely expressed disdain of

Alec's business ability, their mutual respect and affection would be evident through the years ahead.

The Hubbards, after dividing their time between Cambridge and Washington for years, were inclined to settle in Washington, now that fortune seemed at last assured. For that reason and because "I am hardly ever free from colds all winter long in Boston and neither are you, it seems," Mabel proposed that she and Alec become Washingtonians too. Alec had chosen Boston over Washington five years earlier and his preference had not changed, but he yielded without cavil and at once began plotting the new course. Remembering the invitation of five years earlier from the National Deaf-Mute College near Washington, he wrote President Edward M. Gallaudet, who replied at once, "I should be very happy to have you connected with the work of this college should you take up your residence in Washington."

"It is so wonderful [Alec] ever came to me," Mabel had written her cousin and confidant Caroline McCurdy in February. "Every day I see something new in him to love and admire. His wonderful talent and genius is but a small part of him. It is wonderful that he should be so clever, but far more so that he should not only be that but also so utterly without conceit of any kind, so very true, and as thoughtful for others as a woman, far more so than I." In this characterization, of course, we see the poetic license that precedes the marriage license. Such a state of grace could not be unremittingly sustained over a period of forty-five years by any man, including Alexander Graham Bell; and during those years Mabel Bell did not seem either surprised or disconcerted at discovering this. Indeed, she wrote Alec's mother two months before the wedding in a more realistic vein. Quoting Alec as saying that "like a true Briton his spirit all depends upon his having a good dinner," she added that "I am beginning to learn that my happiness in life will depend on how well I can feed him." Nevertheless, the truth in her girlish rhapsody turned out to be much more than a grain; and so to the end she remained deeply grateful, all things balanced, for what life had brought her.

Alec's present, wrote Mabel to her cousin Caroline a week before the wedding, was "an exquisite cross of eleven round pearls, the prettiest he could find in Boston." She could have added that he also made over to her all but ten of his 1507 shares in the Bell Telephone Company, constituting about thirty percent of the total. (He kept the ten shares as a sentimental token.) Her wedding gown, she wrote, "is perfectly simple, all one piece from neck to feet."

On the day of the wedding, July 11, 1877, young Tom Watson and his still younger assistant Eddie Wilson wore white gloves for the first time in their lives. "We didn't know," Watson remembered a half century later, "that at such occasions there was always a room for the men to dress in and as we were ashamed to wear our gloves in the horse car, we put them on behind a tree in front of the house before we went in."

The marriage took place at the Hubbard house on Brattle Street, Cambridge, in the middle bay of the big room with the crimson curtains, the room in which young Professor Bell had first called on Mabel Hubbard and had sung into the piano for her father's edification. Long afterward, when death had parted Alec from her, Mabel remembered the "lovely July evening" of her wedding, and how fragrant the room was with "Madonna lilies which the gardener had saved for the occasion."

From Niagara Falls, the honeymooners went to Brantford. At the insistence of the elder Bells, the Symonds girls observed the old Scottish custom of invoking good fortune and happiness upon the bride by breaking oatcakes above her head as she entered the house. Here for the first time Alec's wife met his mother (having already met his father in Boston). "Mrs. Bell is just as nice and kind as she can be, so bright and quick," Mabel wrote her own mother. Next day there was a party with some thirty people, including Chief Johnson, head of the Six Nations, who talked Mohawk over the telephone from Tutelo Heights to Brantford.

The newlyweds returned to Cambridge by train as the great railroad strike and urban riots of July 1877 convulsed the United States. The failure of the Boston & Maine engineers' strike in February had helped inoculate New England against the contagion, and so they came through safely. In Montreal, however, wrote Mabel, "we were so doubtful about our chances of reaching Boston undisturbed that Alec bought a revolver with ammunition enough to kill a hundred men, he said. I think he was rather disappointed not to have a chance to show it off."

In historical perspective, we can see that Alexander Graham Bell was more than an onlooker in that crisis. His telephone would be a major technological factor in the new urbanism, against which the upheaval of July 1877 was at least in part a visceral reaction. Without the telephone as its nervous system, the twentieth-century metropolis would have been stunted by congestion and slowed to the primordial pace of messengers and postmen. And the modern industrial age would have been born with cerebral palsy.

"The telephones are doing very well," Gardiner Hubbard had written Anthony Pollok on Independence Day, a week before the wedding. "There are nearly 100 in operation [actually more than 200, acccording to Watson later] in many different places, and they have not failed in a single instance. . . . We shall soon begin to reap the rewards of Alec's invention." On August 1, Hubbard as trustee reported to the Bell Telephone Company's board of managers. There were now 600 telephones in operation, still without a failure or complaint. (Watson later set the figure for that date at 778.) Charles Williams had the Boston agency, Gower had the rest of New England, Ponton had the oil regions of Pennsylvania. Thanks to Hubbard's travels on the Railway Mail Commission, agencies had also been parceled out for Albany, western New York State, New York City, Ohio and Indiana, and South Carolina, Georgia, and Florida, with negotiations afoot for Chicago, Pittsburgh, San Francisco, Baltimore, and Washington, D.C. Williams was manufacturing telephones at the rate of twenty-five a day.

In Boston at noon on the day of Hubbard's report, Bell attended the first shareholders' meeting of the Bell Telephone Company. He and the others present — Gardiner Hubbard, his brother Charles Eustis Hubbard (who had ten shares), Thomas Sanders, and Thomas Watson — elected themselves to the board of managers, and chose Sanders as treasurer. Tom Watson became "Superintendent," in charge of manufacturing, and Bell was appointed "Electrician," or what might now be called director of research and development.

In fact, however, it would be Tom Watson who filled the last-named position. For on August 4, 1877, almost exactly seven years after he had arrived in the New World, Alexander Graham Bell and his bride left New York aboard the steamship *Anchoria* for Plymouth, England.

21

Family, Fame, and Foreign Parts

Aboard the *Anchoria*, Bell got up at six every morning, took a bath, and then (Mabel choosing for the moment to tolerate his old habit) luxuriously went back to his bunk until lunch. There was dancing in the evenings, sometimes to Bell's piano accompaniment. He read one of the captain's books on navigation and at once drew up plans for a new steering apparatus, which seems to have sunk without a trace. He set up telephones, and so everyone aboard got telephone fever. The captain, who was tutoring Bell in navigation, made up his mind to have the *Anchoria* equipped with a telephone communications system as soon as possible.

Then came landfall. "I cannot tell you what a longing I have," Bell wrote his mother from shipboard, "to see again the places I remember so well, London, Bath, Edinburgh and Elgin. I don't know how it is, but Elgin bears the palm with me." The Bells proceeded at once to Glasgow, where Alec went "perfectly wild," Mabel reported, at being back in Scotland. He wanted her to like everything Scottish, especially Edinburgh things, almost begging her to like an Edinburgh breakfast roll. When they visited Edinburgh itself a month later, after an English interlude, Alec went first to his father's old house in Charlotte Square and then took his bride on a tour of the Old and New Towns.

A few days later, at the end of September, they went to Elgin and then to Alec's old haunts at Covesea, a row of a half-dozen whitewashed, thatched cottages, in one of which they rented a little room. Alec's plan was for them to rough it, catching and cooking their own fish. "Neither one of us," wrote Mabel, "ever saw a fish cooked, much less ever did it ourselves. Alec appears to think it a very simple operation, but I have an idea that the fish has to be opened and cleaned, and a part of its inside taken out first." Presently they accepted their landlady's offer to do the cooking and gave themselves over to rambling the cliffs and walking the beach. The

Covesea idyll lasted a week. They spent the last day on the beach, Alec being "wild and full of fun, though rather ashamed that the inventor of the telephone should go wading." But Mabel persuaded him "that he should not be the slave of his own position."

The newlyweds had already made what for Alec was a sentimental journey to Bath. It brought him painful memories of Edward's death and left him with a fearful headache. In London he went with Mabel to Highgate Cemetery and mused at the grave of his brothers and grandfather, marked with a gray, moss-grown slab and headstone. In this elegiac mood, Alec complained that most of his old friends were dead or gone. But he exaggerated the toll of seven years. Alexander Ellis was there and glad to see him. James Murray and his wife were little changed from the day Alec had stood up as best man at their wedding, though now they had five children. Susanna Hull was still alive and would remain so for another half century. And one Mr. McBurney came to dinner with his wife, née Marie Eccleston. McBurney was "a little man with a brown beard and ghastly white face." Mabel, who knew all about Alec's old, unhappy flirtation with Marie, serenely recorded his opinion that she "looks if anything younger and less stout than when he last saw her."

More than two months before that occasion, in late August, Mabel had begun to believe herself pregnant, and by early September she was sure. Mabel's condition, and later the baby itself, persuaded Alec to put off their return to America; though they had first meant to come back in November 1877, their sojourn eventually stretched to the end of October 1878. So their daughter was born in London on the evening of May 8, 1878. "Such a funny black little thing it is!" Alec wrote his mother. "Perfectly formed, with a full crop of dark hair, bluish eyes, and a complexion so swarthy that Mabel declares she has given birth to a *red Indian!*" "Alec," Mabel observed, "is at once so fond of it and yet so afraid of the poor little thing, and he hardly knows how to hold it." But he lost no time in discovering "that all its organs of speech and sight and hearing are perfect." After two weeks of consideration, they named the baby Elsie May — Elsie as a Scottish form of Grandmother Bell's name Eliza, and May as both Mabel's nickname and the month of birth. By September, Alec was writing: "I do so love little children, and I like nothing better than being among them. I can hardly wait for Elsie to quit the baby stage . . . [and to be] old enough for me *really* to love her."

They had set up housekeeping early in September at rented rooms in Jermyn Street, London. But Mabel wanted a house of her own, and Alec's habit of striding up and down late at night while cogitating put the other tenants strongly on her side. So late in November they moved into what Mabel called "our own dear little home" on West Cromwell Road in South Kensington, a newly furnished house four stories high with seventeen rooms and a large plot of ground, which rented for £225 a year. There was a room for each of the two servants, one of them being Mary Home, who had been with the Bell family for about forty years, including twenty-five as housekeeper for Alec's grandfather. Here they lived during the rest of their English year, except for a few weeks in the summer of 1878, which they spent at a little country hotel in Middlesex; by then, Mabel's sisters and mother had come over, and the women spent most of the summer days "lying on shawls spread out under the shade of some great tree in the meadow."

The Bells began by living better than they could afford, since Alec thought it wise "to make as good a show as possible." Besides, he was getting from fifteen to twenty-five pounds for lectures, and characteristically concluded that this income alone would support them in style. Then he began to worry and economize — which meant such expedients as traveling third class while engaging a seventeen-room house. Early in December, however, William Reynolds gave him a check for the full amount remaining due for Reynolds's share of the British patent rights. He put it out at interest and seems never again in his life to have felt financially insecure. "Do you know," Mabel wrote him the following summer apropos of engaging rooms, "you have got the reputation among us [Hubbards] of being very particular, much more so than we. Why is that? You never have lived as well as we — a few years ago Papa was a very wealthy man, and we had our saddle and carriage horses and summer house in Newport, and yet we don't mind living cheaply, as you do."

Mary Home turned out to be so helpful that at last she was persuaded to go back to America with the Bells. Mabel's knottiest household problem consisted of routing Alec out of bed at a respectably early time in the morning. At Jermyn Street he had got into the habit of breakfasting at eleven, two hours later than Mabel. "He would complain of sleepless nights and a bad headache," wrote Mabel, "and I never had the heart to pull him out, in fact I couldn't." At West Cromwell Road, she prevailed somehow, and they breakfasted together at eight-thirty. "He often feels cross and headachy when I awake him and begs hard to stay in bed, but if I am firm, after breakfast the headache has quite disappeared, and he is bright and thankful he has been awakened." Yet, she confessed, "it is hard work and tears are spent over it sometimes." The years to come would show that she had won a battle but not the campaign.

So too with the problem of his weight. The slender, six-foot bachelor who took stairs two at a time, executed a Mohawk war dance at moments

Alexander Graham Bell, 1877

of triumph, and impressed lecture audiences with the athletic energy of his bearing had now, as a family man and public figure, begun to sacrifice agility to appetite. Mabel did not approve. "I never saw him look better, but he is growing awfully stout," she noted in London as early as mid-September 1877. Writing his mother later that month from Scotland, she was more circumstantial:

Alec . . . is perfectly happy with his Edinburgh rolls, Scotch oatmeal porridge and red herring. Last night he swallowed a whole dish of finnan haddock which was intended for us both. In fact Alec is growing tremendously stout, and can hardly get his wedding trousers on now. I remember your warning long ago and scold just as hard as ever I can, but it is no use. Alec proposes buying a book, teaching fat men to grow thin!

At Covesea an old friend of Alec's commented on the change. By late October his weight had risen from the 165 pounds of his wedding day to 201. On or about Christmas Day his trousers burst for the third time, and he had new clothes made. When he reached 214 pounds in the following summer, he managed to trim down to 200; but he was doomed never to get significantly below that weight and over the years would wax to 30 or 40 or sometimes even 50 pounds above it. His whiskers grew with his weight. After a series of colds in the winter of 1877–1878, he yielded to Mabel's urgings and as a protective measure allowed his beard to fill in the space between his side-whiskers. With beard and bulk he had become as stately as his father and was well on his way to superseding George Bernard Shaw's characterization of Uncle David Bell as "by far the most majestic and imposing looking man that ever lived on this or any other planet."

Mabel never ceased to rejoice in her husband's impressive public presence. As for his private aspect, she wrote after nearly half a year of marriage: "He is just as lovely as ever he can be, and instead of finding more faults in him, as they say married people always find in each other, I only find more to love and admire. It seems to me I did not half know him when I married him." Alec himself testified two months later that "we are enjoying life about as much as it is possible to do."

Thanks initially to Sir William Thomson's praises, the telephone by itself would have made Bell a notable figure in England from the moment of his landing. In Plymouth, at the 1877 meeting of the British Association for the Advancement of Science, a paper on the telephone drew a large audience. "The telephone is beyond all measure the lion of the Association meeting," a local newspaper observed. On August 20, when the Bells themselves arrived in Plymouth for the meeting, Mabel wrote: "Alec is really the chief person here, everyone seeks to do him honor, he has been introduced to all the great people, Lord this, Sir that, and all are anxious and eager to speak to him. . . . When he went down to Table d'Hote this afternoon the whole large hall full of people turned around to stare at Alec."

But Bell, of course, did not leave it at that. He had already given a telephone demonstration to leading citizens of Glasgow, where the next morning's newspaper posters blazoned the word "Telephone." He gave another, along with a lecture, at the Association meeting. He did not accept all the lecture invitations that showered him thereafter, but over the next four months he gave at least ten telephone lectures in England and Scotland, mostly sponsored by scientific and technical societies, but drawing crowds of the general public to each — in two cases as many as two thousand. As in America, he took care to give due credit to his predecessors and now also to the Providence experimenters. And as at Salem, so many were turned from his lecture before the London Society of Arts and Manufactures that it was soon repeated. His practiced forecasts of what the telephone would mean in everyday life evoked cheers and even outbreaks of laughter, not the laughter of disbelief but that of sudden delight, the exuberance of Shakespeare's Puck, who could "put a girdle round about the earth in forty minutes."

Bell gave additional private demonstrations also, one from down in a Newcastle coal mine to the surface, one in London between divers and the surface of the Thames. Bell himself went down in a diver's helmet for a test and came up with bloodshot eyes and a headache. But the most notable private demonstration was the one for Queen Victoria.

Alec may not have remembered, but his father must have, that thirteen years earlier the Queen had chosen not to witness a demonstration of Visible Speech. Now it was by the Queen's invitation that in January 1878 Alec set up telephone connections between Cowes, Osborne Cottage, and the Council Room at Osborne House. In the Council Room on the evening of January 14, 1878, along with Princess Beatrice and the Duke of Connaught, Queen Victoria appeared in black silk and a widow's cap. Alec described her later to Mabel (who had made a trip to Paris to get a gown for the occasion and then had not been invited after all) as "humpy, stumpy, dumpy," with her ungloved hands as red, coarse, and fat as a washerwoman's and her face also fat and florid. Still he granted that Victoria's face was not coarse, that she was genial and dignified and, all in all, quite pleasing. He heard later by way of one of the royal household that Her Majesty had been much pleased with him personally. Her diary for that day described the telephone as "most extraordinary." The line to Cowes had gone dead, and so the waiting singers and musicians in Cowes and London did not get the royal ear. But conversation got through from Osborne Cottage, and also a song by Kate Field, an American writer and journalist whom Reynolds had hired to publicize the telephone. As Miss Field was about to sing "Kathleen Mavourneen," the Queen happened to be looking away, and so Bell, already habituated to his wife's deafness, touched the Queen's hand and offered her the instrument. This breach of court etiquette did not seem to faze the Queen, and for the onlookers it gave a fillip to the occasion. Within a couple of days, the story had Victoria smiling sweetly as Bell pulled at her arm.

Reynolds had chosen well in retaining Kate Field, who had already earned her pay by writing a score of articles on the telephone and inspiring others. She knew influential and articulate people in English literary and social circles and was herself currently writing regularly for the London *Times* and the *New York Herald*. Mabel liked her: one would "know her for an American a mile off; it's such a pleasure to see one." Though Miss Field was forty, "she seems thirty, has exquisite teeth, but lolls around on chairs in a most unladylike way." Bell's lectures, indeed his very presence, evidently helped keep the telephone before the British public. How much of the telephone's fast-growing notoriety otherwise was of Kate Field's contriving cannot be determined — which was one evidence of her professional skill. At any rate, as early as the end of 1877, *Punch* was including in its list of New Year's resolutions: "To make myself thoroughly acquainted with the Eastern Question in all its bearings, the relations between Capital and Labour, the principle and construction of the telephone."

Kate Field was planning a "Matinée Téléphonique" for the press, in order to "get one general chorus of gratuitous advertising" before the opening of Parliament captured the spotlight. The Queen's well-reported interest in the telephone helped make the affair a decided success. A few weeks later Miss Field put together her own articles and some earlier stories copied from Bell's scrapbook in a little six-penny pamphlet entitled *The History of Bell's Telephone.* An "Intercepted Letter" from a supposititious American lady tourist to a friend provided a thin commentary and connective. It included a speculation that light as well as sound would eventually be transmitted, and "while two persons, hundreds of miles apart, are talking together, they will actually *see* each other!" This was just the sort of notion Kate Field might have got from Bell himself.

Along with Miss Field's paid praise went the more credible flattery of imitation. By the end of February, Mabel found the shops full of "Domestic Telephones" — thread or string "telephones." "Wherever you go, on newspaper stands, at news stores, stationers, photographers, toy shops, fancy goods shops, you see the eternal little black box with red face, and the word 'Telephone' in large black letters. Advertisements say that 700,000 have been sold in a few weeks." "Telephone" had become an English household word; and to some degree, presumably, so had the name of Alexander Graham Bell — or "Graham Bell," as the English chose to call him.

"Business," Bell wrote Gardiner Hubbard in the fall of 1877, "is hateful to me at all times." Besides, it "would fetter me as an inventor." But "never having been engaged in any business in my life, no person can tell what dormant business ability I may possess!" The year that lay ahead would test his ability and confirm his prejudice.

At the time of his marriage, as security against possible misfortune to his and Mabel's other interests, Bell had made Gardiner Hubbard the sole trustee of his telephone rights in Great Britain, Germany, Austria, Belgium, and France. Alec and Mabel were to get the income up to three thousand dollars per year. Any excess was to be added to the principal until it reached two hundred thousand dollars. Thereafter they would receive all the income during their lives, and their children would get the principal afterward. Bell resolved not to meddle with or even inquire about Hubbard's disposition of the trust properties, but this did not inhibit him from trying to advance their fortunes otherwise. Indeed, because the trust was meant eventually for his children, Bell felt it all the more his duty to do so.

Bell therefore went forward in his association with William Reynolds. Pending Reynolds's arrival in England, Bell spoke for both of them to the callers who flocked to see him. When Reynolds showed up early in September 1877 it was none too soon for Mabel. "We have ever so many applications, some very important ones, and I wish he would see about them, and take some of the work off Alec," she had just written. But Bell soon had misgivings about Reynolds's executive ability, kindly and pleasant though Reynolds was personally. There was talk in late October of a company to be capitalized at £200,000, but it came to nothing. As inquiries and callers multiplied, Reynolds arranged for telephones to be manufactured by a rubber company that specialized in telegraph wire insulation. The demand was already so great as to incite infringements.

Prospects improved as 1878 began. Reynolds's retaining of Kate Field helped. As soon as the telephone had been successfully demonstrated to the Queen, Reynolds issued a prospectus for a proposed telephone company. Infringers were selling telephone parts in easily assembled kits on the pretext that these were not telephones as such, but Reynolds put a stop to this by initiating legal action against the more active and stubborn offenders, who yielded without a contest.

Meanwhile the very basis of such legal action was in hazard. Sir William Thomson's description of the telephone at the British Association meeting of 1876 had been published in English periodicals, and this prior publication had threatened to undo Bell's right to a British patent. Fortunately the published drawing had shown Sir William's iron-box receiver with its metallic diaphragm attached to the rim with a screw, and bent away from it (because of damage in transit), as if this were the proper arrangement. The British courts in their turn seized upon this accident to bend the law toward justice, holding that vital feature — the metallic diaphragm — not to have been legally disclosed. In conformity with British law, Bell cleared the way further by officially disclaiming certain nonessential patent claims that had undeniably been published. But several letters to *The Times* charging Bell with theft of the telephone idea from Wheatstone or Reis seized on these pro forma disclaimers as confessions of guilt. And all these troubles dampened the enthusiasm of prospective backers.

So instead of a £500,000 capitalization, which he had projected in January, Reynolds in March sold fifty-five percent of the English patent to a group of capitalists for £10,000 cash and £10,000 in another twelve months. Of this Bell received £500 at once for his expenses and services in England, and the Bell trust got something more than £3000.

Bell's services consisted, in the new backers' view, chiefly of his promotional contributions through lecturing. His principal technical contribution had been in the matter of interference by induction. He had heard the chattering of nearby telegraph messages on telephone lines in the United States and had met that by using a complete wire circuit, rather than a single wire grounded at each end, so that the interference, being induced in both the outgoing and incoming wires of the circuit and thus in opposing directions, largely neutralized itself. In Great Britain, where a larger number of telegraph wires were likely to be found close to a telephone line, the neutralization had to be more nearly perfect. Perceiving that this required the outgoing and return telephone wires each to be precisely the same distance from the disturbing wire, Bell hit upon the successful scheme of twisting them about each other inside a common insulation. He took out a British patent on this device in November 1877. But others as far back as Faraday had, unknown to Bell, proposed much the same remedy for similar interference. The new company felt that the patent could not be sustained against a serious challenge and so declined to buy it. While in England, Bell also experimented with a call bell device, a system of hotel annunciators for telephones, and a galvanometer for measuring the intensity of a telephonic current, none of which turned out to have commercial value.

At first Bell also did his best to make a business contribution. Two days after the entry of the British investors into the enterprise, Bell sent each of them a copy of a letter "concerning the future of the Electric Telephone" and some suggestions "in regard to the best mode of introducing the instrument to the Public." In it he elaborated on the ideas he had been presenting to his lecture audiences over the past year, especially that of a central exchange. Bell urged the leasing of lines, on no account letting control of them pass from the company's hands, and charging each customer either a fixed annual rental or a toll based on duration of calls. To root the system initially, he suggested free trial periods for a few of the principal shops, as an incentive for householders and other shops to acquire telephones. These well-conceived suggestions seem to have drawn no response.

Within the company Bell had an ally in his old London friend Adam

Scott, who had resumed his long-past custom of frequenting the Bell household in London and had also worked his way into Reynolds's favor and employment. By the end of April, Scott was temporary secretary of the new "Telephone Company, Limited." Scott's ideas coincided so exactly with Bell's that for several months Bell paid little or no attention to the company's affairs, unwittingly losing the confidence of his fellow directors by his neglect of meetings, appointments, and letters.

In July, however, the two friends joined in persuading members of Parliament to kill a bill absorbing the telephone system into the government system of mail and telegraphy. At that time Scott alerted Bell to the ineptitude and slovenly ethics of the company's manager, one McClure. Not only had McClure tried to cheat the company, but he had also, among other follies, favored merely making and selling telephones while leaving the lines and exchanges to others. Bell presented his charges against McClure in a privately printed letter to the directors, urging that McClure be fired immediately, and pointedly recommending Scott's ideas and ability. McClure was replaced by another man, though not by Scott. But in mid-October 1878 Bell resigned as a director of the company, because of "the gross mismanagement of the Company's business and the personal discourtesy with which I have been treated by the Board of Directors and by the Acting Manager."

McClure's successor turned out to be honest but incompetent; Reynolds, whose health had been poor all the while, bore out Bell's early doubts of his executive ability; and the directors, though able in their own fields, could not seem to grasp the needs and opportunities of the telephone business. By early 1879, moreover, Edison, inventing to order, had ingeniously bypassed the Bell electromagnetic receiver. Edison's receiver used the variations in the rubbing friction on the surface of a moist chalk drum when a varying electric current was passed through the point of contact. This device, though skittish and inconvenient, served its strategic purpose in forestalling a checkmate by the Bell patent holders in England and thus making the inevitable merger more expensive to the Bell group than had been anticipated.

In the summer of 1879, however, Gardiner Hubbard swooped down on the faltering English company as a trustee of the Bell interest, reorganized and revitalized it, and came home to fatten the trust with $50,000 in cash, as well as stock in the reorganized company worth $100,000 by the following summer.

The German patent had been lost because Bell, for lack of money, had not applied for it until August 1877, apparently too late under German law. This permitted the German electrical firm of Siemens and Halske to manufacture telephones and sell them in Germany and elsewhere without permission or royalty payments. Because in Austria periodic payments were required to keep a patent alive, Bell and Hubbard later believed Austrian rights also to have been lost after 1878; but in 1880 it turned out that the man Bell had given power of attorney there had kept them alive at his own expense.

In October 1877, hearing that Edison was trying to be first in France with the latest stage in his evolving variable-resistance riposte to the Bell transmitter, Bell sent instruments to be demonstrated by a French applicant for an agency, and shortly afterward he followed in person. He found the newspapers full of telephone articles, and the minister of war anxious to get telephones (having heard that Bismarck was using them). Cornelius Roosevelt, a competent and likable New Yorker who had acquired an interest in the French rights from Hubbard, arrived in December to discover the French patent in jeopardy and already being infringed because of late application. But Bell heartened Roosevelt for the struggle to save it, and Bell's acknowledged standing helped win government patronage.

Late in the summer of 1878, Gray and Edison, in the interest of Western Union, published insinuations in France against Bell's honor as an inventor. Frederic Gower by then had found the telephone lecture outmoded by the commercial success of its subject and so had come to France in the Bell cause, but presently he patented a minor variation on the telephone, set up his own "Gower system" in France, and was reputed eventually to have made considerable money. (Gower's later life was short but lively, including a marriage to the noted opera singer Lillian Nordica and a final disappearance aboard a balloon crossing the English Channel.) Despite these setbacks, Gardiner Hubbard in his transatlantic rescue mission of 1879 may have made some settlement with Roosevelt and Gower to the profit of the Bell trust.

A Norwegian civil engineer named Jens Hopstock on his own initiative took out Scandinavian patents in Bell's name after reading about telephones and making some himself. In October 1877 Bell gratefully gave Hopstock a two-year license for Scandinavia with an option to buy the Scandinavian rights afterward. As Hopstock was energetically beginning his promotion, word came that Hubbard, unaware of the arrangement, had licensed another firm in Scandinavia (though Scandinavia was not included in his trust). Someone presently gave Hubbard evidence of "duplicity" by Hopstock, who was soon afterward displaced by Reynolds in the attempt to organize the Scandinavian business.

As for other foreign rights not included in the trust, Bell, like the first Napoleon, parceled out nations as gifts to sundry relatives. Australia, for example, went to his Symonds cousins. His father got three-fourths of the Canadian rights and Charles Williams the remainder, in return for supplying a thousand telephones. Mrs. Hubbard was later given Bell's rights in Scandinavia, then sold them to Mabel for twenty thousand dollars in cash. Italy was saved when a wealthy Boston friend, one of the Sears family, volunteered to pay the patent expenses as a gift; Bell insisted, however, on regarding the money as a business loan. Then he gave Italy to his brother Melly's widow Carrie, now the wife of a Brantford farmer named George Ballachey. Assorted nations went to Uncle David, Bell cousins, and others. As in the case of the trust nations, the ultimate return from these high-sounding interests was variable, sometimes negligible, but the generosity of Bell's intentions was real in any case. Later he made over a number of such foreign rights to the Oriental Bell Telephone Company in return for stock, which he then distributed to members of the family in lieu of the original, more indeterminate boons.

Among Bell's casual dispensations were telephone rights in nations that had no patent systems or in which patent rights had already been forfeited. Meanwhile there were kaleidoscopic organizations, reorganizations, and mergers of national and international companies, involving stock issues, substitutions, dividends, divisions, and exchanges. Considering that, as in the Scandinavian case, even Hubbard did not always have clearly in mind which countries the trust covered, it is not surprising that Bell, despite his resolution of noninterference, grew uncontainably suspicious of the trust's condition. Hubbard himself had planted seeds of doubt in 1878, when in wrestling with what turned out to be his last personal financial crisis he got Bell's permission to borrow from the trust. In the summer of 1880 Bell's pent-up fretfulness burst out and nearly ruptured his relations with his father-in-law. Bell charged Hubbard with mishandling — he came close to saying misappropriating — a trust held for Hubbard's own daughter and grandchildren. Though deeply wounded by this, Hubbard made allowance for his son-in-law's temperament and paternal feelings and also drew upon his own in making a reproachful but self-controlled reply, which refuted the charges and pointed out his substantial salvage from the English venture. The air had been cleared, and no such clash troubled it thereafter. By 1887 the trust fund had grown nearly to its goal of two hundred thousand dollars, and after Hubbard's death in 1897 the family found that he had let it grow further to two hundred sixty thousand dollars.

His thrashing about in the financial thickets turned Bell's prejudice into conviction. A third of a century later he would write, "I am not a business man and must confess that financial dealings are distasteful to me and not

at all in my line." But business perplexities were not to be his only tele-
phone-generated headaches. As early as February 1878, William H. Preece,
a British electrical expert, accurately predicted other trials. "When once
. . . a new thing is shown to be true," he wrote, "a host of detractors
delight in proving that it is not new. The inventor is shown to be a
plagiarist or a purloiner or something worse. . . . Professor Bell will have
to go through all this."

Already a "host of detractors" had sought to transfer Bell's honors to
Gray or Reis or others and even to brand Bell a thief of ideas. Reis himself
was silent in death; and when Bell left for England, Gray had not yet gone
back on his acknowledgment of Bell's probity. But on March 26, 1878,
Western Union and its subsidiaries backed a laboriously assembled collection of claimants in a swarm of interferences filed against Bell's second basic
patent, No. 186,787. All of them challenged Bell's priority; two — Elisha
Gray and Amos Dolbear (of whom more will be said later) — also would
impugn Bell's personal character, or at least that of his associates.

As the summer of 1878 waned, Bell grew increasingly bitter. "The more
fame a man gets for an invention," he wrote, "the more does he become a
target for the world to shoot at." As for patents, "if my ideas are worth
patenting, let others do it. Let others endure the worry, the anxiety, and
the expense." The "feverish, anxious life" he had led since his marriage
already had begun to make him "irritable, peevish, and disgusted with life."

It was not merely the struggle for patent rights from which he turned.
He felt doubly vulnerable as a family man to the calumny that his great
invention had brought upon him. "Please don't be so distressed about that
article in the *Times*," he wrote Mabel.

I am beginning to be quite troubled too, just because you are. . . . Let the press
quarrel over the inventor of the telephone if it pleases. Why should it matter to
the world who invented the telephone so long as the world gets the benefit of it?
Why should it matter to me what the world says upon the subject so long as I
have obtained the object for which I laboured and have got you my sweet sweet
darling wife? And why should it matter so very much to you and to my little
Elsie so long as the pecuniary benefits of the invention are not taken from us —
and so long as you are conscious of my uprightness and integrity? . . . All questions of priority will soon be settled by the Patent Office. . . . Truth and Justice
will triumph in the end. . . . Let others vindicate my claims if they choose but
keep me out of the strife.

In this mood he renounced the subject that had absorbed him so long. "I
am sick of the telephone and have done with it altogether, excepting as a
play-thing to amuse my leisure moments." If he worked any more at telephony, "let it be from a love of science."

What, then, would he work at? Or would he work at anything to any effect, now that he need not? Mabel had relayed a perceptive warning two years before: "Mamma says your mind is so fertile it is always drawn off by every new idea that comes up. . . . But . . . you will never do anything of value that way. She thinks it well you have to support a wife. If I came to you rich, you might think you had nothing to do but please yourself flying about."

Bell saw the danger too. In his nocturnal pacings he may have wondered if he were now only a spent shot on a long, effortless, powerless trajectory to dust. In September 1878 he appealed to Mabel: "Make me *work*, there's a good little girl — at anything, it doesn't matter what, only make me work, so that I may be accomplishing something."

But he was already at work along the lines that would, as it turned out, chiefly engage him thenceforward: scientific speculation in the hope of greater fame, invention for the fun of it, and the education and general welfare of the deaf simply because he could not be happy in withholding any special power he felt in himself to relieve human misery.

As already noted, Bell's dreams of scientific glory had begun during his boyhood in Edinburgh, one of the world capitals of academic science. There he had been seized by boyish enthusiasms for botany and zoology and had assumed the title "professor of anatomy" in his circle's "Society for the Promotion of Fine Arts among Boys." In a dreamy way he had puzzled over the ways of wind and water at Bell's Mill and of birds in the air over Corstorphine Hill. At Elgin, in the ruins of Pluscarden Abbey, he had grappled for the meaning of the universe; and in his Weston House room, with his fork-twanging and throat-tapping dissection of vowels, he had unknowingly paralleled Helmholtz in acoustics. In London he had attended physiology and anatomy classes at the university and had hob-nobbed with eminent philologists. And in Boston, with the cachet of his university professorship, he had entered as an equal into a notable scientific community. Meager as his formal scientific training had been, Bell was not unreasonable in taking his science seriously.

But in his post-telephone years he took a turn that exposed his chief weakness as a scientist. That weakness was in mathematics. At the Royal High School, so far as schoolboy mathematics held any interest for him, it was as a grab bag of diverting riddles. Even the rigor of following a perceived method to its precisely correct result bored him. In science, mathematics must be both earnest and rigorous. As for the direction of his scientific ambitions, his telephone success seems to have pushed him toward physics. Bell had an intuitive sense of where science and technology were heading in his day; and so a recognition of physics as the most fertile field of post-Darwinian science may also have influenced him. But whatever the ratio of chance to choice, he did not come to terms with the fact that in physics, more than in most other sciences, higher mathematics was becoming more and more essential.

Bell's contemporary Thomas Edison, perhaps because his boyhood had not impressed him with the glamor of academic science, all but boasted of his mathematical incapacity. When he needed mathematics, he remarked contemptuously, he could hire mathematicians. Edison accepted his consequent limitations as a physicist. In the privacy of his notebooks Bell might admit his own limitations, but from time to time he would struggle to transcend them. Fortunately his Scottish common sense kept these efforts within the bounds of an occasional pastime and saved him from actually publishing cosmic pipe dreams that could have brought him only humiliation.

The most persistent of his speculations evidently went back well before 1876. In July of that year Bell bought a book reviewing recent advances in physics. "My beautiful new theory over which I have spent so many years of thought, my theory of the nature of attractive force with its explanation of Gravitation . . . is there, dimly outlined, by a man who lived *at the end of the last century!*" he wrote Mabel. "I had looked forward to the publication of these ideas at a future date as a means of placing me high among scientific men. Now I can at best only build upon another's foundation." Bell's "beautiful new theory" had indeed occurred to others, and not surprisingly. It was a symptom of the general craving in the heyday of Newtonian science for mechanical models to explain physical phenomena. Unable to understand gravitational force in mechanical terms, Bell had grandly postulated that it did not exist; and he set out to explain its apparent existence in terms of primordially imparted motion, luminiferous ether, and whatever other cogs and levers he had to invent.

He might have surmised something from the fact that a seed planted "at the end of the last century" had never sprouted. But if he had been so easily daunted he probably would not have invented the telephone. Instead he went on to inform Mabel that "I shall swallow my disappointment as best I can, read up what this man has written, and *work out my theory.*" Some forty years later he would still be working it out, and gravity would still be operative.

Early in his British sojourn, Bell saw Sir William Thomson's laboratory at the University of Glasgow and, as Mabel put it, "went wild over the wonders of science shown to him. . . . He says he never before appreciated his own knowledge of science. . . . He recognized at once everything shown him, though he had probably seen few of them before." Coming just afterward, the atmosphere at the British Association meeting and the respect he was accorded there brought on a notebook spasm of fruitless efforts, using simple algebra, to deduce the effects of the moon's attraction on atmospheric pressure. If he broached this to Thomson, Sir William probably handled him gently. Thomson liked Bell as a man, quite properly respected his scientific learning and ability in the fields of acoustics and vocal physiology, and greatly admired his telephone inspiration. On a second visit to Glasgow, Bell wrote Gardiner Hubbard proudly, "Sir William and I have made some remarkable discoveries with the telephone, and Sir William is to write a letter descriptive of them to appear in the Royal Society's Transactions, the letter to appear in our joint names." Thomson showed intense interest in Bell's phonautograph slides of vowel curves, and the two men debated Helmholtz's ideas as scientific equals. Mabel and Lady Thomson got along well also, Lady Thomson's speech being easy to read, and the two couples enjoyed what Bell called "a delightful week" together. At Covesea, Bell had amused himself by dropping sugar lumps into water to study the bubbles that rose; and Mabel must have thought of this, and been reassured, when she wrote, "Sir William is just like Alec in investigating every little thing that comes in his way. I saw him yesterday at lunch deep in the study of the vibrations of light through a tumbler of jelly!"

Bell's public efforts in science — in November 1877 a lecture on Visible Speech to the Philosophical Society of Glasgow (with Sir William in the chair), in March 1878 a paper entitled "The Natural Language of the Deaf and Dumb" before the Anthropological Society of London — did him credit, for in them he spoke from long experience and sure knowledge. His proudest moment as a savant came when, through the noted philologist Professor Max Müller, he received an invitation to deliver a course of lectures on speech at Oxford University. The lectures were given on four successive days, October 22–25, 1878, the first lecture drawing an unusually large audience for such affairs, and each successive lecture a still larger one. "We only wanted a great many more!" wrote Mrs. Müller to Mabel afterward; "they were so clear that every one could follow them." President Warren of Boston University remarked in his next annual report that the occasion had set a precedent for some arrangement of exchange professorships between American and European universities.

But meanwhile something — perhaps the unsettling influences of fatherhood and spring — in May 1878 had excited another spasm of gravitational speculations. Revealing of Bell's troubles with mathematical concepts was his puzzled observation at this time that the idea of resolving a force into two equivalent forces by means of a triangle of vectors was absurd, because two sides of the triangle must be greater than the third and so could not be equivalent to it. "Must study more," he noted a couple of days later; "no confidence in any mathematical investigation of my own."

Months earlier, Thomson had given Bell a warm letter of introduction to James Clerk Maxwell, one of the world's greatest mathematical physicists, the very type of the modern in physics. On June 4, within a week of Bell's latest return to closet science, Bell called on Maxwell at Cambridge. Bell's notebook and letters are silent on what passed between them, and within a fortnight he was once again at his private war with gravity; so if Maxwell commented on Bell's notions he worked no cure. But in Maxwell's later comments on telephony, there is evident some disdain for Bell as a basic scientist.

Bell took invention far more seriously than he did scientific theorizing. Already behind him were years of frustrations and failures, the price of one historic triumph; ahead were many more years of brain-racking labor, deferred hopes, and only a few triumphs. But Bell's heart never sickened more than momentarily. After all, though fame may have been a spur, the enchantment of the chase was itself worthwhile.

Moses Farmer — and Elisha Gray — could have told Bell about the inventor's bitterest pill: not mere failure, which does not kill hope, but a rival's success, which does. When he first heard of Bell's telephone and realized how simple a thing it was, Farmer told Tom Watson one day, he could not sleep for a week. "Watson," said Farmer with tears in his eyes, "that thing has flaunted itself in my face a dozen times during the last ten years and every time I was too blind to see it."

Bell in his turn felt the same pang when he heard about Edison's phonograph late in 1877. He thought it "perfectly wonderful," Mabel reported. But he wrote Gardiner Hubbard a few weeks later almost in the words of Moses Farmer: "It is a most astonishing thing to me that I could possibly have let this invention slip through my fingers when I consider how my thoughts have been directed to this subject for so many years past." Like some of the telephone claimants, he passed easily from the feeling that he should have thought of it to the conviction that in principle he had. In his telephone lectures he had remarked that if some implement could be made to follow the curves of a phonautograph tracing, it would reproduce the sound that had made the tracing. "And yet in spite of this the thought never occurred to me to indent a substance and from the indentations to reproduce sound."

Bell suggested to Hubbard that the telephone patent might, in conjunction with his lecture remarks, embrace the phonograph principle — "the idea of producing the effects of articulate speech by moving a plate of iron or other material in the way in which the air is moved by the voice." Chauncey Smith, the Bell Telephone Company counsel, agreed that there might possibly be some overlapping of the telephone and phonograph patents. But a company official, Charles Cheever, pointed out the other edge of the sword: as Bell had foreshadowed the phonograph, so had others foreshadowed the telephone. To win the former might be to lose the latter; while the challenge itself, whatever its outcome, would breed bitterness. So Bell gave up the idea.

Bell kept these speculations private. In public he gave Edison full credit for his "most ingenious instrument." Anyway, there was room left for Bell in the phonograph field, just as there had been for Edison in telephony. Edison's diaphragm stylus indented its recording on a sheet of heavy tinfoil wrapped around a hand-cranked cylinder. Each recording ran for only a minute or two; and its quality, poor to begin with, deteriorated to a mindless growl after a few playings. Early in April 1878 Bell and Alexander Ellis spent an hour together testing the phonograph's utility for philology and found it wanting. "The extreme vowels *ee oo*," wrote Ellis, "were scarcely differentiated." "Bite" could not be told from "bout." So far, he thought, the phonograph was "merely a highly interesting & ingenious *toy*." Nevertheless the first step was made, Ellis wrote, and "there are *plenty* to follow."

Bell already contemplated taking those steps, all the more since the Edison Speaking Phonograph Company, which had paid Edison ten thousand dollars and a twenty percent royalty for sole rights to exploit the phonograph as a music box, had been organized in January 1878 by Gardiner Hubbard and other telephone executives. Hubbard, unlike William Orton of Western Union, was evidently flexible enough to join with a man in one field while fighting him in another. Perhaps also he shared Tom Watson's notion that pushing the phonograph might help divert Edison from telephony. The new company's chief income turned out to be admission fees from popular exhibitions of the novelty during the year or so of the "phonograph craze." Then, as with the telephone lectures, the novelty faded. Edison himself, though he still delighted in his phonograph achievement above all others, lost faith in its commercial promise, which he considered to have been mainly as a business dictating machine. In his severely pragmatic way, he abandoned further work on it for ten years, turning instead to the development of electric lighting. The phonograph itself fell into disuse. In time, Bell would seize the opportunity thus offered.

In England, however, Bell's inspiration carried him no further than two notions for the commercial application of the phonograph, one wildly chimerical, the other merely impractical. He proposed the former to Hubbard in the spring of 1878 as likely "to realize a large fortune in a couple of months or so." The vision was of a ratchet noisemaker of the kind then used as a policeman's alarm in Cambridge, Massachusetts, but with a phonograph track that would cry out "Fire" or "Help" when whirled about. The absurdity of this proposition would seem to be instantly evident. So it was to Hubbard and others. If for some reason one did not choose to yell the alarm himself, why not use a policeman's whistle or the aforementioned ratchet? And whence would come so great a demand as to yield a quick fortune? It is difficult to explain Bell's fantasy except by the circumstance that makes his touching of Queen Victoria's hand understandable: the physical handicap of his wife. Mabel was deaf, and her speech was not readily understood by strangers. Such a device could have been useful to her. And Bell, without realizing it, may have identified the welfare of all the world and its multitudes with hers.

That summer Bell also suggested that Watson work out, patent, and manufacture thousands of phonographic "Swearing Tops" as a Christmas toy that would make an amusing outcry when touched with a pencil as it spun. With the profits, he wrote, "we can work at Flying Machines & all sorts of things next year in comfort." Watson pointed out that the top "could hardly speak the shortest word before it had made several revolutions," and that he himself had little spare time to work on the matter — in effect, that the trivial end scarcely justified what it would take to overcome the formidable technical problems, if indeed they could be solved at all. There the matter ended.

Yet along with these phonographic crotchets Bell was giving serious thought to a couple of ideas that he would develop over the years to more notable purpose. One of them was expressed lucidly in a lecture, "Speech," before the Royal Institution in London on May 17, 1878: "It has not long been known that the electric resistance of selenium varies under the action of light. . . . If you insert selenium in the telephone battery and throw light upon it you change its resistance and vary the strength of the current you have sent to the telephone, so that you can hear a shadow." Conscious of Edison's growing success with solids in variable-resistance telephony, Bell already envisioned leapfrogging Edison's carbon button by using a light beam of variable intensity impinging on selenium in a telephone circuit.

The other and more significant inventive goal was the one alluded to in his "swearing top" letter to Watson: "flying machines." Watson may have been as skeptical of that as of the "swearing top," but he could not have been surprised. "From my earliest association with Bell," he later recalled, "he discussed with me the possibility of making a machine that would fly like a bird." The flight of birds had, in fact, fascinated Bell since the boyhood days when he watched them soar over Corstorphine Hill, studied their anatomy, and even wrote a poem on the subject. Walking with Watson on the beach at Swampscott during his Salem days, Bell came across an extremely dead gull and, while Watson kept well to windward, "measured its wings, estimated its weight, [and] admired its lines and muscle mechanisms." With powered flight in mind, he also took an interest in a very light steam engine Watson had built. And before leaving for England in 1877 he got Watson to promise that as soon as the telephone business was well launched, Watson would leave it and join Bell in aerial experimentation.

At Aberdeen on September 23, 1877, by Mabel's report, "Alec . . . went out for a long walk this morning and saw some sea gulls flying and since then has been full of flying machines." A week later at Covesea she was marveling,

What a man my husband is! I am perfectly bewildered at the number and size of the ideas with which his head is crammed. . . . Flying machines to which telephones and torpedoes are to be attached occupy the first place just now from observations of sea gulls. . . . Every now and then he comes out with "the flying machine has quite changed its shape in a quarter of an hour" or "the segarshape is dismissed to the limbo of useless things." . . . Then he goes climbing about the rocks and forming theories on the origin of cliffs and caves. . . . Then he comes home and watches sugar bubbles.

At Pluscarden Abbey, Bell took notes and made sketches while watching the rooks that gathered there. Afterward he drew the outline of a rather modern-looking airplane and set down some general ideas about propeller-driven, heavier-than-air flying machines: for example, two propellers rotating in opposite directions, light fuel such as oil or gas, and a "parachute arrangement in case of accident to machinery." A few days later he was sketching propeller arrangements for vertical takeoff and landing. Though he made no actual experiments, the subject of aeronautics recurred in his notebook for the following March.

Bell made a point of visiting institutions for the deaf in England and Scotland when he had the chance, explaining his own views to them and learning from their experiences. After Professor David Hughes of the University of London announced his invention of the microphone early in 1878, Bell's well-reported interest in the deaf brought him numerous letters asking if the new device could help them. He replied cautiously that it probably could at some time in the future. In Scotland that fall, Bell tried it out on a partially deaf man who "held the telephone attached to the microphone to his ear, and . . . understood everything that was said, even when [Bell] went to the other end of the room, but when he took the telephone away he was helpless." Bell was wise, however, in raising no hopes of an imminent breakthrough in that respect. Not until many years had passed would electronic aids to residual hearing be developed and become basic in educating the deaf.

Bell was recalled to active service in the education of the deaf by a letter from Thomas Borthwick, a businessman of Greenock, Scotland, who had read a newspaper report of Bell's lecture on Visible Speech at nearby Glasgow in November 1877. Heartened by Bell's account of teaching speech to the deaf, Borthwick asked Bell's help for his eight-year-old deaf-mute daughter and about ten other such children in the town. Borthwick had already tried to get the Greenock school board to open an articulation school for them. But the board had rejected the idea as both visionary and beyond its authority. Finding that funds might be raised privately, however, Borthwick now implored Bell to help obtain a teacher trained in the Visible Speech method, or else to come himself and instruct some Greenock teachers in it.

Immediately interested, Bell promised to get a qualified teacher if a class were formed and a salary guaranteed. He began looking to Greenock as a pilot project for Great Britain in his cherished program of public day schools for deaf children. It would, he thought, be "likely to inaugurate a revolution in the method of teaching deaf mutes in Europe." When the teacher Bell tried to bring from the United States backed out belatedly, Bell decided, pending the arrival of a substitute choice, to launch the school himself, with the help of Mabel's beloved former teacher Mary True, who happened to be on vacation from a teaching position in England and had volunteered her services.

At Greenock, early in September 1878, Bell seemed to find himself again. "I have been so happy in my little school," he wrote, "happier than at any time since the telephone took my mind away from this work. . . . I have been waiting for months past for something to do. I have been absolutely rusting from inaction — hoping and hoping that my services might be wanted somewhere. Now I am needed, and needed here." It was a small class after all, comprising only Jessie Borthwick and two younger girls. The schoolroom he confronted on his arrival was a dusty, cobwebbed storeroom in the local academy, cluttered with the effects of a dead writing master. But it was soon put to rights. And in it Bell saw not just three little girls "but the *thirty thousand deaf mutes of Great Britain*." "I shall make this school a success if I have to remain till Christmas!" he wrote almost wishfully. He must have had mixed feelings when the permanent teacher arrived three weeks later to take it over.

Mabel's feelings were evidently not so mixed. "It is a sorrow and great grief to me," Alec wrote her from Greenock when she seemed reluctant to join him there, "that you always exhibit so little interest in the work I have at heart, and that you have neither appreciated Visible Speech nor have encouraged me to work for its advancement." Perhaps she regretted, even while respecting, Alec's conviction that he could not properly be both her husband and her teacher; if so, she may not have been eager to see him long absorbed in teaching others. Whatever the reason for her attitude, Alec's reproach seems to have had its effect. Certainly in years to come Mabel would work almost as devotedly in the Visible Speech cause as Alec himself.

For several years Bell kept in transatlantic touch with the Greenock school. It worked no revolution, perhaps, but it lived and sent out shoots. After a few years it became part of the city school system. The teacher appointed on his recommendation in 1883 trained a number of others who carried the oral method back to their several native towns in England. At last report in the 1960s the school itself was still teaching speech to the deaf in Greenock.

When the Bells returned to America after the Oxford lectures, Bell still felt the pull of the long-interrupted work he had taken up once more in Greenock. It was in the Greenock days that he had renounced the telephone so vehemently. "Do you know, dear," he had written Mabel from there, "I think I can be of far more use as a teacher of the deaf than I can ever be as an electrician." To avoid being drawn back into telephone work he took passage for himself and his family to Quebec, with the intention of going straight to Brantford, thereby dodging the headquarters of the Bell Telephone Company at Boston.

But it turned out that while the Bell Telephone Company could now get along without Bell the "electrician," it could not, in the vital patent struggle then unfolding, find any substitute for Bell the witness. When Bell's intention became known, the company's officials and lawyers held a desperate conference in Boston. It was Gardiner Hubbard who came up with the proper strategy. When Bell arrived with Mabel, Elsie, and Mary Home at Quebec on November 10, 1878, there to meet him was none other than Tom Watson, with a plea from the company in Boston. In effect it said: Mr. Bell, come here, we want you.

22

The Witness

What Bell lacked in business sense he made up in luck, at least in the chances that led him into partnership with Gardiner Hubbard and Thomas Sanders. The two men shared a warm regard for Bell, faith in the future of his creation, and a resultant purpose to make him rich, almost (it sometimes seemed) in spite of himself. Not all nineteenth-century inventors were so fortunate in their business associates. With these attributes in common went sharp differences between Hubbard and Sanders in temperament and therefore in business policies; but even their differences were, in counterpoise, all to the good.

Hubbard was the promoter, energetic and expansive, a man to whom the past was a closed book and the present only a perfunctory preface to the glorious future. He preferred to skip prefaces. Unlike Bell, who sometimes alternated between elation and dejection, Hubbard had a direct-current temperament, never pessimistic, though sometimes a little less optimistic than usual. On his travels with the Railway Mail Service Commission he had been a Johnny Appleseed of telephone agencies; and in the spring of 1878 he had persuaded the general superintendent of the Railway Mail Service, a thirty-two-year-old managerial prodigy named Theodore N. Vail, to quit the government and become general manager of the telephone company. This latter stroke turned out to be one of Hubbard's greatest services to the enterprise, if not his greatest.

Thomas Sanders, as treasurer of the company, saw another side to Hubbard's blithe promotions. "He is not a businessman, does not pretend to be and therefore cannot appreciate the weight of the burden I am carrying and the desperate necessity for capital," Sanders wrote Bell in March 1878. Theoretically, given some capital and credit to start with, the company's strict policy of renting rather than selling instruments would in time finance its own expansion, just as a small deposit at compound interest would grow to a fortune in a century or two. Hubbard, however, seemed to think that faith could not only move mountains but also nullify time, mathematics, and commercial law, that what was theoretically attainable in a distant future might be considered as in the pocket today and spent forthwith; and so he proceeded to order telephones from the Williams shop as fast as customers would rent them, without regard for the question of how to pay for them. That, in Hubbard's view, might be left to Sanders as treasurer.

Fortunately it could — to the extent of Sanders's personal fortune and credit, all of which he threw unhesitatingly into the financial breach, sinking $110,000 before he got a dollar back. "I have exhausted my means and stretched my credit to its utmost, having received no aid from Mr. Hubbard financially from the first," Sanders wrote Bell in his March 1878 letter. Sanders saw only one solution: organizing a national company and selling stock. Hubbard resisted the idea of diminishing his and Bell's proportionate interest in the company. But Sanders took a more realistic view. "It is the worst policy imaginable to refuse to part with such portion of our interest as to make what we have left valuable," he wrote Bell.

Necessity had already, in February 1878, brought the organization of the separate New England Telephone Company, to which the Bell Telephone Company had sold the exclusive right to lease telephones in New England. From this, Sanders told Bell, the original company had received "the first money so far as I can ascertain that has been advanced by any one in any part of the world to forward your invention." Sanders now wanted to enlist the New England company's backers in forming a single national company and raising adequate working capital by the sale of stock; and he told Bell, "I want you to persuade Mr. Hubbard to consent to it."

Perhaps Bell guessed how much weight his business advice would have with his father-in-law; in any case he chose not to venture any, nor even to answer Sanders. And after all, circumstances compelled Hubbard to yield. In July 1878 the Bell Telephone Company was reorganized as a corporation rather than a voluntary association and was able to sell enough stock to stave off collapse, even though the New England company hung back from merger.

Despite the sharpness of their disagreements, Hubbard and Sanders respected each other's contributions. They "used to have little spats," recalled one early telephone employee. "I heard Sanders tell [Hubbard] once to mind his own business. . . . He didn't say anything back. He was always dignified. He never lost his temper." Hubbard wrote Sanders in the critical spring of 1878, "Your letters have been frank and manly and it is absolutely necessary for our success and comfort that we should at all times . . . be open and frank." Sanders testified thirty years later that in all their "spats" Hubbard "cherished no malice." And in the midst of it all, Sanders himself conceded to Bell that Hubbard "is entitled to great credit and receives my unbounded admiration for the energetic, audacious and skilful manner in which he established agencies and organized the inception of the business."

Tom Watson in retrospect saw clearly how valuable was the contribution of each, and they in turn recognized the value of Watson's intelligence, youthful energy, and unmatched telephonic experience, especially for keeping the technical quality of the company's instruments and the technical competence of its agents well ahead of the competition at the crucial time of first public impressions. And when in February 1879 Hubbard reluctantly but inevitably surrendered the absolute voting majority given him by his and Mabel's combined stock, and with it his dictatorial authority, he left Theodore Vail, prototype of the new class of professional corporate managers, in masterful charge of operations. Bell's luck was in the box and driving four-in-hand.

The going would have been rough even if the Bell Company had had the road to itself. It did not. By 1878 the Western Union Telegraph Company, with more than forty million dollars in capitalization and three million a year in net profits, had got into the race.

Since the spring of 1876 the Western Union had been paying Edison to do telephonic research. But the move was at first more precautionary than aggressive. Through most of 1877, Hubbard, Elisha Gray, and President William Orton of Western Union put out occasional feelers toward compromise and combination. Then attitudes began to harden.

The Bell Company drew courage from its lawyers' confidence in the strength and scope of Bell's patents, from general business hostility toward Western Union, from the growing demand for telephones, and from the Bell Company's better instrument quality, earlier occupation of key cities, and longer experience in telephony. Western Union, for its part, had its resources of money and its own experience and massive physical properties in electrical communication though not in telephony. It also had been assembling its stable of Bell's rival inventors. In December 1877 the Western Union set up the American Speaking Telephone Company to hold the combined patent rights and claims of Thomas Edison, Elisha Gray, Amos Dolbear, and others of less account, with a Western Union subsidiary, the Gold and Stock Company, serving as the exclusive agent of commercial telephone operations. On paper there were further corporate ramifications; in fact, however, the Western Union ran the whole affair. Also tending

LIST OF SUBSCRIBERS.

New Haven District Telephone Company.

OFFICE 219 CHAPEL STREET.

February 21, 1878.

Residences.

Rev. JOHN E. TODD.
J. B. CARRINGTON.
H. B. BIGELOW.
C. W. SCRANTON.
GEORGE W. COY.
G. L. FERRIS.
H. P. FROST.
M. F. TYLER.
I. H. BROMLEY.
GEO. E. THOMPSON.
WALTER LEWIS.

Physicians.

DR. E. L. R. THOMPSON.
DR. A. E. WINCHELL.
DR. C. S. THOMSON, Fair Haven.

Dentists.

DR. E. S. GAYLORD.
DR. R. F. BURWELL.

Miscellaneous.

REGISTER PUBLISHING CO.
POLICE OFFICE.
POST OFFICE.
MERCANTILE CLUB.
QUINNIPIAC CLUB.
F. V. McDONALD, Yale News.
SMEDLEY BROS. & CO.
M. F. TYLER, Law Chambers.

Stores, Factories, &c.

O. A. DORMAN.
STONE & CHIDSEY.
NEW HAVEN FLOUR CO. State St.
"   "   "   " Cong. ave.
"   "   "   " Grand St.
"   "   "   Fair Haven.
ENGLISH & MERSICK.
NEW HAVEN FOLDING CHAIR CO.
H. HOOKER & CO.
W. A. ENSIGN & SON.
H. B. BIGELOW & CO.
C. COWLES & CO.
C. S. MERSICK & CO.
SPENCER & MATTHEWS.
PAUL ROESSLER.
E. S. WHEELER & CO.
ROLLING MILL CO.
APOTHECARIES HALL.
E. A. GESSNER.
AMERICAN TEA CO.

Meat & Fish Markets.

W. H. HITCHINGS, City Market.
GEO. E. LUM,   "   "
A. FOOTE & CO.
STRONG, HART & CO.

Hack and Boarding Stables.

CRUTTENDEN & CARTER.
BARKER & RANSOM.

Office open from 6 A. M. to 2 A. M.

After March 1st, this Office will be open all night.

The first telephone directory, published February 21, 1878, about a month after the first exchange had opened in New Haven, Connecticut toward intransigence on both sides was the long-standing feud between Orton and Hubbard.

Events in February 1878 increased the assurance of the antagonists: the patenting of Edison's carbon-button transmitter on the Western Union side, the financial infusion from the New England Telephone Company on the Bell side. Sanders, who as usual had been acting as drag anchor against Hubbard's full sails, came around to Hubbard's view that a fight was inevitable. And so on February 21, 1878, negotiations were formally broken off.

Operating now in earnest, the Western Union cut telephone rates (and sometimes, it was charged, Bell Company wires also), put pressure on district telegraph companies not to enter telephony on the Bell side, and guaranteed its customers against any losses or penalties as a result of patent decisions. Watson and Vail claimed now and then, doubtless with much truth, that Western Union activity helped by advertising the telephone and by stimulating the efforts of Bell agents; and Sanders and other Bell officials shed no tears when William Orton died suddenly in April 1878. Nevertheless the fight hurt, especially later in 1878 when the much louder Edison transmitter began winning customers away from the Bell magneto transmitter.

Luck armed the Bell Company against the Edison transmitter when in February 1878 Charles Cheever came across a German-Jewish immigrant named Émile Berliner, who had filed a variable-pressure transmitter caveat on April 14, 1877, thirteen days before Edison's first carbon transmitter patent application. After trying in vain through the spring of 1878 to develop a variable-resistance transmitter without using Edison's carbon, Watson took a look at Berliner and his instrument; and in September 1878, after some dickering, the Bell Company hired Berliner in exchange for control of his caveat and inventions. It could thus file an interference against Edison's transmitter and prevent the final issue of a patent. Edison won his patent after long-drawn-out proceedings, but not until 1892. By then it no longer mattered. The Berliner interference had protected the Bell Company from an injunction against its use of Francis Blake's carbon transmitter (acquired in October 1878 and in effective use by early 1879), and that immunity had been a factor in the final settlement, which among other things gave the Bell Company Edison's telephone rights.

There was irony in the fact that the Bell Company did to Edison precisely what the Western Union would have done to Bell had Gray's caveat preceded Bell's patent application. But the Bell Company could not afford to settle for savoring ironies. Bell's patents had been fully granted, notwithstanding the salvo of patent interferences touched off by the Western Union in March 1878. The Bell patents were thus the big guns in the company's fight for telephone supremacy. As early as April 1878 Hubbard urged using them in vigorous legal action against the competition.

Sanders succeeded in postponing the move until the organization of the new Bell Telephone Company provided more money for what was bound to be an expensive contest. Then on September 12, 1878, in the Circuit Court of the United States for the District of Massachusetts, the Bell Telephone Company sued for an injunction against Peter A. Dowd. The nominal defendant, as agent of the Gold and Stock Company, had rented out telephone transmitters of the Edison type and also magneto receivers, which the Gold and Stock claimed as the invention of Elisha Gray. The real defendant was the Western Union, which assumed the defense in the case. The record of telephone patent litigation would, in the course of nearly twenty years, pile up to a stupefying height. But this first suit would turn out to be the crux of it all.

In challenging the priority of Bell's inventions, the defense in the *Dowd* case relied chiefly on Elisha Gray, but not exclusively. Of the others, only Professor Amos E. Dolbear of Tufts College had previously entered Bell's life directly. And besides Gray, only Dolbear ever claimed to be the victim of unethical conduct by Bell or someone acting for him. Gray's story has already been told here, but Dolbear has thus far appeared in this account only in the peripheral way he figured in Bell's own mind before the *Dowd* case: as a factor in speeding Bell's second basic patent application and as one of the chorus of belittlers who had recently turned Bell against telephone work. The public airing of Dolbear's story in the course of the legal battle seems at this point to demand of us, as it did of Bell himself, a closer look.

Dolbear's early career resembled that of Elisha Gray. Born at Norwich, Connecticut, in November 1837 — in the same house, as it happened, where Benedict Arnold was born — Dolbear lost both parents as a small boy, and so between the ages of ten and sixteen he worked hard for his keep on a New Hampshire farm. Then followed two years in a Worcester pistol factory, four years of schoolteaching in Missouri, and three years as a machinist in a Taunton, Massachusetts, machine shop and in the Springfield Armory. Like Gray, Dolbear next turned to less physically arduous work for the sake of his health, studying after hours until he could enter a small Ohio college in 1863. After graduation in 1866, he studied for a year at the University of Michigan and then taught for a year at the University of

Kentucky and six years at Bethany College, West Virginia, before settling down for the next thirty-two years as a professor of physics at Tufts College.

Like Gray also, Dolbear enjoyed tinkering with electrical and acoustical devices. At Michigan, for example, he played with the idea that the motions of an armature at one end of a electromagnetic circuit would be reproduced in one at the other end; but in practice he found the generated force too weak to produce the gross effects he had looked for. As late as 1874, he remarked in a letter to Edward Pickering that "I am not acquainted with the literature of electrical work beyond what is in the text books." But then, Bell's own lack of electrical learning had left him uninhibited enough to invent the telephone. Like Bell, the young Dolbear was ambitious. "I have a name to make & *expect!* to make one," he wrote his Michigan mentor at the age of twenty-nine.

By Dolbear's account it was on September 20, 1876, that he made a blackboard sketch of what he thought might be an improvement on the Bell telephone as he then knew it: his model was to have a sheet-iron (instead of a membrane) diaphragm, and a straight-bar permanent magnet (instead of an electromagnet) with a coil wound around one end and in circuit with a similar instrument. A week or two later he asked a part-time student, who came only on Saturdays, to make the models. But the student did not finish them before beginning an absence of several weeks. Busy with other matters, Dolbear did not start making the instruments for himself until after Christmas.

According to Dolbear, he was spurred to action at that time by his friend and Bell's, Percival D. Richards, who on hearing Dolbear's ideas offered to get financial backing and help promote them. "Let us make some money!" Richards wrote. Then, with a view to combining forces, Richards took it upon himself to consult Gardiner Hubbard. Richards came back to Dolbear with the disappointing news that Bell had tried both permanent magnetism and metal diaphragms in his telephones before Dolbear thought of them. Dolbear later charged that he had been discouraged from applying for a patent by Richards's further report that, according to Hubbard, Bell had already patented the ideas; but Hubbard denied having made such a statement. Bell later testified that on hearing of Dolbear's ideas he had dismissed them as well-intentioned but not novel suggestions. Whatever Richards's report was, Dolbear heard it on January 6, 1877. The chance of a patent gone, said Dolbear later, he dawdled over the models thereafter, but by early February finished them and transmitted speech with them. Meanwhile Bell received his Patent Number 186,787, which covered both the metallic diaphragm and the permanent-magnet phone.

After Bell's first Salem lecture in mid-February 1877, Dolbear wrote him: "I congratulate you, sir, upon your very great invention, and I hope . . . that you will be successful in obtaining the wealth and honor which is your due." But in March he learned that Bell's patent had not been applied for until January 15. On March 20 Dolbear wrote privately as to Bell's permanent-magnet telephone, "I *think* that a mutual acquaintance informed him as to what I was doing and before I could get in working order he, by the aid of a skilled electrician and working nights and Sundays, completed his and got it patented instanter, so I suppose that I shall lose both the honor and the profit of the invention."

Strictly speaking, it was almost certainly true, as noted earlier, that what Richards told Hubbard late in December 1876 had contributed to the urgency of Bell's patent application. That fact by itself in no way reflected on Bell's honor. The aggrieved tone of Dolbear's remark, however, seems to insinuate that he thought Bell had appropriated his ideas, which was quite another matter. Yet clear, precise, and circumstantial written evidence existed then and still exists to show that, as also noted earlier, Bell had anticipated Dolbear by a comfortable margin in conceiving and reducing to practice, as well as in patenting, the improvements at issue. From that it follows that Bell did not steal them. Such was the eventual judgment of the courts, as it must be that of history and should have been that of Dolbear.

For some time Dolbear confined the insinuation to his private correspondence. In a letter he wrote Bell early in May 1877, far from airing his suspicion, Dolbear remarked on how common it was, especially in telegraphy, "for two or more persons to hit upon the same idea at the same time, without any knowledge of each other's work." He gave Bell an account of having conceived the new form of telephone "last fall" and then having begun some experiments that "resulted in this present form of the Telephone . . . early in Dec. last" (a chronology inconsistent with his earlier and later testimony). His friends, Dolbear went on, had persuaded him "to keep away from you," perfect the instrument, and patent it. "This I was doing when your invention of the same thing was made public. . . . I did at one time seriously think of contesting your patent, but . . . I hadn't a cent to spend for any such purpose, therefore I beg you not to be uneasy in the slightest degree as to your claim." Dolbear only asked that in speaking of other inventors Bell give him credit for independently arriving at the idea: "I admire both your wonderful ingenuity and persistence, and I would love to count you as one of my friends."

By July, however, brooding seems to have hatched Dolbear's suspicion into a conviction. On July 8, 1877, three days before Bell's wedding, Bell wrote Dolbear that he would be giving a paper on the telephone at the

British Association meeting, and that wanting "to give credit where credit is due," he would "like exceedingly to have the opportunity of knowing what portion or portions of my invention you claim as your own in order to be able to allude to the matter in England."

Dolbear went that day to the Hubbard home and stated his case in an interview with Hubbard and Bell. He began by claiming priority of conception in the permanent-magnet features (but said nothing about the metallic diaphragm, which he later testified "to me never seemed an essential matter"), and he blamed his slowness in reducing his idea to practice on the alleged false report of an existing patent by Bell. According to Dolbear, Bell granted that Dolbear might have got a patent if he had applied for one before Bell. It is not evident why Dolbear claimed this as a concession, since it merely stated a truism, but in any case Bell later denied saying it. Dolbear in his turn later disputed Hubbard's insistence that Dolbear had finally conceded Bell's priority at the interview. Nevertheless, in a letter written that very day at Bell's request, Dolbear carefully avoided including priority of conception in specifying his claims to recognition.

In August, after Bell and his bride had left Boston, Dolbear wrote Hubbard demanding "a share in the profits of this invention." Otherwise he threatened to release "a small treatise on the subject of telephony," which would reveal to the public that "my rights are equal to those of Prof. Bell" and thus "diminish the sale under that patent." He also had "good counsel and some parties in Boston that are somewhat anxious to test the validity of the last patent (that of Jan. 30, 1877)." Hubbard told him to go ahead and take his chances in a formal judicial proceeding, but pointed out that to claim undue credit in a published work "might be an injury to Mr. Bell, for which no adequate atonement could be made."

Dolbear proceeded to publish his small treatise *The Telephone*, stating in the preface that "as the speaking-telephone, in which magneto-electric currents were utilized for the transmission of speech and other kinds of sounds, was invented by me, I have described at some length my first instrument. . . . Steps have already been taken to secure letters-patent." In the text, however, he referred to Bell's membrane telephone, shown at Philadelphia, as "the first speaking-telephone that was ever constructed, so far as the writer is aware," adding that "it was not a practicable instrument." As for his own instrument, the text merely said that he had devised it (date not specified) "without the slightest knowledge of the mechanism which Prof. Bell had used." There is no evidence that these remarks tended to "diminish the sale" under the Bell patent.

As for the "parties in Boston" who wanted to test Bell's patent, they were probably officials of Western Union. At any rate, Dolbear wrote

President Orton sometime during that summer of 1877 and negotiated during August with the Western Union's telephone subsidiary, the Gold and Stock Company. Early in September he signed an agreement to apply for patents on all his telephone inventions and assign them to Gold and Stock, in return for the patent and legal expenses involved, one-third of the net profits of such patents, and twenty-five dollars a week in support of further research for the company. In October he applied for a patent. Thus it came about that Dolbear figured in both the telephone patent interferences and the *Dowd* case.

When Bell arrived at Quebec in November 1878, Watson's most urgent message concerned the *Dowd* case only indirectly. In the Speaking Telephone Interferences before the Patent Office, each party had a limited time in which to file a preliminary statement of claims and dates of work, or else he would lose by default — which in Bell's case would knock the foundations out from under the case against Dowd. Even so, it took Tom Watson's dramatic flair, youthful exuberance, and evocation of high excitement in days past to overcome Bell's distaste for involvement in the telephone controversy. And at that, Bell insisted on first depositing his family at the Bell homestead in Brantford and getting an explicit promise that the Bell Company would pay his expenses to Boston. He dictated his statement while in the Massachusetts General Hospital for treatment of an abscess and with relatively few letters or documents to bolster his memory. But a supplementary statement could be filed later and was. What counted in his statement of November 20, 1878, was that it came in time to keep the fight alive.

As the Bell Company must have hoped, this first skirmish reawakened Bell's own instinct for combat. Within hours of his arrival at the hospital he was writing Mabel for the keys to a bureau in the Cambridge house where she kept his letters to her, "not that I have the remotest intention of publishing any of my love letters! I am only anxious to discover any statements that may help me to fix dates." Afterward he looked up Clarence Blake and others around Boston for the same purpose. By January 1879 he was engrossed in following the development of strategy for the great Dowd battle.

Bell did not presume to dictate that strategy, even when he disagreed with it. He knew it was in the hands of masters. The company's leading counsel in patent litigation was James J. Storrow, a little, dark-whiskered man, precise, alert, and keenly discerning, who had inherited or acquired a bent for technology from his father, a noted civil engineer, and combined it with study at Harvard Law School to become one of the foremost patent lawyers of his time. Storrow prepared for what he may have sensed would be the great work of his life by spending an entire summer at his country home studying electrical physics. Associated with Storrow in the telephone cases was the Bell Company's regular counsel Chauncey Smith, Storrow's equal in acumen and legal eminence but a piquant contrast to him in looks and manner. Born in Vermont and graduated from that state's university, Smith for years had, like Longfellow and Gardiner Hubbard, lived on Brattle Street in Cambridge. He was a scholar of the law as well as one of its leading practitioners in Boston. But like Storrow he also had a respect approaching reverence for technology and its creators. "Mr. Smith's fair round figure . . . completely filled a big arm-chair," ran a newspaper description a few years later; "with his fat ruddy face, dimpled and clean-shaven and bearing a benign expression, his long silvery hair falling upon his shoulders in ringlets, and his bright eyes peeping through gold spectacles, [he] was the picture of Uncle Toby or the good old Vicar of Wakefield."

The picture was usefully deceptive. After seeing Smith in action, Bell resorted to military rather than ecclesiastical metaphor. At the outset, Smith had made Bell ill with worry, so Bell said; for he proposed to base his strategy not on Bell's cherished concept of the undulatory current as a new phenomenon, but on the use of electricity in general for transmitting speech, as exemplified by the instruments described in the patents. After Smith opened the case for the Bell Company on January 25, 1879, however, Bell was "delighted." Just as Smith had anticipated, the Western Union counsel had built their case around the proposition that undulatory currents were not new with Bell, and that Dolbear, for example, had preceded him in consciously achieving that effect. By granting all that, and basing Bell's case on the *application* of undulatory currents, Smith's argument had, in Bell's words, "all the effect of opening up on the enemy a heavy fire from a masked battery . . . [and] from an unexpected quarter."

Smith's strategy was to be of fundamental significance in the outcome. Through it the Bell interests would win control of the basic principle of telephony, not merely of some particular devices in it. But legal defenses are not often taken by storm, and certainly not those involving such enormous stakes. So this propitious beginning was only a beginning. Both sides and their chief candidates for opulence, Bell and Gray, labored through the spring and summer of 1879 to gather, arrange, and present evidence and to study and reply to that of their antagonists. Mabel and her cousin Mary Blatchford spent much of April "wading through" the accumulation of old correspondence and notebooks in the Brattle Street house. "Oh! that those lawsuits were ended," Bell wrote his wife that month. "I am afraid to make more inventions, for fear of being dragged into an interminable business connection with the Company." All day, with an assistant provided by the company, he had been in a Boston library hunting up references to earlier experimenters.

Bell went down to New York to hear Elisha Gray's testimony taken. "Poor Mr. Gray," he wrote Mabel. "I feel sorry for him — for I feel sure he would never of his own accord have allowed himself to be placed in the painful position in which he now is — that is upon the supposition that he is an honourable man." A few days later, on April 7, the Bell side surprised and disconcerted the other side by producing Gray's evidently forgotten letter to Bell of March 5, 1877, in which Gray had written of the telephone, "I do not, however, claim even the credit of inventing it, as I do not believe a mere description of an idea that has never been *reduced* to *practice* — in the *strict sense* of that phrase — should be dignified with the name invention." When Gray was asked to verify the authenticity of the letter, he remarked ruefully to his counsel, "I'll swear to it, and you can swear at it!"

But the effects of Gray's letter were more moral than legal, as were those of a more recent letter from Professor James Watson of the University of Michigan, describing the incredulity Gray had expressed at accounts of Bell's telephone in the Centennial Exhibition. After all, what Gray himself may have thought of his legal claims at any past time had little relevance to the present proceedings. It was what the court would conclude from the established facts that mattered.

Even after his embarrassing letter had turned up, Gray himself wrote bravely to Professor George Barker that "our case is strong enough to stand on its merits." Dolbear told his story in detail. Franklin L. Pope, the Western Union's chief electrical expert, testified to the thorough investigation he had made of Gray's telephone claims at Orton's request in the summer of 1877 and recapitulated the reasons for his report in favor of them. Gray's "wash-basin receiver" was claimed to have essentially anticipated Bell's (though Pickering's tin-box receiver had by the same reasoning anticipated Gray's). Reis, Van der Weyde, and other gropers were portrayed as finders. And still others in the Western Union's string of claimants had their days in court.

All this came up against the impregnable truth that Alexander Graham Bell had conceived, reduced to practice, and patented the basic form of the telephone before anyone else. A telephone counsel of a somewhat later time marveled in retrospect at Bell's "capacity for long-sustained mental effort." He remembered taking a deposition from Bell that began in mid-morning; by two that afternoon, the lawyers mustered the courage to ask

Bell if he cared to pause for lunch. "I don't lunch," said Bell, and proceeded with his deposition to the end of the day. But the performance would have been still more remarkable if Bell had been weaving a fabrication instead of simply telling the truth. Storrow put the case neatly, also on a later occasion, when he wrote Bell, "You know how much I like to work with you because you are quick to catch an idea — but the real excellence of your deposition and its naturalness lies in the fact that in telling your own history you are telling the story of the man who invented and who knew that he had invented, the electric speaking telephone."

And so it was chiefly Bell's own testimony in July 1879, an account filling nearly a hundred pages of the printed record, that overmatched the Western Union with all its money, its experts, its eminent counsel, and its hopeful protégés.

The defense completed its testimony in May 1879, and the evidence in reply was closed in September. The final printed record contained over six hundred pages of pleadings and evidence and as many more of documentary exhibits. But even before the completion of the case for the defense, George Gifford, the Western Union's chief counsel and one of the nation's leading patent lawyers, had anticipated the verdict of the court and of history. Having informed his clients that the Bell patents could not be defeated, he and Chauncey Smith conferred for a week that summer on the terms of a possible compromise ("much to our disgust," Mabel Bell wrote for herself and her mettlesome husband). Smith refused Gifford's suggestion of an equal share for each side in the combined patents of both, and so the Bell Company and the Western Union took up the negotiations directly.

On November 10, 1879, the Western Union agreed to give up the telephone business and to assign all its telephone patents to the Bell Company, in return for twenty percent of telephone rental receipts over the next seventeen years. In view of the fact that the Edison transmitter and some other useful improvements came with the settlement, along with a network of established agencies and customers, it represented a decisive victory for the Bell Company. When Tom Watson got word of it, he took the whole day off to roam the woods and shores of Swampscott and Marblehead, declaiming poetry to the sky. Others may have shared that impulse, for in a matter of days the company's stock nearly doubled in value.

The Western Union had several reasons for accepting a settlement seemingly so advantageous to the other side. The redoubtable Jay Gould's American Union Telegraph Company had made overtures to the Bell Company for an alliance against the Western Union. William Orton was dead.

And the conservative eastern stockholders who controlled the Western Union were led by William H. Vanderbilt, who unlike his late father tended toward appeasement in business struggles.

It does not follow, however, that the Western Union would have been nearly so generous if it had seen any prospect of legal victory, or that George Gifford and its other able counsel did not make as strong a case as they possibly could. The official disposition of the *Dowd* case in 1881 by a consent decree, which upheld the Bell patents and enjoined Peter Dowd, was cited in later litigation as evidence of collusion between plaintiff and defendant, which, it was alleged, nullified the weight of the decision. In answer Gifford made affidavit at length that no effort or expense had been spared in contesting the Bell patent, and that he had not advised giving up the fight until he was sure it was hopeless. And the judge in that case said, "I would attach almost the same weight to an opinion by him, against his own clients, that I would attach to the judgment of a Court." Even without such testimony, common sense by itself would seem to make all this evident.

As for the Bell Company's decision against fighting it out for all or nothing, it stemmed from no weakness in the Bell basic patents, but from the fact that countersuits had been brought against the company by the Western Union for infringement of such later improvements as the Edison transmitter. In these, victory could not be certain, but expense and distraction were. Relief from them was worth the price.

Over a period of eighteen years from first to last, the Bell patents were tested in some six hundred separate cases. Such of the records as were printed filled 149 volumes. The suits were brought by the Bell Company against numerous infringers, whose object was usually to sell as much stock to the public as possible before being enjoined, though a few may have faintly hoped to make enough of a case to get bought off by the Bell Company. Most defendants gave up at an early stage; but of those cases that went to a final hearing, the Bell Company won every one.

Some cases would repay study by a psychologist; others would delight a connoisseur of rascality. The association of Antonio Meucci with Dr. Seth R. Beckwith offers a fine specimen to each. By 1885 Beckwith's professional specialty seemed to have become the doctoring of documents. He had shown rare skill also in kiting speculative telephone corporations and passing the strings to other hands just as the wind failed. Meucci, a seventy-seven-year-old veteran of Garibaldi's army who had found peace in the making of candles and sausages on Staten Island, had filed a caveat in 1871 for what he called a telephone. The caveat never became a patent and never could have become one. It described an acoustic or tin-can telephone, not an electric telephone; and this, of course, was not original.

The name "telephone" in the caveat was enough, however, for Beckwith to smell it out and use it as a basis for infringements on the Bell patents by his Globe Telephone Company of New York, and when that flimsy enterprise fluttered down, by his Meucci Telephone Company of New Jersey. This in turn dropped like a stone under legal pressure. Storrow wrote Bell in 1886 that Meucci's "whole story is a piece of fraud, supported by a forgery," and that Meucci himself "is the silliest and weakest imposter who has ever turned up against the patent." A historian might hesitate to use the superlative in the face of such stiff competition for it, but Storrow's exasperation is understandable. In 1958 Giovanni E. Schiavo, a conscientious historian, made the best possible case for Meucci. Nevertheless, Meucci's own testimony as presented by Schiavo demonstrates conclusively that Meucci did not understand the basic principles of the telephone, either before or for several years after Bell invented it.

Most defendants in infringement cases, however, merely tried to resurrect issues that had already been dealt with in the *Dowd* case, hoping to invalidate Bell's patents by establishing priority of conception for someone else, even for those like Gray or Dolbear whose rights had been acquired by the Bell Company in the 1879 settlement. So with two exceptions Bell was spared the necessity of appearing again as a witness. His *Dowd* case deposition was simply stipulated bodily into case after case, as well as into the speaking telephone interferences that dragged on in Patent Office hearings from 1880 to 1889 before being decided in his favor. The deposition went into the *Spencer* case, the record of which in turn was stipulated into the *Overland* case, which the Supreme Court of the United States decided in favor of the Bell patents. Thus Bell's 1879 depositions had as full a scrutiny and as conclusive an endorsement as American law affords.

Bell's two further appearances as a witness were for four weeks in 1883 in the suit against the People's Telephone Company, commonly known as the Drawbaugh case, and for about nine weeks in 1892 in the so-called Government case.

In 1880 Daniel Drawbaugh made his public debut as one of history's most prolific retroactive inventors. Until that time his career was summed up by a business card of 1874: "Daniel Drawbaugh. Inventor, Designer and Solicitor of Patents. Also Models Neatly Made to Order. Eberly's Mills, Cumberland County, Pennsylvania." With the title of "Designer" all sides were in agreement; as to the others, as with everything else Drawbaugh is re-

Daniel Drawbaugh.

INVENTOR, DESIGNER

and

SOLICITOR . PATENTS.

☞ Also Models Neatly Made To Order.

Eberly's Mills,

Cumberland County, Pennsylvania.

[See Other Side.]

Dan'l Drawbaugh

INVENTOR OF
THE FOLLOWING PATENTS.

Stave, Heading & Shingle Cutter.
Barrel Machinery.

STAVE JOINTING MACHINE, Many in use.
Tram & Red-staff for leveling face of Millstone.
Rine and Driver for running Millstone.
Nail Machinery for Feeding Nail Plates.

PUMPS, ROTARAY & OTHERS.

Hydraulic Ram.

THE DRAWBAUGH Rotary Measure-
ing Faucet, very extensively used.

CARPET RAG LOOPER -- A little
device by which rags are looped quick and firm
without Needle or Thread.

ELLECTRIC CLOCK.

MAGNETO ELECTRIC
MACHINE,

For short line Telegraphing, Fire Alarm,
and Propelling Electric Clocks. It can be
applied to any form of Electric movement.
Gives entire satisfaction USEING NO
GALVANIC BATTERY.
☞ For SIMPLICITY it has NO RIVAL.

corded as claiming, there is room for doubt. On the reverse side of Draw-baugh's business card were listed a dozen or so of the inventions he professed to have made, some patented, some not, none of them noteworthy, one or more copied faithfully from an encyclopedia. His bill letterhead described him as "Daniel Drawbaugh, practical machinist. Small machinery, patent office models, electric machines, etc., a specialty." He was skillful or plausible enough to have prospered, giving his father $1400, investing $2200 in the Drawbaugh Manufacturing Company, stocking his shop with a variety of machine tools, owning a house free and clear. Between 1851 and 1867, he received eight patents, and by the early seventies had sold rights to some for amounts ranging up to $6000.

In 1880 Drawbaugh broke the news that between 1866 and 1876 he had invented the Bell telephone, the Edison transmitter, the Blake transmitter, the Peirce mouthpiece, and indeed most of the successful improvements in American telephones up to 1880 (though, as it turned out, none of the improvements that appeared subsequently). He explained his failure to apply for patents earlier on the grounds of having been too poor to pay the fees and furnish the models.

When asked in court to explain the mental and experimental processes leading to these inventions, Drawbaugh replied: "I don't remember how I came to it. I had been experimenting in that direction. I don't remember of getting at it by accident either. I don't remember of any one telling me of it. I don't suppose any one told me." However, one of the Bell Company's experts made an elaborate series of charts showing that in "specific mode of organization, details of arrangement and peculiarities of form" every one of Drawbaugh's alleged telephone inventions was identical with types well known by 1880.

In the late sixties and early seventies, at his shop three miles from Harrisburg, Drawbaugh had indeed exhibited something he claimed to be an electric speaking telephone (which was already being forecast as a future development) to several score of his innocent neighbors, though he confined the demonstration to the muffled transmission of words from one floor to the next below. Though hazy as to the mechanism, the neighbors swore to having heard this done; and no doubt they had heard it, probably by the mechanical conduction of sound along a hard rod or a taut cord. But until 1880 no one else knew of these alleged inventions, or at least no one took them seriously enough to consider commercial development of them. Drawbaugh did not list them on his card, presumably seeing them as less notable than his "Tram & Red-staff for leveling face of Millstone." He admitted under oath having spent five days with a friend at the Centennial Exhibition and hearing all about Bell's exhibit, yet not mentioning to anyone that he had already invented the first four years of telephone technology in advance. And he had no explanation of this uncharacteristic reticence.

In 1880 certain promoters, mostly businessmen of Cincinnati and New York, purchased Drawbaugh's claims, organized the People's Telephone Company (probably so named to capitalize on rising popular hostility to big business), sold as much stock as they could, began putting out telephones in New York, and were therefore sued for infringement. For lawyers this meant livelihood. The Bell Company's New York counsel alone earned fifty thousand dollars in fees from the Drawbaugh litigation. For historians, the chief benefit was the 125 pages of testimony by Bell during February and March 1883, in response to four weeks of cross-examination on his Dowd deposition. It was this turn on the stand that earned Bell the commendation already quoted from James J. Storrow.

In the end, of course, after eight years of litigation, Drawbaugh's claims were rejected. He himself had meanwhile been reprimanded by the court for his falsifications. For his consolation, however, there was the twenty thousand dollars paid him at the outset by the People's Telephone Company. With that he went back to obscurity until 1903, when he briefly reappeared to announce that he had invented radio before Marconi.

Bell was brought to the witness stand one last time because of an invention by Dr. James W. Rogers of Tennessee. The invention was juridical, not telephonic. Rogers recognized that any trumped-up legal challenge to the Bell patents would protect infringers in fact even if the challenge were ultimately defeated in law, provided it could meanwhile be made the pretext for delay in enjoining the infringers — a delay, say, until the patents expired. In short, whereas others had tried and failed to take the big gun, Rogers proposed instead to cut off its ammunition. Such was the principle. The mechanism was bribery.

Rogers had a son, J. Harris Rogers, who was an electrician, and a friend, the dishonorable Casey Young, who was a former member of Congress from Tennessee. The son provided the inventions, which seemed to be about one percent inspiration and ninety-nine percent tracing paper. They were, in fact, such transparent infringements that the Bell Company could not at first see why anyone should trouble to submit them to litigation. Casey Young provided the brazen shield. As soon as he and his confederates incorporated the Pan-Electric Company, with a paper capitalization of five million (at an actual cost of $4.50 for the use of the Tennessee state seal), he bought the names of sundry illustrious ex-Confederates for large chunks of stock. General Joseph E. Johnston became president of the company, for example, and Senator Isham G. Harris of Tennessee went on the board of directors. Ten percent of the total stock was given to Augustus H. Garland, former governor of Arkansas and currently a senator from that state, who was also appointed attorney for the company. He rendered a first quid pro quo by solemnly assuring the public that the Rogers patents did not infringe the Bell patents. This induced the gullible, and those with faith in the abundance of the gullible, to invest generously in Pan-Electric stock. More money came in from the organizing of subcompanies, the Pan-Electric promoters individually getting blocks of their stock and the Pan-Electric Company receiving cash bonuses in exchange for the questionable right to put out telephones under the Rogers patents.

Already the original promoters were far ahead of the game. And they had no intention of quitting it at that point. In 1884 they got a bill through the House of Representatives authorizing the federal government to bring suit for the vacating of patents under certain circumstances. This failed in the Senate. But the election of Grover Cleveland that year revived the influence of Southern politicians in the executive branch. To head the Department of Justice as attorney general, President Cleveland appointed none other than Augustus H. Garland. The Pan-Electric ring now looked forward to a Golden Age.

They immediately asked their fellow stockholder, the attorney general, to sue in the name of the United States for the annulment of Bell's patents, on the grounds that Bell had obtained them by fraud and was not the first inventor. Propriety or prudence led Garland to go deer hunting in Arkansas, so that the solicitor general, a Virginian, could hastily grant the request while his superior was off in pursuit of the fast buck. Again for propriety's sake, President Cleveland submitted the question to the secretary of the interior, a Mississippian; but the latter upheld his fellow Southerners by the tortuous reasoning that, whether the charges against Bell were substantial or not, allegations of fraud in the matter had now been made against federal officials (in the Patent Office) and should be tested by means of a government suit at government expense against the American Bell Telephone Company.

In January 1887 the Government case began its dreary course. Fortunately for the Bell Company, the scheme to prevent injunctions against infringers fell through. Injunctions were granted against various infringing subcompanies; and faced with the prospect of arguing the merits of the Rogers claims, the parent Pan-Electric Company did not contest the issue. So the Pan-Electric schemers did not gain all they had dreamed of. But the government suit cost them nothing, and in its early stages it made their operations easier. In September 1885 the *New York Tribune* exposed the Pan-Electric stockholdings of the attorney general along with those of other past and current holders of high federal office. Other newspapers joined in the attack. But a congressional committee whitewashed Garland, and Cleveland, though uncomfortable about the matter, retained him as head of the Department of Justice. The Government case dragged on while the Supreme Court upheld the Bell patents, and even after the patents expired. The government's counsel in the case also expired in the summer of 1896, after which the government let his case follow him. It was never formally settled.

The government suit charged Bell with claiming the invention of something already widely known to exist in the form of the Reis "telephone," and also with somehow concealing the existence of the latter from the

Patent Office's expert examiner in that field, Zenas F. Wilber. It charged Bell's attorneys with bribing Wilber to show them what was in Gray's caveat, to let them insert it into Bell's application, and to shuffle the latter into priority over the caveat. And it accused Bell of being an accessory after the fact to this string of alleged misdeeds.

Early in October 1885, Bell returned with his family from a long trip to Newfoundland and Cape Breton Island to find these charges filling the newspapers. He stayed up through all or most of the night for several days in a row to compose a long, coldly furious refutation of them in a letter to Attorney General Garland, along with a stinging indictment of Garland's own conduct. Storrow toned the letter down a little in the latter respect, but approved it otherwise; whereupon it was printed as a nineteen-page pamphlet, and copies were sent to the Justice Department's files, to President Cleveland's desk, and elsewhere.

As Storrow doubtless expected, the letter did little good in the litigation, but it helped Bell to let off steam. He needed that relief, since his chance to testify in the case did not come until the spring and summer of 1892, when he spent nine weeks, at intervals, on the fullest consecutive account of his telephonic work that he would ever give. The Bell Company's historian of the litigation called it "the most detailed and best arranged statement of his telephonic work." Bell went over the whole ground afresh in direct testimony and was cross-examined at length, all parties concerned having the benefit of his previous depositions and the enormous mass of accumulated testimony since the beginning of telephone litigation. Printed in full by the American Bell Telephone Company in 1908 "because of its historical value and scientific interest," the direct and cross-examination of Bell alone fills 445 pages.

But the telephone controversy did not die with the Government case. It merely left the courtroom.

Amos Dolbear did not mellow with the years. He considered himself "outrageously swindled" when he discovered, after selling out his telephone rights in 1878 to the Western Union for $10,000, that it had been ready to pay $100,000. (In the same statement he remarked that he himself had been ready to take $5000.) In 1879 he invented an "electrostatic" or "condenser" telephone, which was essentially a Bell telephone with a condenser instead of a closed circuit. Thus Dolbear had a pretext for denying that what undulated in it was a "current." Nevertheless, in 1881, after the Dolbear Electric Telephone Company had been incorporated and begun telephone operations, the Bell Company enjoined it and was sustained in 1886 by the Supreme Court.

Dolbear was neither a charlatan like Drawbaugh nor an innocent like Meucci. In 1882 he received a patent in wireless communication, after some experiments that gave him a valid place in history as a precursor of Hertz and Marconi. He remained an effective and popular professor of physics at Tufts College, where his students affectionately nicknamed him "Dolly." But an alumnus remembered for half a century the impressive way in which "Dolly" displayed a box to a freshman physics class in 1895 with the declaration, "That is the first telephone that was ever invented. I invented it." The information lodged in freshman notebooks and apparently even in freshman minds. Some alumni many years later affixed a bronze plaque to the brick wall of Ballou Hall at Tufts in commemoration of Dolbear's claim.

Elisha Gray's bitterness seems to have crystallized in 1885. It was then he began claiming that his counsel in the *Dowd* case had put up only a sham defense after negotiations got under way between the Bell Company and the Western Union. Such a betrayal was self-evidently improbable. It would have been irrational for either side to have given away anything irretrievably to the other before the negotiations were settled and the bargain sealed. For Gray to have supposed otherwise suggests that disappointment had at last begun to affect his judgment. Gray's counsel, William D. Baldwin, promptly denied Gray's charge. Today we may add to common sense and Baldwin's sworn statement the weight of the Bell Telephone Company's archives, now open to historical study. They show in detail that it was the progress of the *Dowd* case that affected negotiations, not the other way around as Gray charged, until early September 1879, by which time both Gray and Bell had presented their cases. And then, far from the Western Union's letting up on Bell, it was the Bell counsel who agreed to postpone their cross-examination of the other side during a crucial stage in negotiations.

A second and perhaps more decisive factor in Gray's turn to bitterness developed in 1885 and 1886, when Zenas Wilber (probably liquored up or bribed, or both, by agents of the Globe Telephone Company) made affidavits that he had allowed Bell to examine Gray's caveat in full. Between these affidavits, Wilber made others that directly contradicted them. But Gray chose to believe the worst and brood over it — though even had it been true, it would have had little significance.

Like Dolbear and Bell, Gray went on inventing. His most important later invention, patented in 1888 and 1891, was his "telautograph," which transmitted facsimiles of handwriting. It saw much use in banks and railroad stations. All in all he received about seventy patents and earned more money from them than Bell and his wife ever got from the telephone. But he spent most of it on further research.

In November 1900 an article by one John Paul Bocock appeared in *Munsey's Magazine*. It was appropriately called "The Romance of the Telephone," and its burden was that Bell and his coconspirators had cheated Reis, Dolbear, Gray, and others out of their just claims to the invention of the telephone. "Interwoven in this story," wrote Bocock, "are such marvelous oaths, such charges of corruption and treachery, such tales of ruin and oppression, such accusations against men high in the public esteem, such sacrifices of truth and honor, such disappointments and defeats of the many who have sought to share the reward of the one, that the bare relation of them all, were that possible, would surpass any romance ever written."

The inventory was fairly accurate, but the discreditable items in it were not chargeable to Bell or the Bell Company; quite apart from character, the Bell side had truth with it, which made dishonesty pointless. Nevertheless Bocock's insinuation was to the contrary, and it stirred Elisha Gray to lash out one last time. He wrote a letter to the *Electrical World and Engineer* insisting that, through Wilber's alleged disclosure of the caveat, "I had shown [Bell] *how* to construct the telephone with which he obtained his first results." Gray's letter appeared in print early in 1901, soon after his sudden death near Boston while experimenting with underwater signaling to vessels at sea. Among Gray's belongings was found a scrap of paper bearing the words: "The history of the telephone will never be fully written. It is partly hidden away in 20 or 30 thousand pages of testimony and partly lying on the hearts and consciences of a few whose lips are Sealed. — Some in death and others by a golden clasp whose grip is even tighter."

What Bell thought of it all appears in a letter of thanks that he wrote in March 1901 (but did not send) to the author of a reply to Gray's posthumous charges:

Ever since the commencement of litigation in telephone matters, I have been obliged to keep silence — my counsel always advising me that the *Courts* would look after my reputation and sustain my rights — which they have always done. This was pretty hard to do at first, and I can remember how I used to writhe — in silence — under the unscrupulous attacks which were made upon me. But as years went by I became callous and indifferent as to what people thought or said about me or the telephone. . . . For some time past I have felt that the articles which have appeared in the public press demanded some reply, but I did not care to undertake it myself, and my old defenders have all passed away. . . . I had almost reached the conclusion that the time had come for me to speak out in my own behalf, when the sudden death of Elisha Gray caused me to change my mind.

I had a very high respect for Elisha Gray, and have always had the feeling that he and I would have become warm friends had it not been for the intermeddling of lawyers, and the exigencies of law-suits. . . . Whatever Mr. Gray may have thought of me, I have always had the kindest feelings towards him; and it therefore seemed inopportune that I should say anything in conflict with his claims at a time when we are all mourning his loss.

23

Disconnected

Long before Bell ceased to be a courtroom witness in telephone matters, he became a witness to them in another sense — that of a detached onlooker.

Soon after his return from England, he acted for the last time as a director of the Bell Telephone Company. Late in January 1879, in debt and with its treasury bare, the company faced bankruptcy. During February and March, however, it escaped that fate by merging with the New England Telephone Company into a new National Bell Telephone Company with greater capitalization. As a director, Bell actively resisted efforts of "the monied men" to reduce the influence of the original patent associates in the new corporation. In particular he did what he could to continue Gardiner Hubbard as president, partly out of regard for his abilities, partly from family loyalty, and partly, Bell wrote, as "my guarantee that no arrangement will be made with Western Union until my reputation has been cleared." But the monied men could stomach no more of what they considered to be Hubbard's fiscal irresponsibility, and so the Boston financier William H. Forbes became president. "Poor Papa," wrote Mabel, "felt very badly about it altogether." Still, Bell had already judged Forbes to be "the only man [on the board] who impresses me as possessing marked ability," and so Bell's chagrin was less acute. Besides, both Hubbard and Thomas Sanders were appointed to the controlling five-man executive committee, so the patentees were not immediately frozen out.

Nevertheless the old guard's power was clearly passing. Bell was left out of the new board at his own request. Gardiner Hubbard continued to play some part in the company's affairs. But Thomas Sanders had already resigned as treasurer, and he soon withdrew to the sidelines. Tom Watson found himself rich, overworked, yet still young enough to hanker after "a larger life and new experiences." The "lonely, fascinating, pioneer work" of early days was now parceled out among many workers. So Watson quit the telephone company entirely in the spring of 1881 and turned eagerly to travel and self-education. Bell's position by then was stated in his reply to an inquiry in February 1880: "As I do not attend to the business matters in connection with the Telephone, I have referred your letter to Mr. G. G. Hubbard."

An indirect tie with telephone affairs was also cut. Bell's father, to whom Bell had given seventy-five percent of the Canadian rights in 1877, had found the telephone business "very up-hill work in Canada." He had enjoyed giving mellifluous public demonstrations of his son's invention, but not being harassed by scheming rivals and technical problems. So with some help from Alec, Alexander Melville Bell sold his Canadian interest to the National Bell Telephone Company at the end of 1879 for one hundred thousand dollars, which freed him and Mrs. Bell to move nearer their son. It freed them also from money concerns. "I seem always to have had . . . the necessity of considering ways and means," wrote Mrs. Bell, "and I wonder if I can ever break myself of the habit of doing so." It was a challenge anyone might welcome.

Since returning to Boston from Europe late in 1878, Bell had been drawing a salary — three thousand dollars a year from the Bell Company, supplemented by a thousand dollars each from Hubbard and Sanders — in return for postponing his professional work, making himself available as a witness in court, and assigning the company any new telephonic inventions. After Forbes became president of the new corporation, he and Bell differed on where Bell should live (Forbes urged Boston), on how much he should be paid, and on how much emphasis he should give strictly telephonic research. The agreement was patched up for another year, but finally ended in July 1880.

While it remained in effect, Bell filed six applications resulting in telephonic patents: an arrangement of tubes for simultaneous listening and speaking to one instrument; the twisted-wire metallic circuit he had developed in England; an arrangement for cutting a telephone into and out of the main circuit; an elaborate call bell; and two types of variable-resistance transmitters. Aside from duty, he was probably driven by pride. Mabel had reported talk by company officials that "the discovery of the telephone was an accident, and you can do nothing further." And she urged him to "bring out something no matter what, so it proves that that was not the end of you." The transmitters showed some ingenuity, one using the variation in electrical resistance of contacts sliding at varying speed, the other using the varying separation of conductive particles on a plumbago-coated balloon as it expanded and contracted. But none of the devices, except the twisted-wire circuit, offered enough practical advantage to be used commercially.

Bell did not have the proper temperament for inventing on demand, not even on Mabel's demand.

In 1881, after quitting the company, Bell took out patents on a telephone cable design and on a receiver in which the current vibrated a diaphragm in passing through it, rather than electromagnetically. These also found no commercial use. They were his last telephone patents. "I have not kept up with the literature of telephonic research," he remarked matter-of-factly in 1888. In 1891, perhaps aware of Elisha Gray's "telautograph," he wrote that "my old autograph telegraph should be developed and perfected into a practical machine." But he did not attempt it. In 1895 and 1896 he did tackle the problem of an automatic switchboard, convinced for a while, as Mabel understood it, that his plan "does away entirely with the switchboard in the central office." He never got it into patentable shape. In 1904 he told a newspaper reporter that some day "there may be a system by which the subscriber can move certain buttons and call up whom he pleases . . . automatically." But by then he seemed willing to leave the prize to others.

Thus Bell's telephone work receded into the storied past. A first rush of historical consciousness had come over Tom Watson even before Bell's wedding day, when Watson dismantled the outgrown laboratory at 5 Exeter Place in Boston, its function now being shifted to Charles Williams's establishment. From it Watson salvaged a memento, which he wrapped carefully and labeled with the words:

This wire connected Room No. 13 with Room No. 15, at 5 Exeter Place, and is the wire that was used in all the experiments, by which the telephone was developed, from the fall of 1875 to the summer of 1877, at which latter time the telephone had been perfected for practical use. Taken down July 8th, 1877.

T. A. WATSON

Only a few months later Watson was writing Bell as if an age had passed — as in a sense it had, and not merely for Watson and Bell. "Do you ever think of the days we spent together at Exeter Place?" he asked. "Some were dark and gloomy but I hope we shall see no more such. It seems like a dream to me when I look back and think of the changes and events that have taken place in the last two years."

Bell himself was early mindful of his place in history. During the spring of 1879, engaged in researching the historical background of telephony for the *Dowd* case, he decided to write a scholarly history of the subject as "a means of establishing my reputation upon an enduring foundation." By this he meant his reputation not only as inventor of the telephone but also as a scholar. Mabel remarked in her journal that "however hard and faithfully Alec may work on his book he cannot prevent ideas [for new inventions] entering and overflowing his brain." She was right, of course, and the project never got beyond a bundle of notes.

Bell need not have worried about his fame as inventor of the telephone. Elisha Gray's dissent notwithstanding, the evidence piled up in eighteen years of litigation gives it a historical foundation as massive and solid as Edinburgh's Castle Rock. Bell's place in the public mind was further secured by Gardiner Hubbard's calculated policy of attaching Bell's name to the instrument itself and to the companies that offered it. This was, to be sure, good business strategy at a time of competition from other models. But Hubbard more than once gave his concern for his son-in-law's permanent fame as one motive.

And by chance, the very tones and overtones of Bell's name worked to the same end. Though deaf, Mabel had a good ear for the rhythms of speech, and her insistence on Alec's always using his full name showed this. The four-two-one progression of syllables gives the name a gathering force and culminating impact. The last syllable itself has the initial blow, metallic peal, and liquid diminuendo of the age-old communications device it echoes. The name "Bell," moreover, goes well with the first syllable of "telephone." Its sense is fitting to the instrument. And as Bell himself demonstrated in a drawing at the top of a youthful acrostic sonnet to his father, the name lends itself to a simple, shapely, instantly recognizable emblem, so that Bell's name survives as a device of corporate heraldry, however impersonal the official corporate designation.

Reinforcing these powerful guarantees of celebrity was Bell's eye-catching and ear-seizing personal presence. At first, as he grew stouter and grayer, he did not prize his premature venerability. "I have a very great personal objection to having my photo published," he wrote Edison at thirty-two. But this objection softened as his years caught up to his looks. And whatever his own feelings, the striking contrast of the dark, luminous eyes, mobile mouth, and vibrant speech with the patriarchal beard and imposing figure seemed eminently appropriate to a man of mark. He resisted the publicizing of his private life, but welcomed interviews about the past and future of the telephone, keeping scrapbooks of clippings and by the mid-nineties, if not earlier, subscribing to a clipping service. These clippings show him safely established in the hagiology of the new age. Even if a little less exalted than Edison, he was accepted as a miracle worker:

"Are you the inventor of the telephone, sir?" asked the conductor of a train in North Carolina in 1880.

"Yes," replied Bell.

"Do you happen to have a telephone about you? The engine has broken down and we are twelve miles from the nearest station."

The *New York World* published a Christmas sentiment from Mark Twain in 1890: "It is my heart-warm and world-embracing Christmas hope and aspiration that all of us — the high, the low, the rich, the poor, the admired, the despised, the loved, the hated, the civilized, the savage — may eventually be gathered together in a heaven of everlasting rest and peace and bliss — except the inventor of the telephone." Gardiner Hubbard promptly wrote Twain in good-humored remonstrance and got an equally prompt reply addressed "To the Father-in-law of the Telephone." Twain explained his position as one of personal war with the Hartford telephone system, "the very worst on the face of the whole earth." No man, he insisted, could make a twenty-word message intelligible over it in less time than a week. "And if you try to curse through the telephone, they shut you off. It is this ostentatious holiness that gravels me. Every day I go there to practice, and always I get shut off." Since Professor Bell had invented the instrument in the first place, he could not escape responsibility. "For your sake I wish I could think of some way to save him, but there doesn't appear to be any. Do you think he would like me to pray for him? I could do so under an assumed name, & it might have some influence." But Bell was probably doomed, and so he had better come up and use the Hartford telephone; that would reconcile him to Hell. "Meantime, good wishes and a Merry Christmas to *you*, Sir!"

Mark Twain had as great a reverence as any American for technology in general, including the telephone, which as early as 1877 he had used as the basis for a short story in the science fiction genre. That he took for granted the canonization of Bell is made more, not less probable, by his Christmas joke, the whole point of which is in its tone of playful blasphemy.

Nevertheless, Alexander Graham Bell was, like Edison, a human being, not a saint. Like Edison and other human beings, Bell had elements of vanity, wishful thinking, and mischievous humor in his nature. And these three elements, along with his irrepressible dramatic sense, once in a while made him yield to the temptation of a reporter who was ready to be awed himself and eager to awe his readers. So Bell would obligingly hint at new marvels in the making, marvels that in the end did not come off. To his credit be it said that the hopes were real, even if slighter than he let on, and also that he took care to hedge them with doubts.

As the years and milestones passed, the telephone industry honored its father. One October evening in 1892, at the New York offices of the American Telephone and Telegraph Company, about a hundred telephone officials, politicians, and newspapersmen gathered for the formal opening of telephone service between New York and Chicago. In what may have been an unintended echo of Bell's early telephone lectures, the program opened with a cornet solo. There were formal addresses and other obligatory rituals. But the historic highlight was Bell's conversation from New York with William Hubbard, his Centennial Exhibition assistant, in Chicago. It was Bell who was center stage at the receiver when the photographer took a flashlight photograph for posterity.

The opening of the New York-Chicago line, 1892

In 1911 an association called "The Telephone Pioneers of America" was formed for those who had worked in the telephone industry for at least five years and some part of whose service dated back at least twenty-five years. In a letter inviting Bell to join — which he did with pleasure — occurred the piquant remark that "the earliest date shown is that of Mr. Enos M. Barton, who dates his beginning in 1876. I know that your date will considerably precede this." Barton had dated his telephone service from his association with Elisha Gray.

It was in 1911 also that a new regulatory act made it illegal for telegraph and telephone companies to issue franks to those outside their employ, as the telephone company had long done for Bell in recognition of his historic contribution. Theodore Vail, president of the company, worked out a technically legal dodge to continue the practice, but Bell declined it, "as I do not have any official connection with telephone or telegraph companies, [and] I do not wish to . . . [appear] to evade the law."

Bell's connection with the telephone may have been officially ended, but it was historically imperishable. Over and over to the end of his life, in interviews, in published writings, in addresses, and, it may be supposed, in private retrospect, he relived the great days of the seventies. If it be charged against this narrative that its record of those days is out of chronological proportion, it can be said in reply that for Bell, as for everyone, there were days that filled his years; and a biography that reflects that fact is truer than the most unremitting daily journal.

Nevertheless the chief actor in that recurrent drama was a tall, thin, nervous, dark-haired young man. And so its stately, gray-bearded reviewer could not help remarking now and then, "I have become so detached from it that I often wonder if I really did invent the telephone, or was it someone else I had read about?"

PART THREE

After the Telephone

24

The Uses of Prosperity

In the Bells' rented house in Washington one day in March 1879, while Alec pored over old letters preparatory to a *Dowd* case trip to Boston and Mabel sat on the floor amid the clutter of his packing, they "got into a long discussion on riches." Her sights were set on fifteen thousand a year, a scale of living that would include a "fine house and carriage." He disapproved. Five thousand a year would be plenty. Did she want him to give up scientific work and devote his life to making money?

"No," she said, "only I want the money too, if I can get it."

"So you shall, my dear," said Alec, turning toward her so that she could read his lips clearly.

"Well," said Mabel, "that's all I want to know. I may have it if I can get it."

The fount of this promised fortune had begun as a one-third interest in the original telephone patent. By cutting in Thomas Watson for a one-tenth interest when the Bell Company was organized as a voluntary association in 1877, each of the three original associates was reduced to a thirty-percent interest. When New England was turned over to the newly incorporated New England Telephone Company in February 1878, the Bells received 301 shares or fifteen percent of the new company, since half of its total of 2000 shares were earmarked for raising capital. (The source of the odd share is not clear.) The Bell Telephone Company, incorporated in July 1878 to replace the original Bell Company, earmarked about a third of its 4500 shares for new capital. The Bells received 813 shares as their proportion of the remainder. Mabel sold eight of these shares. Thus, when the two companies were consolidated in March 1879 and their respective shareholders exchanged share for share of the new National Bell Telephone Company's total of 7250 shares, the Bells came out with 1106 shares or a little more than a seventh of the total.

To drop from a one-third to a one-seventh interest sounds like a comedown. In fact, because the newly recruited shareholders represented vivifying capital, the Bells were better off than before. Whereas to bolster Bell's spirits in December 1878 Gardiner Hubbard had assured him that his and Mabel's stock could be sold immediately for $50,000, in March 1879 it was quoted at $65 a share for a total market value of $71,890. Mabel evidently considered so giddy a rise as too good to last. Two days after the "discussion on riches," she wrote Alec in Boston:

Enclosed please find a blank power of attorney to sell seven hundred of my shares at sixty-five dollars each immediately. Mama says Papa and Mr. Morgan both think this is the time to sell, that later on I won't be able to get so much money, so please sell out immediately, please, *please, please, please, please,* *PLEASE PLEASE PLEASE.* Are you sufficiently impressed by the importance of the subject now? If you love me *do* do something right away the moment you get this.

Here was a business duty in which Alec would have gained by being dilatory. How promptly he acted is not certain, only that at the end of August 1879 their telephone stock holdings amounted to about eight hundred shares, so that he could not have sold much more than three hundred by then.

In June, Mabel wrote her mother that what was left of their stock was quoted at $110 a share, "and we feel very rich." Early in September it rose to $300 a share; and Bell cashed in on what he took to be the speculators' fever, selling two hundred shares at that price. The next day shares rose to $337. The Bells now had six hundred shares left. Two days after that the stock stood at $350, and Gardiner Hubbard refrained from selling all his own shares only from the fear that to do so would prick the bubble. He was cheerful anyway. "Eighteen months ago," he wrote his wife, "I knew not where to beg or borrow one hundred dollars. . . . It does seem as though the good time might really come at last. I wished several times last night that we were ten years younger and might have that much more time to enjoy it in."

Early in October, Bell sold fifty of his wife's shares at $500 and another fifty at $525. That left them with about $150,000 in cash and five hundred remaining shares in the Bell Company. Bell now thought it might be wise to hang on to those for awhile. But he did sell a dozen or so at $775 a share in early November. A week later the Bell Company announced its agreement with the Western Union, and Bell shares sold at $1000 each. "We are beginning to realize that we have wealth," Bell wrote his father just before Christmas. "I think that we have too much interest in the Telephone Company. I am not sure but I think that Mabel is still the largest stock holder."

In 1880 the National Bell Telephone Company was reorganized as the American Bell Telephone Company with 73,500 shares, which were exchanged at the rate of six to one for those of the previous company. At the end of that year Mabel owned 2975 of the new shares. Unspent money from previous sales was conservatively invested in bank and railroad stocks. From all American sources, Bell wrote his father at that time, their income was $24,000 a year. "We should be able to live on that."

In 1881 Bell sold nearly a third of their remaining telephone stock, putting most of the proceeds into United States bonds. Nevertheless the telephone stock retained at the end of the year was worth $75,000 more than the stock held at the start. In addition the Bells now had more than a quarter of a million dollars in other investments and bank accounts. An inventory in 1883 showed all their American investments (not including more than $100,000 in the trust fund) as worth $900,000 and yielding an annual income of about $37,000. All assets considered, Bell had by a narrow margin become a millionaire.

With this the Bell fortune seems to have reached a financial plateau. The Bells kept the remainder of the telephone stock as a permanent investment; at the beginning of 1885 it amounted to 2038 shares, which in the preceding year had paid $32,380 in dividends. But Mabel informed Alec at that time that, aside from the trust fund (now approaching $200,000), "my total income and capital from all sources on 1st. January 1885 was $806,475.50." The income from her stock in what now became the American Telephone and Telegraph Company did not approach the 1884 figure again until 1904, when it reached $31,000.

The Bells usually spent all their income and sometimes more. Gardiner Hubbard chided his son-in-law in 1893 for not being "as considerate in regard to your expenditures as you were ten or twelve years ago." They often had to meet current bills by borrowing against expected dividends. "I do try to be careful," wrote Mabel in 1896, "yet there is eight hundred owing on the vase for Mamma's golden wedding. . . . I will not hire the carriage and pair as I intended, but I must get a trap because the carryall Perrin drives is really not respectable." Bell wrote Hubbard a few months later, "I highly approve of your plan of re-investing the income of the Trust as far as possible, it is the only way in which we can save anything." In 1900 Mabel reckoned total income as $37,615.63 and total outgo as $45,415.53.

That level of spending was lavish by the standards of those days. Yet while the Bells, as will be seen, lived well, they did not indulge in riotous waste. Over a period of some forty years, about $950,000 went for Bell's nonpaying researches and experiments. Bell contributions to the welfare of the deaf totaled about $450,000 in the same period, although this included money from permanent endowments by Bell as well as from the Bells' current income. In the 1880s, moreover, the Bells joined Hubbard in financing a large diversified farming operation at Moxee, Washington, which was a losing venture for many years. Its chief yield for the Bells was of jobs for various Bell and Hubbard relatives. So, even allowing for more than $200,-000 from phonograph patents, $200,000 inherited from Mabel's mother in 1909, and what were probably substantial earlier bequests from Mabel's father and Alec's father and mother, as well as such windfalls as an unexpected dividend of $22,500 from the International Bell Telephone Company in 1901, the Bells were in no position to live on the imperial scale of the Morgans, Vanderbilts, and Carnegies.

The public, when they thought of the matter, supposed otherwise. The telephone industry, unlike steel, oil, copper, and automobiles, gave rise to no colossal personal fortunes. Yet it became an enterprise of great magnitude, and some outside it probably assumed that if the Bells had started with a one-third interest in it, they continued to own a third of it. Even some more sophisticated observers grossly exaggerated the Bells' fortune. In 1889 Bell came across an article in *Cosmopolitan Magazine*, "Wealthy Women of America," which credited Mrs. Bell with a share in "her husband's millions" and the expectation of inheriting "several million more" from her father (whose estate in the end amounted to little more than half a million). "I wonder," wrote Bell to his wife, "whether the wealth of the other ladies mentioned is as much exaggerated?"

Mabel had already felt the consequences of "the idea of our great wealth." "Begging letters are easily left unanswered — not so personal appeals, especially from acquaintances for charitable purposes. Tradespeople also are only too apt to overcharge, . . . [to a degree] that is very appreciable on our moderate income." She wrote that in 1885. A quarter century or so later, her small grandson witnessed the practice:

Grandfather's house on Connecticut Avenue was designed in the old-fashioned way with a great circular stairway curving around the center. One afternoon Grandmother came regally down to greet a piano salesman and his boss. As she paused on the bottom step, the young salesman started his pitch. She answered him perfectly so he was not aware of her deafness. The boss, standing off to one side, said loudly, "Don't be a fool, Sam, that's Mrs. Bell. The price *to her* is $100 *more*." The salesman was obviously shaken, so the boss took over. Standing in the wings, I thought this the funniest thing I had ever heard! But afterwards I tipped Grandma off to what had been said behind her back, and there was no sale.

Despite a lifetime of such encounters with greed and trickery, Mabel remained generous in large matters. "You may think me superstitious," she wrote Alec in 1883, "but I feel as if no good can come of our wealth unless we try to do good to others besides ourselves with it. So that although the morbid desire to do something for my fellow deaf-mutes *at a distance* still continues, I do not *dare* oppose or discourage your doing what you can to aid them."

But her experience with cheaters inclined her to balk at her husband's impulsive dribbling away of a hundred or two hundred dollars at a time in response to some pathetic plea or other. "You say I have never been pleasant when you proposed giving away a hundred dollars or so but we have never been even with our accounts," she explained in 1898; "if we were I really would say nothing." Thus in 1905 Bell passed along a request from one of his Clarke School pupils of 1873 for a loan of three hundred dollars to build up a poultry and fruit business on his small farm. "Dear Mabel," Bell penciled on the letter, "Can I give this old pupil $300 on his note? Would like to help him, & the deaf have few friends able to help them. May I cast my bread upon the waters in this case?" But no bread was launched. In 1900, on another letter requesting a contribution toward erecting a statue in Washington of their old Brattle Street neighbor Henry Wadsworth Longfellow, Bell summed up Mabel's role in a whimsical note:

Respectfully submitted to M. G. B.
$250.   ?
  25.00 ?
   2.50 ?
    .25 ?   *AGB*

Longfellow now stands in bronze effigy near a Connecticut Avenue bus stop; but while the Bells' share in this inspiration to passengers and pigeons is not certain, Bell probably anticipated the trend of his wife's response.

Even in such cases, Mabel played a part more of restraint than of interdiction. At Alec's instance in 1881, she lent five thousand dollars toward a fresh start by one of his father's old friends, who had been ruined by an unscrupulous partner. Friends of earlier days in Canada and Scotland could count on loans of one or two·or three hundred dollars. Five hundred went to the self-styled "Joseph Faber," the speaking machine man, in 1885. There were loans and gifts to occasional students, including an American studying opera in Italy and a Japanese studying meteorology at Columbia University. When Percival Richards importuned Bell for help during the eighties, Bell was anguished by his lawyers' strict injunction against giving

Richards any help that could possibly be traced back to him, for fear of prejudicing the Government case, in which Richards's testimony was likely to figure. Nothing stopped him in 1891, however, from throwing at least five thousand dollars, possibly more, into what he knew was a hopeless effort to save Thomas Sanders from going bankrupt through bad investments. Afterward he saw to it, at considerable expense over a number of years, that young George Sanders learned the printer's trade, was established in his own printing office, was given business, was helped out from time to time by loans and gifts, and did not sink into the bleak penury that so often deepened the curse of deafness.

Aside from the money he poured into more general aid to the deaf, Bell was often moved by special appeals. One totally deaf young man whom Bell helped to attend Columbia University ended up by giving him as a reference in an attempt to cash forged checks. But Bell was not dismayed by that. Among other benefactions was one of fifteen thousand dollars for the support of two small children after their deaf-mute father had died penniless.

And of course there were the less personal gifts: to the YMCA, to the Garfield Hospital, to the Walter Reed Memorial Fund, to the relief of political exiles in Siberia in 1891 and of destitute Cubans in 1898, and so on.

For all that, the Bells did not stint themselves. They could and did live where they chose and pretty much as they chose.

Alec dissented for a while as to place of residence. "I am afraid a quiet life in Washington is too good a thing to be hoped for," he wrote Mabel in March 1879, "so I think we all had better settle down together here in Cambridge." That year the Bells summered in the Hubbard house on Brattle Street, the spacious Victorian home where Gardiner Hubbard had become excited over the multiple telegraph, where Alec had courted, proposed to, and married Mabel, where the variable-resistance "hole" in the telephone specifications had been plugged. The now-famous inventor fitted out a workshop there and worked away happily while Mabel's two sisters, two cousins, parents, and small daughter — who was just beginning to walk and speak — kept her occupied. In the summer of 1880 they returned again, and Mabel wrote her mother: "I think it will be long before Alec can get over the feeling that this house of all others is 'home.' He keeps saying how nice it is to be here, and how can I want to live away from it."

But aside from happy memories and personal tastes, Alec had nothing to hold him in Cambridge. If he had thought of teaching again at Boston University after getting through with the *Dowd* case, he gave up the idea when the university discontinued its School of Oratory. Meanwhile Mabel and her parents maintained their gentle pressure in favor of Washington. In 1879, after the Bells rented winter quarters there, Mabel reported that "Alec has about stopped railing at Washington and is beginning to find there are nice and scientific people here." In 1880 Gardiner Hubbard consented to invest some of the trust fund in a permanent home for the Bells in Washington. That settled it.

In the fall of 1882 the Bells moved into what a journalist called "one of the largest and probably the most costly of the new houses in Washington," a three-story brick mansion in the conglomerate style of the seventies. If great architecture is frozen music, this was congealed cacophony, the baying of gables and turrets about a mansard roof — or as its contemporaries saw it, "perfectly built and elegant in every appointment." The architect had been given carte blanche and so he spent over one hundred thousand dollars at the low prices of 1879, as a result of which the original owner found himself unable to afford living in the house and therefore sold it to Gardiner Hubbard (as Mabel's trustee) for a reported ninety-nine thousand dollars. The Bells were said to have then spent twenty-eight thousand dollars remodeling it. The house and its grounds monopolized an entire block on Scott Circle, the official address being 1500 Rhode Island Avenue. Its interior embraced, along with more usual accommodations, an oratory with stained glass windows, a billiard room, a library, a music room with a grand piano for Bell's use, and a big conservatory with walls of enameled brick, "claimed . . . to fully equal that of . . . the tower of Pekin." Even the stable was wainscoted in pine, lit by electricity, and topped by steam-heated quarters for servants, of whom the Bells had at least half a dozen.

The brick was "Pompeiian red," a color that seemed appropriate early one January morning in 1887 when a policeman saw flames breaking out of the mansard roof. The fire had apparently escaped from a grate and browsed awhile among the heavy timbers of the roof. Bell was out of town, but his wife was awakened in time when her little dog jumped excitedly on her bed. No one was injured, and the fire was kept to the third floor. But in chasing the fire through the architectural jungle, the firemen thoroughly drenched the interior and ruined its elegant furnishings. Fortunately the damage was largely covered by insurance.

Bell came home to find his library intact and also that of the late Joseph Henry, which he had bought from Henry's widow. His study, however, was a litter of manuscripts and notebooks. From the hosing in bitter cold, ice had built up inches deep on the floor. The housekeeper had hired some men from a nearby rooming house to begin clearing debris and salvaging what they could. One alert and intelligent young man, an eighteen-year-old black Virginian named Charles F. Thompson, so impressed her that she

Bell in his study, 1884

entrusted him with Bell's study. Many years later Thompson remembered in detail the arrival on the scene of Bell himself, a tall, heavily built man with black hair and beard streaked with gray: "With a genial smile, he shook my hand as if he had known me for years. . . . That handshake electrified my whole being." Bell was delighted to find that young Thompson was able to read his manuscript notes. He impressed on Thompson the importance of gathering up and preserving every scrap with figures, writing, or drawings on it. "I suppose," wrote Thompson in 1923, "it must have been the way I carried out those instructions that started me out over the trail of thirty-five years of service with Dr. and Mrs. Bell." In time Thompson became Bell's chief proxy in coping with the gritty details of domestic life.

The Scott Circle house was refurnished as elegantly as before. Two years after the fire, a reporter visited Bell in his top floor study, where he worked from one in the afternoon to four in the morning. A narrow door opened into it from the book-lined library. In the study, books are everywhere. An easy lounge lies in front of the fire, and a globe stands in one corner. At a common flat walnut desk, sitting on an office chair cushioned with green leather, Mr. Bell works. The desk is covered with books and papers. . . . A porcelain hand with letters pasted upon it lies at one side, and this, I am told, is an invention for teaching deaf children to converse with each other by touching certain spots on the hand, which represent letters.

Bell himself had "the dark complexion of a Spaniard. His face is full and regular, and his forehead very high and whiter than the rest of his face. His hair is thick, and its color is that of oiled ebony. His face is covered with a full, black beard, which curls and twists, and his eyes are a soft, velvety black. He dresses usually in business clothing, and he is democratic in his manners." (The reporter tactfully excised the gray from Bell's beard and hair.)

But on March 1, 1889, a few days after the reporter's visit, the house was sold for ninety-five thousand dollars to Levi P. Morton, the incoming vice president of the United States. Morton and his heirs retained it for half a century; and when they did not occupy it, it was usually rented to such eminent tenants as the Russian ambassador Count Arturo Cassini, Secretary of State Elihu Root, the inventor and engineer John Hays Hammond, and Assistant Secretary of the Treasury Ogden Mills. Meanwhile it was again remodeled, happily beyond recognition. Yet when its present owners, an association of paint manufacturers, bought it in 1940, it fetched precisely what Morton had paid for it, or four thousand dollars less than Hubbard had paid in 1882.

In the summer of 1891, after wintering for two years in a house on Nineteenth Street, the Bells began building a new house at 1331 Connecticut Avenue, just below Dupont Circle and opposite Gardiner Hubbard's lawn-girdled mansion. A side lawn separated it from the house that Bell's cousin Charles J. Bell, who had become a successful Washington banker, was already building. The Bells' new home was a three-story, red-brick and stone residence, far simpler and cleaner in its lines than the Scott Circle extravagance. It was commodious, however, and elegantly furnished in the taste of the times. Over the years the Bells added large Sèvres vases, Italian marble mosaics, paintings, carved teak paneling from India, and other souvenirs of their interests and travels. For his Wednesday evening gatherings of notable men, Bell had a special annex built. In their Connecticut Avenue home, the Bells would remain comfortable and contented for the rest of their lives, during the seasons they chose to spend in Washington.

For Bell, Washington had one drawback to which he could not be reconciled: its summer climate. Heat made him headachy and miserable, and sometimes it compounded his distress with heat rash. During the summers of the early eighties he fled with his family to Cambridge, Newport, the mountains of western Maryland, or the North Shore of Massachusetts. In the summer of 1885 his father proposed a still longer flight to Newfoundland, which for the elder Bell would also be a sentimental journey to the scene of his New World sojourn as a young man. Gardiner Hubbard added the suggestion that they all go by way of Cape Breton Island, at the northern end of Nova Scotia, where Hubbard could pay a visit to the Caledonia coal mines at Glace Bay in which he had sunk so much money. Thus Bell came to the Cape Breton town of Baddeck (accented on the second syllable) on one of the labyrinthine ocean inlets known as the Bras d'Or Lakes.

A book by Charles Dudley Warner describing the charms of the town and its setting led Bell to make a point of seeing Baddeck en route. As he sauntered up the road to the Telegraph House hotel, one young woman on the veranda thought him "the handsomest man I had ever seen." She noticed especially his "wonderful head of black hair liberally sprinkled with white." The same observer found Mrs. Bell "a slender, graceful woman, with the gentlest manners, her sweet sympathetic face framed in the most beautiful soft brown hair." The Bells were just as susceptible to Baddeck. Warner had not overdrawn its appeal. From a hillside rising out of Baddeck Bay it looked across a mile or so of blue water to the red bluffs of a high headland jutting into the lake. The scene reminded Bell of the salt lochs he had left behind in his beloved Scotland; and the very names of Baddeck's people — MacKenzie, McCurdy, McDermid — strengthened his sense of homecoming. When their steamer for St. John's ran aground, the Bells cut short their Newfoundland trip in favor of a return visit to Baddeck.

Their September stay confirmed first impressions. "This morning we drove to the New Glen," wrote Mabel in her journal, and saw "forest-covered hills, undulating valleys with trim, well-kept fields and neat little houses, pretty streams. . . . Baddeck is certainly possessed of a gentle restful beauty, and I think we would be content to stay here many weeks just enjoying the lights and shades on all the hills and isles and lakes." Cool air, salt water, security from casual interruptions and the demands of Washington society, and the look of Scotland — Bell could not resist these. In contrast, Washington greeted him on his return with the infuriating slanders of the Pan-Electric conspirators. The Bells came back to Baddeck the next summer already equipped with a plan of the area drawn up by an able young Baddeck businessman named Arthur W. McCurdy, and that summer of 1886 Bell bought fifty acres at the point of Redhead, the peninsula across the bay. As an interim establishment they rented and later bought and enlarged a cottage on the Baddeck side of the bay. For the next thirty-six years, Baddeck would be Bell's summer refuge.

Piece by piece over seven years Bell bought out all the farmers on the headland. He made sure of controlling the access road and the springs that would furnish water. After consulting a local authority on Gaelic in 1889 he renamed his seagirt fastness "Beinn Bhreagh" (pronounced "ben vree-ah") or "Beautiful Mountain," rejecting such grim possibilities as "Cnoc Mhaisheach." Until the main house was done, the Bells lived in the "Lodge," which Bell had designed, near the point.

The building site was to be at the end of the point, in a sloping field that dropped off steeply to the bay on three sides and ran into a wooded hillside on the fourth. According to family tradition, Bell chose the scenic prospect for each room by previewing it from a swaying two-story scaffolding on a hay wagon, his intrepid wife up there with him despite the poor sense of balance that accompanied her loss of hearing. Plans were drawn up by a Boston firm in 1892 and the building constructed by a Nova Scotian contractor for $22,000. In reporting its completion late in November 1893, the *Halifax Chronicle* headlined it rather hyperbolically as "The Bell Palace at Baddeck," with a subhead, "Said To Be The Finest Mansion in Eastern Canada."

"The general style of architecture," reported the *Chronicle*, "is after the French chateau." It was a hopeless pursuit, however, which stopped with the cone-roofed round towers, one of which swelled out at each end of the two-story main section's water side. Between them ran a long, wide piazza or porch. A wing sprouted from the main section at an obtuse angle, and it also had its big piazza facing the water. The house's other exterior features included towering freestone chimneys, an assortment of gables, dormers, and balconies, and a covering of cedar shingles on roof and sides. The "Bell Palace" did not adorn the general scene. But it was to be of long and happy service to the Bell clan, which, like the house, was large and rambling.

Whatever its architectural merits, the big house managed somehow to accommodate twenty-six people besides the servants during one summer in the later days of the clan's proliferation. There were ten fireplaces in addition to the great one in the main hall or drawing room. If the lavish expanses of ash and cherry paneling were rather somber, they made the blazing fires inside and the panorama outside all the brighter. Bell had a big, book-lined room to work in, where for inspiration or mere respiration he could take his ease on a lounge by a big window with a fine view.

Bell loved scenic grandeur as much as when he had daydreamed as a boy on Corstorphine Hill. Here he could feast on it: not the Firth of Forth or

The Lodge and the Bells

The main house at Beinn Bhreagh

Bell's retreat: the houseboat *Mabel of Beinn Bhreagh*

the low, blue Pentland Hills or the Old Town on its hill, but — not so very different — Great and Little Bras d'Or, the island-dotted St. Patrick's Channel, the low blue range of the Shenacadie hills and Salt Mountain in relief against the western sky, and the hamlet of Baddeck on its green hillside. Shipping had to pass so close to Beinn Bhreagh's point that individuals aboard ship could be seen from it, and in those years there would often be full-rigged sailing vessels with canvas bright in the sun. And when the fire at his back and the view through glass seemed too sheltered and tame, Bell could climb alone to the hilltop behind the house to stand in the wind and look down on the world.

Still, not even the surroundings of Beinn Bhreagh satiated Bell. By himself or with Mabel, he explored the woods and cruised the waters for many miles around. A Washington friend, the noted explorer and journalist George Kennan, after a visit with his wife, bought a summer cottage across Baddeck Bay from Beinn Bhreagh. In 1893 Kennan, an ardent lover of nature himself, discovered a group of mountain lakes in the wilds some thirty miles north of Baddeck and bought government land for a cabin. A year later the Bells visited the spot, built their own cabin half a mile away, and joined the astronomer and physicist Samuel P. Langley in buying up two square miles of land around the lakes to protect them from timber speculators and lumbermen. Every September for a dozen years, the Kennans and Bells spent several weeks there boating, fishing, and exploring the forest. When darkness settled over the lonely lake, Kennan recalled, "the profound stillness was broken only by the soft hooting of owls, the occasional bark of a fox, or the wild, wailing cry of a loon."

For Mabel, the darkness was presumptive silence. When still water and motionless leaves were visible, when there was no pressure of a breeze or faint vibration of earth under her feet, she could positively see and feel silence, or so it seemed to her; and conversely, when life stirred visibly and vibrated perceptibly about her, she knew there must be sound. A lifetime of depending on her sight for more than its normal service seemed to enrich her visual impressions. The sensuous joy Alec found in music came to his wife through color and form. So life at Beinn Bhreagh meant even more to her than it did to him. "This is one of the most absolutely beautiful days I ever saw," she wrote her mother from there one winter.

The glory of it is almost pain. . . . Imagine two perfect suns equally bright, one beneath the other, reflected from the glassy sea. Not a ripple — not a cloud. A sheet of ice thin and brittle as thinnest glass seems forming across the lake. . . . It is afternoon — the afternoon sun shining squarely in from over Washabucket's black shoulder. All up St. Patrick's channel is the same still glassy sea, with the little islands dotting it, with far distant hills palest purple against the pale orange sky, and nearer hills white with snow and dark purple. The sun is like a burning glass. . . . It is too much beauty for one human being to hold alone, and yet I am alone and worse than alone, for Alec sits reading in darkness. . . . This is really the one flaw in our married life — that he cannot rejoice in the sun, or I in the dreary gray cold.

As to the irony of their opposite preferences in weather, Alec was as conscious of it as she was. "Sun this morning would have pleased you," he wrote her from a journey through Maine one spring, "the sky now beginning to fleece up for my benefit." Alec preferred night and storms and Scotch mists. But the difference was not as complete and rigorous as that of the little wooden couple in the weather-house barometer. Beinn Bhreagh gave each a full measure to his special taste, and more besides that both could love. They shared outdoor afternoons and evenings. He had no grudge against sunsets, though he would not have written as she did that "for sunsets Beinn Bhreagh has no equal, and night after night I watch the gold and rose red and turquoise blue of sky and clouds and the purple of mountains and the silver of sea." There is no reason to suppose that he did not enjoy the flowers she gloried in raising and arranging, like chords of color and seasonal fugues, to match what he found in his Beethoven and Bach. "My rhododendrons and azaleas are perfectly glorious now," she would write in June, "and the yellow azaleas are a brilliant mass of color from the foot of the lawn." Or in August, "my garden is a mass of foxglove, yellow lilies, larkspur, and magnificent, perfectly magnificent purple Japanese iris."

To Mabel, Beinn Bhreagh was a home for all seasons, though the family lived in Washington from November to April or May. On returning to Baddeck one September, she wrote of her welcome in "the great hall with its immense logs blazing and a thicket of golden glow and autumn leaves just touched with scarlet. About were great bowls of pink poppies, dahlias, asters, perennial phlox, . . . old-fashioned verbenas, heliotrope, lupines, . . . making with . . . the scarlet Turkey rug and blazing fire such a brilliant, warm scene as must be seen and felt to be realized." "Seen and felt" — those were her keys to life and the world.

The Beinn Bhreagh estate became more and more elaborate. The Lodge had preceded the big house. So had a houseboat, the *Mabel of Beinn Bhreagh*, which between cruises on the Bras d'Or Lakes was drawn up to the shore of a secluded inlet at Beinn Bhreagh and linked to land by a footbridge, thus serving as a solitary retreat for Bell. As he added to its amenities and conveniences, it became the equivalent of another cottage. With the big house was built a gardener's cottage, and other buildings were acquired or built as land purchases continued and operations expanded: a shepherd's house, another for a laboratory assistant, wharves and boat-houses, stables, a dairy, a warehouse, a windmill, and so on. Mabel had fruit trees set out by the dozen, though her father took more interest in that enterprise than did Alec. By 1907 the place had twelve miles of roads.

Despite efforts to make it yield a profit through farming, livestock, and gardening, Beinn Bhreagh's expenses generally exceeded its income. In the depression year of 1895, its running expenses came to about ten thousand dollars. "I am not at all troubled about our financial condition," Mabel wrote her father a little defensively that January, "it is much better than it was this time last year, and our income is large enough for any reasonable family. . . . It is only that we always spend before we get our money." And she added, "I can't see why we shouldn't have a good time with our money while we are young enough to enjoy it."

Mabel's sound precept was faithfully followed. The Bells had money enough and spent it freely enough to feel at home not only in Washington and Baddeck but also the world over. They, their children, and sometimes one or more of the Hubbard family, toured England, France, and Italy in 1880, 1881–1882, 1888, 1891–1892, 1895, 1901 (without Alec), 1907, 1909, and 1920, as well as Mexico in 1895, Japan in 1898, and most of the Orient in 1910. Mabel complained in 1892 that Charles Thompson, her husband's valet, was learning more from the experience than were the Bell children, but it must have enlarged the girls' outlook. At any rate, Elsie was certainly not unprepared when her destiny eventually made her one of the most widely traveled women of her time.

Mabel had a keen eye for geography, and her letters could convey the flavor of a local or regional culture. So it was with her letters from Europe, though the scenes they described had long before been worn thin by in-numerable literary traverses. There is something characteristic and poignant, however, in her view of a Sorrento tarantella, performed, so far as she was concerned, in total silence: "The brilliant scarlet, yellow and peacock blue of trousers or skirts contrasted brightly with the white stockings and shirt sleeves, and I thought I should never weary of watching [the young peo-ple] dance in and out among each other."

Alec also loved travel, not so much for its visual stimulus as for its cornu-copia of curious or provocative things, ideas, and customs. Science, tech-nology, and the treatment and education of the deaf headed his checklist of things to investigate in foreign lands. He also enjoyed evidences of his inter-national fame: the "perfect storm of applause," beyond that given any other, that he received during the award of a "*Diplome d'Honour*" [*sic*] at the International Exhibition of Electricity in Paris in 1881; and the access his celebrity gave him to famous men, scientific and otherwise. Mabel herself wrote in 1898: "There's nothing like coming to Japan to find out what a big man my husband is. For his sake the children and I are received with such tremendous attention that I at least am beginning to think myself a very big personage indeed."

Still, except within the family, Bell was a fundamentally solitary and aloof man, notwithstanding his mastery of showmanship; and the ideas and things he encountered in traveling could also be found in books with more dispatch and less physical and temperamental stress. Age, weight, and impatience with ceremony advanced together in Bell. In Japan he got word that the emperor would receive him at ten in the morning, and that he was to appear in full evening dress, white gloves, and silk hat. When Charles Thompson brought the news to him, he looked at Thompson "with a peculiar expression on his face" and said, "Charles, are you kidding? Do you really mean I am to put on that horrid outfit at ten in the morning? Good Lord!" Thompson managed to rout Bell out of bed at what for him was an unearthly hour, get him into his "horrid outfit," and send him on his way. On his return Bell immediately went to bed without comment, woke at two in the afternoon, and asked when the consul was coming to take him to the emperor. After a while the details came back hazily to his mind. But he did not like what he recalled of the emperor's exaggerated hauteur, traditional or not, and so he commented, "Well, I am glad I was not awake."

In the spring of 1901, Alec wrote Mabel: "Under the circumstances will it be worth while my going to Europe this year? I would prefer to go quietly up to Baddeck when the hot weather comes and remain quietly there until you return. . . . While, of course, I should like to be with you all, traveling is a great interruption to consecutive work, and I should prefer not to go." On earlier occasions Mabel and the children had gone ahead and been joined by him a month or two later; this time, apparently, he did not go at all. It was the only such occasion, but it marked a decline in the frequency of his foreign junkets.

Money thus could liberate him from the bonds of space but not entirely from those of time. Nevertheless, taking all in all, Alexander Graham Bell enjoyed more than four decades of enviable compensation for whatever he may have suffered during those earlier years at the Williams shop, at Exeter Place, and on the witness stand.

25

The Private World of
Alexander Graham Bell

During more than half his lifetime, millions throughout the world knew the name of Alexander Graham Bell, and many would have recognized him at sight. He was a world traveler as well as a world figure. He met emperors and at least one empress. He was lionized in society, cheered at exhibitions, applauded at scientific meetings, and sought out by reporters. Medals, prizes, and honorary degrees were showered upon him. And he looked his part. He bore himself with the majesty of a Moses and the benevolence of a Santa Claus. When he entered a room, he seemed to fill it.

No one word covers all his activities, but the one that covers most is the word "communication." It applies to his work as a young teacher of speech, as a phonetician, as an advocate and teacher of speech for the deaf, as inventor of the telephone, as an organizer and collaborator in development of the phonograph and the airplane, as a frequent and masterful public speaker, and as backer and adviser of key journals in the fields of general science, deafness, and geography.

Furthermore, he and his wife united two numerous and close-knit families, wherefore his houses were built to accommodate and did accommodate platoons of guests. He was nominally a member of more clubs and other organizations than he could recall at any given moment, and he was active in a number of them. And for many years he presided over a brilliant salon of Washington scientists and men of affairs.

Yet his son-in-law David Fairchild said of him soon after his death: "Mr. Bell led a peculiarly isolated life; I have never known anyone who spent so much of his time alone."

Paradoxical though it may seem, Fairchild meant his observation literally, and he made a good case for it. Bell's lifelong habit of working alone through most of the night and sleeping through most of the morning; the designedly limited social activity of his Beinn Bhreagh summers, especially the quarter century and more of weekend withdrawals to the total seclusion of his houseboat, immersed in silence except for the sounds of waves and forest; his nocturnal ramblings in woods or on city streets; his hours of solitary piano-playing after everyone else had gone to bed — these were the evidences Fairchild offered.

But there was a subtler sense in which Bell stood aloof, one that he himself expressed in a letter of 1894 to Mabel: "You are always so thoughtful of others — whereas I somehow or other appear to be more interested in *things* than people — in people wholesale, rather than in persons individual." His other son-in-law, Gilbert H. Grosvenor, reading this letter soon after Bell's death, noted emphatically that it was a "very accurate description of Dr. Alexander Graham Bell; his interest was in mankind '*wholesale.*' " And Bell's wife agreed with his own self-analysis, perhaps more than was warranted, when she wrote him in 1895: "Your deaf mute business is hardly human to you. You are very tender and gentle to the deaf children, but their interest to you lies in their being deaf, not in their humanity, at least only in part."

Testimony to Bell's aloof and solitary ways spans his life, and much of it is his own testimony. His earliest memory was of being alone and lost in a wheat field. His Sabbath meditations, his long afternoons alone on Corstorphine Hill, the testimony of his boyhood friends are all to that effect. "All the stories I have ever heard of father [as a boy]," wrote Mabel to her Grosvenor son-in-law, "show him as quiet — bright enough and clever, but never especially light-hearted." The year alone with his grandfather in London reinforced his aloofness; he emerged looking and acting older than his years, and he would always seem so. This outward aspect in itself tended toward constraint between Bell and his near contemporaries. "Don't get absorbed in yourself — mix freely with your fellows — it is one of your great failings," Marie Eccleston wrote him on the eve of his departure for the New World in 1870. But he did not or could not take her advice. "Since I came to America," he wrote Mabel in 1883, "I have made a great many acquaintances but very few '*friends.*' " "Father is full of fun at times," wrote Mabel to Gilbert Grosvenor in 1906, "but I don't think anyone would call him exactly a 'jolly' sort of fellow or a merry one."

Bell's own rueful self-appraisal was quite as consistent and even more forthright. "I often feel like hiding myself away in a corner out of sight," he confessed to Mabel a month after his Centennial Exhibition triumph. "Whenever I try to say anything I stop all conversation. If there is anything of value in what I say people leave all the talking for me to do — and

I don't like it at all." Bell's arresting voice and presence — and later, of course, his worldwide fame — probably tended to overpower intimacy and dampen the give and take of banter. But there was evidently an underlying element of unconquerable reserve. From the Paris electrical exhibition of 1881 he wrote Mabel that in the crowd "I could not go many steps without hearing 'Monsieur Bell' and without staying for five or ten minutes to chat with some one." Yet in the same letter he also wrote, "I quite agree with you that I ought to exhibit some interest in other people and show that I am as anxious to receive others as friends, as many are to receive me." In 1890, depressed by attendance at the funeral of Thomas Sanders's mother (who during a crucial time had been a sort of acting mother to him), Bell wrote Mabel more desperately:

I feel more and more as I grow older the tendency to retire into myself and be alone with my thoughts. I can see that same tendency in my father and Uncle in an exaggerated degree — and suppose there is something in the blood. My children have it too, but in lesser degree — because they are younger I suppose. You alone are free from it — and you my dear constitute the chief link between myself and the world outside.

"The world outside" is a suggestive phrase.

Every man inhabits concentric spheres of life — the innermost worlds of consciousness and subconsciousness, the private world of home and family, the public worlds of friendship and livelihood, and for a few, the outermost world of celebrity. The innermost worlds are not open to a biographer's sure knowledge, some reaches of them not even to his subject's own understanding. And therefore such questions as *why* Bell was inclined to be aloof and solitary will be left open to the reader's speculation.

The metaphor of concentric worlds is suggested here in order to express a distinct tendency in Bell's post-telephone life. Independently wealthy, secure in his fame, Bell could and did enlarge and elaborate his world of home and family so much that it encroached on his worlds of professional and social life. Especially at Beinn Bhreagh, to which he resorted more and more as the years passed, he created what became a smaller, more amenable, partial substitute for "the world outside." But at Washington, also, the same tendency was evident in the community of Bell and Hubbard family establishments and in a social and intellectual circle of Bell's own organizing and leadership.

Bell did not carry this tendency to the point of secession from the outside world. Thanks in part to his wife, the tendency remained only a tendency. Throughout his life he continued to be eminently active in national scientific, humanitarian, and educational organizations. Still, the preference for privacy was a significant characteristic of his later life. That is why this chapter, as well as the one preceding and another later on, is devoted to what might seem an account merely of Bell's "home life." For him it was more than that — it was a busy and populous private world.

Mabel Bell worked on her husband's aloofness through direct exhortation. "I cannot bear to think of you living all alone, shut in yourself, holding no communication with your neighbors," she wrote him in 1895 from Paris. "*Please please* don't go back to such a life. . . . Please try and come out of your hermit cell. . . . I want you to succeed in your experiments, but not to lose all human interest in the process." And a dozen years later she wrote, "You have lived too much by yourself. You've talked about nature and solitude and all that, but you haven't been in the crowd at all and that's what you need."

She could be very winning in her plea. From France while he was in London she wrote him in 1888, "Accept all the invitations to dinner you get and meet all the great men you can — I want to hear all about them. I always feel as if you were my second self and all the gorgeous people you meet I meet too, and enjoy far more than if I really met them. Never mind a little dyspepsia. We'll go home to Cape Breton and live on bread and milk the rest of the summer."

But probably more effective were the contagion of her spirit and the example of her own delight in social life — in which, as her husband, Alec had little choice but to take some part. In April 1882 she had "a calling dress of ruby silk and velvet for next winter made at Worth's" in Paris, and being proud of it she may have worn it when the Bells called on the James G. Blaines the next spring. ("I think Mr. Blaine was lovely," she commented afterward; "I am afraid he has captured me nearly as much as he has Alec.")

Alec did not always come willingly. Once, to dodge a round of afternoon calls, he hid with his work in the attic. But Mabel tracked him down by his cigar smoke. "Charles, Charles, Charles," he moaned as Thompson helped him dress, "I will have to give up smoking. . . . There is too much work to be done in the world for men to spend their time sipping tea from house to house." On another occasion, facing the prospect of afternoon tea at the Hubbards' home, Twin Oaks, after callers had kept him all day from some projected work for the deaf, he first locked himself in the bathroom and then, when harried out of that stronghold, retreated again to the attic. There Mabel found him "crouching behind rolls of carpets. I got him out, helped him dress and carried him in triumph to Twin Oaks. By that time he was in a very penitent mood, ready to promise to be very nice to the people."

And she had honeyed blandishments for him, all the sweeter for being wholly sincere. "I am always so awfully proud of you when we go to parties," she told him in 1893.

There's not another man in Washington unless it's Justice [Horace] Gray who is as distinguished-looking as you. I don't mean handsome, doubtless there are plenty prettier than you, my husband — I'm nothing if not impartial! — but none that look so evidently somebody. I always expect the listlessly polite face of the hostess to change and light up when you come, and the look of interest to deepen when your name is announced.

Bell did not entirely shun the club life of Washington. In 1880 he became an early member of the Cosmos Club, which drew heavily upon the scientific establishment of Washington for its intellectual distinction. But he felt more at ease in a well-channeled current of conversation. "Just let him talk and don't have other people maintaining private conversation at the same time," Mabel advised their daughter in 1906, by which time Alec had got set in his ways. "He likes what he calls 'general conversation' when each one at table listens to one speaker and is listened to in turn. Even two people talking privately and softly together bothers him."

Bell's fame, money, and personal presence, operating in the nation's political capital, which was also one of its scientific centers, made it possible for him to organize a remarkable social life in his own home. In 1879 Gardiner Hubbard gave his son-in-law a taste of what Washington could offer him with a dinner at which the invited guests were Julius Hilgard, Ferdinand Hayden, Albert Myer, Spencer Baird, and Orlando Poe, all in the top echelon of government science and engineering. Later Bell began arranging his own regular soirees, mostly of men distinguished for talent or intellect. During the mid-eighties John Fiske, the popular historian and essayist, attended several of Bell's Monday evening musical parties, at which Bell played the piano "very finely" in company with a violinist and a cellist. Fiske thought Bell "a very enthusiastic man." After Fiske had sung a couple of lieder, Bell shook his hand and exclaimed, "What a truly splendid singer you are." "I am more impressed by Graham Bell than almost any man I have seen since Huxley," Fiske wrote his wife; "he is a superb object to look at."

By the mid-nineties Bell had begun a series of "Wednesday Evenings," informal receptions for a score or so of men, mostly scientists. Occasionally one or more women of the Bell clan were present. These were not "society affairs," nor were they ever reported in the press, yet George Kennan remembered them as being "more instructive and entertaining than anything of the kind that I have ever known." Bell planned and arranged a program for each one, taking care to find out in advance about the latest activities of each guest and to devise a course of discussion that would engage each in turn. One guest would usually speak at some length or read a paper on the subject of his current interest, and the conversation would take off from that, ranging perhaps from the indigenous races of China to the life history of eels, or from the latest volcanic eruption to cancerous growths in plants. Bell shone in the imagination, charm, and savoir faire with which he steered the discussion away from monologue on one hand and fragmentation on the other.

Guests lists survive for two successive Wednesday Evenings in April 1896. The first included such men from the cream of American scientific society as John Wesley Powell, William McGee, G. Brown Goode, Edward Morse, Alpheus Hyatt, William Brewer, and Samuel Scudder. More piquant to the historian of science is the simultaneous presence of those famous archfoes, the paleontologists Othniel Marsh and Edward Cope. But in addition, engineering was eminently represented by William Sellers, medicine by John Shaw Billings and George Sternberg, and politics by Senator Francis Cockrell of Missouri and Postmaster General William Wilson — all these from a list of only nineteen guests! The second Wednesday Evening included, among seventeen guests, the eminent scientists Benjamin Gould, Ira Remsen, Charles Walcott, Arnold Hague, Henry Rowland, Asaph Hall, Samuel Scudder, and Albert Michelson. Education was represented by Presidents Thomas Mendenhall of Worcester Polytechnic Institute and James Angell of the University of Michigan and former President Andrew White of Cornell (Remsen would later be president of Johns Hopkins); government (aside from scientific agencies) by United States Commissioner of Labor Carroll Wright (later president of Clark University); and politics by Congressman Joseph Cannon (later Speaker of the House). If a group photograph had been taken on either occasion, and a historian were to discover it without knowing its circumstances, he would feel sure it was a composite, so high would he calculate the odds against such a remarkable assemblage under one private roof.

And yet, perhaps because of the very richness of information and experience offered, these symposiums exposed no deep personal feelings and drew out no confidences. Charles Walcott, who eventually as secretary of the Smithsonian had other contacts with Bell, could not recall after Bell's death "any special stories or incidents of note" in their twenty-five-year acquaintanceship. They were both busy men, he explained, and so "when we met we transacted our business quickly and went our ways," having no occasion to speak of "that which was nearest and dearest to us." Marcus Benjamin, who came to the Wednesday Evenings from time to time over a period of twenty years, paid tribute to Bell's "unfailing cordiality and graciousness, which continued to the end." But, he added, "I don't remember that I ever had any serious conversation with him."

There were others much closer to Bell, however. John Wesley Powell, the geologist, ethnologist, and scientific organizer, a man like Bell in his striking physical presence and his penchant for grandiose scientific speculations, and Simon Newcomb, the self-made, Nova Scotia–born astronomer, mathematician, and economist, both frequented the Bell and Hubbard homes in Washington with easy freedom. After Bell's death, his companion in Washington and Baddeck, George Kennan, testified that "he was always a most sympathetic, congenial and lovable man to me, and my friendship with him has been one of the best things in my whole life." Many years later still, Bell's son-in-law Gilbert Grosvenor expressed the opinion that in the final reckoning Bell's closest friend outside the family was the astrophysicist, pioneer in aeronautical design, and secretary of the Smithsonian Samuel P. Langley, who resembled Bell in a different way from Powell, being reserved to the point of shyness or gruffness with most people and yet warm and witty in the company of a select few, including Bell.

Nevertheless, in America the only man outside the family on a first-name basis with Bell was, by Mabel's testimony, Fritz de Sumichrast, scion of a noble Hungarian family that had settled in Scotland. He and Alick (as Fritz still called him) had been close friends in the Royal High School and afterwards, even when Fritz turned frivolous and spendthrift. Eventually, after Sumichrast had been discovered in bed with a prostitute, his wife divorced him. A few months after the Bells sailed for Canada, Sumichrast likewise emigrated in search of a new life. He made one for himself, remarrying happily and putting his cosmopolitanism to use by teaching languages and writing on foreign affairs for Halifax and Montreal newspapers. Then, in 1886, enemies or busybodies dredged up the old scandal and charged him, on that basis alone, with unfitness for his current headmastership of a church-affiliated girls' school in Halifax.

Perhaps fortified by the memory of what his beloved grandfather had endured, Bell rose to the defense of his old friend without equivocation. "I have known him from boyhood," he wrote the lord bishop of Nova Scotia. "He has always been the same upright conscientious man you know him to be now, and there is nothing in his past life to justify the cruelty of reopening an old wound. . . . If I can be of any service to him or his friends please command me." Sumichrast was allowed to keep his position, but the affair had made it distasteful; and so in the fall of 1887 he and his wife made a final fresh start in the United States. Through the Bells, Hubbards, and Scudders, Sumichrast went well recommended to Harvard, where President

Eliot took him on trial as an instructor in French. Two years later, the appointment became permanent, and Sumichrast went on for many years as a respected member of the Harvard faculty. The friendship of "Alick" and "Fritz" endured. "When are you coming down to Baddeck again?" Bell wrote in 1897; "I am sure you know that we will be glad to welcome you here at any time."

Close as the two old friends remained, however, Fritz was an infrequent visitor in the Bell household. From day to day, it was within the clan that Bell found as much warmth and intimacy of personal relationship as his temperament let him accept.

Even in the family circle Bell seemed reserved by comparison with his ebullient Hungarian brother-in-law, Maurice Grossman, who had married Mabel's sister Gertrude. Maurice's death from cancer at the age of forty-one in 1884, after two years of illness, led Mabel to make a diary entry that reveals something not only about Maurice but also about Alec and, by indirection, about herself. Maurice "was so intensely alive," she wrote.

We seem such a quiet ordinary family now. Alec is a man out of the way certainly, but he is more quiet in general life. He never shocks us and takes away our breath with his overflowing spirits and utter disregard of conventionalities as Maurice did sometimes. We are all so quiet-mannered and self-restrained. Maurice was so demonstrative and energetic in all his protestations, . . . hot-headed, impulsive. . . . Every feeling as it possessed him was acted out.

On the morning after Maurice died, Gertrude's daughter asked her, "Mamma, what are you looking at?" "Nothing," said Gertrude, "I have nothing to look at now."

Still, against a Maurice Grossman not only Bell but most other men too would have seemed "quiet-mannered and self-restrained." Within the family, and especially so securely within it as he was at Beinn Bhreagh, Bell let down his guard at least part way. Mabel's sister Grace noticed at Baddeck in 1890 that Alec "is quite a different person here from what he is in Washington. Here he is the life and soul of the party." Mabel added her own touches. "Nothing is done without him; no detail relating to our enjoyment, comfort or safety escapes him. He is forever on the go. At night when all are sleeping he paddles about. . . . When a high wind is blowing or the boat is to be moved he is up, no matter how early the hour, directing, arranging everything."

Of his blood relatives, Bell's favorite among the younger generation was his Uncle David Bell's son Charles, who served for a short time in 1880 as Bell's secretary. According to Mabel, Charlie was "a good, kind, gentle boy

... but not much quicker than I in seeing the point of Alec's puns and jokes." Nevertheless Bell thought him "very clever." Having fallen in love with Mabel's sister Roberta, Charlie Bell recapitulated his cousin Alec's experience in courtship. When he proposed to Berta without a settled means of support, Gardiner Hubbard ordered him out of the house and forbade further communication with "Miss Hubbard." But Mrs. Hubbard cooled her husband's temper. Hubbard finally allowed that Charlie might try again when he had a steady job. Meanwhile he was not to bind Berta in any way nor see or write to her for the present. In due course, as with Alec and Mabel, Charlie and Berta were married; and Charlie became a successful Washington banker.

The old tension between Bell and his father had quite passed away with Alec's worldly success and emergence as a paterfamilias. By 1881 the elder Bells had lived ten years at Tutelo Heights, financially independent by reason of Melville Bell's investments in mortgages and Alec's later gift of a share in Canadian telephone rights. Though Melville Bell on occasion served as an interim professor in Queen's College, Kingston, Ontario, he was affluent enough to decline a permanent appointment out of preference for his Brantford homestead. But his brother David moved to Toronto in 1880, Alec settled permanently in Washington, and Melville himself sold his Canadian telephone interest.

To be near Alec and his family, Melville and Eliza Bell sold the Brantford place in April 1881 and moved to a house on Thirty-fifth Street in Georgetown, a mile or two from their son's establishment. Melville was then sixty-two but had a quarter century remaining to him for life as a Washingtonian. Eliza's three unmarried nieces, the Symonds girls, came to live with them; and soon afterward David and his wife bought the house next door, where their daughter Aileen — noted in the family for having refused George Bernard Shaw's proposal of marriage — kept house for her parents. David and Melville spiced their days with hot arguments over the authorship of Shakespeare's plays, but on Saturday mornings the two white-bearded patriarchs would walk amicably side by side to market with baskets swinging from their arms.

Alec later made a laboratory out of an old stable at the back of his father's property, and thenceforward when in Washington he would come out to the lab almost every afternoon and visit with his parents before returning home. After Mabel paid her father-in-law a visit in 1891, she wrote Alec, "Your father was in his favorite position on the sofa with his head where his feet should have been and his legs high up on the head of the sofa. He wore his usual linen wrapper and black alpaca and looked even more like Santa Claus than usual."

In Washington on February 15, 1880, the Bells' second child was born, another daughter, three weeks ahead of time but healthy and full grown. Alec was absorbed in his latest invention, the "photophone," which he thought more wonderful than the telephone. But, Mabel wrote, "he cannot assert it is more marvelous than this little, living human mite." The baby was christened Marian, after Mabel's youngest sister, but for some reason she was never called anything but Daisy.

On August 15, 1881, while the Bells were summering in Massachusetts, a son was born prematurely and with a breathing difficulty. He lived several hours and, being strong and healthy otherwise, might have pulled through if regular breathing could have been established. During late July and early August, Bell had gone to Washington in an effort to locate electrically the assassin's bullet that was slowly killing President Garfield. A year later, seeing Garfield's successor in the casino at Newport, Mabel wrote Alec wistfully that "but for Guiteau [the assassin] our own lives might have been different. You might not have gone to Washington, but have stayed with me and all might have been well." In all the years that followed she could never wholly shake off the devils of "if" and "why?" In 1883 she punished Elsie for some mischief by withholding candy. "I don't know what more I said," she wrote Alec, "something it was about God's punishing me as I did her when I did wrong. How He had promised me a baby if I would be careful, but I had not been, and He took the little one from me."

Alec repressed his feelings more sternly, but his actions showed them. He cabled Charlie, then in England, a notice for *The Times* of "our little Edward's" birth and death. He also set to work on a "vacuum jacket" machine for artificial respiration. And in Paris during the November after the baby's death, having gone with Mabel and the two girls to Europe for change and distraction from grief, he had a French artist paint a small canvas from a young Rockport artist's portrait of the dead infant in its casket, a portrait Mabel did not know had been made. As Mabel had done, Alec came close to speaking his feelings in a commentary on his living children in 1883. Lest they become callous toward real people, he suggested, they should be led to grieve at the loss of a doll, though not by means of reproaches: "We may take another little one to our arms but it can never take the place of the other. . . . If you were to lose a child through your own carelessness the sting would only be embittered by any reproaches addressed to you by others. . . . Santa Claus might entrust them with another baby — if he felt convinced that they would love it and care for it — and treat it so that it should not die."

"I would like a boy oh so much," wrote Mabel after the birth of a nephew in 1882. She got her wish on November 17, 1883, with the birth of a son, Robert. But again the baby was premature. "Poor little one," she wrote afterward, "it was so pretty and struggled so hard to live, opened his eyes once or twice to the world and then passed away." Three hours later, unaware of what had happened, Alec arrived home from the Hartford convention of the National Academy of Sciences.

In 1887, when Mabel was twenty-nine, a letter from Alec alluded to his having seen her cradling a "plaster baby" in her arms. "I love you very much, my darling little wife," he told her, "and wish indeed you could be blessed as you desire — *with safety to yourself*. I love you too much to risk your life." Mabel's sister Roberta Bell died in 1885, and two years later Charles Bell married his late wife's sister Grace Hubbard. They had a son, whom they named Robert. In 1895, when Mabel was thirty-six, she wrote Alec that "the touch of Robert's little hands on my face last night seemed to set some wheels going inside me that had been stilled a long while." And later that year, apparently having been assured by a physician that child-bearing was no longer inadvisable for her, she wrote Alec, "I can accept things philosophically when convinced there is no help for it, but I must be convinced first. . . . This is my last stone unturned. Let me turn it, if in vain I will submit." But she and Alec had no more children.

Mabel did her best to "accept things philosophically." "I am constantly thankful ever for my little babies' few hours of life," she wrote as the old century neared its end. "I have had them and my whole life is the happier for this consciousness." And early in the new century, with a baby grand-son to delight in, one who "likes to be hugged," she wrote again, "I do not know of anything I am more thankful for than that which is also almost my deepest sorrow, the little boys that were mine so short a time. Their tiny lives, lived so few hours, scarcely even that, have colored mine to its furthest."

She and Alec in any event had their two daughters to love and worry about. Elsie at three delighted her father with her "daring gymnastic feat of walking up Papa to his shoulders, ending in a back summersault to the ground," while she cried out for "more, more," long before she had fin-ished. But she was already evidencing what a noted physician later called "one of the worst cases of chorea he ever saw," a nervous disorder especially common in children. The trouble recurred for almost a decade. At last Elsie was put in the care of the noted neurologist and novelist Dr. S. Weir Mitchell, inventor of the "rest cure," who took her away from her parents and their eventful household for a year or more of quiet treatment. By early 1891, Mabel was "delighted and happy over her improvement. . . . She is decidedly taller than I am, and large but well proportioned." A year later Mabel reported with complete sangfroid that Elsie "is perfectly delighted

The Bells and their daughters Elsie and Marian ("Daisy"), 1885

at having diphtheria . . . [and says] 'I have had everything Daisy has had except typhoid fever and pneumonia.' "

From shipboard en route to Europe in the spring of 1895, when Elsie was seventeen and Daisy fifteen, Mabel reported proudly to Alec that "people seem rather to question my right to be the mother of 'those tall girls.' " With dark hair and eyes like their father's, they were taken for Spaniards and were attracting the attention of men. "Elsie has looked very handsomely and talked right and left in broken French, Italian and English with an ease and composure of manner that amazes me. After all I don't think she will be a wallflower. Daisy is more quiet but she is a dear, thoughtful child." In Paris that spring Elsie blossomed. She "never was more beautiful than she is now," her mother wrote. "Photographs give no idea of her beauty." Like a Henry James heroine, she accepted the interest of Parisian gallants with innocent vanity. "She is so thoroughly frank and unreserved that you see the best and worst of her at once, . . . her absorption in herself, and her pleasure in her own good looks, and at the same time her perfect sincerity and honesty, her great desire to do right, to be a good woman and to conquer her faults." Daisy, her mother thought, was "far more reserved and more ready to argue a point of obedience, but very thoughtful and with a mind so bright and clear and resourceful that it is a delight to watch her meeting any little difficulty."

Even after Elsie's long separation from her parents, her mother remained close to her, as well as to Daisy. In their relationship with their father, however, the girls had to compete with his manifold preoccupations. In a letter written to Alec about 1885, Mabel came down hard on him. "Your children need you," she wrote;

their characters are unfolding and they are a puzzle to me. All our lives we may regret that you were too absorbed in irregular night work — tending where? — to give them the care I cannot. I believe in God. Perhaps the reason our boy was taken from us so early was that we have not done our duty by the children we have. . . . Why was our wealth given us if not to give *you* time to make up to *your* children what they lose by their mother's loss [that is, her deafness].

When he turned his attention to his daughters, Bell was a warm, exciting father. But at other times his enthusiasm turned elsewhere. In the mid-nineties both girls could not help but feel some jealousy of his interest in and affection for his blind, deaf protégé Helen Keller. "Mamma," said Daisy during the European tour of 1895, "do you know I don't think Papa quite appreciates us. He thinks I can't do anything but swim and that is because he taught me himself." Mabel took care to let her husband know that. She prevailed on Alec to join them in France that summer. "At home he always has something in which he is absorbed," she wrote, "but away with nothing to do but be attentive to his wife and children he devotes himself absolutely to that, and they enjoy it."

Four years later, Alec admitted to Mabel that "I do so long for the confidence of my children, but I suppose the fault is my own." He blamed his innate tendency to "retire into myself and be alone with my thoughts." But a letter he wrote Elsie in October 1897, a letter full of anguish and disappointed hopes, suggests another factor besides. He had been conducting a class in science for the children of the Beinn Bhreagh establishment, including — and, his letter confessed, chiefly on account of — his nineteen-year-old daughter. She had submitted with disappointing passivity, showing no inclination toward voluntary, independent experimentation. Then she had suddenly tried a simple experiment on her own. "My heart gave a great *leap* (!) when I found your note in the Library," he wrote her in an excited hand.

It is a little thing that you have done — and yet how big! . . . Suppose that you had been trying for a long long time to stimulate a *corpse* (!) — and had failed after many efforts to get any response — and as you sit with the lifeless hand in yours, grieving over the dear departed — just imagine what your feelings would be — if the supposed corpse *should return the pressure of your hand!* These were my feelings my dear when I found your note. A little thing indeed but a *sign of life!* And in the reality — life — what possibilities lie! A *future* not a past. . . . Perhaps . . . at last my Elsie is to awake to life and usefulness.

How many times since his sons died had the fantasy of the revived corpse risen in Bell's mind? The strange image reveals that what seems cruel was only a cry of despair. Perhaps Elsie understood this.

Elsie indeed had a long, useful, and happy life before her, but not that of a scientist. Mabel seemed to fathom her husband's mind. On Thanksgiving morning, 1904, she wrote him: "How many years have passed since our supreme Thanksgiving morning dawned for us both? I don't want to count, only to be thankful we have had such a happy life together. I can even find it in my heart to be thankful for the loss of our little sons, although the pain seems as heavy today as at the time. At all events I had the great comfort of being the mother of sons, even if they did not live, and how much happier to think what they might have been than the cruel reality of Thomas Edison's son. It is scarcely possible that we could have carried on the tradition of the last three generations of Alexander Bells to a fourth, and after all it is better the name should go out in a blaze of glory."

Notwithstanding Mabel's remark, earlier quoted, about Alec's interest in deaf children more as specimens than as individuals, he seems to have been accurate as well as sincere when he wrote her that "somehow or other I have a tender spot in my heart for children." Some of the qualities of childhood lingered in him all his life. Perhaps for this among other reasons, Bell, as one newspaper reported on his visit to a school for deaf children, seemed to "inspire all children with involuntary confidence." But beyond that, there were several boys and young men over the years — Arthur McCurdy's son Douglas, the young engineer Frederick "Casey" Baldwin, and at last Elsie's own son Melville — toward whom Bell felt almost as he might have toward the two sons he never knew.

Like Alec, Mabel worried about being shut off from their children and from the world at large, but the obstruction in her case was physical, not psychological. "I have always declared I would sooner be blind than deaf," she once wrote, because "the blind through their ability to hear are able to be the centre of everything, whereas it is extremely difficult for a deaf person to be kept or to keep himself in close touch with the intimate family life going on around him." She felt this "very strongly." But she never despaired. "From my power of speech reading," she wrote proudly and justly, "I have been able to overcome much of the difficulty and am, I believe, nearly as much the centre of my home as any hearing mother can be."

Since many elements of speech are determined by vocal organs not visible in conversation, the lip-reader needs a broad vocabulary and a nimble mind to deduce from context, and sometimes also from facial expressions and body movements, which of several possible sounds makes most sense. "One must try for the complete story," Mabel wrote. "One thing obtained you can think back and recognize for yourself what those details must have been." She managed this so well that none of the family thought of her as deaf. Alec never needed to use finger spelling with her as he had with his mother. Lipreading was enough for Mabel. "It is no uncommon occurrence," she wrote in 1895, "for my husband to talk to me perhaps for an hour at a time of something in which he is interested. It may be on the latest geographical discoveries, Sir Robert Ball's Story of the Sun, the latest news from the Chinese war, some abstruse scientific problem in gravitation — anything and everything. Very rarely do I have to ask him to repeat." At dinner he included her in all conversations, thumping the table if necessary to draw her attention to his lips. She needed only a little time to accustom herself to the speech of strangers, and they in turn gradually found themselves at ease in talking with her. Having to keep an appointment after a Hubbard dinner party, Alec once asked another guest to escort Mabel home; years later the guest still remembered a conversation the rhythm of which was dictated by the spacing of lampposts, Mrs. Bell talking in the darkness between them and pausing under each one to watch her escort's reply.

Mabel did her best to maintain unbroken communication with hearing people, both for its own sake and to preserve her facility in speechreading. "I shrink from any reference to my disability and won't be seen in public with another deaf person," she wrote near the end of her life. In this course she had to contend not only with her handicap in understanding others but also with the deficiencies in her speech that made it difficult at first acquaintance for others to understand her. Nevertheless Alec continued to feel that he could not properly combine the roles of husband and speech

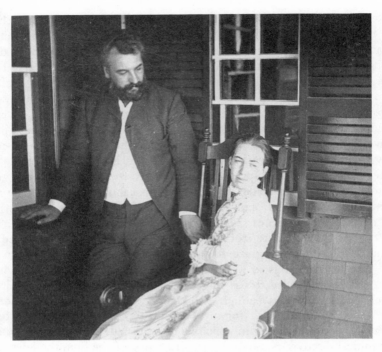

Alec and Mabel, 1883

teacher. "The value of speech is in its intelligibility, not its perfection," he once said apropos of his wife's speech; and whatever trouble it gave new acquaintances, he considered her speech intelligible enough so long as the family and old friends understood it easily.

In their love the Bells found deliverance, Alec from his temperamental constraint, Mabel from her physical handicap. "I *dread* absence from you," he wrote her in 1879 while away testifying in the *Dowd* case. "It is *NOT RIGHT* my darling. Let us stop it *NOW*. . . . Let us lay it down *as a principle of our lives, that we shall be together*." She in her turn wrote him in 1882, "It is wonderful how I miss you the moment you are out of reach." And again in 1884: "It does feel awfully lonesome without you, my big burly husband. You do take up so much room in a house that this feels very big without you, and then when you are here you are the object around which all my life moves, and now you are away it feels empty and object-less. Do take care of yourself, for I love you so very much."

The "principle" Alec had laid down in 1879, however, remained subject to occasional suspension throughout their lives as Alec's experimental and humanitarian interests called him away, or as his growing distaste for intercontinental junketing inclined him to give Mabel and the children a long head start on their travels. Moreover, Mabel recognized that her husband could be absent in spirit though present in the flesh. "I wonder," she wrote him in 1889, "do you ever think of me in the midst of that work of yours of which I am so proud and yet so jealous, for I know it has stolen from me part of my husband's heart. . . . I live in hope that you will not quite forget me, and that we may pass many another summer like the last when we had thoughts and interests in common." And in 1894: "I realize as I see Mamma and Papa, Grace and Charlie together how little you give me of your time and thoughts, how little willing you are to enter into little things, which yet make up the sum of our lives."

But when he was actually away, she felt the difference. "I like you around," she wrote him during one separation, "even if you are a bother sometimes. You are the mainspring of my life, and though when it is gone the other wheels go on by themselves for a time, it is very languidly and more slowly, and I want you back to give me an interest in life." In the spring of 1895, after she and Alec had come back from a Mexican trip for the two of them, Mabel and the children sailed for Europe while he stayed behind at Beinn Bhreagh. There he found a note she had left for him, which said in part: "I want to tell you how very happy I am that we have had these last six weeks to ourselves for ourselves. I had wanted you so much and I felt that the only way I could hope to have you really was to go away with you. . . . You were very, very good to me all the time."

Alec was contrite, as he had been before and would be again. "You have made me very, very happy, my darling, during all our wedded life," he wrote her in reply,

and I too enjoyed these few weeks alone with you in Mexico, although I fear you did not think so. I meant to give you pleasure, but pleased myself instead. I meant to devote myself to you, but the scientific men and old mines, etc., were all for me. I fear that selfishness is a trait of my character. I can see it very clearly in others, but I do not recognize it in myself until too late. It was selfish in me to let you all go to the other side of the world without me, for the sake of my experiments and the Convention.

He had written her half a dozen years before, "Deaf-mutes, gravitation or any other hobby has been too apt to take the first place in my thoughts, and yet . . . you have grown into my heart, my darling, and taken root there, and you cannot be plucked out without tearing it to pieces." And in the last year of the old century he would write again, "I love you, my dear, though sheep and kites, deaf-mutes and gravitation stand in the way. There is a mixture for you!"

In stirring his mixture Bell was rich enough to ignore the conventional working day and to be the night owl his nature seemed bound to make him. But Mabel now and then remonstrated. "Our worst quarrels have always been about that," she noted in her journal less than two years after their marriage. "No," she then added scrupulously, "the front rank belongs to the all important one of getting up in the morning, but this follows close behind." Some years later, in a letter of uncertain date, her reproaches to him had a ring of desperation:

You always say I am scolding and so sulk and will not hear, but I cannot bear to have people think you are doing nothing, and what can you do if you don't work when others work? Please, please, if you love me, work in the daytime and not at night. Alec, I am frightened and don't see what we are coming to. . . . It is so nice being rich, please don't spoil it for me by [the] feeling that perhaps it is a curse instead of a blessing, because if we were poor you would have to work hard and regularly.

In this matter as in others, Alec had spasms of penitence. One December night in 1891, for example, he climbed twice up and down the mountain at Beinn Bhreagh so as to get to sleep by twelve; but at three A.M. he was still wide awake and writing Mabel, "To take night from me is to rob me of *life*. No more useless reform for me as yet." In 1904 he dictated a new resolution to Mabel for solemn embalming in his journal: "For years past, I have formed the habit of retiring at 4 A.M. and I have come to the conclusion that it would be best for us all around if I could substitute early morning work for night work." He would therefore try sleeping from midnight to seven for a month; but if that did not cure him, he announced in advance, he would go back to his old ways, since he had undertaken a major project involving the statistics of deafness. The cure failed, of course.

Between these brief fits of reform he cherished the peace of the small hours, the silence that his wife knew through all her days and nights. "March 3/99 after midnight (chiming clock)," reads a surviving note in a large, indignant scrawl. "If you cannot stop this thing from striking at night, either it or *I* must go out of the house. AGB."

In due course he would go to bed, at which juncture he yielded to his one confessed superstition: a fear of having moonlight fall on him while he slept. He did not attempt to explain the feeling, except to point out the derivation of "lunacy" from the Latin word for moon. Nevertheless he felt strongly enough about it not only to ward it from himself but also to check the rest of the sleeping family on nights of full moon and to pull curtains or place screens so as to shield them also.

Rousing Bell could be as arduous as getting him to bed. His daughter Elsie once looked back over more than seventy years of world travel, wide acquaintance, and life with a numerous clan to adjudge her father "the soundest sleeper I have ever known. . . . He was so hard to awaken that he often stayed up all night in order to be up on time for an early morning engagement. His eyes were very sensitive to light and he used to wind a heavy bath towel around his head to keep out the light." At precisely nine in the morning, Charles Thompson would come in to announce the time. An earlier awakening would give Bell a headache; a later one would upset his confidence in the proper order of things, even though he often blinked the first call and went back to sleep. If this relapse endangered the keeping of an important engagement, Thompson would persist. That failing, Elsie or, as the ultimate force, Mabel would be called in. Thompson remembered mornings when Mrs. Bell would emerge from the bedroom pulling her husband by the hand, like a reluctant child. On one such occasion, after she remanded her husband to Thompson for dressing, Bell sat down heavily in an armchair with a disconsolate "Ugh, ugh, ugh!" Thompson looked at him and smiled. "Young man," said Bell, "it is easy to perceive that you are not a married man. If you were, you would sympathize with me instead of laughing."

In 1903, at one of Bell's Wednesday Evenings, someone (probably the eminent bacteriologist and epidemiologist George M. Sternberg, who was present that evening) commented on the spreading of disease by spitting on sidewalks. This led to the immediate organization of an ad hoc committee to lobby for a municipal ordinance against the practice, Bell being elected chairman. On the morning of the first meeting, Thompson found Bell unusually hard to rouse. At last he sat up crossly and complained, "Charles, you have given me a very bad headache by annoying me so often." Thompson explained gravely that he was to preside at ten over a meeting in protest against expectoration on the sidewalks of Washington. "Let them spit all they've a mind to," said Bell, with which words he retreated under the covers and went instantly to sleep again.

Except on pressing occasions, however, Bell's breakfast — always coffee, two eggs, and Scotch oatmeal with cream and brown sugar — would simply be left on a tray by the bed for him to resort to in his own good time; he did not mind eating it cold. Along with it went one or more morning newspapers, in which Bell would be engrossed for an hour or so. ("I do not see why it should come to an issue between me and your experiments or thoughts," wrote Mabel in 1899; "rather might it be between me and some of your numerous newspapers.") Then he would settle back for a smoke (a cigar up to the late nineties, a pipe thereafter), ring for Charles, dress, and be about the business of the day.

Charles Thompson holding tetrahedral kite cells, about 1907

First on Bell's daily agenda was the morning mail, if he could resist an immediate return to whatever had been occupying him at bedtime. Then he would live everything twice, as his son-in-law David Fairchild put it, by dictating a full account of his and his family's doings of the previous day to Mabel, Elsie, or his secretary. This went into a series of notebooks labeled "Home Notes."

About 1880 Bell hired William Schuyler Johnson as his private secretary, a tall, thin, good-natured, conscientious young man with a deaf brother. Johnson accompanied the Bells on a European sojourn in 1881 and won their hearts, wherefore they took it all the harder when he died in 1883 from some painful, lingering disease. "Is it not hard," wrote Bell, "that a young man just beginning life with bright prospects and a hopeful future should die now?"

In the mid-eighties Bell struck up a close friendship with Arthur Mc-Curdy, the young Baddeck businessman who had acted as his agent in acquiring the Beinn Bhreagh estate. They were well matched in chess and also in their boyish curiosity about nature. Mabel thought McCurdy "a very energetic, wide-awake man. When we were out he could not pass an uncommon stone, leaf or shrub without examining it." In 1888, when a hard winter for the McCurdy store's farmer customers made its accounts uncollectable and sent it into bankruptcy, Bell hired McCurdy as his private secretary. McCurdy held the job for nearly fifteen years before striking out on his own again.

An exchange between Bell and McCurdy in 1896 throws light on Bell's office habits. "Our work," wrote Bell, "is actually in a chaotic condition. . . . This is entirely my own fault, and I sympathize with you in having to work with such an unsystematic man as myself. . . . *My* time did not correspond to any one else's time, and I was *never* willing to face everyday matters in proper season." He outlined an admirably systematic plan of attack on correspondence. "At present there is none, or very little, but this is because we have neglected it," he explained. "People are tired of writing and receiving no answer. The moment we attend to it, it will revive ad nauseam." Six weeks later, McCurdy listed some suggestions for making the grand plan more effective:

1. You must come to the office in some sort of season, and not put off office work until three or four in the afternoon.
2. Don't take letters away from the files of the office and expect me to find them when wanted.
3. Don't take unanswered letters away, and expect me to find them.

Bell could scarcely have been as delinquent in correspondence as all this suggests; his voluminous letterbooks by themselves refute the idea. Perhaps his good resolutions came often; perhaps his secretaries kept him near the mark. Between secretaries in 1914, nevertheless, he was again bemoaning "my usual state of muddle in the matter of correspondence."

In his private study, Bell refused to have a telephone. Indeed, he found the one-sided conversation of someone else at a telephone so distracting that he would not have it within earshot of his work (although otherwise he loved to amuse Mabel and himself by repeating such fragments to her and trying to reconstruct the other side). In his later years these facts were elaborated by newspaper reporters to have the inventor banishing his creation from his house. This fable amused Bell, and he gave it further currency by whimsically repeating it. Sometimes also, when a telephone message disrupted his plans, he would ask jokingly, "Why did I ever invent the telephone?" After his death, however, Mabel grew annoyed at the newspaper notion that her husband had scorned his great invention. "There are few private houses more completely equipped with telephones than ours at 1331 Connecticut Avenue," she protested. "Mr. Bell's one regret about the telephone was that his wife could not use it or follow his early work in sound." (In talking to Alec and others who were used to her speech, Mabel could, in fact, use the telephone with the help of an intermediary, one of the children or grandchildren perhaps, who would listen at the receiver and repeat the other party's words for her to read from their lips while she spoke into the transmitter.) As for the line at Beinn Bhreagh, Bell once remarked that the family could not live there without it.

"I have found by experience," wrote Bell in 1885, "that I can only deal with one thing at a time. My mind concentrates itself on the subject that happens to occupy it and then all things else in the Universe — including father, mother, wife, children, *life itself*, become for the time being of secondary importance." Mabel also found this by experience. "If Alec is well it is by my care," she wrote her mother one October; "he is nearly as irresponsible as a baby. He always was, you know." After a day of experimenting with kites, he had gone to his unheated laboratory and fallen asleep without thought of the chill, until someone came with his overcoat and brought him home. "The thing he was working on," she recalled after his death, "was always the biggest and most important thing to him for the time being. That was a trait that often annoyed people because to them that thing might seem very unimportant compared to what they wanted him to do. And they were often quite right too. But he could not be moved."

His capacity for total absorption in the concern of the moment was one of those traits in Bell which might be called childlike, as might also his ready enthusiasm and his fascination with the world about him. Mabel, exercising her wifely prerogative, sometimes called him inconsiderate. Notwithstanding that, and notwithstanding Bell's own self-accusations, he cannot be charged with either deliberate selfishness or callousness. On the contrary, few men are so kindly, charitable, and easily roused to sympathy as was Alexander Graham Bell. In seventeen years of close association with him, his son-in-law David Fairchild never once heard him criticize anyone harshly, much less unjustly. One of Bell's favorite sayings was that "one should never impute unworthy motives to others." He was unfailingly optimistic and constructive in the face he showed the world, whatever Celtic glooms may have drifted through his mind in the solitude of night. He would not even permit himself to swear out loud, at least not in the presence of others. "In all the years I served Mr. Bell," wrote Charles Thompson, "the only way I have ever known him to give vent to his feelings when vexed was by two or three grunts or as many hisses through his teeth that sounded like 'G e e t h e, g e e t h e, g e e t h e!' Once I remember him hissing three or four times in succession. His secretary looked at him and said, 'Damn it!' 'That is just it,' said Mr. Bell, 'but you said it, sir.' "

There were times when Bell resisted casual intrusions. Usually, however, he was eminently approachable. Tourists who ventured to call at Beinn Bhreagh unannounced and unintroduced often found themselves being conducted about the estate by its master with an affable enthusiasm that went far beyond the requirements of mere politeness. Three classes were especially favored, both in Baddeck and in Washington: reporters, young inventors (or would-be inventors), and anyone concerned in any way with the welfare of the deaf. A reporter who intercepted Bell returning from his Georgetown laboratory in late afternoon was likely to be surprised by an invitation to call for an uninterrupted talk sometime between midnight and four the next morning. "These are my hours for work," Bell would say, "come right along."

An episode that tells much about Bell began at a young people's dance in his Washington home when a guest named William Glasgow Powell twice conspicuously refused to take the offered hand of a young lady. The next evening, upon Bell's inquiry, Powell explained privately that though the girl herself did not know it, her mother was the illegitimate child of Powell's grandfather. On this ground alone, said Powell, the insult had been offered deliberately, and he was not sorry for it. "This is what *STAGGERS ME*," Bell wrote Powell, after pondering the statement. He still could not believe that he had understood Powell correctly. "No blame, of course, can attach to the mother for things that happened before she was born — nor, surely, can blame attach to the innocent girl whose hand you refused to take *because her grandparents did wrong*." Powell replied proudly that he had acted on the "principals" that had "guided my family and connections to prominence in their respective States and in the annals of their country, from Colonial times to the present." Stupidity, cruelty, and arrogance are usually found in combination; but the crudeness with which Powell flaunted all three might have stung a man less gentle than Bell to a harsh reply. Instead, he merely acknowledged Powell's request for permission to show the correspondence to an uncle. "I am glad to give it," wrote Bell, "for you obviously need the counsel of an older and wiser head than your own, one more experienced in the affairs of the world."

In its tone and its revelation of character, the exchange echoed one that Bell had on board a ship bound for America in 1892. As he described the incident to Mabel, a young shipboard acquaintance

in the smoking room talked in rather an insulting and sneering way of the "niggers" of the South. I replied that I thought the negroes were entitled to equal rights with himself. It looked at first as if there might be some sharp words. The other gentlemen, however, so promptly sided with me that Mr. Kean very wisely allowed the subject to drop and devoted himself to making himself agreeable. He seems to be a nice young fellow and can hardly be held responsible for his Southern sentiments.

But intolerance is one of the hardest things to tolerate, and even Bell could not always manage to do it. In 1904 Charles Thompson and his wife were turned away from a hotel in Sydney, Nova Scotia, because of their color. Furious, Bell encouraged the passing of a protest resolution by the people of Beinn Bhreagh and Baddeck. "I anticipate a stirring time, and the matter may even become a National issue," he wrote Mabel. "I think the people here, with Arthur McCurdy at their head, are so fully aroused that it is unnecessary for me to do anything further here. This is a Canadian affair. . . . I can help materially, however, by being interviewed in the United States. Think I better write to the American Consul in Sydney." And he was as good as his word.

Ordinarily, Bell was not politically minded. In 1874 he took out his first naturalization papers in order to expedite patent procedure; he completed the process and became a full-fledged citizen of the United States on November 10, 1882, under the spur of a prospective nomination to the National Academy of Sciences. In 1896 he told a vote-hunting Canadian politico that "I am an American citizen and deem it my duty to abstain from all participation in Canadian politics." But as a resident of the District of Columbia he could not vote in the United States either. "I am glad I haven't a vote," he wrote Mabel in 1894, "for on the question of Tariff (how do you spell it?) I'm sadly perplexed and declare I cannot decide whether I am a Republican or a Democrat. I'm a Mugwump, that's just what I am — a cross between the two."

Yet when he did recognize social justice and compassion as political questions, Bell was capable of expressing decided, even radical, opinions. In 1894, during the great depression of the nineties, Mabel seemed a little uneasy in reporting to Elsie that

Papa rather sympathizes with what he conceives to be the purpose of Coxey's Army, namely to force the Government to give them work. . . . I think Papa rather holds that as the Government is of the people by the people and for the people, it is bound to provide for its starving members. . . . Papa thinks there is something wrong in a system which feeds and gives criminals work, but refuses it to honest labor. Papa [sees] the Government as the organized instrument of the people.

His Grandfather Bell would surely have agreed.

In the circle of his day, evening brought Bell back again from his several other worlds to that of his family. Little seen by the rest of the household during the day, he would appear at the dinner table for an interlude of easy sociability before retreating to the solitary world of night.

In his early years of married life he had been annoyed at dinner by a set of English china decorated with vividly accurate pictures of various insects. In France he finally acquired a cream-colored, gold-edged set bearing only a simple monogram, after which he confided to Thompson that at last he could enjoy his dinner without imagining a caterpillar or moth in every spoonful of soup. Beginning in the mid-nineties he further enhanced the pleasure of his dining by using glass straws or tubes, thick ones for imbibing soup, thin ones for less viscous liquids. Mabel considered them unsanitary, being hard to clean; but she tolerated his using them at Beinn Bhreagh. The refusal of others to adopt the idea seemed to nonplus Bell, who perhaps did not reflect that their intake, unlike his, did not have to negotiate a hair-raising passage between mustache and whiskers.

With such amenities established, Bell could give himself over to the enjoyment of his victuals and the enlargement of his girth. Neither his appetite nor his conviviality needed the stimulus of alcohol. He was not an abstainer on principle, and as a "social function" he sometimes drank wine or beer, though nothing stronger. He simply had no desire to tipple, and he chose not to cultivate one, having found that a mere glass of wine or a single teaspoonful of whiskey or brandy gave him a flushed face and a touch of fuzziness. Such a dose he found beneficial only when suffering from a nervous headache or when exhausted by a long tramp in the woods.

Bell's stimulants were conversation at dinner and music afterward. Elsie remembered many years later that "we never sat down to a family meal together but some experiment was tried or somebody was sent to look something up in the dictionary or encyclopedia, or that my father did not pull out his little black notebook and tell us some amusing story of something he had read in the daily paper."

After dinner, Bell often would play the piano and, in his fine voice, lead such of the family and guests as cared to join him in a medley of folksongs and traditional favorites. Scotland would be summoned up in "Annie Laurie," "Loch Lomond," "Scots Wha Hae," "The Laird of Cockpen," and others, but there would be Negro spirituals too, and such American rousers as "Little Brown Jug." "As the family grew sleepy and drifted off," remembered his grandson Melville Grosvenor long afterward, "he'd switch to a piano recital with me turning the pages. So long as I could stay awake and turn the pages, I was allowed to sit up and help him, for father felt this was an extraordinary experience for me. . . . Sometimes he played into the wee hours, and we left our doors open so we could dream to his beautiful music."

Melville's mother, Elsie, remembered the latter phase a little differently, as a mother would. Her father, she once recalled, "would often become so interested in playing pieces by Mendelssohn and Beethoven that he would sit up until two or three o'clock in the morning, regardless of the fact that his grandchildren were trying to sleep."

Sought or unsought, sleep would come at last to all of the family but one, and silence would rule over the house, except for the whisper of turning pages or the scratch of a pen in Bell's study (even those solitary sounds presumably being stilled now and then as Bell reflectively watched his blue pipe smoke eddy upward). At last, in the hush before dawn, he would rise and, if the moon were up, make his rounds to shield his sleeping family from its cold, uncanny light. Then he too would go to rest.

26

". . . And He Will Go On Inventing"

In February 1879, Bell bought a set of the *Encyclopaedia Britannica* with the intention of reading it through from start to finish. Whether or not he finished it, his appetite for miscellany persisted. Early in 1892 he reported to Mabel that he had spent the evening reading Jules Verne, then had turned to "my usual night reading, Johnson's Encyclopedia. Find this makes splendid reading matter for night. Articles not too long — constant change in the subjects of thought — always learning something I have not known before — provocative of thought — constant variety." Like his Wednesday Evenings, Bell's encyclopedic browsings demonstrated his omnivorous curiosity. That trait impressed his son-in-law David Fairchild. So did Bell's capacity for seeing the commonplace as fresh and exciting. Once, when a grandchild came to kiss him goodnight with a toy balloon in her hand, he looked up at it and said to Fairchild, "Isn't it wonderful! See how it rises!" "Wondering to him," recalled Fairchild, "was almost a passion."

It was this compulsion that Bell had chiefly in mind when he told a meeting of inventors in 1891 that "wherever you may find the inventor, you may give him wealth or you may take from him all that he has; and he will go on inventing. He can no more help inventing than he can help thinking or breathing." But in Bell's case, at least, there were reinforcing motives.

One, implied in a further remark on the same occasion, was the enduring satisfaction of achievement, of realized ambition. "There is no truer test of ownership than the fact of creation," he said. "That which a man makes himself, he feels he owns." He put it more plainly during the great days of 1876, when he wrote Mabel (in extenuation of tardiness in applying for his Centennial medal) that "the real reward of labour such as mine is *success* . . . a medal far more valuable than any made of gold . . . a medal that you may wear around your heart — and that will wear as long as history itself!"

Another motive, once he had won fame as a young inventor, was pride. "I can't bear to hear," he wrote Mabel in April 1879, "that even my friends should think that I stumbled upon an invention and that there is no more good in me."

As if to reassure himself, Bell followed up the last-quoted words with descriptions of two ideas that had just occurred to him. One, a "magneto-scope," was not significant; the other, "a new method of deep sea sounding," was destined to be useful in other hands.

The latter idea had come to Bell only the preceding night, and so he could trace its development clearly. Someone's reference to sound in water had suggested to Bell the notion of an underwater foghorn, its sound to be picked up by a submerged telephone. This in turn had reminded him of a submarine microphone he had devised earlier and of the suggestion it had elicited from Julius Hilgard of the Coast Survey: that a shell be arranged to explode on contact with the ocean bottom and the depth then calculated from the time taken for the sound to reach the surface. Bell's new idea was that "there must be *echoes* in water as well as in air, and if a sound were made in the water at a short distance from the surface, the sound would travel to the bottom, be reflected up, and reach the submerged ear (or apparatus) . . . as an echo. . . . The computation of the depth would be easy when once the velocity of sound in water is known."

John Tyndall, one of that generation's best-known students of acoustics, was "much struck by the idea," Bell reported in November 1880, "and says the experiment should be made." But though Bell made the suggestion publicly on several occasions over the next thirty years, he never got around to making the experiment.

Tyndall, like Bell, evidently considered the idea a new one, and in proposing a telephonic listening device it was. But long before Bell's proposal, the French scientist Arago had suggested measuring ocean depths with sound waves; and the American oceanographer Matthew Maury had tried doing so by exploding shells underwater, though he did not realize that a direct connection was necessary between the water and the ear, and so he heard no echo. After Bell's suggestion, there remained several technical obstacles, which Bell may have chosen not to grapple with. How much impetus, if any, his suggestion gave to the work of others in this direction is uncertain. At any rate, not until the early twentieth century were all the technical elements in hand for a practical sonic sounding system: a microphone to detect faint sounds, an oscillator to generate underwater sounds, a hydrophone to locate their direction, and a sufficiently accurate timing device. The last item came along in time to permit the first conclusively successful test of deep-sea sonic sounding in February 1922, only a few months before Bell's death.

Whether or not the obvious technical problems deterred Bell, the *Dowd* case left him no time for bouncing echoes off the sea bottom during the spring and summer of 1879. Furthermore he had already opened a pursuit more to his taste for the spectacular. He even had a name for his new quarry: the "photophone."

In England six years earlier, it had been discovered that the electrical resistance of the element selenium varied with the intensity of the light that fell on it. When the newly wed Bells arrived in 1877, English journals still carried occasional reports of experiments with that strange property. Mindful of Edison's success with variable-resistance solids in telephones, Bell saw in selenium a way to outflank his rival's carbon-button patent. He divulged his line of thought (it may be recalled) in his speech of May 17, 1878, before the Royal Institution: "If you insert selenium in the telephone battery and throw light upon it you change its resistance and vary the strength of the current you have sent to the telephone, so that you can hear a shadow."

Bell brought back some selenium and, before the *Dowd* case engrossed him, tested the samples at MIT. But their resistance was too great for "the experiment of hearing light," and so he put off serious work until after the *Dowd* case. Perhaps his long courtroom reliving of the telephone quest suggested a pattern of attack. Certainly the subsequent photophone campaign evokes a sense of déjà vu.

Bell began it in Cambridge on October 15, 1879, by hiring a counterpart to Tom Watson on terms much like those arranged with Watson in 1876: fifteen dollars a week, a one-tenth interest in such of Bell's ideas as the assistant helped reduce to practice, and a one-half interest in any inventions the assistant himself made during the life of the contract. Twenty-five-year-old Charles Sumner Tainter, three months younger than Watson, had developed his mechanical skill in his father's carpenter shop and, though he dropped out of school at an early age, continued to learn much through his father's subscription to the *Scientific American* and from scientific and technical books borrowed from the Watertown Public Library. When Tom Watson came into Charles Williams's shop in 1872, young Tainter had been there two years. Soon afterward, however, in order to get on a projected expedition to observe the transit of Venus, Tainter took a job making chronographs at the telescope firm of Alvan Clark and Sons in Cambridge. The plan worked. Tainter circled the globe, in 1874–1875, as the Transit Expedition's instrument maker. After three more years with the Clarks,

Tainter set up his own shop to make experimental apparatus for Harvard professors, a chore the Clarks had done as a favor but were glad to let their young graduate take over.

Early in December 1879 the Bells rented a house in Washington, and Bell set up a laboratory nearby, where he and Tainter promptly went to work. The photophone at first had competition from the other variable-resistance transmitters which Bell was developing under his contract with the telephone company, as well as from his meditations on switchboards, phonographs, and sound-ranging systems. But on January 22, 1880, Bell and Tainter inscribed a declaration of purpose in their lab notebook: "We recognize that an Electric Photophone if perfected would probably be found of less practical utility than many of the other ideas we have discussed; but we are both so fascinated by the scientific prospects opened up, that we have determined to make the electric photophone our great object of search."

Bell seemed deeply committed to the new emprise, and his emotional intensity communicated itself to Tainter. On January 28 they tried unsuccessfully to get a long-distance effect from selenium, and more than a week later the notebook recorded that "the failure of the experiments produced so great a reaction that we have both been quite unwell ever since." Only about three weeks after that, however, Bell noted that "the problem of the reproduction of speech by the agency of light was . . . solved by Mr. Sumner Tainter and myself in my laboratory No. 1325 L St. Wash. D.C. on Thursday, Feb. 19, 1880."

The means was simple and the range limited. The knottiest technical problem had been to develop a light-sensitive selenium cell of sufficiently low resistance. That achieved, Bell and Tainter placed the cell in a telephone circuit and by a system of lenses threw upon the cell a beam of sunlight reflected from a voice-vibrated plane mirror. The voice distorted the mirror's surface and thereby varied the intensity of the reflected beam. This in turn varied the resistance of the selenium cell and thus made the battery current "undulatory," as required for telephonic speech. At that stage of experimentation, the mirror and cell were within earshot of each other inside the laboratory, and so the experimenters listened to the effect through a telephone receiver in another room.

Their early success seemed to reassure Bell that his life thenceforward was not to be merely a long postlude to the telephone. His elation can be seen in a letter he wrote his father on February 26. "Only think!" he exulted, apropos of Marian's birth on the fifteenth, "Two babies in one week! . . . Both strong, vigorous, healthy young things, and both destined I trust to grow into something great in the future." He gave his dramatic

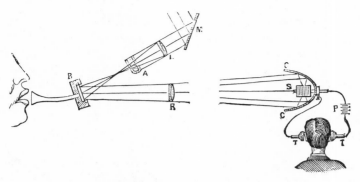

The photophone, 1880

instinct full play. *"I have heard articulate speech produced by sunlight! I have heard a ray of the sun laugh and cough and sing!* . . . I have been able to *hear a shadow,* and I have even perceived by ear *the passage of a cloud across the sun's disk."* Entreating his father to hold the letter in confidence, he cast aside the caution of January.

Can Imagination picture what the future of this invention is to be! . . . We may talk by light to any visible distance without any conducting wire. . . . In warfare the electric communications of an army could neither be cut nor tapped. On the ocean communication may be carried on . . . between . . . vessels . . . and light-houses may be identified by the sound of their lights. In general science, discoveries will be made by the Photophone that are undreamed of just now. . . . The twinkling stars may yet be recognized by characteristic sounds, and storms and sun-spots be detected in the sun.

He closed happily, "You are the grandfather of the Photophone and I want you to share my delight at my success."

At the Smithsonian Institution early in March, Bell and Tainter deposited a sealed tin box with a model of the device and an account of its development thus far, so as to establish priority without revealing their line of research prematurely. A hoax story about a "diaphote" for seeing by telegraph had happened to appear in the *Boston Transcript* in February and was copied by American and English newspapers — even including the august British journal *Nature* — in succeeding weeks. To Bell's amusement, news of his mysterious box, appearing alongside the "diaphote" hoax, touched off indignant claims by several inventors that he had stolen the idea of "seeing by telegraph" from them; and some divulged their own schemes for grids of small selenium cells which, by affecting magnetized shutters or polarized light, would produce pictures on a screen.

In time, Bell himself would toy with that idea, but just then the photophone seemed wonder enough. Early in March 1880 he wrote a scientist friend in England that his new invention (still unrevealed) would "prove far more interesting to the scientific world, than the Telephone, Phonograph, or Microphone." Next day he remitted two pounds to his English cousin Chichester ("Chester") A. Bell, one of Uncle David Bell's sons, to buy selenium in small lots at various places. "As I expect the price will rise when my invention is made public, I want to be sure of having a supply on hand."

"Red letter day for Photophony!" Bell noted on March 26, after getting audible effects over a distance of 82 meters. Then on April 1 Tainter transmitted a message 213 meters from the top of the Franklin School to a window of the L Street lab. Another sealed box was bestowed on the Smithsonian, and Bell made solemn note of the historic utterance: "Mr. Bell, if you hear what I say, come to the window and wave your hat." Several years later he relived the occasion in a talk to engineering students at Cornell. "It is unnecessary to say," he said, "that I waved with vigor, and with an enthusiasm which comes to a man not often in a lifetime." Such moments, he told them, are "worth a lifetime to live for."

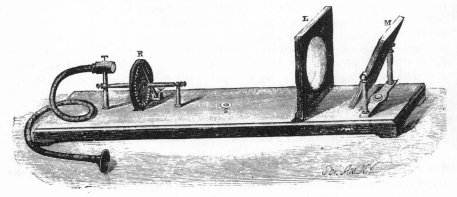

The intermittent-beam sounder, 1880

Not only did Bell and Tainter achieve their original goal, but they also stumbled on an unexpected effect that Bell found entrancing. To test the carrying power of the light beam, Bell had resorted to a device analogous to one of his early harmonic telegraph transmitters: a whirling, perforated disk that made the light beam intermittent, like a make-and-break current, and so produced a simple telephone tone for testing. Bell discovered that such an intermittent beam, falling on a diaphragm, produced an audible tone directly, without the intervention of any electrical apparatus. He went on to produce such tones from all kinds of substances, including a test-tube of cigar smoke (although he got no response one day in July from a fried egg and a stick of sugar candy).

By August he and Tainter had devised a variety of selenium cells, a parabolic reflector at the receiving end to concentrate the light on the cell, and more than fifty methods of varying a light beam, including magnetic fields that affected polarized light, lenses of variable focus, variable apertures, and adaptations of Koenig's manometric capsule. Bell and Tainter also hit upon a simple way of getting selenium into the required crystalline state in a few minutes, instead of the elaborate accepted method, which took forty to sixty hours. Bell summed up their work in a paper, "received enthusiastically," before the American Association for the Advancement of Science on August 27. Echoing his basic telephone patent, he emphasized as "the fundamental idea" in the photophonic transmission of speech his conception of "what may be termed an undulatory beam of light in contra-distinction to a merely intermittent one." And he presented the photophone as being "greatly due to the genius and perseverance of my friend Mr. Sumner Tainter of Watertown, Mass." Indeed, the Washington experiments were subtitled "Researches of Sumner Tainter and Alexander Graham Bell," and the ruling pronoun was "we."

As early as May, Bell had considered the photophone far enough along to be offered to the National Bell Telephone Company. He asked only two thousand dollars in return for expenditures directly attributable to its development. President William H. Forbes accepted for the company. But in the tone of Forbes's acceptance can be heard the first small cloud crossing the sun: "Whether this discovery ever approaches the telephone itself in practical importance or not, it is no less remarkable and a thing which we should be glad to possess."

Bell either missed or shrugged off that note of doubt as to the photophone's "practical importance." In June he assured his father that "when the Photophone is once divulged, there will be hundreds of persons to work it up into practical usefulness, and I will then take 'a new departure' in a different direction." "The first practical application" of the device, he told Mabel several times that month, would be to supersede the Blake carbon transmitter.

But the note of doubt sounded again in a somewhat mocking *New York Times* editorial of August 30, commenting on the paper read to the AAAS by "Prof. Bell, who invented a good deal of the telephone." The *Times* pictured the puzzlement of "the ordinary man" as to how sunbeams might be used in lieu of wires. Would there be "a line of sunbeams hung on telegraph posts?" Would it be "necessary to insulate them . . . by a thick coating of gutta-percha?" This raillery was neither kind nor fair. Yet it did suggest the formidable problems remaining: interference by fog, smoke, or rain, and the limited range of transmission thus far achieved.

The *Times* pinprick may have passed unfelt, in Bell's elation just then at being given the French government's Volta Prize for his invention of the telephone. Established by the first Napoleon in honor of the Italian scientist Alessandro Volta, the Volta Prize for scientific achievement in electricity had been awarded only a few times before. Among all the formal honors that were to come to Bell in his lifetime, he would regard this one as the greatest. Early in September he left for Paris to accept it and the fifty thousand francs, equivalent to ten thousand dollars, that went with it. On his way home in November he spent a day with Tyndall at the Royal Institution experimenting with intermittent light beams. "It was delightful," Bell wrote home, "to see another man get excited over my experiments." And a few days later he lectured on the photophone before the Society of Arts in London, being introduced as a man who had "made for himself a world-wide reputation in all places claiming civilization."

Further to brace himself for a return to photophone development, he asked his cousin Chichester Bell to join him in the work at Washington. He and Chester had always hit it off well, and Chester had charmed Mabel three years before in England — "a thorough gentlemen every inch of him," she thought, with "a beautiful navy black beard as black as Alec's." Several times since then, the Bells had urged their bachelor relative to join them in America.

A year younger than Alec, Chichester Bell was a man of parts, an adept at both the piano and the old Irish hockey game of hurly-burly, an amateur boxer with a broken nose as a prize, an engaging companion. Chester's wit and musicianship had made him a close friend of young Bernard Shaw, who used him as a model for the chief character in a play, *The Doctor's Dilemma*, Chester having been a practicing physician for a time. "We studied Italian together," Shaw recalled, "and though I did not learn Italian I learned a good deal else, mostly about physics and pathology." It was Chester Bell also who introduced Shaw to Wagner's music and made Shaw "the Perfect Wagnerite." More to Alec's purpose in 1880, however, was Chester's knowledge of chemistry. He had worked for two years as chemistry tutor to the Duke of Marlborough, had studied chemistry in Germany, and had served as a professor of chemistry in Dublin and London. Alec must have looked forward to a new turn in photophone matters when he wrote Mabel that Chester was to leave for America early in December.

Before Chester left England, he and Alec planned a "campaign against

Chichester A. Bell

selenium, tellurium, and sulphur" making use of Chester's chemical expertise. Back in America, Tainter was meanwhile unsettling their strategy by discovering that lampblack could be used instead of selenium in the photophone. For some weeks after returning, Alec was, as he put it, "in the doctor's hands . . . on account of functional derangement of the heart brought on by too much Photophone." But Tainter's discoveries and the prospect of Chester's collaboration revived Alec's zeal along with his health. He decided to use his Volta Prize money as working capital for a permanent, self-supporting laboratory, to rely chiefly on Tainter and Chester, but to be augmented from time to time, he hoped, by "distinguished specialists of different kinds." The establishment would be called the "Volta Laboratory."

By mid-April 1881, the new chemical orientation had produced what Bell reported to his father as "great discoveries." "We have baptized our new Laboratory baby 'The Spectrophone,' an instrument for spectrum analysis by means of sound." He described it immediately in a paper before the Philosophical Society of Washington. An intermittent beam passed through the substance to be analyzed and then through a prism to produce the absorption spectrum. The spectrum was scanned by a diaphragm with a slit parallel to the light and dark bands. As the bands were passed by the slit, a

341

photophone receiver on the other side translated them into intervals of sound and silence. The instrument's chief use, Bell suggested, would be in the invisible part of the spectrum, "where the eye is useless, [but] the ear is invaluable."

The "spectrophone" turned out to be the last trophy of photophony. A few days after the spectrophone paper, the Volta Laboratory hunters started other game and never returned to the old spoor.

Its shortcomings unremedied, the photophone fell into neglect. It could not challenge the Blake transmitter in simplicity and ruggedness. "Quite a wonderful thing, though not an invention of recent date," commented a kindly visitor to the Philadelphia electrical exhibition at which Bell exhibited his photophone in 1884. "As far as we have learned, no practical application of Professor Bell's invention has as yet been made. Its importance is altogether theoretical. Yet we have no doubt that the time is not far off when even this discovery will find its use in the everyday walks of life." Time passed, but that faith was not vindicated. The *Scientific American* in 1890 praised Bell's spectrophonic exploration of the infrared spectrum, but doubted that the spectrophone had any other use. In an American Telephone and Telegraph Company exhibit at the Columbian Exposition in Chicago three years later, the photophone — now called the "radiophone" by Bell — transmitted speech a hundred feet to a lampblack receiver, from which hearing tubes passed into the listeners' booth. The limit of transmission was still no more than six or seven hundred feet.

In February 1896, Bell chose "Radiophony" as the subject of a talk to the students of Illinois College, suggesting it as an uncrowded field of scientific study. He himself seems to have retained his enthusiasm for the invention, though not for serious work on it. That September he told a Washington reporter that recent advances in electric lighting, combined with proper lenses and reflectors, might significantly extend its range and eliminate the need for telephone wires. An AT&T engineer began experimenting with the photophone or "radiophone" that year and in 1897 pushed its range to several miles by using an arc light, responding to slight variations in current, as the transmitter instead of Bell's mirror diaphragm, and by improving the efficiency of the selenium cell receiver.

But it was also in 1897 that Marconi first succeeded in sending radio signals several miles.

Asked by an interviewer in 1899 if Marconi's radio might be used in the Boer War, Bell pointed out the possibility of what would now be called jamming the signals. "It is as easy as cutting the wires," he remarked. "Any electrician can do the work easily." As he probably meant to suggest, that fact, along with radio's vulnerability to interception, gave his photophone some advantage in military operations; and such was indeed to be the chief application of the photophone principle. Eventually the achievement of coherent light in the laser beam would greatly extend its possible range. For all that, the photophone concept still remains one of limited utility. Nor is its scientific significance great — or even perceptible.

Yet, Marconi notwithstanding, Bell in 1898 held the photophone to be his greatest invention; and in 1921, less than a year before his death, and in an age of intercontinental radio communication, he could still tell an interviewer that "in the importance of the principles involved, I regard [the photophone] as the greatest invention I have ever made; greater than the telephone." The explanation of that statement is more likely to be found in psychology than in physics.

For all of Bell's brave assertions and his upsurges of euphoria, he brought himself to admit now and then, as he had at the very outset of the quest, that the photophone was not likely to be a money-maker. Such an admission came soon after the debut of the spectrophone, when on May 31, 1881, he recorded another declaration of purpose under the laboratory notebook heading of "Historical jottings": "We fully decided . . . to devote our time to something that would pay . . . so that I might have a self supporting laboratory."

He and his two associates had a project in mind, one that had an appealing historical symmetry to it, though they did not say so. Bell having neglected his variable-resistance principle in telephony, Edison had taken it up and made it dominant in the field. But in so doing, and in going on to electric lighting, Edison had in turn neglected his own inspiration, the phonograph, which in 1881 remained a laboratory toy, still too crude and limited for everyday use, its records brief, poor in quality, short-lived, and not subject to mass duplication. Now Bell and his associates proposed to take their turn at leapfrog. In 1878 Bell had begun to think about such possibilities as the electroplating of records. In the spring of 1881 the Volta Laboratory associates came up with the idea of impressing on the record a permanent magnetic field, varying from point to point, which would produce sound as a pickup of some kind traversed it — the embryo, in short, of modern tape recording.

The phonograph enterprise overlapped and then displaced the photophone between mid-March and late May of that year. Though trials of the magnetic recording idea did not succeed, new ideas kept interest alive. Of "a great discovery" by Chester early in June, apparently involving an improved stylus, Alec wrote, in a wry allusion to Mark Twain's fictional get-rich-quick artist Beriah Sellers, "There's millions in it!"

343

But another manifestation of the Gilded Age suddenly interrupted phonograph work. On July 2, 1881, the crazed Charles J. Guiteau, whose delusions had been given a political turn by the meaningless factional struggles of the recent presidential campaign, shot the new President of the United States, James A. Garfield, in the back as the President walked through the Washington railroad station. The bullet did not kill Garfield. That work was left to the doctors, a procession of whom began almost at once to insert unwashed fingers into the wound, scornful of Joseph Lister's fourteen years of well-supported insistence on antisepsis in surgery. One homeopathic physician pushed his fingers deep into the wound and said, "My God, General, you ought to have surgical advice!" Garfield replied, "There are about forty of them in the adjoining room — go and consult with them." The repeated finger pokings did not locate the bullet, nor did surgical probes. But Garfield's comparative youth (he was forty-nine) and his physical stamina helped him rally in spite of the doctors. By mid-July the general expectation was that he would recover.

Anesthetic and surgical techniques had not yet advanced far enough to permit with safety the formidable exploratory operation that would have been needed to find the bullet. A. Nelaton of Paris had developed a metallic gunshot probe tipped with a small porcelain bulb, which was characteristically marked by lead when contact was made. A bullet could thereby be distinguished from bone. But the Nelaton probe required some pressure, possibly dangerous. On July 6, with the President's life seeming at stake, Bell's astronomer friend Simon Newcomb suggested to the attending physicians the possibility of locating the bullet through some sort of electrical or magnetic effect, such as retardation of a rapidly revolving magnetized needle, though Newcomb doubted that such an effect would be detectable. The Bells had forsaken the heat of Washington for Massachusetts in mid-June, but Bell read of Newcomb's suggestion in the newspapers and promptly wired him an offer of assistance, Boston being a likelier place than Washington to have such a device properly made.

Bell himself had already begun thinking about possible ways to locate the bullet. He remembered hearing in Paris of a patient's having been lit up like a Chinese lantern by an electric light in his stomach that faintly revealed the shadow of a tumor; and he tried the idea on his accommodating secretary William Johnson to the jack-o'-lantern extent of a small light and a bullet held in the mouth. The bullet was indeed shadowed on Johnson's cheek. But Bell also happened to have experimented in England in 1878 with less awkward techniques for locating hidden metal.

He had been led to them through his device of canceling out the interference of induction on a telephone line by means of a complete wire circuit with the outgoing and incoming wires twisted together. In one outdoor bypath of this research, he used a rheotome grounded at points in the earth some distance apart, and a telephone receiver with one ground fixed and the other movable from point to point; those points at which the movable ground did not pick up the tone of the rheotome lay on equipotential lines. In another offshoot, he found that when two overlapping flat spiral coils were so positioned as to silence the induced tone, the passing of metal across the overlap (or vice versa) disturbed the balance and made the tone audible. Back in America in 1879 he had written out his experiment with the former device at the request of Gardiner Hubbard, who saw in it a possible way to detect valuable metallic deposits in the earth; and he had presented the second in a paper before the American Association for the Advancement of Science.

As soon as he understood Garfield's condition, Bell made some preliminary experiments of that sort at Charles Williams's establishment in Boston, using as the most promising form the induction balance invented in London in 1879 by his friend David Hughes. The results were encouraging, and a reference to them in his wire to Newcomb led the latter to switch support to that approach. George M. Hopkins of the *Scientific American* independently thought of the Hughes induction balance, tried it, and published his hopeful results in the *New York Tribune* on July 11. Three days later Bell and Summer Tainter arrived in Washington and set to work desperately at the Volta Laboratory.

Efforts to adapt the induction balance to the President's need went on simultaneously at the Volta Laboratory, the Williams shop in Boston, and an electrical shop in Baltimore. By telegram Bell asked for suggestions from Hughes in London, Henry Rowland at Johns Hopkins, and John Trowbridge at MIT. George Hopkins, out of ideas, yielded the field to others, and the Baltimore shop soon did likewise.

The urgency of the work was heightened on July 23, when the President's condition suddenly deteriorated. The temperature he had been running since early July rose to 104 degrees, and next morning a drainage incision was made below the wound. Bell and Tainter, frantically trying various coil designs, battery powers, and other arrangements, had meanwhile managed to extend the effective range of the balance from two inches to three. On July 25, Professor Rowland called at the lab and suggested the use of a condenser in the primary circuit. That modification pushed the range another half inch further, which the doctors considered good enough to warrant a first trial on the President.

Early on the evening of July 26, Bell and Tainter brought their apparatus into the White House through a private entrance, managing to elude the reporters who had got wind of the proposed experiment. Garfield being asleep, Bell was allowed to survey his sickroom to see how wires might be brought in. As Bell described the scene to Mabel, a woman, perhaps a nurse, sat by the bed fanning the sleeping patient, who

looked so calm and grand he reminded me of a Greek hero chiselled in marble. He has a magnificent intellectual-looking head, as you know, with massive forehead. As I remember him of old, his florid complexion rather detracted from his appearance, giving him the look of a man who indulged in good living and who was accustomed to work in the open air. There is none of that look about him now. His face is very pale — or rather it is of an ashen gray colour which makes one feel for a moment that you are not looking upon a living man. It made my heart bleed to look at him and think of all he must have suffered to bring him to this.

By the time Bell and Tainter had their apparatus set up, Garfield had awakened and his wound had been dressed. Something seemed to be wrong with the balance — it turned out later that the condenser had been connected to only one of the two primary coils — so that a sputtering sound was heard, and the range seemed impaired. But the doctors insisted that the experiment be tried at once. Garfield was turned on his side with his body exposed as far as the thigh. First one doctor, then another moved the instrument over Garfield's body, but "that horrid unbalanced spluttering kept coming & going in an irregular sort of way," and the experiment had to be ended for fear of tiring the President. "I feel woefully disappointed & disheartened," Bell wrote Mabel that night; "however we go right at the problem again tomorrow."

Bell in fact was prostrated until July 29, but on that day he returned to the lab and tried the device with the condenser properly connected and using the earlier plan of flat, overlapping coils, instead of conical coils. This gave a range of fully five inches, at which distance a bullet gave a clear tone. "By forced exertions," the arrangement was given portable form in a wooden case, and on the afternoon of July 31 the doctors at the White House agreed to a new trial on the following morning.

At about nine P.M. on July 31, in response to Bell's hearty "Come up and see us" of that afternoon, a *Boston Herald* reporter called at the little two-story brick Volta Laboratory building, half-hidden amid trees and unkempt shrubbery in the middle of a large lot on Connecticut Avenue, half a mile from the scene of Garfield's agony. "The little mansion was brilliantly lighted," wrote the reporter. "Every room was in use, and all the windows were opened, so that the light streamed out as the air streamed in. A courteous colored serving man bowed his master's guest into the shadowed hall."

346

Bell led the visitor into the main room, where they found Bell's father and Sumner Tainter.

In cabinets, on tables, chairs and floor were coils of wire, batteries, instruments and electrical apparatus of every sort. The light from the jets, burning brilliantly in the centre of the room, was reflected from a hundred metallic forms. It was reflected, too, from the smiling faces of the great electrician and his assistant, who saw success almost within their grasp. . . . The room was full of metals, which disturbed the tone of the balance, and outside and inside, attracted by the light, "annoying insects," as Prof. Bell called them, kept up a monotonous monotone.

Bell's high spirits were dashed by the White House trial of August 1. Instead of the sharply localized sounds detected in trials with bullets both before and after, he heard a feeble sound over a considerable area of Garfield's body. Newspaper accounts deemed the test a success, but Bell and the doctors knew better. On inquiring at the White House next day, Bell found that the mattress of Garfield's bed was supported by steel springs. But he experimented with a duplicate of the bed and found that the springs did not materially affect the hearing distance, nor did they explain the large but limited area of sound. The latter remains a mystery.

A few days later Bell returned to Massachusetts, where until the birth and death of his son Edward on August 15 he commuted daily from Pigeon Cove on the North Shore to the Williams shop in Boston to supervise further induction balance experiments. There he invented a telephonic needle probe. This consisted of a telephone receiver in series with two electrodes, one being a metal plate in contact with the patient's skin, the other a fine needle all of which but the tip was insulated with shellac. A distinct click sounded in the receiver when the needle tip touched a bullet experimentally imbedded in raw beef. In that fashion Tainter demonstrated the needle probe with complete success to some of the doctors in Washington. But they did not choose to try it on Garfield, who had weakened considerably and whose weight had fallen from 200 to 120 pounds.

In Washington on the night of September 19, Bell heard a newsboy shouting outside, "Extra Republican — Death of General Garfield!" "Poor Garfield has gone," Bell reflected; "I hope indeed that there may be an immortality for that brave spirit. . . . If prayers could avail to save the sick, surely the earnest, heartfelt cry of a whole nation to God would have availed in this case." The autopsy showed that the bullet itself was too deep for Bell's induction balance to have detected it. It was, however, completely harmless in its final lodgment. Garfield had died from the infection of his wound and the rupturing of an aneurysm in the splenic artery (which may have been originally injured by the passage of the bullet or, more likely, by the surgeons' probes).

Early in October Bell successfully demonstrated both the induction balance and the telephonic probe to a group of doctors assembled in New York City by Dr. Frank H. Hamilton, one of the late President's consulting surgeons. Five years later Dr. John H. Girdner, who had witnessed the demonstrations, began using and recommending the telephonic probe. As interest in the probe grew among physicians, Girdner assured Bell that his name would "hereafter be coupled with Jenner, Wells, and Harvey as suffering humanity's greatest benefactors," and in Girdner's first paper on the technique he acknowledged it as "the invention of Professor Bell." To Bell he wrote, "I have never nor could I lay any claim to the originality of these inventions." Girdner's next paper a few months later, however, omitted any mention of Bell. By the early nineties Girdner was advertising "Dr. Girdner's Telephonic Bullet Probe," and his obituary in 1933 proclaimed him "the inventor of the Girdner telephonic bullet probe, which was used universally for the removal of bullets before the development of the X-ray."

Bell knew about Girdner's appropriation of credit, but gave no sign of being disturbed by it. Girdner notwithstanding, the University of Heidelberg in 1886 gave Bell an honorary M.D. degree for that contribution to surgery; and as late as 1915 a distinguished British surgeon, who had first heard of the probe from a report of Bell's AAAS address of 1882 on the Garfield case, gave Bell his full due in the *British Medical Journal*. Bell's main concern appears in a letter he wrote Mabel from Washington in September 1881, asking her to look in his study for "the names and addresses of the poor people who want to have bullets located. I would like to help them if possible before leaving [with you for Europe]. One especially is from the father of a little boy who was shot last year." In the Sino-Japanese War of 1894–1895 and the Boer War of 1899–1902, Bell's telephonic probe reduced suffering and saved lives; and even in the First World War, in cases where X-rays were unavailable or inconclusive, it continued to do so. Since Girdner's promotion of the telephonic probe had done much to bring it into merciful use, Bell probably did not begrudge the enterprising doctor his small unearned increment of glory.

It was in the tragic summer of 1881 that Bell, because of the death of his newborn son from respiratory failure, invented a mechanical device for administering artificial respiration. He called it a "vacuum jacket" — an airtight iron cylinder surrounding the patient's torso up to the neck and fitted to have air forced out and in by a suction pump. By reducing the air pressure around the body, the device permitted the atmospheric pressure outside to force air through the mouth and into the lungs. This was an anticipation of the "iron lung" later so vital to victims of poliomyelitis.

The vacuum jacket

Bell constructed a working model in England, where it was displayed at the Physiological Society early in 1882. In America that summer, Bell followed up his AAAS paper on the Garfield experiments by showing and explaining a small model of the vacuum jacket. Bell and his contemporaries viewed it, however, as a means of reviving those rescued from drowning, an emergency in which the old recourse of external manipulation would usually be the quickest available. For this reason, perhaps, the vacuum jacket principle remained unused until others saw its value, when provided with engine power, for cases requiring continuous and indefinitely prolonged treatment.

Early in October 1881 when, to counteract the sadness of little Edward's death, Bell took his wife and daughters for a long European trip, he left his bachelor associates of the Volta Laboratory to their phonographic devices. By then, despite the summer's interruptions, phonograph research was well under way. Two basic goals had been set: fidelity in recording and durability in playing. Ideas had been developed. And progress had been made in reducing them to practice.

The way to the first goal had been conceived and noted down by Sumner Tainter as early as March 28, 1881. Edison's recording needle, as it rose and fell on his tinfoil, did not remove material but simply displaced it, each impression thereby distorting the previous one. The tinfoil's elasticity, such as it was, also worked against an accurate record. Tainter's idea was not to indent or emboss the recording material but to cut, incise, or engrave an impression on it. Not tinfoil, but wax or some other easily carved material was to be used. Edison himself had included wax among the recording materials he had initially proposed. Perhaps he failed to specify the cutting effect of a stylus on wax simply because he thought it went without saying. In the end, however, the courts held otherwise; and Sumner Tainter's explicit formulation, as ultimately set forth in a patent, became the most valuable single product of the Volta Laboratory.

The Volta trio at first saw as another way to prevent distortion the use of a lateral or (as they dubbed it) "zig-zag" cut instead of the up-and-down or hill-and-dale impression, which they referred to as "Edisonian." The lateral cut, they reasoned, would control the vibration of the needle throughout, rather than in one direction only. In time, however, they returned to the hill-and-dale method. Looking back after many years, Tainter concluded that their zigzag grooves had been too large and their pickup too heavy for the energy of the sound waves. It remained for Emile Berliner, whose solid variable-resistance telephone transmitter caveat had stymied Edison's carbon-button patent application, to take up and commercialize the lateral cut some years later.

To reduce wear in playing the record, the experimenters looked for ways to eliminate direct contact with the record surface. Hence the early notion of magnetic action. Bell's own distinctive conception had actually antedated Edison's basic invention. It was the imparting of undulatory motion to a jet of air, which Bell had casually envisioned as a means of reproducing sound during his pretelephone studies of the phonautograph. On September 25, 1881, before Bell left for Europe, an instrument of this sort was tried with promising results. From a reservoir of compressed air, a fine jet was directed by a small nozzle into the lateral-cut wax record groove, and the recorded sounds could be heard: "T-r-r T-r-r There are more things in heaven and earth Horatio, than are dreamed of in our philosophy. T-r-r I am a Graphophone and my mother was a Phonograph." (The experimenters had decided to call the recording apparatus a phonograph, the reproducing apparatus a graphophone, and the record a phonogram.) The graphophone, the phonogram, a description of the invention, and a careful transcript of the recorded message all went into a sealed box and then, on October 20, 1881, to the Smithsonian Institution.

During the six months of Bell's absence, his cousin Chester and young Tainter fully earned the salaries they drew from the Volta Prize money. While Chester busied himself with chemical matters and an effort to adapt static electrical charges to sound reproduction, Tainter, whose living quarters were over the laboratory, applied himself to magnetic and jet instruments. On reading Bell's letters of encouragement and mild prodding, Tainter occasionally waxed testy. "At least *three fourths* of my time," he complained to Bell in February 1882, "is consumed in [making apparatus] that we could hire a man to do for $18 per week if we had a place for him to work in." And in rejoinder to a remark by Bell about "side-shows," Tainter wrote that "with the exception of the time put into that miserable induction balance, I believe that about three days is all I have wasted on them during the past nine months. If the other members of the Association stick to the main question as closely as I do, I don't think the side-show business will ruin us."

More than ten years later, Tainter retained a vivid, if somewhat chaotic, recollection of what he and Chester Bell were up to in those months and years. In response to the questioning of counsel he said:

We made many experiments with etching records and also with electroplating or building up of the record; with various arrangements for bringing in some auxiliary power to work the cutting style; with various methods of reproducing the sounds without contact with the record (magnets, air-jets, and radiant energy were used for this purpose). Different shaped cutting styles and those not adapted to cutting were tried. Much time was devoted to electrotyping, moulding and making copies of records. Records were made by means of photography, and by sensitive jets of liquids. We experimented with stearine, stearine and wax, stearine and paraffin, wax and oil. We tried many forms of reproducers making records of music by direct transfer of the vibration of the sound board. We made records by jets of semi-fluid substances [including maple syrup] deposited on the recording surface, also records on narrow strips of paper coated with wax-like composition; records on wax-coated paper, disks and tubes, with many forms of machines for using wax-coated disks and cylinders, with many forms of recorders and reproducers for wax-coated cylinders and with various forms of speaking and hearing tubes.

At that point the hearing adjourned.

Neither the air-jet nor the zigzag approach had yielded dependable results by February 1882, when Tainter came up with an improvement that turned the experimenters back to direct-contact, hill-and-dale reproduction: a flexible pickup or floating stylus, which maintained a slight, steady, but easily yielding pressure against the record despite any irregularities in the record's surface. This not only reduced the wear of a direct-contact pickup and improved its accuracy, but also made possible finer grooves, which in turn increased the recording capacity of a given surface. The associates continued to experiment with and eventually patented magnetic and jet pickups (Chester seeming to be especially fascinated by the jet approach), but the direct-contact floating stylus held its primacy.

Bell returned in May 1882. Later he testified that "while in Washington [I] was in the laboratory almost daily. I witnessed experiments made by my associates, and showed them my own; and every few days we discussed together the experiments that had been made, with the object of ascertaining the most promising line to be pursued." But his absences from Washington were frequent and prolonged; in 1882, for example, he soon left again to summer with his family in Newport. Even while in Washington he tinkered with his induction balance and vacuum jacket, testified in the Drawbaugh case, worked on the statistics of the deaf, and otherwise diluted his attention to phonograph development.

By the end of 1884, Chester Bell and Sumner Tainter had been working more than three years on their phonographic permutations and combinations. The Volta money was running low; and Chester, increasingly impatient with his American cousin's evident lack of interest in the day-to-day work, decided he could do better back in England. Alec Bell himself felt that the Volta Laboratory's models were ready for the market. Alec had foreseen the difficulties that might arise at this juncture between him and "two other men of independent and original minds." From the outset, therefore, as he explained privately to his son-in-law twenty years later, it had been "agreed that none of us should claim as his own, ideas that had not been reduced to writing. We were all provided with scribbling-books and our claims to invention were to stand or fall by our written notes. The result of this process is seen in the fact that my name does not appear in any of the patents covering the Graphophone. I was the most delinquent in the matter of written notes, and my work which was largely that of directing and guiding the various lines of experiments did not come out in any patentable form."

In 1886 (upon applications filed in 1885) he, Chester, and Tainter did in fact jointly patent two of the noncontact reproducing devices, which never came into use. Chester was sole patentee of four telephone transmitters using jets, and Tainter of two phonographic devices. But the crucial Patent No. 341,214, which embodied the engraving or incising specification, was issued in the names of Chichester A. Bell and Charles S. Tainter only.

An understandably resentful memorandum by Tainter years later explains plausibly enough the adding of Chester Bell's name to a patent for what all conceded was Tainter's idea. The shortage of funds for patent expenses, wrote Tainter, had led him to agree that their attorney, Anthony Pollok, might consolidate some patent applications. When Tainter discovered that the cutting or incising patent had been issued jointly to him and Chester Bell, he protested but was persuaded to go along on the grounds that delicate negotiations, promising rich returns to Tainter as well as to the others, might otherwise be jeopardized. Later still, Tainter asserted, he learned that the promoters involved had deliberately had "the Bell name hitched on to the patent, for it improved the outlook from the business man's standpoint. The justice of the matter did not trouble them."

The share of profits each of the three associates received from the combined patents was not affected by the individual names attached to each patent. Still, the official acknowledgment counted for something to a touchy young inventor. Tainter, however, did not accuse Alexander Graham Bell of being privy to the name-shuffling; and Bell's record of scrupulousness in these matters, as in others, argues convincingly that he was not. It was probably during Bell's summer and fall in Newfoundland and Nova Scotia that the deed was done. In 1888, moreover, Bell cheerfully gave Tainter "exclusive credit . . . for the instrument in its present condition."

While the patent applications were pending, the Volta associates negotiated for a joining of forces with Edison. In patent terms, they now were to him as he had been to Bell in the case of the telephone, their patent controlling a feature important for the commercial development of his basic invention. In business terms, the situation differed, since the Edison Speaking Phonograph Company, control of which Edison had bought back from Gardiner Hubbard and his associates, had long been moribund, in contrast to the vigor of the Bell Telephone Company during the telephone struggle. But after a short period of dickering failed to win his side a three-fourths interest, Edison decided to turn from his electric light enterprises, now well established, take up his neglected phonograph in earnest, and out-invent the Tainter-Bell group, whom he presently referred to privately as a "bunch of pirates." Edison did not hesitate to pirate Tainter's floating stylus in his turn, though developing a number of useful phonographic refinements all his own.

Meanwhile the Volta associates dissolved their original agreement and in

353

January 1886, with Chichester's younger brother Charles and another backer, incorporated the Volta Graphophone Company under the laws of Virginia as a holding company for the patents. A group of Washington court reporters, much impressed by the utility of the Volta Company's slender, wax-coated, cardboard cylinders for dictation, some time later led in organizing the American Graphophone Company, which acquired the patents by exchanging part of its shares for those of the Volta Graphophone Company. The three Volta Laboratory associates divided their American Graphophone shares into four equal parts, one to each as coinventor and the fourth to Bell as financial backer.

Sumner Tainter's shares preserved him from financial worry for the rest of his long life. He married and went on with his phonographic work for a number of years, in 1887 patenting a device that a recent historian considers his best idea, although the industry never adopted it: an arrangement for increasing the rate of rotation of a disk record as the needle approached the center, so as to maintain a constant speed along the groove. About the turn of the century, his health broke down from overwork, and from 1905 on he lived in retirement in San Diego, working on minor inventions in a home laboratory. To his gold medal for photophone development, received at the Paris Electrical Exhibition of 1881, he added the John Scott medal of the Franklin Institute in 1900 and a gold medal at the Panama-Pacific Exposition of 1915, both for phonograph contributions.

Chichester Bell, who soon returned to England, also enjoyed financial independence for the rest of his life, even after marrying in 1889 and raising six children. In England, according to a son, he "spent the rest of his life experimenting in a small way." He died in Oxford in 1924.

Some idea of what the Volta Laboratory associates made from their efforts may be gleaned from Alexander Graham Bell's experience. Having received as financial backer the same amount of stock as each of the other two, he sold it soon afterward for $100,000, which he put in his father's hands as a trust fund for researches relating to the deaf. He had received an equal amount as a coinventor, which, like his former partners, he held for some time longer. When he sold the greater part of this to augment the trust fund, the fund was more than doubled. So the two younger men ought to have done well financially.

The deaf certainly profited, even though of all Americans they were the least able to use the technological wonders that had generated the money.

27

The Realm of the Birds

In the spring of 1879 Bell remarked that he wanted to meet Thomas Edison and see his "celebrated laboratory at Menlo Park." Years later the two men did meet. But when Edison moved from Menlo Park in the eighties to a laboratory ten times bigger, Bell had still not made his pilgrimage, probably not having been invited. Bell's interest in Menlo Park, along with his own Volta Laboratory, suggests an intent at first to follow Edison's approach to invention. The profits from the graphophone project ought to have confirmed that aim. Yet Bell's desultory attention to graphophone research and the donation of his profits to helping the deaf stood in contrast to Edison's lifelong obsession with technological research and his readiness to keep reinvesting all his time and money in it. Bell, in short, did not share and would never share Edison's total commitment to technological research and development.

In the spring of 1883 Bell trailed terminals from the bow and stern of a small boat on the Potomac, sent a strong intermittent current between them, and succeeded in picking up the signal on another boat a mile or so away, using a telephone receiver with similar trailing terminals. (This was not radio transmission nor even an approach to it.) No other significant experiments on that or any other invention were recorded by Bell until 1889. In 1885 it occurred to him that an adjustable array of narrow metal strips might be used to cast the shadow of any given sound-wave form, and the sound itself then be reproduced by an adaptation of the photophone. The aim, though not the method, foreshadowed modern electronic sound synthesizers. "If the above invention can be perfected," he wrote in his notebook, "it will be a greater achievement than the telephone, photophone, or spectrophone." But he never tried to do it — which was probably just as well. "I have got work to do," he wrote that year, "so much work on so many subjects that I want many more years of life to finish it all." And first on his list during the middle and late eighties was work with, for, and about the deaf.

"I doubt the wisdom of having a workshop or laboratory at all," he was writing Mabel by the spring of 1887, when the Government case, the statistics of deafness, and teaching methods for the deaf were taking up most of his time.

Washington is no place in which to carry out inventions. If we only lived in the neighbourhood of a large city, I could have apparatus made in a large workshop, as I did at Williams' in Boston, at any time. In a small workshop with one workman it takes forever to have the slightest thing done, and ideas cool before anything is accomplished. If on the other hand a staff of workmen should be engaged, then they must all be kept in work and that means giving myself up to the laboratory exclusively. . . . Don't urge me to work any more for money. . . . The subjects I long to work at are nearly all unremunerative in their nature. . . . Our income is good enough.

Edison had plainly ceased to be his model.

Yet he never quite lost his fascination with science and technology, and Mabel encouraged him to indulge it. In Washington late in 1889 he hired a young man named William H. D. Ellis to help him wind up certain statistical work, and soon afterward he reopened his laboratory with Ellis and Arthur McCurdy as assistants. "I would be so glad," wrote Mabel, "if you are really going back to experimental work." He began again to dictate his thoughts and experiments regularly to McCurdy for recording in two series of notebooks, the "Lab Notes" and the "Home Notes" (distinguished from each other not by date or subject, but simply by the place where each was kept). His telephone travails having taught him the possible importance of full, dated records, he would eventually fill 210 volumes with them. ("Fugitive thoughts," he wrote years later, "are better preserved in a book than on separate sheets of paper, for most of them are of little or no value and would [otherwise end in] . . . the fire or the waste-paper basket; but some of them are of value [as] . . . the stepping-stones to other ideas.")

By 1889, Bell had moved his laboratory to a small, two-story, red brick building, formerly a stable, back of his father's house in Georgetown. A reporter described it in 1895:

[After] passing through a workshop containing benches and machinery, [we] came into a large room walled with shelves and filled with models and instruments of all kinds. . . . Filling up nearly the whole floor was what at first sight seemed to be a model of a new threshing machine. . . . It was a type-setting machine for the instruction of the deaf . . . by which words could be put upon a blackboard and the letters distributed again. On the shelves in the walls at the left were perhaps fifty models of telephones, and among them the first one that Mr. Bell ever made. . . . Beyond this were scores of cylinders used in the experiments upon the graphophone, little bottles of selenium . . . many scientific instruments, inventions illustrating new and yet unexplained theories as to the property of matter, originated by Mr. Bell . . .

and many other things.

Despite his demurrer to Mabel, Bell resumed his experimental work with a money-making objective: a practical method of copying phonograph records. After working through the winter of 1889–1890 and briefly in late spring, Bell recorded that "a new invention has been born," a method of molding phonograph cylinders from master records in a waxy material Bell called "ozokerite." But the invention was stillborn; the copies could not be counted on to come out of the mold properly. Bell passed immediately to a plan for using agate cement to take a printing impression from photographs. "Here at last is a subject at which we can work together," he wrote Mabel in June 1890. His previous inventions had been to her "all sealed things, relating to the world of sound, but this . . . will bring us together in a common interest." This scheme also could not be made commercially practicable and was soon abandoned.

Aside from unproductive efforts at intervals from 1895 to 1897 to develop an automatic telephone switchboard, Bell made no further serious attempts to invent for the market. Briefly during the early nineties he reverted to the photophone — or as he now called it, the radiophone — the vacuum jacket, and the induction balance. In 1895 and 1897 he tinkered with simple devices to be stored in lifeboats, for condensing fresh water from fog or moist sea air. In 1891 he coiled wire around his head in circuit with a coil around Ellis's to see if it might facilitate thought transference; it did not. In 1895 and 1896 he speculated that a boat with a revolving hull might encounter less water resistance at high speeds, but he did not test the vagrant notion. He daydreamed about "seeing by electricity" with the help of selenium in 1891, and the advent of Roentgen rays and fluorescent screens in 1896 induced further "fugitive thoughts" of that sort; but he did not try to implement them.

Bell's notebooks during the nineties, at least after the short-lived "ozokerite" and agate cement projects, lack the urgency, the expectancy, the exuberant conviction of impending glory, that had imbued his quests for the harmonic telegraph, the telephone, and even the photophone. He applied for no patents. The notebooks would have been few and slim, might even have ended in "the fire or the waste-paper basket" in token of farewell to tinkering, had it not been for the ambition that seized him blithely but tenaciously in the spring of 1891: to develop a heavier-than-air machine that would fly by its own power.

It was Bell's crony Samuel Pierpont Langley who revived Bell's old dreams of mechanical flight and turned them toward action. A large, florid man thirteen years older than Bell, the bachelor physicist Langley hid his innate shyness behind a show of gruff dignity. But to his few intimate friends, including Bell, Langley revealed great charm, wit, and warmth. He was Boston-bred of distinguished Massachusetts lineage, an omnivorous reader from early youth, a leading astronomer, and since 1887 the secretary of the Smithsonian Institution; yet he had no college training. The public libraries of Boston had been his university. As director of the Allegheny Observatory and professor of physics and astronomy at the Western University of Pennsylvania for twenty years, Langley had made the precise measurement of solar radiation his specialty, in 1878 inventing the bolometer, sensitive to a millionth of a degree rise in temperature.

Langley told a reporter in 1896 that he could not remember a time when he had not been interested in flight. Like Bell, he had watched birds as a boy and wondered what kept them up. "I seldom saw a bird flying that I did not think of it," he said, "and even lately I have watched them for hours, trying to understand how they could move about through the air, rising and falling, soaring up and sailing down without any motion of the wings." He had begun his serious experiments on the problem shortly before leaving Allegheny Observatory. There and in Washington he made elaborate measurements of the lift and drift of moving plane surfaces, building long-armed whirling tables and other devices for the purpose. The mere fact that so reputable a scientist took the notion seriously, shown by his publication of results in 1891 as part of the *Smithsonian Contributions to Knowledge,* encouraged others to work in a field stigmatized since the times of classical myth and satire as the resort of fools and madmen.

Mabel Bell may have transmitted the fever to her husband, or at least fostered it, with a letter from Washington in April 1891: "I wish you were here if only to attend the National Academy meetings and to hear the discussion on Professor Langley's flying machines. Of course the papers treat him more respectfully than they would anyone else, still they cannot resist a sly joke now and again." By late May, Bell himself had begun testing the lift of a propeller. In mid-June he was back in Washington, giving as among the reasons "flying machines and Prof. Langley." Langley by now was building and testing models. "Langley's flying machines," Bell wrote in high excitement on June 15, 1891; "they flew for me today. I shall have to make experiments upon my own account in Cape Breton. Can't keep out of it. It will be all *UP* with us someday!"

Bell did not soar immediately into the new project. Not until the end of 1891 did he spend a week or so testing helicopter models, including an unsuccessful trial of one powered by steam jets at the rotor tips. From 1892 through 1897, however, he carried on more extensive experiments, usually at Beinn Bhreagh from May to November or December. This was not a crash program of the Edisonian sort. There were no frenzied stretches of seventy-two hours without sleep. Bell slept his fill, read his newspapers, carried on correspondence about the deaf, tended to various other matters. Still, the notebooks record a substantial effort, and he had frequent spasms of concentrated work. "Alec is simply gone over flying machines and his papers on signs, and there is very little good to be got out of him," Mabel wrote her mother in the spring of 1894. And, experiments or not, the mystery of flight was never far from his mind. A stray note survives from a train ride through Mexican mountains in 1895: "Soaring buzzard's wings turned up at tip . . . surface air almost calm . . . moved in spiral . . . horizontal velocity not less than 30 feet per second."

There was an aimless quality in Bell's flying experiments, as in those of other men during that decade. He had no such rock of original insight as had given him a fulcrum in the telephone enterprise. So he puttered with gunpowder rockets for propulsion, with pinwheel helicopter rotors, with spring-powered models, with monoplanes. Rockets misfired or ignited the model, free-flying rotors did eccentric flip-flops. But Bell shrugged off the setbacks. "The more I experiment," he wrote in 1893, "the more convinced I become that flying machines are practical."

After all, little machines had indeed flown, not only Langley's models but also the flying toys, both helicopter and winged, invented by the Frenchman Alphonse Pénaud in 1871. Bell knew about the latter, made his own similar toys, and even contemplated marketing them to support the research. If small, weak machines would fly, why not large, powerful ones, notwithstanding the stubborn fact that weight increased as the cube of a model's linear dimension? Bell wanted so desperately to surmount that mathematical obstacle that privately (fortunately for his reputation) he seized on the analogy of the heavier, yet more effective, telephone diaphragm to speculate that the heavier an object was the better it would fly. When his imagination leaped to flying battleships, he awoke to the absurdity of the notion.

Late in 1894, Bell began laborious tests of wings and propeller blades on whirling arms like the devices Langley had used. These went on in season

Samuel P. Langley

Bell's snapshot of the Langley model
in flight, May 9, 1896

during 1895, 1896, and 1897. In May 1895 Langley's forbiddingly precise standards, perhaps even some hint of skepticism by Langley as to Bell's experimental technique, may have given Bell pause. At any rate Mabel wrote him from Paris that "if Mr. Langley has changed your ideas, why then I can't see why you should not come over." A month later, however, she wrote more positively, "I do so want your name associated with successful experiments in flying machines. I don't like to think of your stopping them unnecessarily." Whether or not as a result of Mabel's urgings, he went on.

In May 1896 Bell accompanied Langley to Langley's houseboat on the Potomac at Quantico, Virginia, to see the first trials of a steam-powered, propeller-driven airplane model. Langley had invited Bell as the only man he knew whom he could bear to have as witness to a possible failure. They got up at five on the morning of May 9. After preparing the sixteen-foot-long model to be catapulted from the houseboat, Langley retreated alone to a little pier, too nervous to be any closer to the test, while Bell had a boy row him out on the bay, from which point he managed to photograph the machine in flight. Three days later Bell wrote an exultant letter to the editor of the journal *Science*, likening the spectacle to that of "an enormous bird," soaring in a great spiral to a height of a hundred feet, traveling half a mile, and then, when the steam gave out, settling unhurt on the water "as slowly and gracefully as . . . any bird." A second trial was equally successful. "No one could have witnessed these experiments without being convinced that the practicability of mechanical flight had been demonstrated." On the day of the flights, Bell wrote Langley a note of congratulation ending with what, coming from the inventor of the telephone, was strong language: "I shall count this day as one of the most memorable of my life."

Both Langley and Bell went back to their experiments with renewed enthusiasm. But Bell found success stubbornly elusive. Somehow he could not achieve the combination of precisely measured observations, systematic yet imaginatively progressive experimental variations, and sure grasp of unifying principles that might have unlocked the mysteries of flight. "I am finding out in the laboratory," he wrote in October 1896, "that a great deal has yet to be learned concerning the best way to combine aero-planes or aero-curves — so as to gain the full benefit of the surfaces." For the ebullient Bell, this was a modest, even pessimistic, report. It was during these years of 1895–1897 that, as if to hedge his research bets, he worked now and then on an automatic telephone switchboard.

It is possible to read too much pathos into Bell's aeronautical frustrations. He was not a driven man like Langley, backing off from the scene of a trial as though the shock of failure might shatter his heart. Bell could take a setback with a few hissings of "g e e t h e" and perhaps a steamy climb up the mountain, forget it all in a rousing pianistic gallop with "The Laird of Cockpen" that evening, and next day come back as sanguine as ever.

Whatever Langley thought of Bell's aeronautical science, he found sanctuary and support in Bell's resilient enthusiasm. At Bell's warm invitation, Langley spent several summer vacations with Bell on Cape Breton, sometimes cruising alone with him for days at a time on Bell's houseboat. Charles Thompson set down a revealing anecdote of that relationship. "I have seen them sitting on deck under the awning for hours and hours, neither of them uttering a sound, but both of them eagerly watching the seagulls soaring about the boat. I remember one day Professor Langley said suddenly, in a raised voice, 'Isn't that maddening!' 'What's maddening?' said Mr. Bell. 'The gulls!' said Professor Langley. 'I was thinking they were very beautiful,' Mr. Bell replied. They both eyed each other for a moment and then laughed heartily. Professor Langley always appeared buoyed up in spirit when leaving after these summer visits."

Bell supported Langley publicly as well as privately. "I have not the shadow of a doubt," he told an interviewer for *McClure's Magazine* in 1893, "that the problem of aerial navigation will be solved within ten years. That means an entire revolution in the world's methods of transportation and of making war. I am able to speak with more authority on this subject from the fact of being actively associated with Professor Langley . . . in his researches and experiments." To a *Cincinnati Enquirer* reporter early in 1896 Bell said, "I am at present engaged in trying to solve the problem of aerial navigation. . . . I believe that it will be possible in a very few years for a person to take his dinner in New York at 7 or 8 o'clock in the evening and eat his breakfast in either Ireland or England the following morning." And a few weeks later Bell's enthusiasm and the weight of his prestige emboldened the *New York World* to give an extended and glowing account of Langley's Quantico success. "The problem of the flying machine has been solved," the story began categorically. "Those who read this article are reading of the fulfillment of a world-old dream." And it closed by remarking that "fifteen years ago a man who had the temerity to deliver a serious lecture on the prospects of navigating the air would have ruined his professional reputation by the indiscretion. Now the much-derided 'cranks' are having their innings."

Bell's effectiveness as herald or pitchman for the age of flight was made apparent by the man from *McClure's*. "Professor Bell has the happy faculty of expressing great ideas in simple words. . . . He is as enthusiastic as a school-boy thinking of the kite he will make as big as a barn-door. His black eyes flash, and they seem all the blacker contrasted with his white hair; the words tumble out quickly, and those who have the good fortune to listen are carried away by the magnetism of this great inventor."

The weight of opinion against which Langley, Bell, and other pioneers of flight contended may be gauged from Bell's encounter with Sir William Thomson, who had become Lord Kelvin, at Halifax during Kelvin's American tour of 1897. Mabel found Kelvin "greyer than when I last saw him twenty years ago, but otherwise the same kindly, loveable, simple man." It was presumably Kelvin's kindliness that prompted his first words to Bell, expressing "regret that he was going into aeronautics. Alec took issue, and Lord Kelvin . . . drew him aside and plunged right into scientific talk right there in the midst of the crowd." Mabel took it upon herself next day to write Lord Kelvin in explanation and defense of her husband's work. But Kelvin was not persuaded. After returning to England, he replied belatedly and soothingly:

I was quite sure that your husband would not go on in respect to flying-machines otherwise than by careful and trustworthy experiment. Even if the result is to demonstrate to himself that a practical useful solution of the problem is not to be found. . . . When I spoke to him on the subject at Halifax I wished to dissuade him from giving his valuable time and resources to attempts which I believed, and still believe, could only lead to disappointment, if carried on with any expectation of leading to a useful flying machine.

Even as Kelvin wrote his letter, the Wright brothers had begun combing the Dayton, Ohio, Public Library for information on the subject.

If Kelvin's warning disheartened Bell, he did not admit it. Bell's experiments went on after Kelvin left, and when Bell read and forwarded Kelvin's letter to Mabel, he made no comment. Anyway, the Spanish-American War had just begun, and the search for Cervera's fleet in the Caribbean turned Bell's thoughts to aerial reconnaissance and bombardment. "I am not ambitious to be known as the inventor of a weapon of destruction," he wrote, "but I must say that the problem — simply as a problem — fascinates me, and I find my thoughts taking more and more a practical form." From late 1897 to mid-1899, it was not disparagement but distraction — sheep breeding, travel, the deaf — that kept him from all but scattered experiments.

The year 1898, however, also brought a turn in Bell's aeronautical experiments that started him up a blind alley. As early as 1896 he had given some thought to kites as an experimental device, including the Hargrave or box kite. These ideas recurred occasionally in 1897. Then, in June 1898, he wrote Mabel that "the importance of kite flying as a step to a practical flying machine grows upon me." Kites would permit experiments involving equilibrium or stability in which cost would forbid the risk of engines, and conscience the risk of men.

The kite idea made sense. In August 1899, for the same reasons, the Wright brothers began their experiments by using a large biplane kite. In their first experiments at Kitty Hawk in September 1900, they flew their man-carrying glider as a kite. But to them, kite flying was transitional. To Bell it would become an end in itself, so obsessing him as to be a fatal drag, figuratively and literally, on forward motion.

"The Laboratory Annex was so filled by the big kite that there was no room for experiment," he wrote Mabel from Beinn Bhreagh in May 1899. "Just fancy a kite 14 feet 7 inches long by 10½ feet wide and 5 feet 2 inches high! A monster — a jumbo — a 'full-fledged white elephant.' " The ambivalent language echoed the disquietude of other remarks in the same letter. "I am no longer young," he wrote — he was fifty-two — "and the experiments on which I have been engaged for years should be completed sufficiently for publication, so that younger men can take up the thread of research. . . . Don't take me any more away from my work until it is finished — or I am!"

A few days later he reported "looking over the records of the multitudinous experiments I have made relating to [aeronautics] and asking myself the question 'Where am I at?' " To read his answer is to have a sinking feeling. In what may have been an instinctive last pulling away from the kite entanglement, Bell suggested that air drag was so fundamental a concern as to call for eliminating wings altogether except for guidance, and obtaining lift simply by angling the propellers upward. Modern helicopters demonstrate that the idea has its uses, but for heavy loads at fast speeds, wings are with us still. The significance of Bell's notion (a notion he quickly rejected) lay in showing his readiness to jettison his years of inconclusive airfoil tests, as he soon did in taking up kite-flying.

On the other hand, the kite obsession led him to his last notable original idea in technology, a concept he would not otherwise have had and one probably more important than any he would have come up with in the mainstream of aeronautics.

As for Bell's contributions to aviation up to this point, it can be argued that he helped to keep Langley at it; and that Langley's example, his writings, and the literature supplied and recommended on request by his Smithsonian Institution helped in turn to set the Wright brothers going. The history of flight, like all history, is a web so complex and finespun that no strand can safely be called redundant.

So as to catch up with Bell's other doings in the last two decades of the nineteenth century, this narrative will reserve till later its accounts of Bell's technological work after 1899. But enough has been said so far to warrant a general assessment now.

A natural tendency is to measure Bell against Edison. The two men were born within a few days of each other; they are both famous for inventive concepts astonishing in their combination of simplicity, originality, and social impact; and they even leapfrogged each other in their inventive work. Bell, like Edison, retained fertility of technological imagination well into later life. But unlike Edison, Bell did not show it to the public. He recorded his ideas in his private notebooks and moved on without devoting himself to cultivation and harvest, whereas Edison never ceased to push his technical ideas through development and patenting to commercial use.

What differences between the two men explain the lapse of nearly twenty years between Bell's graphophone patents of 1886 and his tetrahedral patent of 1904, while Edison was patenting and promoting his (and some of his assistants') inventions at a spectacular rate?

The key is in a paradox: Bell's assets in the telephone triumph became his liabilities in later invention.

In the beginning Bell was driven by various pressures and perhaps by innate temperament to achieve something great — to make himself heard. The need of money on which to marry gave him added incentive. The specific goal of electrical speech transmission was latent in a number of simple, well-known physical principles; but these principles were in fields not usually conjoined, chiefly electricity, acoustics, and physiology. To bring these together and fuse them required imagination and a wide-ranging curiosity, both of which Bell had to the point of extravagance. Bell also chanced to be well grounded in the very assortment of special fields that converged at the crucial point. And with all these attributes, the inventor of the telephone needed to be armored by stubborn self-persuasion against the doubts and denials of supposed experts. Bell, it should be plain by now, had a full suit of that quality.

But the very success of the telephone tended to satisfy Bell's craving for fame and fortune. He married Mabel, outshone his father, and became a permanent worldwide celebrity. So we find him thereafter resisting the suggestion that he invent for money and — so far as public acclaim was concerned — gradually contenting himself more and more with the durable glory of his great invention, especially as age and the comparative meagerness of his later results weakened his assurance.

Bell's multiplicity of interests spread his time and thought over too many fields for productive cultivation of his inventive ideas. He preferred leaving it to others to develop and promote those ideas, good or bad, proved or purely speculative, concrete or impressionistic: the photophone, sonic sounding, the vacuum jacket, "seeing by electricity," and so on. In 1890 he tried to work out a conscious strategy or methodology of invention, but his reluctance to commit himself exclusively to any long-term project gave him little occasion for strategy.

Bell's impatient, easily diverted imagination required.a direct dash at a problem. He not only lacked a mathematical mind but also a mathematical temperament, the readiness to plot a combination of indirect approaches with painstaking thoroughness and precision. The vacuum jacket concept, not to mention the telephone, was a fine prize of Bell's dashing approach; but such goals as an automatic switchboard or so complex a concatenation of interdependent elements as an airplane resisted Bell's frontal assaults.

Related to this characteristic was Bell's reliance on analogy. The analogy of sound waves with electrical fluctuations was magnificently fruitful in the telephone. But we have seen examples of the pitfalls of analogy in Bell's notion that as a heavy telephone diaphragm turned out to be sensitive, so a heavy phonograph pickup or a heavy flying machine would work better than light ones. His gravitational will-o'-the-wisp arose from a mechanical analogy of atoms and atomic forces with the collisions of billiard balls or similar gross mechanical phenomena. Once he even asked himself, "Are odors vibrations? If so, may they not be vibrations between sound & heat?" (This, to be sure, was only one of his "fugitive thoughts." Even so, it might better have got clean away.)

And in much of his later work, he seems to have tried to recapitulate his telephone experiences, even to particular turns of phrase. That pattern resembled a series of extended analogies, usually misleading.

Bell's independence of thought, his indifference to the objections of experts, and even his resilient optimism after experimental failures, all these qualities turned to his later disadvantage.

Bell did not deliberately eschew study — he was aware of Bernoulli's law, for example — but his reading was haphazard and spasmodic. In 1882 he wrote a Boston bookseller that he already had about four hundred books on science and technology, but wanted the dealer to get him more on physics and sound and to send as a standing order everything new about electricity, sound, and general physics in any language. Presumably he read at least some of these works. His notes specifically recorded at one time or another that he had read Oliver Lodge on electricity, LaPlace's *Celestial Mechanics*, and most current journal articles in his fields of interest, as well as the same readings that the Smithsonian recommended to the Wright brothers at the outset of their study.

And yet, though he recognized what it cost him, Bell seemed reluctant to accept, absorb, and proceed from what he read until he had reworked or rethought it all for himself. The cavilings of experts were not always as wrongheaded as that of Lord Kelvin against aviation, but Bell tended to ignore them anyway. And certain pet notions, like "thinking by telegraph," would keep resurfacing despite total lack of experimental support. All this was probably abetted by his tendency to turn inward upon the private worlds of Beinn Bhreagh and family life.

Bell's qualities as an inventor compel respect even when viewed as faults. His doggedness especially excites sympathy and a measure of admiration. Once in a while — a great while — Bell would let out his feelings to Mabel. "Sometimes feel away down," he wrote her in 1896, "things go so slowly here. . . . Many important things have actually been waiting *years* to be tried — and I suppose never will be tried now. One man can do so little." But he kept trying, literally to the last day of his life.

And she did what she could to sustain him. "I do so appreciate all the wonderful, unfailing, uncomplaining patience that you have shown in all your work and the quiet, persistent courage with which you have gone on after one failure after another," she wrote him in 1901. "How many there have been, how often an experiment from which you hoped great things has proved contrary. How very, very few and far apart have been your successes. And yet nothing has been able to shake your faith, to stop you in your work. I think it is wonderful and I do admire and love you more as the years go on. But oh how I wish that you may have success at last."

This catalogue of Bell's difficulties as an inventor would be exhaustive without being true, if it were not viewed in perspective. Three points may help in that.

First, Alexander Graham Bell invented the telephone. If he had done nothing else, if he had tried nothing else, he would still rank as one of history's most notable inventors.

Second, the pattern of youthful achievement and then early abatement of creative power is common among inventors. Most who produced one great invention produced no other commensurate with it. In some respects, Edison was an exception to the rule. Edison's continued effectiveness in calling forth original ideas from his younger colleagues, making those ideas commercially practicable, and organizing and directing team assaults on specific technological objectives amounted to genius, certainly by his own famous definition of it as one percent inspiration and ninety-nine percent perspiration. And at a level somewhat below the highest order of inventive originality, Edison himself remained prolific for many years. Nevertheless, even Edison came up with no great original concept after 1880. Yet Bell, as will be seen, achieved a fundamental technological insight as late as 1902, at the age of fifty-five.

Third, invention was only part of Bell's life after 1881. When asked, he consistently gave his occupation as that of "teacher of the deaf." On those grounds alone, it would be unreasonable to measure him against Edison, who had no other occupation than the development of technology, and scarcely any other interest. For Bell, invention may have been something more than a hobby, but it was certainly less than a calling. The conclusion of an interview with a Washington reporter in 1902 stated the case as Bell saw it. " 'It is pretty hard and steady work,' said Prof. Bell when he spoke of his researches and experiments. 'But then,' and he smiled as he gave his hand, 'it is my pleasure, too.' "

28

Scientific Circles

In the spring of 1879, while Bell eyed the example of Edison, he did not foreclose an alternative career: that of professional scientist. "Science, adding to our knowledge, bringing us nearer to God," he told Mabel solemnly, was "the highest of all things." The zoologist Spencer Baird had succeeded the late Joseph Henry as Secretary of the Smithsonian Institution in 1878. Bell now aspired to succeed Henry as physicist at the Institution. He worked up a paper, "Vowel Theories Considered in Relation to the Phonograph and Phonautograph," for the National Academy of Arts and Sciences, in part "to show the scientific men here that I have a special familiarity with the subjects that were Prof. Henry's chief objects of study." (The paper, published in the *American Journal of Otology*, reported his experimental verification of "the ideas of Helmholtz as expressed by Ellis.") It turned out, however, that no money would be available for such an appointment without a special appropriation by Congress, and Bell's code of honor compelled him to refuse Gardiner Hubbard's offer to lobby one through.

So as a scientist, Bell kept his amateur standing. His most substantial published contributions to science would be those dealing with the inheritance of deafness (to be described later). In August 1879 he read a paper before the AAAS, "Experiments Relating to Binaural Audition," later published in the *American Journal of Otology*. These experiments used separate telephone receivers for each ear and showed, among other things, that direction of sounds could to some degree be discerned with one ear. Aspects of his work with the photophone, spectrophone, induction balance, telephonic probe, and vacuum jacket were presented in scientific papers or published communications, as were some results of his sheep-breeding work. But except for the sheep experiments and other studies in heredity, Bell's serious publication in science virtually ceased by the mid-eighties.

Still, Bell's wistful ambition to be once more an active, publishing physical scientist never quite died. He was probably flattered to see himself so termed in print, though he avoided making any such claim himself. (Referring to Edison, Gray, and others, Mrs. Bell wrote him in 1893, "They are inventors and you are a scientific man. Please let me say this for I know it is true.") In 1895 he thought of taking up meteorology because of Cape Breton's position in the track of every large storm traversing the continent, but Mabel presently described him as "recovering from weather on the brain." News of Roentgen rays, later known as X-rays, interested him in 1896; and when he heard in the summer of 1903 that they were useful in treating external cancers, he suggested inserting radium sealed in thin glass tubes into deep-seated cancers. His doctor put the idea, properly credited, into print, and it attracted considerable attention. It appears that a French physician had anticipated Bell by a few months, but Bell's suggestion must have helped initiate the treatment in America.

Bell's yearning for scientific glory found a private outlet in his grand, nebulous theories about gravitation, conservation of energy, molecular forces, and other keys to the universe. Though in 1889 he thought of reading a paper on the conservation of energy before some scientific society "where its conclusions could be discussed, and established, or controverted," he did not do so — fortunately, since more than a year later he confided to his notebook, "Have not got a proposition clearly formulated. What is it I want to prove?" Not a year of the nineties passed that he did not festoon his notebooks with long strings of algebraic calculations, based not on experiment but on circular reasoning, so far as can be determined. His like-minded friend John Wesley Powell egged him on in this, but there is evidence that the mathematically sophisticated Samuel Langley did Bell the favor of holding him back from publication — or more likely, that he seconded Bell's own good sense in the matter. Thus Bell was spared the ridicule heaped on Powell when, after his retirement as director of the Geological Survey, Powell brought out the first volume of a projected trilogy intended to order and correct all human knowledge (in unfortunately eccentric terminology). To the end of his life Bell could happily resort to his private hobby as a hook on which to hang his dreams but not his reputation.

If Bell felt certified as a scientist because he belonged to scientific societies, he did not show it. Among those that claimed him as a member at one time or another were the Washington Academy of Sciences, the Philosophical Society of Washington, the Anthropological Society of Washington (which suggests the influence of John Wesley Powell), and the American Association for the Advancement of Sciences, of which he was a

Fellow. But on receiving a dues bill from the American Forestry Association in 1899 he wrote his secretary, "Didn't know I was a member — Can't you get me out of a few of these things. Just drop this unless Mrs. Bell wants me to keep it up."

Bell took his membership in the National Academy of Sciences more seriously, as he should have, since it was the most distinguished and exclusive of American scientific honorary societies. Over the years after his election in 1883 he submitted five papers to it, all but one ("On Apparent Elasticity Produced in an Apparatus by the Pressure of the Atmosphere") dealing with some aspect of heredity. He also served on six ad hoc committees. But he declined election as treasurer of the Academy in 1897.

In 1898 Bell was appointed a regent of the Smithsonian Institution, by then under the direction of his good friend Langley. This imposing office, he found to his disappointment, gave him little more to do than to pass upon vouchers. But there were exceptions. As a regent, he took the lead in having James Smithson's remains brought to Washington in 1904 from their quarry-threatened resting place by a cliff-edge near Genoa, and he and Mabel went there to oversee the operation. That, subconsciously at least, Bell did not regard the enterprise as earthshaking except in the narrowest sense is suggested by Charles Thompson's anecdote of the morning he went to rouse Bell for the final official delivery to the Smithsonian. Bell, who had not turned in until four that morning, asked grumpily, "Why am I to get up if I don't want to?" "To escort the remains of James Smithson to their last resting place," Thompson replied. "Nonsense, what are you talking about, Charles, he has been dead fifty years." "Can't help it, sir, he is in Washington now and you brought him here." There was a brief silence. "What did he come back here to bother me for?" Only the sight of Mrs. Bell speechless with laughter at the foot of his bed startled Bell into wakefulness and so got him to the Navy Yard on time.

As a regent, Bell's most significant contribution was in the case of Charles L. Freer, a fastidious, art-loving, Detroit bachelor who had retired in 1900 at the age of forty-four after having made millions as a railroad car manufacturer. Freer had begun serious art collecting in the early eighties with the etchings of James McNeill Whistler. He went on to acquire the largest existing collection of Whistler's art and along with it the personal friendship of Whistler himself, who steered his affluent admirer into becoming one of the world's great collectors of oriental art. Freer's adviser in America was the noted authority on Japanese art Ernest Fenollosa, son of the Salem musician whose piano had been used by Bell in his first attempt at variable-resistance transmission of sounds. It was not this link, however, that brought Freer into the life of Alexander Graham Bell.

In 1904 Freer offered to leave the Smithsonian his art collection and half a million dollars to maintain it. He suggested that a committee of regents come to see the collection. Secretary Langley, old, sick, heartbroken by the recent failure of his last flying machine, and in any case no orientalist, showed little enthusiasm for the offer. But Bell's daughter Marian, or "Daisy," who had studied art for a time with the sculptor Gutzon Borglum, aroused her father's interest in the collection — though Bell himself had apparently exhausted his appreciation of oriental art at the Centennial Exhibition, now preferring strict realism. A year after the offer was made, a committee consisting of Langley, Bell, Senator John B. Henderson, and James B. Angell set out for Detroit.

The committee's average age was over seventy, its feeling for oriental art was zero, it ignored Freer's suggestion that it add Fenollosa to its roster, and it submitted ungraciously to Freer's showing of more than five hundred works of art and a thousand specimens of pottery piece by piece. Fortunately, at Bell's insistence, Daisy came along as art's advocate and helped prevent a rejection on the spot. In Washington, President Theodore Roosevelt emphatically (that is, with table thumping) informed the still doubtful regents that if they did not accept so magnificent a gift, he would find some other way to save it for the people. Daisy's wire was the first to reach Freer with the news that on Bell's motion the regents had unanimously accepted the collection; Bell's own wire was second. By 1923, when the collection finally opened in a million-dollar building, it was appraised at seven million dollars; and by 1965 Freer's endowment of it had grown to more than seventeen million.

Socially as well as institutionally, at the Cosmos Club and especially at his own Wednesday Evenings, Bell throughout his later life hobnobbed regularly with eminent scientists. Daniel C. Gilman, retired president of the Johns Hopkins University and the Carnegie Institution of Washington, wrote Bell a note of thanks in 1906 for a Bell dinner at which he had heard "the progress of American science set forth by most competent and most interesting leaders. You have the rare gift of calling out men of mark in talk about the subjects in which they are specially interested." Bell met Helmholtz on the latter's American visit in 1893. Some years later H. G. Wells, not a scientist but a prophet of science, called on Bell and doubtless heard much about aeronautics (which Wells thought would come to pass, but without much effect on transportation).

The three men closest to Bell in Washington outside of the family — Langley, Newcomb, and Powell — ranked high in American science. Bell and Powell went for regular horseback rides on which Bell, tall and massive, seemed oddly matched with the small, rotund Powell. Powell's seamed, weatherworn face with a rather bulbous nose, his scraggly gray hair and unkempt gray beard streaked with tobacco stains and the remnants of its youthful red, might also have seemed incongruous with Bell's kingly visage, had it not been for Powell's striking likeness to the traditional representation of Socrates, a likeness doubtless emphasized by Powell's visible intelligence and force.

Langley and Newcomb happened to be at Beinn Bhreagh together in the summer of 1901. Since they seldom agreed, Bell found enlightenment and some mischievous amusement that season in raising scientific questions and listening to the two men debate — or squabble, as sometimes happened. One such moot question was the ability of a cat, dropped with its back down, to right itself in free fall. When Newcomb denied the possibility, mattresses were commandeered, cats drafted, the veranda manned, and Langley vindicated. The contrast of austere scientists with cascading cats, and perhaps also the presence of the journalist Mark Sullivan, who put the story in the papers, made the episode a standard family anecdote.

Bell's money let him be a benefactor as well as a friend of scientists, individually and collectively. As individuals, scientists were apt to be too prudent to need charity or too proud to take it. But when in 1883 Bell paid a good price (five thousand dollars) to Joseph Henry's widow and daughters for Henry's library, no charity had to be imputed, though the money was welcome to its recipients. And when, beginning in 1907, Bell and nine other well-wishers at William James's instigation joined in paying an annual pension of five hundred dollars to Charles S. Peirce and his wife, they did so anonymously. Peirce, who was not only old, sick, penniless, jobless, impractical, and eccentric, but also a scientist of ability and a philosopher of genius, was able in the seven years left him to produce some of his greatest work.

Bell gave generously toward dinners and other ceremonial expenses of the chastely unmonied National Academy of Sciences. His gift to the Smithsonian was larger and more productive. In 1890 he and another Washingtonian each gave five thousand dollars to start the Smithsonian Astrophysical Observatory, a project close to Langley's heart; a year later Congress began appropriating funds for the new bureau's work, destined to be notable and long-continuing. With the approval of Bell and the regents, Langley used most of Bell's five thousand dollars for aeronautical experiments.

In rate of scientific return for money invested, Bell's support of Albert A. Michelson ranks first.

Since the late sixties, Simon Newcomb had wanted for astronomical purposes a more accurate determination of the speed of light than Foucault's revolving mirror method had yielded. Early in 1878 he began planning to do the job himself. Then Michelson, a young ensign teaching physics at Annapolis, wrote him to report preliminary results from his own improved version of the Foucault method. Newcomb congratulated him on "a triumph" and corresponded with him enthusiastically In 1879 Michelson published his new determination, the first made in America, and the scientific world at once joined Newcomb in recognizing the young man as a brilliant experimenter and an inspired technician. Newcomb, an astronomer at the Naval Observatory and the Nautical Almanac Office, arranged for Michelson's transfer to help him with further measurements of the speed of light. But a few months later Michelson went on leave to study in Europe with Helmholtz and other leading physicists.

From Berlin in November 1880, Michelson wrote Newcomb that he had conceived an experiment of fundamental importance in physics: to determine "the motion of the earth relative to the ether." In those days scientists postulated that something they called "ether" filled all space in the universe. It was supposedly neither solid, liquid, nor gas, yet with attributes of each. But the only evidence adduced for its existence was the assumption of that mechanistic age that the ether had to exist as a medium for transmitting light and gravitational force. Like Bell with his gravitational speculations, scientists could not accept the notion of force being exerted without material contact, nor of light waves traveling where there was nothing to undulate.

Michelson believed in ether too; he merely wanted to measure some specific effect of it. His apparatus was ingenious, but its principle was simple. A light beam would be split into two beams diverging at right angles, which after traveling identical distances would be reflected back to a common path. If scientists were right in believing that the earth moved through "stationary" ether, one of the recombined halves of the light beam, because of a different speed relative to the ether, would arrive at a slightly different time than the other half, thus producing a visible interference pattern (the out-of-phase waves neutralizing each other at certain points).

Helmholtz liked the proposed method, Michelson wrote Newcomb. But there was an "unexpected difficulty, which I fear will necessitate the postponement of the experiments indefinitely, namely, that the necessary funds do not seem to be forthcoming." Perhaps it was Newcomb who put Michelson's case to Bell. At any rate, Bell made the historic experiment possible by offering Michelson five hundred dollars out of the just-received ten-thousand-dollar Volta Prize. Michelson's conscience made him point out that he was "young and therefore liable to err" and that his reputation as a scientist rested on a single research that had happened to succeed. But Bell — perhaps with all the more fellow feeling — insisted on putting up the money anyway.

On April 17, 1881, Michelson wrote Bell a startling report: "The experiments concerning the relative motion of the earth with respect to the ether have just been brought to a successful termination. The result was however *negative*, . . . showing that the ether in the vicinity of the earth is moving with the earth; a result in direct variance with the generally received theory of aberration." Only two hundred dollars of Bell's grant had been used; the balance was therefore at Bell's disposal. Bell replied with one of his rare understatements: "I think the results you have obtained will prove to be of much importance." And he told Michelson to keep the money for any further experiments he wanted to make.

The results were of more importance than either Bell or Michelson realized. Some leading physicists immediately questioned the very existence of the "ether," though Lord Kelvin, whose mind required a mechanical model of physical phenomena, did not. Michelson himself, as his letter to Bell indicated, retreated only so far as to hypothesize that the earth carried the ether along with it, at least near the surface. In 1887 he and Edward Morley collaborated in repeating the experiment with much more elaborate equipment and precautions against experimental error. They proved that if the ether existed it had no effect on the velocity of light in any direction near the earth's surface. Michelson still clung to the "ether drag" explanation, until he himself eliminated it with still another experiment ten years later.

By then, subtle mathematical physicists had advanced other explanations for what Michelson reported. These in turn played a part in Albert Einstein's revolutionary theory of physics, of which the first major installment appeared in 1905. In 1907 Michelson became the first American to win a Nobel Prize in science, not for any contribution to the newfangled notion of relativity, but for his experimental techniques and their results in the study of light. Michelson was not a good enough mathematician ever to follow Einstein's reasoning. But he was a great experimenter and in due course accepted the verdict of experiment: that Einstein was right, and that there was no such thing as the ether.

In July 1883, after Michelson had resigned from the navy and become professor of physics at the Case School of Applied Science, he spent several days at Bell's home in Washington. Telephone testimony had kept Bell there while his family took refuge from the heat at an ocean resort. "It is perfectly delightful," he wrote Mabel,

to meet a man with whom I can talk my thoughts. . . . I have a very high respect for his abilities. . . . Michelson is one of the few *young men* [he was then thirty and Bell was thirty-six] towards whom I have felt drawn naturally. . . . We have been discussing the affairs of the *Universe!* with a vengeance. Gravitation — Electricity — Magnetism — Meteorology — Chemistry — Moleculars, Atoms, "Points of Matter" — Education — Evolution and Religion — are a few of the topics.

The episode brings out what Michelson's later Nobel Prize tends to obscure: that Bell and Michelson had much in common. Like Bell, Michelson was an excellent musician — a violinist — who found restoration in his music. Like Bell, Michelson was mathematically unsophisticated, even though more highly trained. As Bell's intellectual life was focused on sound, so Michelson's was focused on light. Like Bell, Michelson owed his fame to invention rather than theory, even though his inventions were scientific instruments. And like Bell, Michelson did his most significant work before he was thirty — at least if his experiment with Morley is considered an extension of or sequel to the "ether drift" experiment Bell had financed in 1881.

Bell's largest financial benefaction to science began soon after his grant to Michelson.

In 1880 John Michels, a foreign-born journalist in his mid-forties, had begun publishing a weekly magazine named *Science*, encouraged by Thomas Edison's willingness to make up the deficits for a time. In the spring of 1881 Bell sent the new magazine a copy of his National Academy paper on the production of sound by radiant energy. Michels, eager to cultivate a possible new backer, managed to print Bell's paper a few days ahead of the *American Journal of Science*, and in July 1881 he went on to suggest that Bell form a small stock company to take over publication of *Science*. But Bell, just then deeply involved in the effort to save President Garfield's life, apparently did not reply. And when Edison withdrew his support in the following winter, Bell was abroad.

Michels tried again with more effect in June 1882. His advertising prospectus had made an ambitious claim: "SCIENCE is essentially the medium of communication among the Scientists of America. . . . SCIENCE enters into competition with no other scientific journal; it was established to fill a void that has been long felt." Bell knew better than to take the first claim as established, but like other scientists he recognized the fact of a "void" in American scientific periodical literature, now that the *American Journal of Science* had become more and more specialized and increasingly directed toward geology. America had popular science periodicals, but none that spoke seriously for and to the general scientific community as did the journal *Nature* in England. Perhaps Bell also liked the idea of taking up where Edison had left off, as the Volta Laboratory was then doing with the graphophone. At any rate he and Gardiner Hubbard took Michels up on his offer.

Bell and Hubbard had meant to keep Michels on as editor, but scientists advised them to get a man of solid scientific reputation. So they soothed Michels's feelings with five thousand dollars for the name and good will of his foundering journal and persuaded Samuel H. Scudder to take it over as editor, guaranteeing him a salary of four thousand dollars a year for three years. Scudder happened to be related to the Hubbard family, but he was also known as "the greatest scholar and most charming writer among the American entomologists." Moses King of Boston agreed to serve as publisher.

Though Bell undertook to bear most of the expense, Hubbard did the work of organizing the enterprise. Late in 1882 he incorporated the *Science* Company in Massachusetts with a capital of $25,000, all of it put up by Bell and himself. Bell had been slated for the presidency, but because of the danger that *Science* might be mistaken for a weapon in the Bell Company's telephone litigation, Bell persuaded Daniel Gilman to take that office. As it was, while Bell's scientific friendships helped bring significant contributions to the magazine's pages, his and Hubbard's telephone associations raised some suspicion of ulterior motives, despite their care to confine their role to that of financial backers.

Publication had been suspended after mid-1882, but it resumed under the new management with the issue of February 9, 1883. To a potential contributor, Scudder had written: "The aim of the journal will be to increase the knowledge of our people, to show our transatlantic friends our real activity, to gain among intelligent people a knowledge of the true aims and purposes of science, and to elevate the standard of science among scientific men themselves." Such aims went beyond even those of John Michels, but so did the journal under Scudder — although in justice to Michels it should be said that besides Scudder's higher scientific standing, Scudder had Bell's authorization actually to pay for articles.

Scientific need and editorial competence did not, however, add up to profit. Subscriptions fell far short of the six thousand necessary to break even. With a subscription list of only about two thousand, advertisements brought few results; and so advertising revenue fell off pitifully. Moses

King quit as publisher before the first year was out. Since a $19,000 deficit at the end of the first year left only $6000 in capital, Bell and Hubbard sank another $10,000 in the venture. The business manager forecast the next year's expenses as at least $30,000, the receipts as no more than $5000. Through the eighties, Bell's expenditures on *Science* were a major drain on his and Mabel's income. Despite Bell's pleadings, Scudder quit in discouragement in 1885 and was succeeded by the hardworking but professionally undistinguished N. D. C. Hodges. By the end of 1891 Bell and Hubbard had transferred ownership to Hodges, but they continued to subsidize him. All in all *Science* cost Bell and Hubbard some $80,000, of which about $60,000 came from Bell.

In the end, the sacrifice was not in vain. Hodges maintained the professional quality Scudder had brought to the journal, and like Michels before him he persevered in wooing the support of the AAAS. At last, in 1894, the AAAS promised an annnual subsidy; and Gardiner Hubbard wrote Bell, "I think it looks as if *Science* might now go & be good for something." That year Hodges transferred the ownership of *Science* to the shrewd, energetic, and enterprising psychologist James McKeen Cattell. For some years after 1895, the AAAS reneged on its subsidy, but Cattell got the Macmillan Company to assume publishing costs in return for advertising space. Then in 1900 the AAAS designated *Science* as its official journal and the dreams of Michels, Scudder, Hodges, Bell, and Hubbard began to come true. Cattell in 1899 wrote Bell to express "the great indebtedness of American men of science for your great gift to science in America." The thanks do not seem extravagant, especially when set against what has since grown from that gift — the official organ of the world's greatest scientific community.

29

The Barricades of Deafness

In the spring of 1879, Bell contemplated a career in science or invention or both. Then, late that summer, the imminence of triumph in the *Dowd* case sent telephone stock soaring. The need to earn a living no longer restricted his plans. So his thoughts turned to the establishment of "a small private school for deaf children, employing teachers to carry on the work, the proceeds of the school to be devoted to advancing the cause of the general education of the deaf by means of articulation." He was to do that and much more during forty years' work in that cause. And looking back from his seventieth year, he would write that "recognition of my work for and interest in the education of the deaf has always been more pleasing to me than even recognition of my work with the telephone."

Bell's concern for the deaf had several springs. The needless isolation of the deaf touched his compassion and sense of justice. In breaking down that isolation, he relished the challenges of research, writing, and debate. Family circumstances also must have moved him. After realizing that, in seeming defiance of "vocal physiology," lipreading or speechreading worked, Bell was haunted by guilt at having failed to discover the fact earlier and persuade his mother to learn the art; perhaps his work for others helped lay that ghost. Mabel's uncanny skill in speechreading sustained his faith, and her father's zeal gave it added force.

An address by Bell in 1887 suggests a still deeper personal involvement. In describing the situation of the so-called "deaf and dumb" in the past, he said:

Who can picture the isolation of their lives? When we go out into the country and walk in the fields far from the city we think we are solitary; but what is that to the solitude of an intellectual being in the midst of a crowd of happy beings with whom he can not communicate and who can not communicate with him. I know that the most lonely place on the face of the earth is the heart of the city of London. I have stood on the sidewalk and seen hundreds and thousands of people pass by me, and not known a soul, and the sense of loneliness in the midst of so many is oppressive. How then must have been the loneliness of the deaf child in the ages that preceded the efforts of the great doctors in England and in other countries?

In Mabel's contribution to his involvement with the deaf, there was an irony that Bell himself probably never fully comprehended. Not until the last year of his life, and even then not to him but to their son-in-law, did Mabel divulge it. However noteworthy her success in mixing with hearing people, she wrote:

I have striven in every way to have that fact forgotten, and to so completely be normal that I would pass as one. To have anything to do with other deaf people instantly brought this hard-concealed fact into evidence. So I have helped other things and people . . . anything, everything but the deaf. I would have no friends among them. . . . To say a child was deaf was enough to make me refuse to take any public notice of it. If help had to be given, given it was from a distance. Of all people I hated [to meet] a teacher of the deaf. I was always on the lookout for a little difference in their manner of addressing me, which would reveal the fact that I was a "case" in their eyes. . . . Above all things I antagonized my husband's efforts to keep up his association with them and to continue his teaching of them. Well, this is a confession of great selfishness. The only excuse is that it is just the spirit enlightened teachers of the deaf wish to see manifested in their pupils. They don't want them to herd together and become a "peculiar people."

Mabel's "selfishness" seems more like self-sacrifice, since none of her many letters to Alec (except for one in their first year of marriage) mentioned any such feeling. Presumably, therefore, she did not speak of it to him either. Nevertheless Bell must have taken note of her aloofness from the work. This may have kept him from devoting himself exclusively to the deaf. But then again, his temperament was not that of a narrow specialist, and he enjoyed his inventive work.

Mabel's self-reproach is hard to square with her conspicuous contribution in 1894, when she sent a paper, "The Subtle Art of Speech-Reading," to a convention of speech teachers in which Bell took a leading part. The standard history of speechreading in America justly describes her essay as "brilliant and absorbing . . . so far ahead of the time that it may still be read today as a modern and authentic exposition of speech-reading." Instead of the vogue then current of concentrating on each successive vocal position, she emphasized the richness of vocabulary and the practiced intellectual agility, which together would enable the speech-reader to reconstruct the essential meaning of an extended remark from occasional clues, the general context, and perhaps some movements and facial expressions of the speaker. "How do you like my paper?" she wrote her husband eagerly; "I hope you will telegraph. . . . Will they like it, what will they say, does it contain any new information? . . . Take care of yourself, my dear old man."

They did like it. The president of the association wrote Gardiner Hubbard that "every deaf person who has seen it wants a copy to keep. They say it has given them great encouragement and stimulation." The *Atlantic Monthly* published the paper in 1895, after which it was translated and published in fourteen foreign countries. Ten years after the paper was published, the nation's leading teacher of speechreading paid tribute to its influence on his work and dedicated his textbook to Mrs. Alexander Graham Bell.

Bell's own ideas about the education of the deaf were formed early. They coincided precisely with those set forth in the 1873 report of the Boston School for Deaf Mutes, the report that described his success with the Visible Speech method. Miss Fuller's school did not teach the sign language and used the manual or finger-spelling alphabet of Dalgarno only temporarily with the younger children, depending otherwise on instruction through speechreading. "A day school like ours, which is believed to be the first of its kind in this country," said the report, "appears peculiarly adapted for instruction in articulation, on account of the advantages enjoyed by children who live at home, where they are surrounded by hearing persons. . . . As they become able to communicate with those about them in a common language, they gradually cease, both in feeling and in fact, to belong to a peculiar and unfortunate class, shut out by their infirmities from the world." Bell added a corollary to this proposition in a lecture on Visible Speech in 1874. "He deprecated the tendency [of the deaf] to intermarry, thus perpetuating a race of deaf-mutes," reported a correspondent of the *Silent World;* "by enabling them to associate with the hearing and speaking, Visible Speech helped to lessen this tendency." Aside from the emphasis on Visible Speech in teaching articulation, these statements expressed Bell's basic creed to the day of his death nearly half a century later.

Livelihood, ideas, friendship, and love — things without which the world is nothing — depend on communication. For all the deaf, Bell wanted the limits of communication pushed to their utmost; and with Mabel as inspiration and evidence, he believed that for more of the deaf than was commonly supposed those limits were wide as the world. Bell was literally a fighter for freedom of speech, though in a sense so elemental as seldom to be used. But he himself did not put it that way. Instead, his watchwords were articulation and speechreading.

As late as 1884, Bell felt that the public at large still needed to be disabused of the notion that the deaf were physically or mentally unable to speak. But by then, teachers of the deaf knew better, largely because of Gardiner Hubbard, his allies, his prime exhibit, Mabel, and the successes of Bell himself; and that part of the public closest to the question, the families of the deaf, were also generally aware of the articulation movement. Having admitted the possibility, no faction among teachers of the deaf could well deny that every deaf child ought at least to try learning to speak. They disagreed in practice, however, as to how determined the trial should be; and so Bell never considered even that phase of the struggle finally closed.

To read the speech of others was still more demanding. Bell recognized that not everyone had the aptitude and perseverance necessary for it. But he believed that persistent example, encouragement, and practice might uncover latent aptitude and fortify perseverance. Here again the issue among teachers of the deaf was the degree of persistence.

The fiercest debates, however, concerned the medium and the environment of teaching, rather than the subject matter. Since writing, finger spelling, and sign language were all easier for the deaf to master and more reliable in use, Bell feared their tendency to displace speech; and the use of any of them in the process of teaching would, he believed, tend that way. The "oral method" or "pure oralism" used nothing but speech in teaching, even before writing had been taught. Bell was not so rigorous. Writing, though slow and laborious, left open the door to communication with the hearing world. Finger spelling, while easier and quicker than writing, closed that door, but maintained the habit of thought and expression in the English language. Except with the semideaf or semimute, therefore, Bell was willing to use both finger spelling and writing in the early stages of teaching, especially with the congenitally deaf (though even in those cases he felt that articulation should be taught from the start). This flexibility, he wrote privately in 1915, served also "to avoid antagonizing the manual alphabet people, so as to unite them with the oralists in opposition to the sign language."

Bell conceded the sign language to be aesthetically beautiful (though Mabel thought otherwise) and easiest of all for deaf children to acquire and use among themselves. But in that very ease he saw danger. Sign language was quite different from any spoken language, being ideographic rather than phonetic, limited in precision, flexibility, subtlety, and power of abstraction, and therefore, so far as it became a medium of thought, a narrow prison intellectually as well as socially. Early in his work with the deaf, Bell became adept in sign language and readily used it with any of his adult deaf friends who needed or preferred it. At dinners and banquets attended by the deaf, Bell would, for their benefit, translate the speeches of others into sign language; and the only deaf man among half a dozen participants in one impromptu lawn discussion remembered with gratitude, years later, that Bell had kept interpreting for him in sign language until the group broke up at two in the morning. But Bell set himself firmly against luring children into that prison of the mind, as he viewed it, whether by their own laziness or by the example of playmates and teachers.

For many years Bell served as the recognized leader of those who opposed the use of sign language in teaching, or — to put it more positively, as he preferred to do — of those who wanted deaf children to be encouraged by every possible means to think and communicate in the English language. He faced a formidable adversary in Edward M. Gallaudet, leader of the other side.

In certain respects the two men were well matched. Though ten years older than Bell and physically slighter, Gallaudet resembled Bell in charm, eloquence, and capacity for enthusiasm. Like Bell, he had a deaf mother. Also like Bell, Gallaudet owed much to the influence and example of a famous father. Thomas H. Gallaudet, founder at Hartford of the first permanent school for the deaf in America, died when his youngest son Edward was fourteen, but continued in memory, as did Bell's father in life, to be a spur to his son's ambition. Bearing a distinguished family name, dedicated to his father's vocation, active and intelligent, Edward Gallaudet, even at the age of twenty, impressed Amos Kendall as the right man to head the Columbia Institution for the Deaf and Dumb when Kendall established it at Washington in 1857 under a congressional charter. Young Gallaudet vindicated Kendall's confidence during fifty-three years as head of the school, which became the National College for Deaf Mutes in 1864 and thirty years later was renamed Gallaudet College after the elder Gallaudet.

Like Bell, Gallaudet had initially scoffed at the notion of speechreading and then had changed his mind on the evidence presented by the Clarke School and Gardiner Hubbard. After a tour of European schools in 1867 confirmed his new views, Gallaudet recommended that children in the college's primary department be taught speech and speechreading as soon as possible and for as long in each case as seemed profitable to the individual pupil. With his usual energy and eloquence he campaigned to open all

Edward Gallaudet

American schools for the deaf to such instruction, at least to that extent. He publicly anticipated Bell's fears that sign language might be a dangerously easy, though inferior, substitute for English.

But as time passed, perhaps partly in reaction against the dogmatic enthusiasm of the pure oralists, Gallaudet backed off to a position of tolerance, even sympathy, toward the use of signs, both as an adjunct in teaching and as a convenience among the pupils outside class time. Like Bell, he wanted to break down the deaf child's isolation as completely as possible. Unlike Bell, he believed that the narrower but more comfortable and reliable range of communication by signs and finger spelling worked better to that end than the far wider but more difficult and uncertain reach of articulation and speechreading.

Gallaudet called his policy the "Combined Method," signs being labeled the "Natural Method" and speech — rather invidiously — the "Artificial Method." Bell in his turn sometimes called the Combined Method the "Sign Method" because of its distinctive feature. He often remarked that it was only a cover for sign language. And Gallaudet did indeed say in an

1893 speech that "every school employing a single method" should be "compelled" instead to use the Combined Method, in which "the sign language is indispensable."

Bell and Gallaudet admired each other's abilities and respected each other's devotion to helping the deaf. For a time in the seventies, it may be recalled, they had discussed the possibility of Bell's joining the college faculty. After Bell settled in Washington, they belonged to several of the same Washington clubs, including the Cosmos, of which Gallaudet was a charter member. Gallaudet had known the Hubbards since 1867, and beginning at least as early as January 1880 he dined with the Bells now and then. The two men did not conceal their differences in private or in public. One evening in 1881 Gallaudet's diary recorded a good-humored verbal "sparring match" between them at the home of a mutual friend. Nevertheless it was Gallaudet's college that in 1880 awarded Bell the first of his several honorary degrees.

Late in 1884, Bell wrote Gallaudet in a mood of mingled hope and apprehension. "Not only do we find all shades of opinion represented in the ranks of the profession," he pointed out, "but we also find a very general feeling of respect and friendship existing among those who entertain the most opposite views." Yet, he went on, "we are still engaged in discussing & rediscussing the questions that were discussed & rediscussed by other teachers before we were born. The experience of the past indicates that these discussions & controversies may continue . . . to the end of time *without settlement* unless some new element can be introduced into the problem." In the spirit of Baconian science, Bell urged that the disputants turn their energies to gathering the facts, from which the theories would emerge naturally.

The spirit of hope prevailed for a while after the summer of 1886, when the quadrennial American Convention of Instructors of the Deaf, including all factions, unanimously resolved that "earnest and persistent endeavors should be made in every school for the deaf to teach every pupil to speak and read from the lips, and that such effort should only be abandoned when it is plainly evident that the measure of success attainable is so small as not to justify the necessary amount of labor." But this, of course, did not touch on the use of signs in teaching; and what it did promise failed to come about. The percentage of pupils taught speech actually fell off during the following year.

A sharper note sounded in a letter from Bell to Gallaudet early in 1887.

I wish you could realize, as I do, how important even imperfect articulation is to a deaf person. I have daily — hourly — experience of its value in my own home, and my heart bleeds for the speaking young men at the College who are placed under deaf teachers with deaf companions. . . . It is the boast of the advocates of the Combined Method that [it] *includes all methods;* and yet the National College . . . excludes from its curriculum the subjects of speech & speech-reading — while the majority of its graduates are semi-mutes! Take away . . . this undoubted reproach. . . . Excuse my warmth of utterance. I can only write as I feel.

Not Gallaudet himself, but some of his supporters uttered even warmer reproaches to Bell, calling his views "pestilential," "ranting," or "the vaporings of an idle brain." Though Bell made no reply to such outbursts, he was privately grateful to his defenders. "I don't know why it is that these attacks make so little impression upon me," he wrote Mabel in 1891. "I suppose I have grown callous and thick-skinned. . . . I cannot realize that I am the person attacked, but look upon myself somehow as a third person." Yet in the same paragraph he confessed that a letter of encouragement had "touched me more than all the abuse." He tried to regard the personal attacks on him as "evidences of victory. . . . You don't throw mud until your ammunition has given out."

Gallaudet in 1886 and Bell in 1888, both armed with elaborate exhibits, traveled to England to testify before the royal commission investigating the condition and education of the deaf and the blind. Though originally allotted only one day, apparently as a mere pro forma rebuttal to Gallaudet, Bell rose to the occasion masterfully, was invited back for another day, then another, then a fourth. In the end he answered more than six hundred interrogatories knowledgeably, temperately, lucidly, and persuasively. His testimony remained the fullest single exposition he ever made of his views on the education of the deaf. When in his justifiable pride he had it printed and bound for distribution to institutions and libraries, he asked Gallaudet's permission to include the latter's testimony also, for the sake of fairness. "Honest controversy never hurts truth," wrote Bell, "and that is what we are after — both of us — I am sure." To which Gallaudet replied in giving his permission, "We *are* seeking to arrive at the *truth* & to use a little current slang I think we shall both 'get there' ultimately."

The first head-on clash between Bell and Gallaudet came over Gallaudet's request in January 1891 for a five-thousand-dollar congressional appropriation to establish a teacher-training school in the college. Bell had told the Royal Commission more than two years earlier that he "would not advocate the establishment, by the Government, of a special training school for teachers of the deaf, for the reason that it will tend to the perpetuation of some one method." He felt that the various methods should prove themselves in free competition. By Gallaudet's account, Bell took a more tolerant view of the normal school proposal some months before it went to Congress. But further thought probably did, as Gallaudet surmised, persuade Bell that "with both College and the National Normal School here, it would be hard work to make headway against such a citadel of the Combined System." Worse still, Gallaudet's bill, as Bell read it, implied that students of the college — by definition, deaf themselves — would be admitted to the normal school. The deaf might teach the deaf many things, Bell felt, but not articulation. Though Gallaudet promised him privately that no deaf students would be admitted, Bell had little faith in the longevity of that assurance and none at all in its binding effect on Gallaudet's successors. He saw the bill, in short, as a thrust to the heart of his hopes for the deaf.

The ensuing controversy turned bitter. Bell opposed the appropriation before the congressional committee on the grounds that it would eventually lead to the deaf teaching the deaf. Gallaudet took this to be giving him the lie publicly, and in that temper, after Bell's testimony had defeated the bill, he publicly replied in kind. "Bell was not satisfied with the muddle he had already made by meddling in affairs that did not concern him," Gallaudet was reported as telling an audience of deaf students in Minnesota, "and wrote letters to all the schools and institutions in the country, that were so worded as to misguide those not informed, and also contained several untruths, that were untruths to his knowledge [that is, as to admission of deaf students]."

Charles Thompson remembered vividly in later years that during the struggle before the congressional committee, one of Bell's supporters suggested privately to him that he prepare the minds of certain congressmen by inviting them to dinner. Bell was furious at the idea, accepted practice though it was in Washington then as now. "If the facts presented to these gentlemen do not convince them of the merits of this case, they can go to blazes," he said, bringing his fist down on the table. Bell was not the sort of man to win his point by lying or any other sort of deceit; and when Gallaudet cooled down, he admitted as much — though, to Mabel's intense annoyance, only in private. Furthermore, Bell supported an increased appropriation for the hiring of articulation teachers by the college, and this was passed. Gallaudet then set up his Normal Department anyway, committing it to the Combined Method. Still, possibly held more firmly to his promise by the fuss Bell had raised, he refrained from admitting deaf students to that department. And he did bring in articulation teachers.

Bell, yielding to mutual friends and to Gallaudet's private retraction, made peace with his sharp-tongued adversary. Gallaudet invited Bell to the 1893 commencement exercises. But the peace was only a truce. "The hatchet is buried," wrote Gallaudet in his diary, "but I know where it is." He dug it up with cool deliberation in 1895, after the breakdown of efforts to merge Bell's association for the promotion of articulation with the American Convention of Instructors of the Deaf. "It seems a great undertaking to 'do up' Professor Bell, but I think it must be done," Gallaudet wrote an ally. At the Convention's meeting in Flint, Michigan, that summer, he delivered a long paper rehashing the Normal Department affair, the merger failure, and two minor squabbles over the nomenclature of methods and the control of a World's Fair exhibit, the general intention being to stigmatize Bell as an unprofessional, narrow-minded, despotic propagandist and intriguer for pure oralism. Further than that Gallaudet did not go, but his more excitable adherents construed the paper as "a scathing denunciation" that declared "war to the knife."

Privately, Bell (who had not been present when the paper was read) speculated that Gallaudet might be suffering from "monomania against myself" (or what now might be called delusions of persecution). When Bell appeared before the Convention to reply, however, he simply denied Gallaudet's interpretation of his motives and policies, and deprecated the spirit of discord that had grown up among workers in the common cause. "He spoke as if inspired," wrote Sarah Fuller; "not a word of retaliation, not a thought of anything but entire truth. . . . He lifted the entire audience into a broader, better, clearer atmosphere." Notwithstanding that, Gallaudet was elected president of the Convention and retained the office to the day of his death twenty-two years later.

The Flint knife cut the social ties between the Bells and Gallaudet for several years. To Mabel, who had requested the break, Gallaudet wrote accepting her decision with regret, affirming his regard for her and her husband, and denying that his Flint speech had been intended as a personal attack. At last in 1900, at Bell's invitation, Gallaudet joined with him in support of a better treatment of the deaf in the decennial census, and the next day Gallaudet went to a Literary Society meeting at the Bells' home. "I hope the hatchet is finally buried now," noted Gallaudet in his diary. It was.

Later that year Gallaudet and Bell, in their old spirit of friendly disagreement, figured prominently at a Paris congress on deafness. Bell had "a high old time" and easily turned back Gallaudet's drive to promote the Combined Method in Europe. At Bell's request during a luncheon for delegates, Gallaudet translated Bell's speech into French. "He seems to be an orator in French as well as in English," wrote Bell. "I must do him the justice to say that his translation was admirable and bettered the original."

Three years later, with regard to exhibits at the coming St. Louis World's Fair, Gallaudet wrote Bell that "if any partisan feeling has existed anywhere, I trust it may pass away."

Up to the present, however, Bell seems to have come closer to the mark with his 1884 warning that "these discussions & controversies may continue . . . to the end of time *without settlement*." A division as to practical emphasis rather than monistic principle or ultimate aim, one charged with emotion and involving a wide gradation of cases, is not likely ever to reach a "settlement," whatever that may mean. Those who have followed after Bell still fear the tendency of signs to displace speech. In reaction to that fear, the successors of Gallaudet still fear in their turn an unreasonable and harmful suppression of signs. The suggestion in both views is of an unstable equilibrium, like that of a pole standing on end. In that analogy, the two forces pulling in opposite directions are like guy wires, together being essential to the structure's stability. Such a balance of tensions is quite in the pragmatic tradition of American politics and society.

Bell's plan of 1879 to start his own school was delayed by the demands of invention and court testimony. The school he had established in Greenock, Scotland, kept up his interest in the subject, however. Its teacher wrote Bell in June 1881 that the plan of holding classes in the regular school building, "where the pupils would associate with hearing children [outside classes], . . . is accomplishing all you had hoped for it." In the spring of 1882, during the European trip that followed his son's death, Bell inspected the Greenock school and made an enthusiastic report to a large lecture audience in Greenock. "A little day school like this one," he said, "is to my mind the ideal." He drew applause with the canny reminder that it was cheaper than supporting the pupils at a boarding school.

By January 1883 Bell was laying happy plans for the little private day school that he finally opened in Washington on October 1 of that year. After a few days in temporary quarters it moved to a small brick building Bell had rented on Scott Circle, old-fashioned but ringed about by the freshness of a lawn and garden. Consciously adopting the Greenock model, Bell gave over the first floor rent-free to a local kindergarten school so that his little class (limited to six) could play with hearing children during recess. Bell wanted his second-floor classroom to be a paradise for children, and by the standards of that day it was. It mustered no grim drill of bolted-down desks, only a low table around which the children could cluster. A soft rug covered the floor. Pictures and ornaments, games, toys, and books, and a bay window looking out on greenery, all appealed to the senses of sight and touch on which the pupils depended for their grasp of the world.

Bell and the teacher he had engaged, Gertrude Hitz, daughter of the long-time Swiss consul general John D. Hitz, brightened the room still more with a device of their own contriving — "whiteboards" of ground glass backed by white cloth, on which they wrote with charcoal. School, teacher, children, and proud director all charmed the *Washington Star* reporter who appeared there soon after classes began; and so Bell's views on education of the deaf were set forth in detail to Washingtonians.

Gertrude Hitz, a young woman of evident ability, reported success to a convention of articulation teachers in June 1884. She and Bell had developed "a museum full of common things," each labeled with its name in Visible Speech and conventional writing. In games, conversations, and lessons, speech had gradually supplanted word cards. Bell's theories seemed verified in all respects. Bell also formed a class for parents, friends, and any others interested, both to spread the word and to help the children's adult circle preserve and exercise what the children had learned.

But at the end of that first year, Miss Hitz left to be married. (Her son, Harold Hitz Burton, eventually became a Supreme Court justice.) And her successor resigned a year later. Bell had counted on having a permanent teacher who would grow with experience. Instead the school demanded more personal attention than he could give it. In his words, he "sacrificed every occupation that could interfere with . . . devoting all my time and all my thoughts to my school." The final blow was the sudden attack on his character and fortune by the Pan-Electric conspirators and their government allies in the fall of 1885. With this new and paramount claim on his time, he tried to organize a couple of mothers and an untrained new teacher into a stopgap teaching force. But they proved to have neither ability, conviction, nor dedication in the work. So in November 1885 he gave it up.

The closing of his little school was one of the great disappointments of Bell's life. He told Mabel that he felt as if his whole life had been "shipwrecked." He asked her to go with him to Boston on Government case business in order "to keep him from an accident," and because, as she noted in her journal, he "feels as if I am all he has left." Mabel had earlier pointed out the school's expensiveness and, more immediately, the indifference and incompetence of the makeshift teachers. Apparently for that reason, she later blamed herself in a measure for the school's closing and near the end of her life confessed that she was "always sorry" about it. But evidence contemporary with the event indicates strongly that the decision was Alec's own and had been inevitable in the circumstances.

Because Bell was no longer an "instructor of the deaf" in the narrow sense of the term, Gallaudet was able to deny him membership in the American Convention of Instructors of the Deaf after the break at Flint in 1895. This rankled. But Bell had already founded his own association.

As early as the fall of 1883, Bell considered forming a national society to promote articulation teaching and help set up day schools. By January 1884 he had decided to confine it to the articulation goal (since most teachers opposed day schools at that time) and to lay the groundwork cautiously, by first organizing a section of speech teachers at the next Convention meeting. But at the 1886 Convention the unexpectedly enthusiastic resolution in favor of speech teaching led Bell and his allies to put off their oral section move, lest the new mood be soured.

At the August 1890 Convention not only was the oral section organized as planned, but on the next day the Bell forces also announced an independent organization, the American Association for the Promotion of the Teaching of Speech to the Deaf. (Gallaudet, in the later time of his open hostility, once amused a deaf audience by writing that sesquipedalian title with flourishes and choreography along a stage-wide blackboard.) It was formally incorporated under the laws of New York State in September, and Bell endowed it with $25,000.

From that beginning until his death, Bell held the AAPTSD faithful to its initial statement of purpose, which committed it to promoting speech teaching but declared its neutrality as to the several teaching methods — oral, manual, and "combined." Not only were oral teachers welcome to join, but also sign teachers interested in teaching speech, and indeed all other advocates of that object, teachers or not. During the thirty years of Bell's active influence, each of the several methods had its adherents among the members and on the board of directors, though Bell's personal sympathies were with oralism. Bell also consistently blocked an official Association commitment to day schools (despite his own decades-long crusade for them), or any other new cause that might overextend the Association's energies or divide its members. Some of its original members resigned for that very reason. But Bell's policy forestalled a renewal of open hostility between the Association and the Convention once the Bell-Gallaudet break had been patched.

As president, Bell got the Association off to a brisk start with a ten-day meeting at Lake George in July 1891. Hope and harmony prevailed, and Gardiner Hubbard helped enliven the occasion with his dry Yankee wit. A hundred and fifty teachers and administrators from a score or more schools attended lectures on aspects of articulation theory and teaching, along the lines of the Chautauqua assemblies then flourishing. It was a "glorious success," Bell wrote his wife midway through the meeting; "I feel that a great work has been inaugurated here." Mabel came herself for the last days of the affair.

Still more attended the Lake George meeting of July 1892. The Chicago meeting a year later, taking account of World's Fair distractions, addressed itself to socializing rather than instruction. But the summer school approach was resumed in the Chautauqua meeting of 1894 and in five more summer meetings at longer intervals between 1896 and 1912. In 1899, however, the Association started a periodical, the *Association Review* (later the *Volta Review*), which served much of the purpose of the summer meetings and used most of the funds available for them. The Chautauqua approach seemed to be losing its appeal anyway. So the 1912 meeting was the last to offer formal classes in Bell's lifetime.

Bell's fame and striking appearance made him conspicuous in any gathering. In the earlier summer meetings, Bell's parents also attended, as well as Mabel and her parents. One teacher vividly remembered two tables being joined in the hotel dining room to accommodate the clan, which also included two Bell cousins and Mary True. But on general principles and to fend off the sort of criticism Gallaudet eventually leveled at him, Bell did not want the AAPTSD to be or seem to be his personal instrument. In the intervals between meetings, moreover, aeronautics and other matters engrossed his mind; and so, as Mabel put it, "the absurd spectacle is presented of me urging Alec on. Generally I do all I can to get him away from it, but I feel that so long as he is president the Association can't be allowed to be a failure."

In July 1893 Bell resigned the presidency in favor of Philip Gillett, an experienced teacher and administrator well known to be a friend of the Combined Method. This helped emphasize the Association's neutrality as to teaching methods. Gallaudet himself publicly endorsed Gillett, though some of Gallaudet's adherents labeled Gillett a turncoat. During Gillett's presidency, the Association and Gallaudet's Convention dickered over terms of union; but the negotiations foundered in 1895 on the fear of each party that the other would gobble it up.

Bell remained active in planning and speaking at the AAPTSD meetings. Beginning in 1899, when Gillett resigned because of failing health, Bell resumed the presidency for five more years, during which time membership was more than doubled to about 350. In 1904 he gave up that office for good. But his interest remained keen. And in addition to his original endowment, he contributed another $25,000 to the Association over the years.

When Bell organized the AAPTSD in 1890, only about forty percent of all deaf pupils were being taught speech. By the time of his death some thirty years later, the proportion had risen to more than eighty percent.

In his crusade for day schools, Bell acted as a private individual, though his influence and contacts in the AAPTSD surely helped. His enlistment in that cause went back at least as far as Miss Fuller's Boston school, perhaps even as far as his first teaching of the deaf in Miss Hull's London school, where the little girls had looked forward so eagerly to seeing their parents again.

Bell took care to put day schools forward as complementing boarding schools, not supplanting them. With most boarding schools, especially those well disposed toward speech teaching, he maintained good relations, visiting oral classes, giving advice when asked, recommending candidates for teaching vacancies, now and then helping out with a check. In 1898 he succeeded the late Gardiner Hubbard on the Clarke School board of corporators. In 1901 he even managed to save the Wisconsin state boarding school from an imminent attack by one of his own overzealous allies in the day school cause. Nevertheless he saw urgent and widespread needs to which boarding schools could not well respond.

Day schools, Bell pointed out, would encourage — indeed, compel — pupils to live in the hearing world more than did boarding schools. The children's incentive to use and read speech would be correspondingly greater. So would their range of personal contacts and later employment, while the likelihood of their eventually intermarrying and thus perpetuating congenital deafness would be lessened, Bell believed. Beyond these benefits, he saw still others. Children would not be taken from home life. Boarding schools might carry on training outside of class time; but on the other hand, day school teachers could more readily advise and encourage parents in extracurricular guidance. The advantage of smaller classes could be gained at less cost per pupil than in boarding schools.

In any case, whatever the comparative merits of day schools, Bell proposed them as at least a partial solution to the tragedy revealed by the census reports: that about half of all deaf children were growing up with no formal education whatever, because there was no room for them in boarding schools, or because their parents did not want to send them away, or because they were too deaf for regular public schools and yet not deaf enough for institutions.

In Bell's view the ideal day school arrangement, on the Greenock model, would be for deaf children to have separate classrooms in a public school building, mingling with hearing children outside of classes. The schools should be state-supported, but controlled by local school authorities. Local control, he felt, would permit more experimentation.

To identify the semideaf, Bell devised what he called an "audiometer."

Conceived in 1879, it consisted of two flat induction coils, one carrying an intermittent current of precise frequency, the other in circuit with a telephone receiver. As the coils were brought nearer on a graduated scale, the sound of an induced current grew louder in the receiver. Standard sound frequencies of well-calibrated intensity could thus be used to compare the hearing abilities of individuals. (Bell's name eventually entered the language in the standard measure of relative differences in sound intensity, the decibel.)

The audiometer revealed unsuspected vestiges of hearing in many previously classed as totally deaf. Once recognized, even a slight hearing capacity was useful in developing articulation and speechreading. The device also detected hearing impairment in many public school children whose handicap had been taken for stupidity or inattention. In exhibiting his audiometer to the National Academy of Sciences in 1885, Bell reported that of seven hundred pupils whom he and an assistant had tested with it, more than ten percent had some hearing impairment. A year later, while testing more children, he told a reporter that one percent were found to be so deaf as to need special classes. Such findings gave still more weight and urgency to his day school campaign.

Bell advanced the day school cause most directly and conspicuously as a lobbyist for state support. His initiation into that work came in 1885 with the help of Robert C. Spencer, a Milwaukee businessman of whom Bell would say years later, "I doubt whether anyone in the country has done as much to promote the teaching of speech to the deaf." In 1879 Spencer had led in organizing a state society for that purpose. When Bell came to address the National Education Association at its 1884 convention in Madison, Spencer won his promise of help in a campaign for state aid to public day schools.

So in February 1885 Bell returned to Madison and spent two weeks testifying before committees, helping draft reports, writing an irresistibly persuasive letter to the legislature, and in general demonstrating a notable gift for the conversion of politicians. Gardiner Hubbard himself could have done no more adroit and effective a job.

The bill was passed; and with Spencer to watch over its workings, it made Wisconsin a model for other states. Encouraged by the state subsidy, day classes for the deaf began in Milwaukee and several other Wisconsin cities. Through Spencer, Bell kept in close touch with their progress. In 1895 and again in 1896 he made strenuous speaking tours of the state, arranged by Spencer in behalf of a more liberal state program. Bell fretted about his impresario's "distribution, before my face, to all these audiences, of pamphlets containing my picture and commendatory notices of myself that make me feel like a Dime Museum Freak." But he stood it for the cause. By 1900 Wisconsin had fifteen day schools for the deaf, and Bell had proclaimed the "Wisconsin system" to be "the most important movement of the century for the benefit of the deaf."

Over a period of thirty years, Bell kept track of day school movements in other states, gave advice and encouragement, wrote letters, and often made special trips to testify. He set two limits on his lobbying activities: he would not oppose the establishment of any school for the deaf, however contrary its methods might be to his personal preferences; and he would not intrude his views unless directly invited to do so by legislative committees or other responsible bodies. But such invitations were easily arranged by day school proponents.

Thus in Michigan over a period of several years he had much to do with drafting a day school bill, appeared several times by invitation before legislative committees, and gave newspaper interviews that helped arouse public interest and support. He made rescue expeditions also to Maine in 1894, Connecticut in 1897, California in 1898, and elsewhere. After the passage of an Ohio day school bill in 1898, an Ohio legislator wrote that "much credit is due to yourself and Mr. Spencer of Milwaukee in furnishing me with an abundance of literature which I freely distributed among both Senate and House members." Letters asking advice on lobbying tactics came frequently from day schools throughout the nation, and since Bell had both intelligence and experience, his replies were astute. At sixty-seven he even promised the governor of Minnesota that he would come there against doctor's orders if a legislative crisis required his help; fortunately he did not have to go.

In his battle for day schools, Bell found allies in the parents of deaf children. He had always recognized their potential service to the cause as individuals. And so he was on hand in Boston in 1894 when, at Sarah Fuller's suggestion, a number of parents met to organize a permanent association. Chartered in 1895, the Boston Parents' Education Association for Deaf Children stated its purpose in its first annual report: to "bring together parents and teachers having a common interest in the pupils." Bell had already moved to form similar associations elsewhere. The movement had to contend with the small number of such parents in most localities, and also, especially among working-class parents, with ignorance, apathy, or lack of time. By 1899, nevertheless, fifteen such associations were in being at various places; and they had already proved themselves as auxiliaries in legislative battles.

Bell took a lively interest in the Chicago day school movement. He gave its backers shrewd advice on the strategy and tactics of organizing, lobbying, cultivating public support, and drafting bills. He also lent the weight of his personal presence. By 1899 the Chicago parents' association had won state aid for day schools, had replaced the superintendent of the state institution with an abler and more sympathetic man, and had installed their candidate as head of the Chicago day schools. In 1918, the city built a fine school to accommodate deaf and hearing children in separate classes but mingle them at play — the arrangement Bell had urged for nearly forty years. The new school was fitly named the Alexander Graham Bell School. Of all the honors Bell had received to that time, insisted one of his closest associates, this was the one he prized the most.

30

A Hand for Helen Keller

Though Mabel Bell once accused her husband of regarding deaf people as "cases," her own sensitivity surely colored her judgment. Only a callous man could have been untouched by the appeals for help and advice that came to Bell. The letters were eloquent not so much in their rhetoric as in their simple statements of fact. "I have been slowly growing deaf for three years and now cannot hear general conversation," wrote Miss Margaret Stark of Delaware, Ohio; "I am forty-three years old." Ellery P. Ingham had managed to establish himself as a lawyer in Laporte, Pennsylvania, but had been growing very deaf over the last four or five years. "I am yet a young man and a long life of *silence* and poverty makes me draw a long breath when I think of it," he wrote. The slow, inexorable closing in of silence, the noiseless decay of lifelong hopes, pervaded these plain tales like a nightmare. "Cases" or not, Bell reached out his hand.

In the earlier years of his fame, Bell sometimes got inquiries about the possibility of his inventing an electrical device to aid in hearing. He had come to recognize his limits as an electrical inventor, however. Before he died, the Bell System's manufacturing affiliate Western Electric had begun developing such hearing aids. But Bell himself could only expose fraudulent gadgets and advise his correspondents as to speechreading courses. In time, as his campaigns for articulation and speechreading began winning public notice, the appeals tended more often toward those subjects to begin with. After Mabel's paper on speechreading was published in 1895, he often enclosed a copy of it with his reply.

In the letters of adults about whom deafness had begun to close, there was the terror of those who knew what they were losing. In letters from parents of deaf children, there was the eagerness of those who had come to terms with despair and now suddenly heard a whisper of hope. In advising them, Bell did his best to steer between deadening caution and false optimism. He stressed the importance of speech training at an early age, if possible as early as that at which hearing children began learning to speak.

Even after Bell's death, Charles Crane of Chicago vividly remembered the long-ago shock of discovering that his eighteen-month-old daughter Josephine had been deafened by illness. Hearing of Bell's interest in the deaf, he had gone at once to see Bell in Washington. When he telephoned for an appointment, "the secretary replied that Dr. Bell was too busy to see anyone, but when I proceeded to state that I was the father of a deaf baby and wished Dr. Bell's advice, the secretary replied that Dr. Bell always dropped everything [in such cases] . . . and for me to come to the house at once." Bell recommended a speech teacher, outlined a course of study, and came to Chicago several times to follow up the child's progress. Little Josephine Crane learned to speak and to read lips so well that few who met her realized that she was deaf. Eventually she married happily and raised a large family.

Bell must have welcomed letters from deaf children, especially those he had helped rescue. He preserved a file of such notes. "I saw a picture of you recently," wrote Belle Munger of Eureka, Kansas, in 1892, "and it brought our journey two years ago, to my memory. You asked me then to write to you. I remember how you amused me all day with your queer watch, opening the case when I blew upon it; then you made it walk and dance. You also told me about your children and how much fun you had with them in the summer, rowing and climbing. . . . We live eight miles from town on a farm. . . . Last Sunday I was nine." "His dominating passion is his love for children," Helen Keller once wrote of Bell. "He is never quite so happy as when he has a little deaf child in his arms."

Bell loved to visit classes, and the children loved to see him. His own account of a visit to a Providence school for the deaf in 1893 recaptures that warm communion. "Some of the youngest children in the school somehow got the idea that I was no less an individual than Santa Claus himself!" he reported to Mabel.

I accepted the situation and described in graphic terms my driving over the tops of the houses. Pandemonium reigned for a time, and the children were much puzzled to know how so big a body could come down so small a chimney. I taught them the word "squeeze" so that they will never forget it!!! I'm afraid that half of the school will write to me before Christmas and I shall have to visit the school in appropriate costume!

Sarah Fuller's pioneer public day school in Boston had prospered. In 1877 its name was changed from the Boston School for Deaf Mutes to the

Horace Mann School; and as its enrollment grew, it moved to larger quarters, until in 1890 it settled in a new building in the Back Bay. But institutional growth did not mean individual prosperity, and so Bell sent Miss Fuller money year after year to buy better clothes for children who needed them, and otherwise to make life easier for them.

Bell also did what he could to help young deaf people support themselves. "There are very few positions in life which cannot be occupied by deaf persons," he told the public. "Nearly all the arts and industries are open to them, and many of the professions." But he was not in a position to give them work directly. The only time on record when Bell asked a special favor of the Bell Telephone Company in the name of his contribution to the industry was in 1883, when he urged President Forbes to employ a few deaf people. Bell specifically recommended a former pupil of his. But there was no vacancy suitable for the young man, nor does it appear that the company subsequently reserved any openings for the deaf.

Bell's interest in young George Sanders was more personal. As there are hearing people who have no ear for music, so there are deaf people who have no eye for speech. George seems to have been one of these. But Bell found the fact hard to accept. He conceded the benefits to George of enrollment at fifteen in the National College for Deaf Mutes in 1882, but he was troubled by the college's lack of courses in speech and speechreading. George "is dearer to me than he or any one else knows," Bell wrote George's maternal grandmother in 1884. "I have no son — not even a nephew — to stand between him and my heart — and I long for the confidential companionship of a young man such as he is growing to be." He hoped that George would visit "often enough to feel at *home*." But George probably sensed Bell's disappointment at his inability to read speech. At any rate, the relationship does not seem to have been as close as Bell had hoped it might be.

At the funeral of Thomas Sanders's mother in 1890, Bell found that George "has grown to be a manly fellow and everyone likes him." George had fallen in love with a deaf girl in whose family a history of deafness ran back through four generations. Bell warned Lucy Swett that marriage to George, who like her was congenitally deaf, would increase the likelihood that her children would be deaf. But Lucy felt that George would surely marry a deaf girl anyway, and she wanted to be the one. "She is a very sweet girl," Bell wrote Mabel, "and I do not wonder that George is in love. . . . Will lovers ever consider the good of those that will come after them?" He regretted that the answer seemed to be no. Yet, he wrote, "I can understand it too."

In 1891, after George and Lucy had married, Thomas Sanders lost nearly all his money. Through most of the nineties, Bell saw to it that the young couple got along. He gave George work as a printer in the bureau he had set up for the collection and dissemination of information about the deaf, and from time to time he gave or lent George substantial sums. By 1898, George Sanders was costing Bell about two thousand dollars a year. But in 1899 George began making good in a small printing business near Phila-delphia, and that drain on Bell's finances ceased.

Among the deaf were those who, like Mabel Bell, insisted that they would rather live sightless but warmed by voices in the dark, than encased in the cold, bright solitude of deafness. Among them also were those for whom even the solitude was dark. They were the deaf-blind.

Bell knew the deaf-blind, too. In the crucial month of February 1876 he had managed to attend a memorial service to the late Samuel Gridley Howe, head of the Perkins Institution for the Blind, and there he had "quite a little talk" by means of finger spelling with Howe's most famous pupil, Laura Bridgman. At two she had lost sight, hearing, even most of her senses of smell and taste. Little remained to make her living body more than the sealed tomb of her mind. But Howe touched that mind and found it responsive. So Laura discovered the existence of the world and learned something of what it held. At the services for her dead teacher, she cried. "The whole scene was one I shall long remember," wrote Bell at the time.

Early in 1887, Captain Arthur H. Keller, a former Confederate officer who had become a newspaper editor in Tuscumbia, Alabama, brought his six-year-old deaf-blind daughter Helen to Bell in Washington. Helen was a healthy child, excited to something like happiness by what she felt of the novel journey. Bell may have seen irony in the contrast of her eager gropings with her father's sadness. Yet in her well-shaped face, for all its intimations of dormant intelligence, there seemed to be an indefinable, chilling emptiness. Bell listened to the story of the illness that had left Helen completely deaf and sightless at nineteen months. Something in his touch, Helen remembered years later, gave her an impression of tenderness and sympathy. She sat on his knee and felt his watch strike. He understood her rudimentary signs, and she knew it and loved him at once. "But I did not dream," she wrote in later years, "that that interview would be the door through which I should pass from darkness into light."

According to Helen, Bell unlocked that door with the suggestion that Keller write Michael Anagnos, director of the Perkins Institution. As it happened, Anagnos was already prepared. A friend of Keller's studying at the Lawrence Scientific School in Cambridge had spoken to Anagnos about Helen's case months earlier, perhaps at the instance of Helen's mother, who had read about Laura Bridgman in Dickens's *American Notes*. Then, on the strength of a tentative inquiry from Keller himself in the summer of 1886, Anagnos had alerted one of his star graduates to the possibility of such a call. She had since been studying Howe's carefully recorded methods in the case of Laura Bridgman and spending much time with Laura. Presumably Bell's opinion in February 1887 rekindled Keller's interest or settled his doubts about Helen's educability. At any rate Keller wrote again to Anagnos and thereby initiated the astonishing life work of Annie M. Sullivan.

Annie was then twenty years old, still haunted by the horrors of her four childhood years in the Tewksbury, Massachusetts, poorhouse, still suffering from the effects of trachoma which had once and would again make her blind, but soon to be called by Mark Twain and others "the miracle worker" and by Helen Keller simply "Teacher." It was on March 3, 1887, that Annie Sullivan arrived in Tuscumbia. That day was to be cherished by Helen Keller as her "soul's birthday." It also happened to be the fortieth birthday of Alexander Graham Bell.

"A miracle has happened," wrote Annie on March 20; "the wild little creature of two weeks ago has been transformed into a gentle child." On April 5 came Helen's famous breakthrough to the understanding that things had names, and three months later she was writing letters. Bell followed the Tuscumbia "miracle" with wonder, as did the public after Michael Anagnos sounded the trumpet. Bell himself helped to spread the news, furnishing a New York paper in 1888 with Helen's picture and one of her letters to him. He saw a wider good coming from the dazzling emergence of her mind. "The public have already become interested in Helen Keller," he wrote in 1891, "and through her, may perhaps be led to take an interest in the more general subject of the Education of the Deaf."

In one respect, Bell stood alone among Helen Keller's admirers and celebrators. He insisted that what Annie Sullivan and Helen Keller between them had done was not a miracle but a brilliantly successful experiment. "It is . . . a question of instruction we have to consider," he wrote, "and not a case of supernatural acquirement." He interviewed Helen himself to measure that acquirement and pressed Annie Sullivan for explanations of it, especially of Helen's command of idiomatic English. From what Annie reported, he found the key in her constant spelling of natural, idiomatic English into Helen's hand without stopping to explain unfamiliar words and constructions, and in her encouragement of Helen's reading book after book in Braille or raised type, with a similar reliance on context to explain new language. This, as Bell pointed out, was the equivalent of the way a hearing child learned English. And it supported his long-standing emphasis on the use of the English language with deaf children, including the use of books. Indeed, he saw the importance of books in the early stages of educating the deaf as "the chief lesson, I think, to be learned from the case of Helen Keller."

At the 1891 AAPTSD summer meeting, Bell gave each member a copy of a handsomely bound "Helen Keller Souvenir," containing accounts of Helen's education by Annie Sullivan and others, among them Sarah Fuller, who had recently given Helen her first lessons in speech. At the Association's expense, Helen and Miss Sullivan came in person to the 1893 meeting in Chicago, and Helen "saw" the World's Fair through the hands of Bell and her teacher; the tour included an exhibit of Bell's Centennial Exhibition telephone. Teachers of the deaf met her and, it was reported, "saw and heard enough to remove all their doubts." A year later, at the AAPTSD Chautauqua meeting, Annie Sullivan delivered — or rather, out of last-minute shyness, asked Bell to deliver for her — an eloquent, yet objective, account of her work and relations with Helen. And in 1896, Helen herself proudly addressed the AAPTSD. "If you knew all the joy I feel in being able to speak to you to-day," she said, "I think you would have some idea of the value of speech to the deaf. . . . One can never consent to creep when one feels an impulse to soar."

Helen Keller and Annie Sullivan were, however, much more to Bell than phenomena or specimens. They were his friends, and he was theirs. "It was an immense advantage for one of my temper, impatience, and antagonisms to know Dr. Bell intimately over a long period of time," said Annie in retrospect.

Gifted with a voice that itself suggested genius, he spoke the English language with a purity and charm which have never been surpassed by anyone I have heard speak. I listened to every word fascinated. . . . I never felt at ease with anyone until I met him. I was extremely conscious of my crudeness. . . . Dr. Bell had a happy way of making people feel pleased with themselves. He had a remarkable faculty of bringing out the best that was in them. After a conversation with him I felt released, important, communicative. All the pent-up resentment within me went out. . . . I learned more from him than from anyone else. He imparted knowledge with a beautiful courtesy that made one proud to sit at his feet and learn. He answered every question in the cool, clear light of reason . . . [with] no trace of animus against individuals, nations, or classes. If he wished to criticize and he often did, he began by pointing out something good I had done in another direction.

When asked long after Bell's death what, aside from her feeling for Helen,

Bell with Helen Keller and Anne Sullivan, July 1894

had enabled her to keep at so exacting a task for so many years, she replied, "I think it must have been Dr. Bell — his faith in me."

Bell's own daughters felt a touch of jealousy at his feeling for Helen Keller. For her part, one of her early letters, written a few months after her teacher first came to her, was to "Dear Mr. Bell," and it said among other things, "I do love you." And more than thirty years later, when he was seventy-one, she wrote him, "Even before my teacher came, you held out a warm hand to me in the dark. . . . You followed step by step my teacher's efforts. . . . When others doubted, it was you who heartened us. . . . You have always shown a father's joy in my successes and a father's tenderness when things have not gone right."

More than once in those thirty years, things went wrong for Helen Keller, and Bell was there with a helping hand. A short story, "The Frost King," which she wrote in 1891 at the age of eleven for Anagnos's birthday and which Anagnos then published, was found to echo the plot and wording of a children's fairy tale published nearly twenty years earlier, a story unknown to Annie Sullivan and not in the books available to Helen. It turned out to have been read to her at the home of a friend in Annie's absence more than three years earlier. At the Perkins Institution a solemn committee (Mark Twain in his outrage called it "a collection of decayed human turnips") cross-questioned the bewildered and frightened child at great length, with Annie Sullivan sent out of the room, before concluding that Helen had unwittingly summoned up the story from her remarkable memory rather than her imagination as she supposed. The ordeal crushed Helen's spirit and her joy in books for months and shook her confidence in her own originality for years.

The kindly author of the original story, Margaret Canby, wrote that Helen's version was no plagiarism but "a wonderful feat of memory" and an improvement on the source. "Please give her my warm love," added Miss Canby, "and tell her not to feel troubled over it any more." Mark Twain was more emphatic, recalling the time he himself had unconsciously plagiarized a passage from Oliver Wendell Holmes. "To think of those solemn donkeys breaking a little child's heart with their ignorant damned rubbish about plagiarism!" he wrote. "I couldn't sleep for blaspheming about it last night." Bell, who had helped Annie Sullivan trace Helen's exposure to the story, saw further than either Twain or Miss Canby. Like them, he pointed out that "we all do what Helen did," that "our most original compositions are composed exclusively of expressions derived from others." But he also observed that Anagnos had "failed to grasp the importance of the Frost King incident," and that "a full investigation will throw light on the manner in which Helen has acquired her marvelous knowledge of language — and do much good."

After a long talk with Helen in 1894, Bell heartily seconded her "strong desire" to be educated in a school for normal students rather than a special school for the deaf or the blind. Bell reminded Captain Keller that his daughter would need a special interpreter in any case, so that a school for the handicapped could offer her no practical advantage. He promised to rally Helen's friends to the underwriting of any expenses. Thus Helen went on to achieve what throughout her life would be one of her chief consolations and sources of pride: acceptance as an intellectual and social equal by people who could see and hear.

In 1897 Arthur Gilman, headmaster of the Cambridge School at which Helen was preparing for Radcliffe College, decided that Miss Sullivan was endangering Helen's health by pressing her too hard in her studies. Having temporarily persuaded Helen's mother of this, he tried to separate Helen from her beloved teacher. Gilman did his best to win Bell's support for the move. But Bell had boundless faith in the wisdom and dedication of Annie Sullivan, and when she appealed to him for help he dispatched his assistant, the venerable John Hitz, to investigate. Afterward Bell wrote Gilman that nothing could justify parting Helen and Annie except evidence that Annie was in some way unfit for her charge; and as to that, his free conversation with Helen had revealed her to be a "living testimonial to the character of Miss Sullivan." Mrs. Keller hurried to Massachusetts and, finding Helen in excellent health and determined to stay with Annie, agreed with Hitz and Bell that Gilman was wrong. Never again was it to be suggested that Helen and Annie Sullivan should be parted.

Three years later, just as Helen entered Radcliffe, a well-intentioned friend nearly persuaded her to withdraw and, with Annie Sullivan, to start and direct a school for deaf-blind children. Bell's decided opposition to the scheme, along with that of other friends, kept Helen in college and out of what would surely have been a fiasco.

Bell's doubts of his own business acumen led him to decline the suggestion that he administer a trust fund set up for Helen in 1896. Nevertheless he took a leading part in organizing the arrangement and contributed a thousand dollars to it. Before and after, he helped out on special occasions, sending Helen $400 when her father died in 1896, $100 toward a country vacation in the summer of 1899, $194 so that Helen could surprise Annie with a wedding gift when Annie married the writer and critic John Macy in 1905. Financial as well as moral support may have led Annie to write early in 1898 that Bell "will never know how deeply grateful I am to him for one of the richest and fullest years we have ever known."

Among Helen's friends and admirers were those who were richer than Bell and less deeply committed to the support of other causes. In dollar terms their gifts to Helen outstripped those of Bell. But he gave her things they could not match with money. "More than anyone else, during those [early] years," wrote a friend who knew Helen in later life, "it was Alexander Graham Bell who gave Helen her first conception of the progress of mankind, telling her as much about science as Phillips Brooks told her about religion." Bell thrilled her with stories that paralleled the Greek epics she loved, Promethean tales like that of the laying of the Atlantic cable. One day he placed her hand on a telephone pole and asked her what it meant to her, then explained that the wires it carried sang of life and death, war and finance, fear and joy, failure and success, that they pierced the barriers of space and touched mind to mind throughout the world.

But Bell's mind, and Helen's through his, responded to nature too. Once, beneath an oak, he placed her hand on the trunk, and she felt the soft crepitation of raindrops on the leaves. For years after that she liked to touch trees in the rain. Then, on another day, he went with her to Niagara Falls and put her hand on the hotel window pane so that she could sense the thunder of the river plunging over its shuddering escarpment. He drove with her and Annie from Washington into the springtime countryside, where they gathered wild azalea, honeysuckle, and dogwood blossoms.

More than once Helen visited Beinn Bhreagh. She spent one night with Elsie and Daisy on the houseboat, from which they all climbed down by a rope ladder to swim in the moonlit lake. There in the fields overlooking the Bras d'Or, Bell told her of his kite flying and his hope of giving wings to mankind. "He makes you feel that if you only had a little more time, you, too, might be an inventor," she wrote. One windy day she helped him fly his kites.

On one of them I noticed that the strings were of wire, and having had some experience in bead work, I said I thought they would break. Dr. Bell said "No!" with great confidence, and the kite was sent up. It began to pull and tug, and lo, the wires broke, and off went the great red dragon, and poor Dr. Bell stood looking forlornly after it. After that he asked me if the strings were all right and changed them at once when I answered in the negative. Altogether we had great fun.

Back at Radcliffe that summer of 1901, she wrote Mabel that "the smell of the ocean and the fragrance of the pines have followed me to Cambridge and linger about me like a benediction."

Now and then Bell thought about Helen's future course in life. As she made her way through college, he began to feel that "with her gifts of mind and imagination there should be a great future open to her in literature." Later he wrote her, "You must not put me among those who think that 'nothing you have to say about the affairs of the universe would be interesting.'" But Helen was more realistic about the limits put upon her direct apprehension of the world, about her inescapable dependence on the words of others for learning what eyes and ears tell most people. She knew also that to the public her blindness was her foremost characteristic, though she agreed with Mabel that deafness was the heavier cross. So her work came to be more and more that of helping the cause of the blind. And because Bell's work lay with the deaf, he and she saw less of each other as the new century wore on.

Each missed the other. When he tried his hand at a letter in Braille, while she was in college, she praised him for not making a single mistake. "It seemed almost as if you clasped my hand in yours and spoke to me in the old, dear way," she wrote him. In 1907 he wrote her, "I often think of you and feel impelled to write but — as you know — I am a busy man, and . . . have always lots of back correspondence to make up." Now and then he wrote again in Braille, but not often enough for it to be easy. He spent a few days in Boston once and tried for a long time one day to telephone Helen's Wrentham home, but Mrs. Macy heard the ringing too late. "We seem bound every time to miss seeing him," Helen wrote John Hitz on that occasion. As public figures, each knew in a general way what the other was doing. "I suppose," wrote Helen in 1902, "Mr. Bell has nothing but kites and flying-machines on his tongue's end. Poor dear man, how I wish he would stop wearing himself out in this unprofitable way — at least it seems unprofitable to me." But six years later, she sent him a note of congratulation on his successes in aviation, to which he replied in proud detail.

In January 1907 Helen wired Bell, "I need you." She was to speak in New York at a meeting for the blind; but Annie, who usually repeated her speech for those who might have difficulty understanding it, had come down with a cold. Bell left Washington at once and lent his matchless voice to the occasion.

In the summer of 1918, Helen asked Bell to play himself in a motion picture of her life. He was then seventy-one, in uncertain health, more susceptible than ever to summer heat, and had "the greatest aversion to appearing in a moving-picture." Still, her letter touched him deeply. "It brings back recollections of the little girl I met in Washington so long ago," he wrote her. "You will," he reminded her, "have to find someone with dark hair to impersonate the Alexander Graham Bell of your childhood." But he promised to appear with her in a later scene, when the hot weather

407

was over, if she wanted him to. To his great relief he was not called upon — which was just as well, since the film was a grotesque failure, both as drama and as history.

The drama of Helen Keller's rescue and rise had, after all, been given a far more enduring form in her own autobiography, *The Story of My Life,* fifteen years before. Supplemented by her own and Annie Sullivan's letters, it both recounted and attested to one of history's most moving triumphs. And it began with the words:

To

ALEXANDER GRAHAM BELL

Who has taught the deaf to speak
and enabled the listening ear to hear
speech from the Atlantic to the Rockies,
I Dedicate
this Story of My Life.

31

The Chain of Generations

Carrying a line of notable achievement into the third generation, Alexander Graham Bell had reason to be conscious of heredity. In the early seventies, over his parents' objections, he stood up for Darwinism, to which heredity was central. As a university professor in mid-Victorian Boston he entered a social setting known for its high incidence of forefather-fixation. All this may have predisposed his mind. But his engagement in the study of heredity arose more directly from his concern with the deaf.

In 1878, soon after Bell returned from England, the Massachusetts State Board of Health enlisted him in its effort to gather statistics on inherited defects, so that the laws of inheritance might be better understood. Given a big bundle of circulars and stamped envelopes, Bell plunged into an investigation of hereditary deafness. He wrote the superintendents of institutions and appealed for data through the *American Annals of the Deaf*. But returns were apparently neither full nor prompt. At any rate, the reports of the state board through 1881 said nothing of Bell's findings.

Bell kept on anyway and accumulated data from the Hartford and Illinois institutions on the marriages and offspring of their graduates. In 1880 the federal government took a more elaborate census than ever before and assembled a distinguished group of authorities to direct the several reports. From the report on the "Defective, Dependent and Delinquent Classes" Bell mined enough additional material to produce a paper for the National Academy of Sciences in November 1883: "Memoir upon the Formation of a Deaf Variety of the Human Race."

Bell's paper appeared at a time of rising interest in the study of human heredity. In England, Darwin's statistically-minded cousin Sir Francis Galton had begun in the sixties by analyzing the pedigrees of famous men, then had studied the inheritance of more clearly definable and objectively measurable traits. In 1875 Richard Dugdale published the most influential nineteenth-century American work on heredity, a frightening catalogue of criminals, paupers, and degenerates in successive generations of a family he called the Jukeses. But most of the traits Dugdale counted could not be measured objectively nor established as congenital, a shortcoming that Dugdale (though unfortunately not all of his readers and imitators) explicitly recognized.

Congenital deafness varied in nature and severity. Still, it was a well-defined physical trait, bore less of a stigma than most inherited defects, was therefore less likely to be expunged from family tradition, and usually went on public record anyway through the requirement of special schooling. So Bell could base his paper on more precise and meaningful data than Dugdale's.

In his paper, Bell emphasized the incompleteness of his data and called for more research. Nevertheless his findings made certain disquieting conclusions inescapable. Deaf parents had a much higher proportion of deaf children than did the population at large. Of 2262 congenital deaf-mutes in his survey, 54.5 percent had deaf relatives. Also, marriage records showed that, unlike those with other defects, the deaf strongly tended to marry the deaf. From those discoveries it followed that deafness could be inherited, and that it tended to propagate itself — whence the paper's title. Bell opposed the idea of prohibiting marriage between the congenitally deaf. Instead he advocated warning them of the risk beforehand and broadening their opportunities for friendship (and so for marriage) with hearing persons through day schools and speech training.

Bell also saw in his findings the more general implication that "if we could apply selection to the human race we could produce modifications or variations of men." It was in that same year of 1883 that Sir Francis Galton's *Inquiries into Human Faculty* introduced the term "eugenics" to designate improvement of the human race through purposeful breeding.

Wisely, Bell did not try to work out precise laws of inheritance for deafness. In the 1860s the Austrian monk Gregor Mendel had done that for simple traits in pea plants and had published his findings; but no one had taken notice, and so Mendelian genetics had to be rediscovered in 1900, whereupon it revolutionized the field. Even after that, certain aspects of the inheritance of deafness could not be explained. A congenitally deaf couple could not be told on Mendelian grounds how many, if any, of their children were likely to be born deaf, let alone how many were sure to be. People are more complex than pea plants, and deafness is not a trait of simple, uniform nature or origin.

Finding that hearing persons from a deafness-prone family could transmit deafness to their children, Bell faced the question of how the tendency could be countered without somehow cutting short all family lines in which it appeared. That anxiety may have led him to his implicit hypothesis that marriage into normal families would tend to submerge or extinguish a hereditary tendency to deafness. In short, he came close to independently formulating the Mendelian distinction between dominant and recessive traits.

The phrase "a deaf variety of the human race" in the title of Bell's paper turned out to be ill chosen. Bell doubtless meant to dramatize a dangerous trend by projecting it to extremity, but instead came near reducing it to absurdity. Worse still, the title antagonized many of the deaf by seeming to imply that they were forerunners of an inferior species. It was not Bell's fault, however, that a newspaper reporter plucked from its context a suggestion Bell had raised in order to strike down: that marriages tending to transmit deafness might be prohibited by law. Periodicals for the deaf repeated the distortion. In 1887 Bell seized a chance to set the matter straight in a lecture before the Literary Society of the National College for Deaf Mutes. Edward A. Fay, professor of languages at the college and editor of the *American Annals of the Deaf*, provided a running sign language translation and afterward helped to publicize corrections of the original misinterpretation. But vestiges of it persisted for years.

A year later the editor of a periodical dealing with charities and corrections began gathering statistics on marriages of the deaf. In 1889, after a number of replies had suggested Fay's *American Annals* as a fitter vehicle of publication, Fay consented to take over the survey in what would become the greatest work of his life. The questionnaire was refined and expanded, and it drew far more replies than had been expected. As soon as Bell heard of Fay's project, he gave Fay all his own data on the subject and put at Fay's disposal some of the income from his endowment for research on the deaf. Fay devoted himself to the work without compensation, but the funds from Bell paid for assistants. Over a period of six years an enormous amount of supplementary research was done. In 1895, again with financial support from Bell, Fay published *An Inquiry Concerning the Results of Marriages of the Deaf in America*.

Fay's work, based on reliable information about 4471 marriages in which one or both partners were deaf, was essentially a more complete and detailed extension of Bell's 1883 investigation. Fay concluded that marriages of the deaf produced a far higher proportion of deaf offspring than did ordinary marriages, ranging from 2.3 percent of children with two adventitiously deaf parents to 25.9 percent of those with two congenitally deaf parents; that the proportion increased when one or both parents had deaf relatives, but that even when both were congenitally deaf they ran much less risk (only about 4 percent) if neither had deaf relatives; and that consanguinity of parents greatly increased the risk, even without deaf relatives. All this tended to substantiate Bell's findings, while offering some comfort to the deaf who had no deaf relatives. What may have given Bell pause was Fay's discovery that the deaf were about as likely to marry the deaf whether they attended day, boarding, oral, or Combined Method schools. Sympathy and common problems seemed to bring them together as effectively as segregated schooling. Still, the matter of marriage was only one of Bell's arguments for day and oral schools.

All in all, Bell's paper, along with Fay's follow-up, stands as the soundest and most useful study of human heredity produced in nineteenth-century America. By that token it may also be reasonably counted Bell's most notable contribution to basic science, as distinct from invention.

Meanwhile Bell had gone on during the eighties to trace deafness through several generations in certain families, confining his scrutiny to New England as being richest in long-continued genealogical records. Within that region he and his assistants researched the history of every family reported to have produced two or more deaf children. He followed the Lovejoy family, for example, from its Massachusetts progenitor in 1644 to its numerous branches in his own day, some of them clearly demonstrating the inheritance of deafness. On the island of Martha's Vineyard, he discovered an isolated community in which a fourth of the population was deaf, while elsewhere on that island he was delighted to find a passionately dedicated amateur genealogist with thousands of notes tied up in little muslin bags, one bag for each old Vineyard family. So Bell was able to track the course of deafness from an early settler down to the silent citizens of Squibnocket.

In the mid-eighties, Bell hired a professional genealogist, an energetic little Bay Stater named Annie E. Pratt. She shared his ardor for the chase. Having located a ninety-six-year-old Lovejoy in 1887, Bell wrote Mabel that "Mrs. Pratt is to start off at once to the boundaries of Maine and New Brunswick to interview the old lady while there is yet time," while he followed another branch to Halifax. Mrs. Pratt later did much research for the exhaustive history of articulation teaching in America, which Bell published serially in the *Association Review*, 1900–1904.

In 1886, as Bell's sweep of New England genealogy proceeded, he hired John Hitz, the gentle, white-bearded father of the first teacher in Bell's little Washington day school, to serve as librarian and research assistant. "No one is better fitted to help me and encourage me in this work than yourself," Bell wrote Hitz in 1888. "There is labor enough in this work to afford you employment for a long time to come." Hitz did in fact serve in that capacity until his death in 1908. (During those years, after converting Helen Keller to a lifelong belief in the religious doctrines of Emanuel Swedenborg, Hitz maintained a fuller and more continuous correspondence with her and Annie Sullivan than did Bell himself.) Transcripts of deeds and vital records, published genealogies and town histories, reports of schools for the deaf, census cards, periodical files, and other data mounted higher and higher in a room of Bell's Volta Laboratory in Georgetown. At Hitz's suggestion in 1887, Bell christened Hitz's office the "Volta Bureau" and on his own motion designated Hitz its "Superintendent." The trust fund was renamed the "Volta Bureau Fund" and earmarked for the bureau's support, except for twenty-five thousand dollars transferred to the AAPTSD on its organization in 1890.

Bell's father and mother took a special interest in the Volta Bureau, which was, after all, literally in their backyard. By giving it a name, their son had given it a life. He himself saw it now as a center "for the increase and diffusion of knowledge relating to the deaf." (The phrase "increase and diffusion," he noted in a letter to Hitz, "is commonly used in the foundation papers of learned societies and institutions — like the Smithsonian.") In this broader role, its collections outgrew the little ex-stable. Bell's father felt that the inevitable new building should be impressive in appearance for the greater prestige of the bureau. When his son demurred at architectural show, the elder Bells contributed fifteen thousand dollars toward that feature. And so in the spring of 1893, across the street from their home, they watched twelve-year-old Helen Keller turn the first spadeful of sod for what became a neoclassic yellow brick and sandstone library with stone steps rising high to a portal framed by massive columns.

John Hitz lived up to his trust and his temple, steadily enlarging the collections by purchase and the receipt of gifts, making the bureau, through correspondence and publication, a major source of information about deafness and the deaf. By the time of his death in 1908, the Volta Bureau had become one of the world's leading centers of such information; and after its merger with the AAPTSD that year, it continued to serve the world as the Association's headquarters and library.

From using census returns, Bell's statistical work drew him into shaping census policy. In 1886 he discovered the original census forms since 1790 strewn loosely over the floor of a dark vault under the Patent Office, more than a thousand volumes in such disarray as to be unusable, whereupon he persuaded the secretary of the interior to have them properly arranged and stored. In 1889, having found them "invaluable" for genealogical research, a purpose for which they had apparently never before been used, he called that use to the attention of the president of the Massachusetts Historical Society, who was much struck by the notion and passed it along to other societies. Since that time, the records have been a mainstay of genealogical research.

During the mid-nineties Bell protested strenuously against a proposal in the House of Representatives to destroy or sell for wastepaper all census population schedules, past, present, and future, as soon as results had been tabulated. He predicted that scientific and other learned bodies would "naturally regard" such an action "as an outrage." The published tables did not use all of the raw data, nor did they give information for particular individuals. "The schedules for 1900 will be as interesting to the students of history in 2000 as those of 1800 are to students at the present day. . . . When new questions are raised, and an answer to them is sought, it will be necessary to go over the former records again, in order to secure the proper basis for comparison." So far as Bell's protest may have helped to save the records, historians — especially those of the quantitative persuasion — are in his debt.

Bell granted that 1880 had brought "the best census of the deaf the world has yet seen." But some trivia had been included, some important questions omitted, some errors made. Questions were so framed as to invite hostility or evasion. All who had become deaf before they were sixteen were arbitrarily classed as deaf-mutes whether they could speak or not. Younger deaf children were often missed, though they above all needed to be located promptly for speech training.

Late in 1888, at the request of a Senate committee, Bell submitted forty-two specific suggestions for making the 1890 census of the deaf more useful and accurate. He protested without avail against the proposed lumping of "the Insane, Feeble-Minded, Deaf and Dumb, and Blind" in a single report. Nevertheless, encouraged by the appointment of Edward Fay as special agent in charge of that category, the first time it had been assigned to a recognized expert on the deaf, Bell helped work out the standard forms. And after the 1890 census, he rescued important supplementary records from disposal as wastepaper and deposited them in the Volta Bureau Library.

During the nineties Bell got his statistical exercise by abstracting and tabulating data from annual reports of schools for the deaf and from occasional questionnaires. As the time approached for the census of 1900, he betrayed so lively an interest that he was offered the job Fay had taken in 1890, and conscience made him accept. He got Congress to designate his report as "The Blind and the Deaf." By the time it appeared in 1906, Bell had frequent occasion to curse his conscience. As Fay had discovered before him, the work absorbed much thought and labor over half a decade. "The question of Census versus Laboratory is ever before me," Bell complained late in 1904. "I hate this census with a personal hatred," Mabel wrote her son-in-law; "I feel it is taking from Father time and strength he cannot spare." Bell dodged involvement with the 1910 census, and the deaf category for that census did not go to an expert. There may have been a twinge of conscience in Bell's private judgment of the report on the deaf that came out in 1915: "a perfect fizzle, not at all comparable to any former census."

In 1889 Bell devised a sorting machine for punch-coded census cards "based upon the principle of turning the cards around an axis by the motion of a rod in a slot punched out of the card at different angles," only to find that the Census Bureau had decided to use Herman Hollerith's new electrical punch-card tabulating system (the Adam of modern data-processing machines). Bell accepted the decision with good grace. The Hollerith system "seems to be an ingenious and practical method," he wrote; "I do not propose to push my own method." More than ten years later, as processing of the 1900 census got under way, Bell examined a Hollerith machine in the Census Bureau and conceived a system of counting electrically, "without the necessity of sorting the cards," which drew on his earlier work with an automatic telephone switchboard. He called his system "combinatorial branching," but it pointed toward the modern binary system of electrical computation. Perhaps his nonmathematical mind would not have carried him much further. In any case, something else must have come up, for his notebook entries on the matter end abruptly with the words, "Have no more time to make notes." The next entry, several days later, deals with ways of condensing fresh water from fog.

As a student of heredity, Bell could not resist moving beyond statistics to experimentation. During the mid-eighties, on a cue from Darwin's writings, he looked into the reported proclivity to deafness of blue-eyed white cats and somehow located several specimens. The three he had seen by 1884, he reported in *Science*, were indeed deaf. He kept a male and female for several years; but while the relationship of the afflicted pair was not platonic, neither was it fruitful.

With his purchase of Beinn Bhreagh, however, Bell acquired a flock of sheep. Thomas Sanders, who for many years had been active in the breeding of fine horses and sheep on his Vermont farm, helped Bell procure some choice Vermont Merinos in 1889 and probably encouraged Bell's venture into systematic stockbreeding. Bell was fascinated by such oddities as the fact that sheep have no teeth in their upper jaws. He also observed that half the lambs born to his flock in the spring of 1890 were twins, though lambs normally come one to a birth, and that what seemed a significantly higher percentage of twins were born to ewes having more than the normal two nipples. Eventually he would have reason to wish he had settled for blue-eyed cats.

Bell set out to determine whether the extra nipples, mere vestiges in that generation, could be made functional and hereditary by selective breeding, and whether ewes thus equipped would bear and raise a significantly higher proportion of twins. If lamb production could be increased, not only would the farmers of Nova Scotia profit, but also the whole mutton-eating, wool-wearing world. "He has thrown himself into these breeding experiments with all his characteristic interest and absorption and thoroughness of detail," Mabel was presently writing. Every sheep was numbered, identified by coded ear punchings, and weighed at frequent intervals; and every detail of its life and pedigree was entered into tables. Bell devised instruments for testing wool strength, processes for extracting oil from the wool, freeze-proof watering troughs, prefabricated sheephouses for winter, contrivances for measured feeding, an octagonal sheep barn, and specially designed and measured sheep runs. On top of Beinn Bhreagh appeared a village of sheephouses, which Bell named "Sheepville."

Nearly every day for several years Bell led his sometimes reluctant family up the mountain for a visit to Sheepville. When possible he was on hand to supervise the critical work of fall mating and spring lambing, once even coming back from a European tour for the purpose and then recrossing the Atlantic. Gradually he entrusted winter care of the flock to shepherds, who on one occasion lost most of it by leaving it out overnight in a blizzard and on another by failing to guard it from wild dogs.

Bell succeeded in developing a strain of ewes with at least four milk-producing nipples. But his tables multiplied more dependably than his sheep. In 1904 he reported to the National Academy of Sciences that "the multi-nippled sheep have not proved to be more fertile than normally nippled sheep; and the proportion of twins born has been quite small." Rather than give up at that point, he proposed to begin breeding especially for the twin-bearing trait. But ten more years of effort still did not yield a multi-nippled stock that consistently bore twins. So he transferred the best of the flock to a younger farmer (who got rid of them by 1920) and auctioned off the rest. Next day, Bell discovered that through a proxy bidder Mabel had bought back the nucleus of a new flock for herself and proposed to continue the experiments with or without his help. In the sheep barn that morning, surveying what would thenceforward be known as "Mrs. Bell's flock," he was heard to say furiously, "I thought I was THROUGH with those damn sheep!" Such, at least, was the story told his secretary by one of the men. So far as Mabel and the rest of the family ever knew, however, Bell was delighted and grateful; and quite likely, whether or not the secretary's hearsay was true, Bell would have missed his hobby of twenty-four years' standing. Certainly he took up the work again with undiminished zeal, and by the time of his death in 1922 had a multinippled flock that he thought might be "a true twin-bearing stock" — though the flock had not grown large enough nor been bred long enough to prove the tendency hereditary.

After studying Bell's meticulous records, the Harvard geneticist William E. Castle concluded in 1924 that the extra nipples had indeed become strongly hereditary in the flock, but were not clearly associated with either an increase in total milk production or a tendency to bear twins. About then, the University of New Hampshire agricultural experiment station took a ram and thirteen ewes from the flock (the rest being sold off) and carried on the breeding experiments until 1941, after which the United States Department of Agriculture continued the work for another three years before disposing of the flock. In its final report, the department held that twinning was not so strongly correlated with extra nipples as to make the latter worth breeding for, nor did the extra nipples furnish more than a negligible quantity of milk. In short, the report concluded, "the multinipple character has no practical value in sheep production."

Bell's sheep breeding brought him into the American Breeders' Association, founded in 1903 by university biologists and breeders of plants and animals. This connection, along with his years of advocating marriage between the deaf and the hearing on genetic grounds, in turn drew him into the new eugenics movement as it entered its period of greatest influence on American thought.

Bell's own published suggestion of 1883, the year Galton coined the term "eugenics," that the human race might be modified by conscious selection, made him a precursor of the new movement. It was launched in 1901 by the aged Galton himself on converging currents of Darwinism, social reform, racism, elitism, and Mendelian genetics. In 1907 the American Breeders' Association, which had become a prime mover of Mendelian research in America, set up a number of committees on specific breeding problems. One was a Committee on Eugenics, the first American group using that term, "to investigate and report on heredity in the human race" and "to emphasize the value of superior blood and the menace to society of inferior blood."

Bell declined the chairmanship, which went to the distinguished biologist and chancellor of Stanford University, David Starr Jordan, but he served on the committee with such notables as the plant-breeder Luther Burbank, the sociologist Charles Henderson, and the geneticists William Castle and Charles Benedict Davenport. Bell, along with Jordan and Henderson, accepted reappointment to the committee in 1909. As secretary of the committee, Davenport persuaded a number of illustrious scholars and scientists to head various subcommittees, Bell serving as chairman of the one on hereditary deafness. In 1913 Bell collaborated with his son-in-law David Fairchild, the new president of the American Breeders' Association, in drawing up articles of incorporation for the society, which soon after was renamed the American Genetic Association; and Bell contributed occasional essays on eugenics and sheepbreeding to the Association's *Journal of Heredity*.

The decline of the eugenics movement was already beginning with the infiltration of racists like Madison Grant, who, aside from the moral stigma they brought to it, thoroughly corrupted its scientific quality, which was already tainted by naive oversimplification of human traits and the crude forcing of them into a Mendelian pattern. The racists used the movement's fading prestige, along with the xenophobia and political reaction of the postwar period, to help slam the door on immigration in the early twenties. But responsible scientists had already begun to dissociate themselves from the eugenics movement. Fortunately the American people, though still susceptible to the pseudoscience of racism, were too tolerant, optimistic, and ethnically varied to stomach racism's social and political corollaries. In Germany, encouraged by perverted eugenics, racism culminated in the incomprehensible horror of Nazi genocide.

It is not easy to look back across that abyss to its sunny approach and see the early eugenics movement as the benign application of science to humanitarianism that claimed the sympathy of men like Bell, Galton, and Jordan. But justice to Bell requires the effort. It also requires a look at his position in the spectrum of eugenics thought.

Bell did yield to the assumption, which all those around him took as axiomatic, that ethnic groups somehow differed inherently in temperament and intelligence, as well as in superficial physical characteristics. But he considered such presumed differences irrelevant to the inheritance of deafness, which was his chief concern. And to the end of his life he escaped the fatal delusion of more and more eugenists that they knew just what those supposed ethnic differences were, quite without benefit of scientific study, and could sort them out as "desirable" or "undesirable." Bell never singled out any specific ethnic group as "undesirable," though it was commonplace in his day for self-styled eugenists to stigmatize the Italians, Jews,

Slavs, and others. In his published writings on eugenics, he alluded only vaguely and casually to restriction of immigration on ethnic grounds, and then only to the extent of insisting that careful, objective studies ought to be made before any groups were presumed to be "undesirable" by heredity and therefore shut out.

Bell's reason might by itself have made him thus properly cautious about racism (though many who seemed otherwise rational succumbed to the disorder). In his case, consistency reinforced good sense. The myth of Anglo-Saxon superiority could not have been palatable to a black-haired Scotsman. As the founder of the Volta Bureau, as the friend and admirer of Marconi, and — at the very height of the eugenics movement — as the most prominent American disciple of both Enrico Forlanini, the hydrofoil boat designer, and Maria Montessori, the educational reformer, Bell could scarcely have supported the racist fringe in its assertion of Italian ethnic inferiority. Anti-Semitic claptrap would have reminded him of Emile Berliner, the Jewish immigrant from Germany who had helped save the Bell Telephone Company, or of Albert Michelson, the Jewish immigrant from Poland, in whom Bell had found as congenial a spirit as any he had ever known. Negrophobia, racism's most vicious American manifestation, had no place in the mind of a man who on his third day in the United States had commented sarcastically on the white American prejudice against interracial marriage and who had ever since stood up for Negro social and political equality. Bell, in short, was too sensible, experienced, and decent a man to be a racist. The race that concerned him was the human race.

As early as 1915, Bell betrayed some uneasiness about what he called "our eugenic cranks." His own ideas remained consistently positive. He opposed any interference with the marriages of "undesirables" (as determined on an individual, not an ethnic, basis). Not only did he object to such interference on grounds of both principle and practicality (perhaps remembering his own demonstration of love's contempt for locksmiths), but he also denied that such marriages could significantly affect the quality of the human race. Eugenics, to him, meant scientific research and discovery, the dissemination of which might encourage those with "desirable" heredity to marry one another for the sake of their own posterity as well as the improvement of the human race.

In his later years, Bell's own research in eugenics dealt chiefly with the innocuous question of whether longevity could be inherited. In 1914 he established the Genealogical Record Office of the Volta Bureau for "the collection and preservation of genealogical records pertaining to long life." Its secretary picked up clues to long-lived individuals from press clipping services, obituaries, and return-addressed postal cards sent to subjects. Bell himself made a study of the long-lived Hyde family, stemming from seventeenth-century Connecticut, and published his findings as a pamphlet in 1918. He concluded that long-lived people had a greater number of children than could be explained simply by the duration of their marriages, and that those descendants also lived longer than the average. Longevity, he surmised, was not heritable in itself, but probably was related to some heritable qualities of vigor or resistance to disease.

As a eugenist, Bell must have been especially glad to see his parents live long and his children marry well.

When Alexander Melville Bell brought his bride to see his maternal grandfather on their honeymoon, the old man drew him aside to predict that "she'll never make old bones." But in the summer of 1894, when Eliza Bell was eighty-four, the big house at Beinn Bhreagh blazed with electric light, Chinese lanterns glowed from trees, bonfires crackled, fireworks scored the night sky, and a military band blared away for the elder Bells' golden wedding celebration. Alec, Elsie, and Daisy, their only living descendants, presented them with a golden loving cup to the applause of two hundred guests. Two years later, the elder Bells saw Gardiner and Gertrude Hubbard pass the same milestone.

Eliza Bell had been born in the first decade of the century that her son had since helped to make so memorable. Now its successor cast a shadow in advance. "What times we seem to be living in," wrote Mabel to her mother from Beinn Bhreagh in 1895. A new world power, its memories of civil war softened by time and poets, was confronting an old empire in the Venezuela crisis. "The stir and excitement has come even here," Mabel reported. "Alec wants to offer his services to the American Government. He is thoroughly heart and soul with President Cleveland's message." An Anglo-American war was averted, but the martial spirit had shown itself. Alec spoke of "inventing lots of fearful machines," Mabel wrote; "is not this strange talk for Christmas time?"

Mabel had already encountered fearful new machines. "There is an exhibition now in progress of horseless carriages," she wrote from Paris in June 1895, "and in a few days there is going to be a race of them from Paris to Bordeaux and back. Steam, petroleum, and electricity are the means of impulsion. . . . I have seen one or two of these carriages in use about the streets."

But Eliza Bell was spared the age of empire and exhaust fumes. On January 5, 1897, she died quietly in Georgetown.

In December of that year, death also took Gardiner Hubbard. He was seventy-five. In 1884 Mabel had written in her journal: "Alec said apropos

420

of his liking few men, that is feeling that they are sympathetic with himself, that Papa was one of those few men; they had really very little in common, but Papa was such a remarkable man and had such a broad and clear grasp of things." By the time he was seventy, Hubbard had reversed old roles by entrusting his son-in-law with entire charge of his money affairs. Two years later, partly on the basis of Alec's generalship in the campaign for speech teaching, Hubbard recanted his original low opinion of Alec's executive ability as emphatically as he had first delivered it. "I know of no one whose judgement in business matters is better than yours when you give your mind to it," he wrote his possibly astonished son-in-law. "If you had been a businessman, you would have made a great success in that line as in your own special lines of thought & work."

A year or so after Alec's mother died, his father, at seventy-nine, married a fifty-four-year-old widow, Mrs. Harriet G. Shibley, whom he had arranged to meet after her photograph in the home of an Ontario friend caught his eye. Two years later Alec reported his father "in fine condition, stepping out briskly like a young man." His voice remained surprisingly strong, clear, and resonant. The elder Bell was working on a lecture on Visible Speech to be delivered at Columbia University; his son thought it "one of the finest things he has ever written." As for the second Mrs. Bell, wrote Alec, "she is just as sweet and good as she can be, and I am perfectly satisfied that he should have married in his old age."

The Bell household continued to feel the tremors of a changing world. When the United States went to war with Spain for the salvation of Cuba in 1898, the indomitable Clara Barton urged eighteen-year-old Daisy Bell to join the Red Cross as a relief worker. "Here is a chance for Daisy," Mabel wrote Alec; "let her take it unless we find the danger is too great." But Daisy had never had yellow fever, and so her father dissuaded her. The war ended a few weeks later anyway, to be followed by the nation's descent into imperialism, which Bell opposed. "He says he wishes Dewey had come right home after the battle of Manila," wrote a *New York Sun* reporter who interviewed him on Christmas Day, 1899, "but so long as he cannot change destiny, he accepts everything as every true American should."

In that same interview, Bell expressed himself as "pessimistic regarding the future of the British Empire and the English people." The British century was nearing its end. When it did pass, Queen Victoria followed promptly. "We had good seats for the procession at the Queen's funeral and saw everything splendidly," Mabel wrote Elsie in February 1901.

At the slow, slow tread of the soldiers marching with guns reversed, all stood so still and motionless and presently all heads were bared. Daisy said you could almost have heard a pin drop . . . in all that great multitude. The soldiers passed, the gun carriage rattled [by], the King followed grave and anxious-looking with the Kaiser reining in his horse so that the King could ride a neck ahead. . . . And then a brilliant mass of variegated colored uniforms passed which we knew covered the persons of almost every reigning sovereign or sovereign's heir in Europe.

Bell marked epochs by technology. "We are in the 20th century & no mistake," he noted in July 1905, "for after dinner, Mr. Hubbard took Mabel and myself for a little drive of *about 40 miles*(!) in his automobile." Bell embraced the new age with characteristic ardor. Within two months, Mabel was writing Elsie that "Papa seems to have thrown every other thought aside for the moment and gone in for automobiling as he goes in for everything he does, with his whole soul. Our machine is a Rambler, and so far it has gone splendidly without accident, or stoppage of any kind. We came over fifty-three miles today, from Gloucester."

Between those two automobile rides, a personal era ended for Bell. With his father, as with his father-in-law, Alec had experienced a reversal of roles. "It is really touching, his dependence on Alec," Mabel wrote from Beinn Bhreagh early in the new century. "Mrs. Bell says it is the first question in the morning, constantly repeated throughout the day, 'Where is Alec? Is he well? When will he come?' And Alec reads to him for hours every evening, or he likes to listen to Alec playing and singing with Mrs. McCurdy." But on August 7, 1905, when he was eighty-six, death came for Alexander Melville Bell. His end was "quiet, and without struggle," wrote Mabel. He "held up his hand and then spread it down as he does when a thing is finished, the characteristic elocutionist's gesture, marking the last fluttering breath."

Only Mabel's mother now remained of that generation. In the spring of 1909, on an automobile tour of Wales, she seemed younger than ever. She was riding along Connecticut Avenue in Washington in October when a streetcar crashed into her chauffeur-driven automobile from the rear and threw it against a pole. The driver lived, but at Garfield Hospital two hours later, Gertrude Hubbard died of her injuries. She was eighty-two.

The new century brought a new generation. Gardiner Hubbard had indirectly helped start a branch he never lived to see, when he led in founding the National Geographic Society at Washington in 1888. Though he committed himself to it as its first president, the society remained a small, sedate, and largely local group during his lifetime; and its soberly bound, drily written journal, the *National Geographic Magazine*, repeated the vain struggles of *Science* in its quest for readership. Among the lecturers Hubbard brought before the society in 1897 was Professor Edwin A. Grosvenor of Amherst College, author of a lavishly illustrated two-volume book on Constantinople, where he had taught history in an American college for more than twenty years. As a guest at the Bells' home, Grosvenor talked proudly of his identical twin sons, Edwin and Gilbert.

By her late teens, Elsie had begun to attract suitors. The Grosvenor twins themselves visited the Bells and were taken by her charm. So was a young Washington architect named George Totten, who was in a better situation to court her. But Elsie did not respond to Totten's feeling, though the Bells, who liked and respected him, reminded her that her mother had not at first responded to her father's courtship. Elsie and Daisy did, however, gladly accept Mrs. Grosvenor's invitation to attend the twins' graduation from Amherst in the summer of 1897. The Grosvenor boys, both of whom graduated with high honors, turned out to be the stars of that commencement. Even *Harper's Weekly* noted their prominence. When George Totten showed up at Beinn Bhreagh soon after, Mabel seized on the twins' minor celebrity as grounds for inviting them also, her private motive being to give Totten's suit the test of competition. "The Grosvenor boys . . . are certainly fine fellows, clever and manly with no nonsense about them," she wrote at the end of their stay. George Totten had also proved to be "really a nice fellow, kind and gentle and up to anything. . . . Alec likes him." He had "won all hearts except the only one he cares about." Mabel commented, "I have entirely gained my object."

Later that year, as Bell told it in retrospect, he was, "by the death of Mr. Hubbard, forced to become the president of the [National Geographic Society] in order to save it." For more than a year, however, Bell's experimental work absorbed his time and thought, while the society's membership dropped from fifteen hundred to one thousand, the life membership reserve melted away, and debt accumulated. At that point, Bell turned his energy and his talent for communication to the problem. The key, he decided, was in transforming the magazine into a bright, interesting, nontechnical periodical, which would recruit members from the general public. To that end, it needed the full time of a vigorous, intelligent editor, whose monthly salary of a hundred dollars Bell offered to underwrite for a year. The board accepted the offer, whereupon Bell (aided by Elsie's presence in Washington) persuaded young Gilbert Grosvenor to quit his New Jersey teaching job and on April 1, 1899, become the National Geographic Society's first full-time employee.

Many years later, near the end of his life, Gilbert Grosvenor maintained that his father's book, with its more than two hundred photographic illustrations in the new halftone process, had chiefly inspired his immensely successful emphasis on profuse photographic illustration of the *National Geographic Magazine*. But if Bell did not inspire that policy, he certainly encouraged it. The magazine should have "a multitude of good illustrations and maps," he wrote Grosvenor in July 1899. And early in 1900 he was more specific: "*Motion* interests. . . . [Run] *more dynamical pictures* — pictures of life and action — pictures that tell a story 'to be continued in our text'!!"

Bell's dramatic instinct by itself could account for this insight. But he may have had an additional cue from an episode of 1894. In that year, Samuel S. McClure, struggling to keep his new *McClure's Magazine* alive through the depression, hit on the idea of a series about Napoleon (in whom there was just then a curious revival of interest), to be illustrated with the new photoengraving techniques. For fourteen years Gardiner Hubbard had been assembling one of the nation's largest collections of prints of Napoleon and his intimates. When McClure heard about it, he rushed down to Hubbard's Washington home and with Hubbard's help ecstatically laid out a complete pictorial history of Bonaparte's career. A few weeks later he brought a talented but still unknown writer, Ida Tarbell, to look over the pictures and write the text. Miss Tarbell, impressed by the spacious gentility of Twin Oaks, apologized to her hostess for McClure's unceremonious enthusiasm. Mrs. Hubbard smiled. "I am accustomed to geniuses," she said. The pictures and Ida Tarbell's articles gave *McClure's Magazine* its first major success and sent it on its way to journalistic greatness.

In a number of letters during the summer of 1899, Bell offered young Grosvenor other ideas for achieving a circulation "of thousands instead of hundreds." The scientific agencies of the federal government, conveniently located near the magazine's headquarters (which occupied half of a small rented room), could provide much material at no cost, but the best writers must be paid well. Hence advertising revenue must be increased, and that required a large circulation. The magazine's content should therefore be of popular interest as well as scientifically reliable.

Timeliness would help too. There should be a stockpile of appropriate material ready at hand as the spotlight of world attention shifted to new places. "Things are working up in the Transvaal. It would be a great thing if you could get up a good map of . . . [what] seems destined to be soon the theater of war." Over his own signature, Bell wrote the publishers of leading magazines and newspapers in August 1899, recommending the *National Geographic Magazine* as a source of "reliable and timely items relating to all the geographic topics that may be occupying the public mind." He commended Grosvenor on his subscription to a press clipping service "to show the extent to which the magazine has been quoted in the public press."

Other markets lay waiting. "We must have today in the United States at least thirty millions of persons under the age of twenty-one. . . . It should be possible to make the magazine . . . of so much value to the schools that it would be used as a sort of text book of current geography." The society might also publish "popular books on geographic subjects." And on the definition of "geographic subjects," Bell delivered his most expansive opinion: "THE WORLD AND ALL THAT IS IN IT is our theme."

Bell also urged that the society be made national in fact as well as name by abolishing the distinction between active and corresponding members, and by including in every issue of the magazine at least one printed form for the nomination of new members. This proved highly effective.

Grosvenor wrote after Bell's death that he had kept Bell's letters of 1899 and 1900 before him "as a chart in all the ensuing years." Grosvenor's unparalleled success as an editor over more than half a century suggests that he might have evolved most of those policies for himself in any case. Furthermore, merely thinking of a policy is far from carrying it through, and the latter was Grosvenor's achievement — in the beginning, literally single-handed. Bell promptly made plain his admiration for that success. "Our list of corresponding members is rapidly increasing — nearly three hundred have joined in the last three months," he wrote less than six months after Grosvenor started work. "The Magazine has been much improved, it comes out ON TIME, and the clippings . . . show it is being extensively noticed." Nevertheless, however much or little of the magazine's success is credited to Bell's ideas, they stand as striking evidence of his publishing insight.

Even in the first frenzied spring and summer of his editorship, Gilbert Grosvenor had found time to court Elsie Bell. Early in May 1899, the young couple joined others of the Bell family for a picnic excursion upriver to Great Falls and Cabin John. "Gilbert really needed the outing," wrote Mabel. "He said he was . . . 'played out.' I think Elsie is seriously considering him." A few days later, Mabel told her husband that "the more I see of the boy, the more I am impressed with the idea that he will be a successful man. I like the way he talks. . . . He is wholly absorbed in making a success of the paper." What puzzled her was that Bert could be "so perfectly self-possessed when his lady-love is near him." "I am sure you weren't," she remarked to her husband. But that was Bert's way, and Elsie seemed not only to understand it but also to emulate it. "Elsie says

The clan at Beinn Bhreagh, 1901. Beginning on left, standing: Mrs. Edwin A. Grosvenor; Gilbert H. Grosvenor; Edwin A. Grosvenor; Alexander Graham Bell. From the left, seated: the second Mrs. Alexander Melville Bell; Mrs. Alexander Graham Bell; Alexander Melville Bell. Leaning over her grandfather on right, Mrs. Gilbert H. Grosvenor

Mabel Bell, Marian Bell, Gilbert Grosvenor, and Elsie Bell at Cabin John, May 1899

she thinks she will probably decide to marry Gilbert, but she is so perfectly matter of fact about it that I am sure she is not a particle in love with him," Mabel wrote Alec in the spring of 1900. Mabel was mistaken. On August 30, 1900, Elsie accepted Bert's proposal; and that October in London, he and she began a happy marriage that would end only with her death sixty-three years later.

Their firstborn, Melville Bell Grosvenor, was a child of the twentieth century, born in his grandparents' Connecticut Avenue house on November 27, 1901. Soon afterward (he was told later) his Grandfather Bell brought him down to show to dinner guests, bent close to him, and shouted "Baaaa!" As Melville was rushed upstairs bawling, Bell announced triumphantly, "He has perfect hearing!" Such grandfatherly antics soon led Bert Grosvenor to move his family into a small rented house nearby. In the years that followed, Melville acquired reinforcements to the number of five sisters and a brother. The latter, born at Beinn Bhreagh in 1909, was christened Alexander Graham Bell Grosvenor at a ceremony attended by President William Howard Taft, a first cousin of Gilbert's mother.

Meanwhile Gilbert Grosvenor had come into his own as one of the most successful magazine editors of the twentieth century. After the first year of his editorship, the society's membership had doubled to twenty-four hundred, and the board added eight hundred dollars a year to the twelve hundred Bell still contributed toward his salary. "Congratulations on the new magazine," Bell wrote him as the new century began in January 1901. Grosvenor brightened the format and shifted from a solid page to two columns. He turned back the insistence of S. S. McClure, by now a recognized authority on magazine policy, that the magazine should play down its sponsorship by the society and take to the newsstands. On Bell's urging, Grosvenor developed and tested his own talents as a magazine writer (though he much preferred editorial work) by getting a number of his articles accepted by other magazines. In 1902, while as always calling for still more illustrations "of the dynamical order," Bell congratulated his son-in-law on the development of the magazine. "Your hand shows more in it than usual," he wrote approvingly of the latest number.

In 1898, after Bell became president of the society but before Grosvenor became editor, the magazine had published a photograph of a comely Filipino woman in a topless costume. The magazine's prim cover, if not the subject's, seemed to establish the display as being all for anthropology. By 1903, when another such photograph came up for Grosvenor's editorial consideration, the magazine had become more general in its readership, and the young editor hesitated. But Bell gave his approval; and so prudery yielded to geographic truth, then and thereafter.

427

Whatever that policy may have done for the circulation of the *National Geographic Magazine* and its readers, the breakthrough into mass membership did not come until 1905. One desperate morning in December 1904, lacking eleven pages of copy for the January issue, Grosvenor received a providential package from a Russian explorer. It contained fifty splendid photographs of Lhasa, the first ever taken of the mysterious Tibetan capital. Grosvenor filled the eleven pages solidly with this bonanza. In February he learned from his cousin, Secretary of War Taft, that the magazine could have free use of photographic plates from the department's forthcoming report on the Philippines, and so the April issue ran thirty-two pages of them. It had to be reprinted to accommodate the flood of new members. That year, membership rose from 3400 in January to more than 11,000 in December. Bell could now end his annual $1200 subsidy. "There is no reason why the membership should not reach ten times ten thousand," he wrote his son-in-law, "in which case the magazine would become one of the most important and influential journals in existence." The cardinal principles, he emphasized again, would be accuracy, general interest, and "dynamical" pictures. The hundred-thousand mark was passed by 1912.

Late in the summer of 1899, Bell had raised the question of "a building — a library, worthy of the Society, and headquarters a little more respectable than . . . the two little lumber rooms in the Corcoran Building." A year later, the society bought a lot at the corner of Sixteenth and M Streets; and in the spring of 1902 Melville Bell Grosvenor, aged five months, held a trowel at the cornerstone laying of what was to be Hubbard Memorial Hall (toward which the Charles Bells contributed half the cost). In the spring of 1903, over Mabel's remonstrances, Bell resigned as president of the society. The new building was nearly done, Bert had taken firm hold of the magazine, and all in all, Bell felt he had done his duty.

In the fall of 1903, Gilbert Grosvenor heard of an Agriculture Department botanist named David Fairchild, who had been traveling throughout the world for the Office of Seed and Plant Introduction, and who had just returned from a trip up the Persian Gulf to Bagdad. When Grosvenor met Fairchild to arrange a lecture for the National Geographic Society, he was startled to find that the plant explorer did not have a long, gray beard, but was only thirty-four; Grosvenor had confused him with another botanist. Bell attended Fairchild's lecture and soon afterward invited Fairchild to one of the Wednesday evening dinners and discussions. Fairchild was won by Bell's "vigor and kindliness." Many years later he remembered feeling "immediately at ease, as one does with any really great and simple character."

Late in November 1904, Elsie Grosvenor invited Fairchild to dinner, where he sat beside Marian ("Daisy") Bell, just back from studying art with Gutzon Borglum. "I was fascinated by her," he remembered later; and when Elsie somehow mentioned in parting that her sister was not engaged, he left with his mind "in a whirl." David and Daisy saw much of each other that winter, the same winter in which Daisy helped save the Freer Collection for the Smithsonian. In the spring of 1905, the courtship accelerated. Daisy had an electric automobile, and she and David drove through the countryside "at incredible speed, twelve miles an hour." They were married late in April, and lived with the Bells until their house was built a year later on a forty-acre tract of woods in Maryland.

The Fairchilds eventually added a grandson and two granddaughters to the roster of Alexander Graham Bell's descendants. Alexander Graham Bell Fairchild came first, on August 17, 1906. "It does look like the little picture of our own baby," Mabel wrote Alec. The future entomologist was "a regular Bell baby . . . no great beauty, but a nice plump, strong, lusty baby, thoroughly satisfactory." Sandy, as he was eventually nicknamed, was to be "scientifically brought up, and so can't be coddled when he cries, which may be all right for the youngster, but is hard on his grandmother." Thus, once again, began the counterpoint of heredity and environment.

It had been morning at Beinn Bhreagh when Charles Thompson roused Bell and read him David Fairchild's telegram announcing the birth of Alexander Graham Bell Fairchild. Still befuddled with sleep, Bell listened patiently and then asked, "Can it fly?" Later he explained that he had been dreaming of kites.

32

Castles in the Air

By the time Sandy Fairchild was born in 1906, his grandfather must have dreamed of kites many times. Since 1899, said Bell in a 1903 *Geographic* article, "I have been continuously at work upon experiments relating to kites. Why, I do not know, excepting perhaps because of the intimate connection of the subject with the flying-machine problem."

"Why, I do not know" — the phrase suggests something deeper than reason. Nowadays we look first to childhood for the roots of such impulses. In his article Bell referred to the string-tailed, diamond-shaped paper toy of "my younger days" and pointed out defensively that "in Asia kite-flying has been for centuries the amusement of adults." His childhood recollections elsewhere mention no kites, however. The most suggestive image among them seems to be that of a rather solitary boy, lying in the heather on Corstorphine Hill and watching birds soar far above him.

Kite flying as a pastime had obvious attractions. It gave Bell moderate exercise in bracing weather amid soul-satisfying scenery. It made sporting demands on his skill and ingenuity. It had the excitement of unpredictability. And because Bell wanted his kites to show up clearly in photographs validating their performance, he covered them in bright red silk. This enhanced the drama of their evolutions, and Bell had an eye for drama.

But Bell also had an explicit and plausible rationale for his kite flying. Like the Wrights and other aeronautical pioneers, Bell chose to experiment with light, wind-supported prototypes before risking men and engines. Mindful of Otto Lilienthal's death in a glider crash, Bell — unlike the Wrights — delayed even introducing a pilot into his craft until he had achieved a high degree of safety through trials with towed or tethered kites. Stability was his concern before all else, including speed. Indeed, as his mind fastened on the sine qua non of safety, a corollary goal became that of *reducing* the speed necessary to maintain altitude. His ideal in 1906 was an airplane that would not have to land at all, given a moderate breeze, but could be tethered a few feet above the ground like a kite, while the pilot climbed down a rope ladder. In the event of an engine failure, the pilot could thus "cast anchor" and dismount with dignity. Bell's near obsession with stability and gentle descent was the premise that made logical what would otherwise seem an irrational indifference to wind resistance or drag. It might be objected that Bell's goal was already being achieved more simply in the dirigible (that is, steerable) balloon. But Bell felt that lighter-than-air craft could not so well contend against adverse air currents.

Having jettisoned his laborious airfoil experiments of the nineties, Bell began his kite experiments in 1898 with a variety of fanciful designs: "spool kites" having cylindrical cells with a long rod as axis, "kites with radial arms" having a starlike cluster of two or more blades radiating out from each end of the axis, and so on. Then he shifted to the box kite, which the Australian Laurence Hargrave had developed in 1892 for the purpose of achieving manned flight. The Hargrave kite resembled a long, narrow box with its ends and a wide section of its middle removed, thus forming two open-ended box cells separated by an empty framework. The wider the space in the middle, the greater was the kite's stability. By the spring of 1899 Bell had built one of these — the "monster" mentioned in an earlier chapter — with end cells the size of small rooms. No wind ever swept over Baddeck with enough force to raise the monster from the ground, which was fortunate for Baddeck.

One reason why the monster remained earthbound was suggested by Simon Newcomb in an article called "Is the Air-ship Coming?" which appeared in the September 1901 issue of *McClure's Magazine*. Newcomb answered his own question in the negative, arguing that if the scale of a working model were enlarged, the wing surface would increase only as the *square* of the linear dimension, whereas the volume, and hence the weight, would increase as the *cube*. Newcomb's premise, as Bell correctly observed, was false, since it assumed that every element in the craft would be increased proportionally in each dimension — for example, the thickness of the wing covering. Newcomb also assumed no change in design as the weight and speed of the airplane were increased. Neither assumption made sense. Still, as Bell also perceived, Newcomb did underscore the fallacy of supposing that a model would work just as well if it were simply duplicated on a larger scale, as in the case of Bell's monster Hargrave kite.

Newcomb could not resist the traditional sneer at windmill-tilters. "No builder of air castles for the amusement and benefit of humanity could have failed to include a flying-machine among the productions of his imagination," he remarked. Not only Bell but also his closest friend, Samuel Langley, were thus publicly awarded honorary degrees in aerial architecture. Many years later, Gilbert Grosvenor remembered that article as having cooled Bell's regard for Newcomb thereafter.

Even before Newcomb's article appeared, Bell had recognized and begun circumventing the supposed obstacle by flying several smaller Hargrave kites linked by a single line. From this he had passed to connecting several small kites by a rigid framework, which is to say building multicelled kites. After Newcomb's article, Bell became more explicit about the advantages of the multicelled kite, pointing out that when cells were joined edge to edge, they could share one piece of framework where two had been used before. Thus a many-celled kite actually weighed less in proportion to the area of its surfaces than did a separate cell.

Meanwhile Bell considered the possible forms of a unit cell. In August 1899, on the subject of triangular kites, Bell had written in his notebook: "There is a good deal in equilateral forms. Quite independently of this — a triangle is *braced*." By the summer of 1901 he made use of this fact by giving the kite cell the shape of a triangular prism rather than a box. The open ends, being triangles, needed no bracing. In contrast, the Hargrave box cell required diagonal bracing across its open ends to prevent distortion, and this increased its weight and wind resistance. In theory the triangular type should not have flown as well as the box shape, but in practice its gain in strength and its lessened wind resistance or drag more than made up for its less effective conformation. Bell was so taken with this form that in March 1902 he jotted down an admonition to himself: "Avoid rectangular elements — let everything be built up of equilateral triangles."

A few days later, at his Washington home, Bell amused himself by making tetrahedral cells out of folded paper and fitting them together to see what kind of forms they produced. His interest remained so casual that he did not look up the geometric name for that shape. Not until August 25, 1902, at Beinn Bhreagh, did the form suddenly capture his full attention. What he wrote down then in a rush of inspiration may well have been maturing for some time in his subconscious mind. To Bell's conscious mind, nevertheless, the insight evidently came with exhilarating suddenness.

A "triangular stick" had been made in his lab that day and then trimmed into what he gropingly called "a perfect equilateral cone each of the four faces constituting equilateral triangles." The word "cone" seeming a misnomer, he tried again. "A figure composed of 4 equilateral triangles having 4 triangular faces bounded by 6 equal edges. Wish I could describe this solid form properly as I believe it will prove of importance not only in kite architecture — but in forming all sorts of skeleton frameworks for all sorts of constructing — a new method of architecture. May prove a substitute for arches — & bridge work generally." He tried to sketch a view but noted impatiently, "Can't draw it." Then he turned his imagination loose again. "Whole structure so solid & so perfectly braced by its construction that it may be treated as a solid body. Only needs support at the three extremities of its base. Structures of this sort may be used in place of arches for bridges — ceilings of large buildings &c. It lends itself to metallic structure. All the parts can be made of metal — & made cheaply." And he sketched standardized joints, parts of which "could be stamped out of sheet metal if desired." At the bottom of the page, a hastily sketched locomotive puffed confidently over a bridge of the new construction.

Presently Bell learned that the form was called a tetrahedron.

About 1900, Mrs. Bell had come across a jaunty jingle in a newspaper, typed a copy, illuminated the border, and hung it in her husband's laboratory at Beinn Bhreagh:

> If things seem a little blue,
> Keep on fighting.
> Stay it out and see it through,
> Keep on fighting.

And so on for several stanzas.

Bell's spirit seemed lifted whenever he read that simple incantation. But the inspiration of tetrahedral construction was an even better tonic for him in the fall of 1902. The tetrahedral kite cell excitingly combined simplicity, strength, and lightness with aerodynamic stability. By November, Mabel found her husband "continually more wrought up over his kite experiments than I like." He had just constructed a large, H-shaped kite, each of its members constructed of tetrahedral cells, and the next day he photographed it soaring like a stubby dragonfly against the sky. A few brief experiments emboldened him to unveil the tetrahedral principle of construction and its aeronautical promise in a paper for the National Academy of Sciences in April 1903. After a few more experiments that spring, he published a copiously illustrated version in the *National Geographic Magazine*.

By the fall of 1903 Bell felt sure that if he himself did not get a motor-driven aircraft off the ground within a year, someone else would; and

433

1902 Aug 25 — Monday — at Bodg.

Triangular stick was made in Lab,
today as shown on p.119 and section
cde was cut out as suggested. It
proved to be a perfect *equilateral cone* ~~composed~~
~~each face~~ each of the four faces ~~constituting~~
equilateral triangles.

A figure composed of 4 equilateral triangles
having 4 ~~its~~ triangular faces bounded by
6 equal edges — With I could describe
~~this~~ solid form properly as I believe
it will prove of importance — ~~it not~~
only in Kite architecture — but
in forming all sorts of skeleton
frameworks for all sorts of constructive
architecture.
— A new method of
May prove a substitute for ~~trusses~~
arches — & bridge work generally.

apex
a   b
elevation.

apex···
a   b
Plan or view
from above.

Can't draw it

Base abc.
Whole structure so solid & so perfectly braced
by its construction — that it may be treated as a
solid body, Only needs support at ~~the~~ the three
extremities of its base (abc) — Structures of this sort

1902 Aug 25 —— Monday — at 1205.

may be used in place of arches for bridges ceilings of large buildings &c. ——

It lends itself to metallic structure — all the parts can be made of metal — & made cheaply.

This part can be cast solid

It needs no solid core.

Can we not try it by casting it in lead?

Flat parts could be stamped out of sheet metal if desired,

he made his prediction public through a direct interview in the *Boston Herald* and articles by Gilbert Grosvenor in the *New York Herald* and the *Popular Science Monthly*.

Bell's confidence did not rest on his own progress alone, but also on that of his friend Langley. After the spectacular success of his steam-driven model in 1896, Langley had proposed to leave practical development to others. Then in 1898 the War Department decided to allot fifty thousand dollars toward the development of a man-carrying plane, and Langley was persuaded to undertake the work. Donating his own time and twenty thousand dollars or so of his own money to the cause, Langley also engaged a young engineer named Charles M. Manly to design and make a gasoline engine for the aircraft. Manly's five-cylinder, water-cooled, radial engine of fifty-two horsepower, weighing only 125 pounds, has been called the world's first modern aircraft engine. In August 1903, a quarter-sized model of both engine and plane flew successfully.

Langley, hypersensitive to ridicule, tried to keep the press away from his trials but succeeded only in alienating it. When on October 7, 1903, the full-sized model with Manly aboard "slid into the water like a handful of mortar" from its houseboat launching platform, the air was filled at least with journalistic catcalls. By the time of the second trial on December 8, 1903, Langley's "Buzzard," as the press dubbed it, was already a national laughingstock. The second dunking of the "Buzzard," caused perhaps by faulty launching, perhaps by collapse of the wings, set off what Mark Sullivan called a "triumphant ecstasy." After the mishap, while Langley was looking after the pilot Manly, a clumsy tugboat crew tore the plane to pieces in grappling for it.

The accident did not hurt Manly, but many believed that in time it killed Langley. Newspaper ridicule not only ended any further financial aid to Langley but also, wrote Bell afterward, "broke his heart . . . [and] contributed materially to the production of the illness that caused his death." Bell delivered a eulogy at Langley's burial in Boston early in March 1906. "His flying machine never had an opportunity of being fairly tried," he said, but "the man and his works will permanently endure."

Langley had lived long enough, however, to see his dream realized by others and to be recognized on his deathbed as "a pioneer in this important and complex science" of aeronautics. For in an irony of technological timing reminiscent of Elisha Gray's alleged independent invention of the telephone, Wilbur and Orville Wright at Kitty Hawk, North Carolina, had made history's first powered, man-carrying, heavier-than-air flight on December 17, 1903, only nine days after Langley's fatal failure.

Meanwhile the Langley tragedy muted Bell's optimism, at least in public, and census work interfered with his aeronautical experiments. In an interview with his longtime friend, the journalist Frank Carpenter, in 1904, Bell explained that he was merely trying to determine the best forms and structures for flight. Only after that would he "be ready to attempt to invent a flying machine." He had, however, "ascertained the shape of the cell . . . the brick, as it were, out of which the flying house must be made."

By 1903 Bell had shifted from black spruce sticks to aluminum tubing for the framework of his tetrahedral cells. Baddeck girls were hired to sew the red silk covering over the frames; later, it was cemented on. After trying several larger sizes, Bell settled on a small, ten-inch cell as the standard unit. One great advantage of a multicelled kite was that the center of pressure on each small cell moved only a fraction of the cell's length as the angle of flight changed. In a single-celled kite of the same overall dimensions, the pressure center would move by a comparable fraction of the whole kite's length and therefore many times as far by absolute measure. So a smaller cell unit gave Bell's kites more stability.

Unfortunately these massed cells sacrificed some lifting power and incurred enormous drag. Bell steadfastly aimed for stability first and lift decidedly second. He considered drag to be of no consequence at that stage. But his immediate goal was a kite that would carry a man and additional weight equivalent to that of an engine. So he found himself building huge kites and waiting impatiently for strong winds.

Within his self-imposed confinement to massed tetrahedral cells, Bell labored by trial and error to find the best possible configuration. By the end of 1904 he broke away from the Hargrave pattern of identical forms at front and rear, joined by an empty framework in the middle, and instead built a single multicelled wingspan with a short horizontal tail. In flight it suggested a great soaring bird, and so Bell called it the *Oionos*, the Greek word for a bird of omen. Later he installed a pendulum-governed mechanism to correct the flight angle automatically, through adjustment of a hinged tail.

Various other configurations were tried at much expense in time and money. After one failure late in 1904, Bell wrote his wife that "we derive great comfort from the laboratory motto 'Keep on Fighting' and etc. Mr. Ferguson, in the depth of his despair, quoted it — and was willing to try again." In the fall of 1905 Bell built a new kite house for what was to be the season's masterpiece, two banks of cells side by side, with a space between to carry a man. Then he waited anxiously for a wind strong enough to lift it. A storm came at last in mid-November. But it made the bay impassable, and the men from Baddeck chose not to come around by land. Bell arrived at the kite house full of hope and found it deserted. "He looked gray when he came home," reported Mabel, "wrote a short note dismissing the staff and closing the laboratory, turned his face to the wall and never spoke again that day or night." The storm continued the next day, however, and Bell managed to collect enough men for a successful test of a smaller model. All the men were hired back in short order.

That December, Bell put all available cells, thirteen hundred of them, into one closely packed bank and called it the *Frost King*. Shortly after Christmas he managed to photograph 165-pound Neil MacDermid dangling nervously from the *Frost King*'s line at a height of about thirty feet in a mere ten-mile breeze. As Bell reported to the Washington Academy of Sciences a year later, MacDermid's unscheduled flight satisfied him that he could build "structures composed exclusively of tetrahedral winged cells that will support a man and an engine in a breeze of moderate velocity," and so it "brought to a close my experiments with kites." In the intervening year, he said, he had looked into the question of motors and propellers, and his assistant Hector McNeil had devised techniques for the mass production of cells and connections. In fact, he added, McNeil "is now taking up the manufacture of tetrahedral cells as a new business."

Since inventing the telephone, Bell had tended more or less consciously to re-create the pattern of events that had led to it, particularly in seeking another Thomas Watson, some able young technician who could help sharpen his ideas and reduce them to practice. As it happened, the way Bell came upon tetrahedral construction did repeat a pattern, but an unintended one. Just as the telephone had grown naturally out of Bell's harmonic telegraph experiments, which themselves yielded nothing of practical importance, so tetrahedral construction grew naturally out of Bell's kite experiments, which in their turn would lead to nothing useful.

There was another similarity. Like the concept of the telephone, that of tetrahedral construction was entirely Bell's own. It had to be, and for the same reason: its fundamental principle was so beautifully simple that it could scarcely have been divided between collaborators.

In America, where timber was cheap, and time and labor were dear, bridge builders had long relied on the two-dimensional truss, a rigid combination of small timbers in which each member sustained tension or compression but not bending. Such trusses consisted of variously interconnected or overlapping triangles, the only polygons that cannot be deformed without changing the length of one or more sides. Bell's tetrahedron was a truss lying in more than one plane. As he put it in his National Academy paper of 1903, "it is not simply braced in two directions in space like a triangle, but in three directions like a solid. If I may coin a word, it possesses 'three-dimensional' strength; not 'two-dimensional' strength like a triangle, or 'one-dimensional' strength like a rod. It is the skeleton of a solid, not of a surface or a line." As in the case of the telephone, the principle was common knowledge. Even the form had appeared in the braced timber trestlework common on Western railroads. But Bell seems to have been the first to see it as the distinct unit of construction now called a "space frame."

What was also remarkable, he saw its practical potentialities clearly from the moment of its first formulation, as already quoted. And he told the world about them, or at least as much of the world as he could reach through the National Academy of Sciences and the *National Geographic Magazine,* in his 1903 article on tetrahedral kites. The tetrahedral cell, he wrote,

is applicable to any kind of structure whatever in which it is desirable to combine the qualities of strength and lightness. Just as we can build houses of all kinds out of bricks, so we can build structures of all sorts out of tetrahedral frames, and the structures can be so formed as to possess the same qualities of strength and lightness which are characteristic of the individual cells. I have already built a [sheep] house, a framework for a giant wind-break, three or four boats, as well as several forms of kites, out of these elements.

In one respect the analogy with the telephone story did not hold. Whereas Gardiner Hubbard had regarded the telephone as a mere diversion from the harmonic telegraph, his daughter Mabel at once grasped the significance of tetrahedral construction. She got her husband's patent attorney Philip Mauro to see him about taking out a patent. "Alec would never have done anything more than talk about it, I am pretty sure," she wrote later. Perhaps nonplussed by the simplicity and sweep of the idea, the Patent Office delayed its decision. But on September 20, 1904, more than a year after the application, Bell received Patent No. 770,626 for "Aerial Vehicle or Other Structure." This and a kite patent granted earlier that year were the first patents Bell had received since a graphophone patent in May 1886 and the first in his name alone since an unimportant telephone patent of 1881.

As in the first telephone patent, the space frame patent's most significant claims were almost buried by what turned out to be inconsequential addenda, in this case concerned with "aerial vehicles." But its first claim specifically covered "a structure whose framework is composed essentially of skeleton tetrahedral elements combined with means whereby the adjacent elements are directly connected at two or more of their corners."

Mabel urged her husband to start commercial development of the new construction system. In October 1904 he proposed that his cousin, the Washington banker Charles Bell, set up a patent-holding corporation like the one Charlie had organized for the Volta Associates' graphophone patents. Mabel seconded the motion. "Here," she wrote Charlie,

is surely a matter worth encouraging by practical men who have no faith in flying machines. Alec has already constructed an arch with his tetrahedral cells . . . certainly a beautiful-looking thing. Alec says that it is portable — can all be taken to pieces & carried in a handbag — put together again, used & then taken up. Certainly such a bridge would be of enormous value now to the Russian or Japanese armies. . . . Only such details as always arise for consideration in the making of practical commercial structures require attention. The invention is now ready for demonstration and practical application.

But after twenty years as a banker, Charlie Bell had grown wary of aerial conveyances, figurative as well as literal.

Early in 1906 Mabel secured the opinion of an able consulting engineer. He did not think Bell's system an improvement over that used for heavy bridges. But he thought it could be "usefully and profitably applied in a wide range of lighter construction," such as footbridges, light highway bridges, and water towers, and as reinforcing in concrete arches, beams, and columns. He also pointed out the advantages in the mass production of standardized structural members. An engineering cousin of Mabel's "talked with a lot of expert bridge and concrete men and they all speak well of the system." Charlie still hung back. By mid-May 1906 it was apparent to Mabel that, corporation or not, a man was needed to work out minor details of construction as they came up, take out supplementary patents, and push the system wherever possible. Hector McNeil was an able workman but not an engineer. The time had come to find her husband another Tom Watson. And within a month she had found him.

Bell's first son was born and died in 1881, his second in 1883. The young engineer Mabel found, Frederick W. Baldwin, was born in 1882. His Irish forebears had settled in Canada in 1798, and his grandfather had been prominent in Canadian politics. Since his preparatory school days, young Baldwin had borne the nickname "Casey," an allusion to the popular verses of "Casey at the Bat" and a tribute to his baseball prowess. Casey Baldwin was a fine all-round athlete, not big but quick and strong. As an engineering student at the University of Toronto, he captained the 1905 football team, one of the greatest in the school's sporting annals. Sailing, however, was his lifelong passion. At sixteen he had shipped before the mast on a timber-laden windjammer bound for Liverpool, and in 1905 he was a crew member of the challenger for the Canada's Cup in yachting. Meanwhile he began reading the growing body of technical literature on aeronautics.

The Bells had once wanted to adopt Arthur McCurdy's second son, John Alexander Douglas, whose mother had died before he was two; and he remained a special favorite of theirs. Mabel wrote Douglas McCurdy regularly after he enrolled as an engineering student at Toronto. In the spring of 1906, as the time drew near for Douglas's summer return to Baddeck, she asked him to bring back any bright young student he encountered who might be able to help her husband. Soon afterward the sociable Casey Baldwin, just about to graduate, dropped in on Douglas, was invited to Baddeck, and agreed to vacation there for a couple of weeks. He ended by making Beinn Bhreagh his home until his death forty-two years later.

Even before Casey arrived, Mrs. Bell had written Douglas McCurdy of her delight at what he had told her of his friend. "He is exactly the kind of fellow from your description that I want associated with Mr. Bell." Since 1903 Bell had wanted to build a full-sized observation tower as a prototype of tetrahedral construction. He wanted to put a motor in his

Frederick W. ("Casey") Baldwin

kites. He wanted to set up an electric power plant. And he was developing an interest in motor boats. Casey Baldwin's newly earned degree was in mechanical and electrical engineering.

From the day they met him, the Bells took Casey Baldwin to their hearts. He was all they would have wanted their sons to have been: modest, cheerful, lively, intelligent. Best of all, he liked them too. Casey's young mind had absorbed and incorporated the engineering discipline and knowledge that Bell had never brought himself to acquire. Yet the young engineer admired the older man's great achievements, responded enthusiastically to the nuggets of insight that Bell's imagination still yielded, and let the tailings pass with good-humored tact.

As if to round out his preparation for what turned out to be a life's work, Casey took some shop courses in the 1906 summer session at Cornell University. When he came back that fall, he set to work on the observation tower. Well versed in the strength of materials and the calculation of stresses, Casey planned and erected a tripodal tower on top of Beinn Bhreagh. The structure itself formed a gigantic equilateral tetrahedron

442

The opening of the tetrahedral tower, August 31, 1907

with seventy-two-foot legs, each assembled from four-foot tetrahedral cells. Some 260 cells were used, all made of half-inch iron pipe. Casey tested one cell in compression up to two tons without observing any excessive strain. To raise the tower, two legs were assembled and joined at an angle on the ground, and their vertex was then gradually lifted as the third leg was built up cell by cell from underneath, so that all the work was done on the ground. That especially pleased the safety-conscious Bell.

The finished tower, complete with observation platform and an access stair inside one leg, weighed less than five tons. Its very appearance was eloquent of lightness and strength. During construction, one seventy-two-foot leg had been supported by its ends in a horizontal position and had been found to sag only three-eighths of an inch at its midpoint. Casey pointed out that besides having remarkable lightness, strength, and rigidity,

443

the tower had been easily assembled from mass-produced parts without skilled labor, and that there was enough redundancy in its members for any one of them to be inspected and replaced without reinforcing the structure meanwhile. When, having outlasted its purpose, the tower was taken down more than a dozen years later, it had passed through the storms of those years without needing any repairs at all.

Bell did his best to fulfill the tower's purpose of demonstration. On August 31, 1907, he presided with his usual dramatic flair over an elaborate opening ceremony, assisted (under some duress) by his small, shy grandson Melville Grosvenor. Scores of guests climbed to the flag-bedecked observation platform to view the Bras d'Or Lake six hundred feet below. Bert Grosvenor kept his camera clicking, and the *National Geographic* ran a generous portfolio of the results. And Casey Baldwin published an article on the tower in the *Scientific American* that October.

Yet for some reason the idea's time had not come. The example seemed lost on structural engineers. Perhaps the emphasis on the vertical had obscured the significance that Baldwin's horizontal deflection test held for the building of wide roof spans and light domes. (Mabel's consulting engineer in 1906 had missed that point, it may be noted.) At any rate, Bell understandably gave up when nothing came of all the publicity. Not until the late twenties, several years after Bell's death, did the great towers of the George Washington Bridge over the Hudson reintroduce the space frame into American structures, and the designer of those towers apparently did not take his cue from Bell's work. Nor did the engineer Konrad Wachsmann, though his space frame system of the 1940s, which helped bring space frames into their own for roof spans, claimed the same principal advantages as Bell's: strength with lightness, and easily assembled, mass-produced parts.

In short, Alexander Graham Bell turned out to be the Gregor Mendel of space frame architecture.

In 1901 Bell asked his cousin Chichester Bell to join him again in "work upon some subject . . . that will pay." "Neither you nor I will ever again be as young as we are now," wrote Alec, "and if we do not make this effort now, *we never will.*" If the proposed reincarnation of the old Volta Association came up with nothing of value within two years, he promised, they would acknowledge "that the vigor and energy of our youth has departed for good" and would shut up shop. But Chester declined to uproot his English household in pursuit of "some subject" unspecified; and after renewing the appeal in 1902 and 1903, Alec gave up.

Mabel had been told of the proposal and was touched by her husband's admission of encroaching age. "What he is trying so hard to do now," she wrote Bert Grosvenor in 1904, "is creative work, which is essentially young people's work. He is now in the prime of life, but it is not more than this, and he has nothing more to look forward to." She was therefore delighted by the affinity that grew up between her husband and young Casey Baldwin in the fall of 1906. Later that winter she proposed that they form "a Tetrahedral Association like the Volta Association." Her idea, she wrote in her journal later, was "*not* the production of a flying machine, but the reducing to commercial use . . . of the Tetrahedral System of Construction in *any way*. Its use in the manufacture of flying machines was to my mind and purpose but one of many possible ways."

But the success of the Wright brothers had not loosened the grip of aeronautics on Bell's mind. He accepted the news with remarkable good cheer and more readily than most of his contemporaries. After their historic first powered flight at Kitty Hawk in December 1903, the Wrights had decided to protect their advantage in dealing with their own and other governments by withholding the details of their machines and discouraging public observation of their test flights. For more than two years, therefore, stray reports of their doings were ignored or downplayed by the press as tall talk. Bell knew better. Octave Chanute, the Wright brothers' mentor, gave him unimpeachable testimony. David Fairchild attended one of Bell's Wednesday evening dinners with Newcomb, Langley, and Chanute.

"What evidence have we, Professor Chanute, that the Wrights have flown?" asked Bell.

Chanute replied, "I have seen them do it."

Fairchild never forgot the "tremendous impression" those words made on himself and the others present. A guest at a Wednesday Evening in the fall of 1905 likewise remembered vividly Bell's rising to announce in "an unforgettable voice," "I hold in my hand a telegram telling me that the Wrights have been in the air for thirty minutes. *Thirty minutes*, gentlemen!"

Early in 1906, about three weeks before Langley died, a *New York World* reporter interviewed Bell. "The impossible [has come to pass] in aerial navigation," said Bell, "and I am proud of the fact that America leads the world in that matter. To the Wright brothers, of Ohio, belongs the credit." This was largeness of spirit in a man who had seen his dearest goal achieved by others. The Wrights themselves seemed a little surprised by the warmth of Bell's friendliness when they met him in Washington later that year.

Bell's graciousness may have been fortified by his feeling that the Wrights had left him something important to do for aeronautical stability. He had

445

a clear mental picture of a tetrahedral airplane hovering steadily and then settling to the ground as gently as a butterfly. Often in the small hours, that alluring vision must have floated before him as frail and elusive and yet as real as a moth in the moonlight. An earthbound tetrahedral tower made a satisfyingly dramatic effect, to be sure, but nothing like that of an airborne structure. Anyway, the Beinn Bhreagh tower could now stand up for itself. So Bell remained air-minded.

By the fall of 1907 two more young recruits had joined Baldwin and the newly graduated McCurdy at Beinn Bhreagh. Twenty-five-year-old Lieutenant Thomas E. Selfridge, a 1903 graduate of West Point, had raised his professional sights from artillery to aviation. When he sought out Bell in Washington in the spring of 1907, Selfridge's aeronautical enthusiasm won him an invitation to see the kites fly at Beinn Bhreagh. There his quiet, courteous, good-humored manner charmed Mrs. Bell. Her husband got President Roosevelt to detail Selfridge to Beinn Bhreagh as an observer, beginning in September 1907. The lieutenant and the two engineers introduced a new precision to Bell's meticulously recorded but poetically calibrated measurements. Now wind velocity was read from an anemometer, not the look of the waves; altitude from a clinometer rather than mere squint and guess; pull from a spring balance rather than the snapping of a manila rope or the pulling loose of a nailed cleat.

September 1907 also brought twenty-nine-year-old Glenn H. Curtiss, whose father, a Hammondsport, New York, harness mender, had died when Glenn was six. The boy had no brothers and only one sister, who became deaf. Curtiss, shy and reticent as a boy, dour as a man, found release in machinery and speed. After opening a bicycle shop, he won every race at numerous bicycle meets and county fairs for three years. At the turn of the century, after his first bicycle defeat, he began making and racing motorcycles. In 1903 he won the national championship and in 1907 rode the fastest mile that any man had traveled till then, averaging 136 miles per hour. Meanwhile he incorporated the G. H. Curtiss Manufacturing Company. Among his customers for light motors were several makers of dirigible balloons. In 1906 Bell, preparing to motorize his kites, ordered a motor from Curtiss. But Curtiss, who till then had little respect for such schemes, took his time about the order and even so did not produce a satisfactory model. Bell ordered a larger one and offered Curtiss twenty-five dollars a day and expenses to deliver and demonstrate it in person. A visit to Beinn Bhreagh in July 1907 changed Curtiss's mind about heavier-than-air flight and inspired a liking in him for the place, the people, and the plans. Hence his return in September as the group's engine expert. During his succeeding three months there, Curtiss seemed rather aloof from the camaraderie of the other young men, who were family-proud college graduates; but something, perhaps in part the deafness of his sister, gave him a bond with Mrs. Bell.

The five-man association Mrs. Bell envisioned that September would, like her earlier proposal, have been confined to tetrahedral construction; but Bell insisted that its object be to develop a practical flying machine without restrictions as to design. He further suggested that his tetrahedral man-carrying kite be finished first, after which each associate in turn, with the help of the others, should build a flying machine of his own design. The others agreed, and so the Aerial Experiment Association began its formal existence on October 1, 1907, for a term of one year. Mrs. Bell was to furnish capital up to $20,000, and Bell was to make his laboratory and other facilities available without charge. Bell would serve as chairman without salary, Baldwin and McCurdy would each get $1000 a year, and Curtiss would draw pay at the rate of $5000 a year when on the scene and half that when away. Selfridge, already on full pay as an army officer, declined a salary.

"My special function, I think," noted Bell shortly before the official beginning of the AEA, "is the coordination of the whole — the appreciation of the importance of steps of progress — and the encouragement of efforts in what seem to me to be *advancing* directions." It was a role much like the one he had seen for himself in the old Volta Association. In this case, being less occupied with other matters than in the eighties, Bell filled that role with more continuity and absorption. He contributed significantly to the step-by-step improvement of successive aircraft and especially to the analysis and remedy of failures. He moderated discussions and rallied enthusiasm. He was, in short, an ideal chairman — or perhaps, in the case of that sports-minded team, an ideal coach.

True to the agreement, the associates completed a tetrahedral kite, the *Cygnet,* an assemblage of 3400 red silk cells looking something like a gigantic wedge of honeycomb. After an unmanned trial, Tom Selfridge on December 6, 1907, crawled into a central space in which he could see very little and do less; and a steamer towed the small boat bearing the *Cygnet* and its 175-pound inmate out on the whitecapped Little Bras d'Or Lake. Lashings were cut in a twenty-five m.p.h. wind, and the *Cygnet* rose gracefully at the end of its towline to a height of 168 feet, where it hovered steadily for seven minutes until the wind dropped. Then, to Bell's delight, it descended slowly and settled toward the water "gently as a butterfly," just as a cloud of the steamer's smoke shrouded it. But Selfridge could neither see his descent nor feel the touching down of the kite's pontoons, and so he did not disconnect the line in time, nor was it cut from the steamer, the kite's position being obscured. So the *Cygnet* was dragged to pieces, Selfridge diving clear in time to be picked up unhurt.

At Beinn Bhreagh months earlier, David Fairchild had already sensed that Baldwin and McCurdy felt little more than polite tolerance for Bell's vision of a tetrahedral aircraft, and that their real enthusiasm was for the biplane designs of Chanute and the Wrights. This must have been plain to Bell at the post mortem on the *Cygnet* that evening, when the others raised questions of control, speed, and wind resistance. He accepted the majority decision to experiment next with gliders of the Chanute type, while resolving on his own account to have more cells made for a powered version of the *Cygnet* to be tried the next summer. But the doubts of his boys must have depressed him temporarily. He read the "Keep on Fighting" verses to his laboratory workers. A few days after the rise and fall of the *Cygnet*, Mrs. Bell wrote her mother: "Father is jolly again tonight, yesterday he said he was an old man."

Far from sulking, however, Bell joined with his customary gusto in the work of those great days that now began for the Aerial Experiment Association. For the sake of a milder winter climate, the AEA at the beginning of 1908 moved its headquarters to the Curtiss shop at Hammondsport, New York. Because Mrs. Bell fell ill just then, Bell stayed with her in Washington. But he kept in constant touch through correspondence and occasional visits. After two months of manned glider trials, the four at Hammondsport undertook each in his turn to design and build an airplane.

In recognition of his daring in the *Cygnet* flight, Lieutenant Selfridge was given the first chance. Already well read in aeronautical literature, he made a last, thorough survey of available information. Chanute's writings were the associates' chief reliance, but Selfridge also requested data from the Wright brothers, who referred him to their patent of 1906 and an article published in France in 1903. If Selfridge obtained the latter items before completing his plane, he neither used them nor made a point of informing his associates. When on March 12, 1908, his light biplane (called the *Red Wing* because Bell's red silk was used as its wing fabric) was pushed on its runners out over the ice of Lake Keuka, it lacked the Wrights' wing-warping mechanism for lateral control. The approach of spring was threatening the ice; and so instead of waiting for Selfridge to come back from a trip to Washington, the others delegated Casey Baldwin (who had neglected to wear skates that day and so could do little on the ground) to take the *Red Wing* up for a trial.

The public had been invited in advance, and a good many had come. In an article later on, Octave Chanute therefore credited Casey Baldwin with "the first public exhibition of the flight of a heavier than air machine in

America." Before a tail strut buckled and forced him down, Casey had flown about a hundred yards at an altitude of ten feet. Five days later he tried again, but a slight wind tipped the *Red Wing* sideways, and it crashed, wrecking both plane and motor, though not injuring Baldwin.

To meet the obvious need for lateral control, Bell wrote Baldwin on March 20 that the tips of the upper wings should be hinged and moved in opposite directions to each other. This concept amounted to the important device called the aileron. A shoulder attachment, he added, could operate the flaps in the proper directions as the pilot instinctively leaned away from a dipping wing. As with sonic sounding and radium insertion, Bell's aileron turned out to have been anticipated by the French. Nevertheless, his independent conception of it demonstrated significant inventive capacity despite his sixty-one years. Baldwin also conceived it independently, his family later maintained.

The next plane, called the *White Wing* because the red silk had run out and been supplanted by white cotton, was Casey Baldwin's progeny. It embodied ailerons, a tricycle undercarriage, and other improvements suggested by experience. Between May 18 and 23, when it was wrecked in a crash, Bell saw each of the other associates take it up from a nearby meadow, Curtiss setting its distance record at a thousand feet.

Unlike the Wrights, the AEA welcomed public notice, and Bell's fame helped win it. Indeed, Bell made a note just before the *White Wing*'s flight: "Must get something ready for Associated Press, otherwise everything done by Baldwin et al. will be ascribed to me." But it was the third AEA plane, Glenn Curtiss's *June Bug*, that climaxed the summer's public triumphs. Experience had led to further improvements in design and detail, including the sealing or "doping" of the porous wing fabric with a sort of varnish. In the last week of June 1908, Curtiss made several flights, one of which exceeded the one-kilometer distance recently stipulated for the *Scientific American* trophy, the first one offered for heavier-than-air flight. No one had yet taken it, so Curtiss and the others at once arranged a flight in formal competition for it on July 4. Bell was at Baddeck, but David and Marian Fairchild joined the thousand or so who watched Curtiss and his "strange, white, flying apparition" whir through the air to capture the prize.

Bell at once wired his patent attorneys to go over the *June Bug* for patentable features. After watching a flight and scrutinizing the machine, their man stressed the ailerons, shoulder control, tricycle undercarriage, and combination of steerable ground wheel and rudder. Before the month was out, Orville Wright complained that the ailerons infringed the wing-warping claim in the Wrights' 1906 patent. Bell pointed out that by warping each wing in a different way, the Wrights' technique created a

449

The *Red Wing*, March 9, 1908

The *White Wing*, May 18, 1908

The *June Bug*, July 4, 1908

difference between the drags of the wings and thus required corrective rudder action, whereas the AEA's ailerons avoided this complication. The Wright patent did claim "any combination whereby the angular relations of the lateral marginal portions of the aeroplanes [that is, wings] may be varied in opposite directions." Despite this claim and the prior French use of ailerons, the latter were included in Patent No. 1,011,106 for a "Flying Machine," granted on December 5, 1911, to all five AEA members as joint inventors. (Another patent on the same day went to Baldwin alone, by agreement of the others, in recognition of his special contributions.)

After the *June Bug* triumph, Bell prevailed upon Baldwin to help him get his tetrahedral "aerodrome" ready at Baddeck while the others helped McCurdy with his plane at Hammondsport. (Out of loyalty to Langley, Bell long adhered to Langley's word "aerodrome," coined at the suggestion of an authority on classical Greek, for what the rest of the world by 1908 was calling an "aeroplane" or "airplane.") McCurdy went ahead with a more powerful engine and improved ailerons. He "doped" the wings with a silvery compound used to rubberize balloon cloth, which led to his naming his craft the *Silver Dart*.

That September, ordered back to Washington for the army trials of the Wright machine, Selfridge volunteered to be Orville Wright's passenger on a required two-man flight. At Fort Myer, Virginia, on September 17, 1908, the plane crashed. Orville Wright was badly injured; Thomas Selfridge died of his injuries that night. His death was history's first airplane fatality. "I can't realize it, it doesn't seem possible," Mrs. Bell wrote her husband; "I miss the thought of him so. . . . Give my love to them all, and let's hold tight together, all the tighter for the one that's gone. Casey called me the 'little mother of us all,' and so I want to be." She offered to put up another ten thousand dollars for a six-month extension of the AEA, and the others agreed.

So McCurdy finished his *Silver Dart*, tried it briefly at Hammondsport in December 1908, and then shipped it to Beinn Bhreagh. On the afternoon of February 23, 1909, McCurdy and his frail, silvery contraption rose from the ice of Baddeck Bay before a sizable crowd and flew about half a mile at the rate of forty miles an hour. This was the first heavier-than-air flight in Canada, and the first by a British subject in the British Empire. The next day McCurdy covered more than four miles in a flight around the bay, the best flight yet made in an AEA machine. During the next couple of weeks he went on to make cross-country flights of up to twenty miles. After much trouble with the Curtiss engine, McCurdy made the last AEA flights on March 29, 1909, the final one being a full circle of three and a half miles at fifty feet.

The *Silver Dart*

After the flight of February 23, the jubilant Bell had invited the whole crowd to Beinn Bhreagh for a gargantuan spread of sandwiches and coffee. The Aerial Experiment Association had brilliantly achieved its goal; and though it did so without benefit of tetrahedral cells, Bell gloried in it nonetheless. The AEA had been more than a business arrangement to him and the other members. It had been a band of comrades on a great adventure — for all they knew, shaping man's future by their nightly discussions in Bell's tobacco-fogged study. It had been a circle of good friends, happily relaxing with fencing and billiards, songs and stories, or tea with Mrs. Bell before the great fireplace after a hard day's contest with wind and water. And for Bell it had been almost like having four brilliant sons to share his life and work. Now it was over.

Late on the last day of its specified term, March 31, 1909, Bell, Baldwin, and McCurdy, the only members on the spot, met nostalgically before the fireplace. As Bell pictured the occasion, "Casey moved the final adjournment, Douglas seconded it, and I put it formally to a vote. We hardly received the response 'aye' when the first stroke of midnight began. I do not know how the others felt, but to me it was really a dramatic moment."

Charles Bell was made trustee of the AEA's patents and other assets. The total investment in the AEA had consisted of thirty-five thousand dollars by Mrs. Bell and three thousand spent by Bell to issue a mimeographed *Bulletin* of the association's work. As payment for the latter debt, Bell took the *Silver Dart* and then turned it over to McCurdy and Baldwin. He gave Curtiss the *June Bug*.

Without regard to the AEA, Glenn Curtiss had just organized his own aircraft company. During the years that followed, he went on to fame as an aviator and wealth as an aircraft manufacturer. In his long-drawn-out litigation with the Wrights, he did not choose to rest his defense on the AEA patents; and so Bell was spared the ordeal of extended patent testimony in his old age. Bell did testify briefly as to his part in and knowledge of the Hammondsport doings, but declined to appear as an expert witness in 1914 on the sufficient grounds that he was no expert. The AEA patents having been declared free of interference in 1915, they were sold to the Curtiss Company early in 1917 by the trustee for $5899.49 in cash and $50,000 in Curtiss stock. Soon afterward, in order to resolve the continuing patent suits and thereby expedite aircraft production for war, the United States government bought the patents of both the Wright and the Curtiss companies. The Curtiss patents, among which the AEA patents were the only ones of real value, were reported to have fetched two million dollars.

After the AEA ended, Casey Baldwin and Douglas McCurdy stayed on at Beinn Bhreagh, where Bell persuaded them to form the Canadian Aerodrome Company and begin making planes for possible sale to the Canadian army. In the summer of 1909, the *Silver Dart* was joined by two more biplanes, *Baddeck No. 1* and *Baddeck No. 2*, embodying a number of improvements. But the army authorities insisted on holding the trials on an undulating cavalry field full of ridges and knolls. On August 2, 1909, after three miraculously successful flights, the *Silver Dart* headed into the sun for a landing, caught the edge of one knoll, slewed into another, and ended in a total wreck, out of which McCurdy and his passenger Baldwin crawled intact, except for a little skinning here and there. Official faith in the prospects of military aviation suffered much more heavily. When at the same field a few days later *Baddeck No. 1* had trouble with a new type of front elevator and sustained some damage, the chances of a government sale died.

There were more experiments and trials that fall and winter. Melville Grosvenor many years later recalled his excitement as a boy of seven watching them in a field a few miles from Baddeck, after which "we drove home at night under the stars, Grampy and the AEA boys singing all the way." But in the spring of 1910, Casey Baldwin and his wife Kathleen, whom he had married two years before, decided to go along with the Bells on a round-the-world trip; Douglas McCurdy went barnstorming with Curtiss; and the Canadian Aerodrome Company sideslipped into history.

Bell had not given up on tetrahedral aircraft. The death of Selfridge in the summer of 1908 confirmed his determination to make "aerodrome" crashes impossible. In August 1908, with Casey back from Hammondsport to help him, he put up a kite which flew "as though glued in the sky." This, he insisted, "rammed home" the lesson of stability. He leaned on Casey's training, judgment, and frankness in working out the proper configuration of cells for his long-planned self-propelled "drome." But Casey soon became absorbed first in the designing of pontoons for water takeoffs and then in hydrofoil boats, and the engine built by Curtiss for Bell's *Cygnet II* turned out to be much heavier than planned. The engine and propeller were installed, but McCurdy, as pilot, could not get *Cygnet II* to rise from the ice of Baddeck Bay in February and March 1909.

From July 1909 to February 1910, Bell worked on a new model, the *Oionos*, which represented a visible concession to the biplane principle, being a triplane with large triangular-prism cells in the spaces between the layered wings. The fact that even with a better engine it could not quite be got off the ground suggests that he did not concede enough.

Finally, in the winter of 1911–1912, Bell encouraged McCurdy, just back from his year with Curtiss, in a last dutiful trial of the tetrahedral approach. *Cygnet III* consisted of a bank of cells mounted on the standard running gear of a Curtiss plane, using the Curtiss front and rear control and tail, with a monoplane elevator in front. In March 1912, just after Bell's sixty-fifth birthday, McCurdy made several attempts to take off in *Cygnet III* from the snow-covered ice of the Bay. On the last trial the "drome" collapsed; and repairs obviously could not be made before the ice went.

A few days earlier, on March 9, 1912, McCurdy had got *Cygnet III* rolling along at about forty miles an hour through four inches of wet snow cover when he "felt some squirming" and cut the motor. A break in the track of the wheels showed that the "aerodrome" had been airborne for about one foot. On the strength of this, McCurdy wired Bell, "Congratulations on technical flight of tetrahedral aerodrome 'Cygnet III' this morning." But Bell was evidently not impressed. *Cygnet III* turned out to be his aeronautical swan song after twenty years of work and hope.

Bell and his associates customarily took pains to record every important trial photographically, and it would not have been strange if McCurdy had snapped a picture of that twelve-inch stretch of inviolate snow which attested the all-time world distance record for powered tetrahedral flight. But apparently he did not; and so, as the season passed, the evidence vanished with the snow and the dream.

33

Portrait of a Patriarch

The seasons of life passed like the seasons of the year. "It's cold here," Mabel wrote Alec from the White Mountains late in the summer of 1904, "the reddening sumac is to the fore and the coming season casts its shadow altogether too much ahead. Why can't the year be young forever, why can't we be?"

The aspect of autumn had come early to her husband with the premature graying of his hair and beard. When he visited the Deaf Mute Institute of Arkansas in 1895 at the age of forty-eight, a local reporter complimented him in print on being a "well preserved man" who "carries his seventy odd years well." He and his sprightly father were sometimes taken for brothers. But having achieved that plateau of venerability, he had at least the consolation of changing little in his last quarter century.

Mabel worried about her husband's weight. "I am always so happy to hear of your taking exercise," she wrote him in 1891. And for his own good she begged him ten years later, "Please, please do walk. . . . It seems to me that to practically die from fat is disgraceful." She herself, blessed with a heritage of Yankee leanness, remained slim and graceful in middle age, a picture of elegance on formal occasions with her ecru lace, moonstone pendant, and narrow gold bracelets against the smooth whiteness of her skin. Her less fortunate husband protested in the spring of 1902 that he had walked himself stiff and sore and kept honestly to a meager diet, and yet after a week, like the legendary chameleon that lived on air, still weighed the same 252 pounds.

In the matter of his smoking, it was not the state of his health but the size of his cigar bill — $122.50 in two months — that troubled Mabel in 1898. "Of course your guests have smoked the best part of that amount," she granted, "but still you do smoke a lot and I do think you might economize on that." Soon afterward he largely gave up cigars, only to take up pipes. A fixture of his desk was the set of two pipes, one to be smoked while the other cooled, and the tin box with parallel bars across its middle against which he knocked out his pipe and by the frequency and amplitude of the taps unconsciously signaled his mood of the moment. He would counter any suggestion that moderation in tobacco might prolong his health by pointing to the incessant smoking of his vigorous father.

True enough, such ailments as afflicted Bell had no apparent relation to his smoking. His left ankle had bothered him occasionally ever since he injured it in Washington during the hectic month of January 1877. His headaches now and then laid him low in response to the state of his nerves and the weather. In 1897, at a summer convention of teachers of the deaf, he suffered something like a sunstroke; and after that, as he wrote in 1902, "I fear the States in the hot weather." His sensitivity to hot weather, which gave him heat rash as well as headaches, made him a public advocate of artificial cooling for houses. In 1901 he suggested a central source of compressed air to be pumped into houses for cooling through expansion. In the summer of 1911, back from a world tour that had included the tropics, he drained the swimming pool in the basement of his Washington house, furnished it as a study, and ran a cold-air duct into it, at first cooling the air with ice and later with a refrigerating plant. But Beinn Bhreagh was a more spacious alternative.

Mabel felt that at Beinn Bhreagh her husband did not frolic in the outdoors as he ought to have, that his canoe paddling and woods tramping of earlier times had been crowded out by his laboratory and study. Still, the afternoons of hillside kite chasing must have served his health. Mabel's cousin Mary Blatchford, the Yankee duenna who had once shielded Mabel from the courtship of young Professor Bell, had long since reversed polarity; and her letters bear admiring witness to Bell at strenuous play. One midnight in 1891 an uproarious storm tempted Bell to quit his study and strike out up the mountain in his bathing suit. Three hours later, wrote Cousin Mary, "there was a tap at the window, and there stood Mr. Bell dripping like a merman and looking as handsome as an Apollo, with his grey curls wet and shining, and his white arms and legs. . . . He was so big and strong that it seemed as if the house would stand stronger with him in it." On another visit twenty years later, she found him "not at all well." He and Mabel passed her door one afternoon, "he in a thick white wrapper, looking like an enormous polar bear, with his heavy grey beard and shock of hair, and she so slim and delicate with her arm around him as if he were a big child and she his little mother." But along came another wild storm and "the excitement cured Mr. Bell of all his trouble. His cold went, and carried all his fractiousness with it. He had been in the houseboat all day, but toward night he came here looking like the spirit of wind himself, and he was as full of joy and excitement as a boy. It is that temper which keeps him young."

"He is a magnificent figure of a man," Cousin Mary wrote a few weeks later,

and his dress — always the same — becomes him wonderfully. He wears long, grey, coarse knit stockings with knickerbockers — or knee-breeches — of grey tweed with a loose jacket plaited and belted. For dinner he simply changes the grey jacket for a white waistcoat and black velvet jacket, and behold him resplendent! The other evening he danced for us, and whether it was a Scotch jig or a Highland fling, or an original-made-up-on-the-spot caper I do not know, but it was a *great show*.

This was the side of Bell that on his world tour in 1910 led a Vancouver reporter to credit him with "the energy of a man two score years his junior."

It was in Bell's family more than in his appearance that the passage of time made itself evident. With the departure of the older generation and the arrival by installments of a numerous younger generation, Bell had become the undisputed patriarch of a large and lively clan.

For Bell as well as the others, Mabel was the heart of it all. "Everything here begins with her and works round again to the same," reported Mary Blatchford. Alec at Beinn Bhreagh wrote Mabel at Washington in 1909, "I am so accustomed to hear a voice call me in the middle of the night at the proper time to go to bed that I forgot all about time till broad daylight." She wondered on occasion if she did not watch after her husband more than was good for him, but she never stopped doing it. "I feel dreadfully worried because you did not have your light overcoat," she wrote him once; "I am sure you have taken cold and I am sure you are feeling miserable and oh I wish I were with you to take care of you." After one parting she wrote him that "it actually was almost as much as I could do to keep from crying when the train moved out and I couldn't see your white head any more." And in 1917, when he was seventy, she confessed from a visit to New York that "with you for company, to plan for and take care of, I don't want anybody else very much, and people realize that and I am left alone now."

In Washington, Alec and Mabel went for a regular midday drive; and in the evening there was often the theater, where Bell, to the fascination of those who saw him, would turn his face toward his wife and silently repeat the dialogue so that she could follow it. Later they became regular, almost

457

The clan, 1918. Back row, left to right: Dr. David Fairchild; Dr. Mabel H. Grosvenor; Mrs. David Fairchild; Mrs. Gilbert H. Grosvenor; Melville Bell Grosvenor; Dr. Gilbert H. Grosvenor; Mrs. Samuel Gayley (née Gertrude H. Grosvenor). Seated: Alexander Graham Bell Fairchild ("Sandy"); Mrs. Alexander Graham Bell; in her lap, Gloria Grosvenor (Mrs. Torfinn Oftedal); partly behind Mrs. Bell, Lilian Grosvenor (Mrs. Joseph Jones); Alexander Graham Bell, with Nancy Bell Fairchild (Mrs. Marston Bates) on his right; Carol Grosvenor (Mrs. Walter Meyers); at Bell's feet, Barba Lathrop Fairchild (Mrs. Leonard Muller)

nightly moviegoers; and then the tables were turned. Mabel could not only follow the dialogue titles like everyone else, but could also read speeches at startling variance (especially during love scenes) with what the captions purported.

Mabel was "Gammie," and Alec was "Gampie" or "Grampie" to the Grosvenor and Fairchild children, of whom Mary Blatchford in 1911 counted three in the big house and five at the "Lodge" (which Bell had given to Gilbert Grosvenor in 1902). "All the plans, the hopes, and the ambitions that have lain buried in the graves of my own little sons," wrote Mabel, "sprang to life with the coming of each one of my three grandsons." The recurrent tragedy of the line was repeated in 1915 with the death of five-year-old Alexander Graham Bell Grosvenor; but Melville Grosvenor and Alexander Graham Bell Fairchild were destined to long, successful careers.

For their five Grosvenor and two Fairchild granddaughters, the Bells may have had other ambitions, but their love knew no distinctions. The earliest conscious memory of the eldest grandchild, Melville, was of sitting on his grandfather's lap and, on instructions, tweaking the nose of Alexander Graham Bell to produce a dog's bark, pulling his hair for a sheep's bleat, and by way of climax, tugging his Santa Claus beard for the deliciously fierce growl of a bear. As each newcomer, boy or girl, reached the age of nose-tweaking discretion, the routine was impartially repeated. Over the years Bell told his grandchildren a continuing story of "the rubber man," whose rubber suit had got inflated and carried him away in a hurricane to endless adventures — a tale of flight, of rotundity rising, that may have helped soothe Bell's aeronautical frustrations. While Bert Grosvenor was absorbed with *Geographic* affairs, his children sometimes enjoyed extended stays at Beinn Bhreagh. "They run to meet me when I come home from the Laboratory," Bell wrote his son-in-law in 1906, "and it is really a pleasure to feel that they love me as dearly as I love them." In 1911 little Alec and Mabel Grosvenor were always waiting for their grandfather by the great hall fire when he came in from the laboratory to have his cup of tea.

It was with Melville, however, that Bell came closest to playing the role his own grandfather had played with him. David Fairchild may have preempted that role with Sandy, whose strong bent for serious science gave his professional scientist father an advantage over his amateur scientist grandfather. By the time Melville Grosvenor was ten, on the other hand, he had to compete with a brother, four sisters, and the *National Geographic Magazine* for his father's attention. Perhaps Melville's name, that of Bell's dead brother, touched a chord of memory in the old man. Melville's shyness might have reminded Bell of his own boyhood. And in any case, Melville had a two-year head start over the other grandchildren.

Bell's notebook for 1911 records his wrestling with the problem of how to discipline Melville, who had been unruly the day before. "I am opposed to corporal punishment," he declared to himself. "The willful infliction of pain upon a little child is to me an unthinkable form of punishment, that degrades the punisher as well as the punished *to the level of the brutes.*" The proper way was to "treat him as a reasonable being, and . . . talk it over with him." The whole matter, including the projected dialogue, was somehow submerged in the shared fun of baking little clay bricks and figurines in the houseboat stove. But Melville's character evidently did not suffer.

In Washington, Melville would often hop on a streetcar at Dupont Circle and ride out along P Street on the front platform with the motorman to the vicinity of his grandfather's Georgetown laboratory. There Bell would help the boy with his homework and encourage him to make simple experiments, such as a shoebox boat propelled by the pinhole steam jet from a water-filled eggshell set in the box over a lighted candle. The two of them would walk back to Connecticut Avenue for dinner, later perhaps going to the movies with Melville's grandmother. During their afternoon strolls one winter the old man and the boy laid elaborate plans to play Crusoe and Friday in the woods at Beinn Bhreagh. When the time came, the pair plunged into the woods dressed only in bathing suits, without food or tools. A cold Scotch mist powdered the air, no game appeared, the dampness made green boughs impossible to break for building material and fire impossible to start by friction. "Well," remarked Bell, "Robinson Crusoe was lucky. His island was tropical." So as a fair equivalent they headed for the houseboat, lit the stove, and heated up a satisfying can of beans.

Bell looked forward eagerly to the summer and fall of 1914, when Melville was to stay at Beinn Bhreagh. Bert Grosvenor felt that he himself had not been all to Melville that he ought to have been, at least in the boy's early years. "The first child is always horribly handicapped by the ignorance of the parents," Bert confessed; "I have tried to make it up to Melville, but realize how lamentably I failed . . . especially in those early days." If Bell had needed rallying to the education of Melville, this would have served. "I am almost jealous of the attention Mr. Bell is giving Melville," Mabel wrote Bert that fall. "He did not devote himself so regularly to his own daughters. But he says there is a difference — Melville is a boy. I wonder what would have happened if his own boys had lived."

Bell's report to his son-in-law that December might have been a letter from Harrington Square half a century earlier. "Melville . . . has been a

Melville Grosvenor and his grandfather help fly a tetrahedral kite at Beinn Bhreagh

great comfort to me here and quite a companion . . . although it must have been very lonely for him without any young people of his age to play with." Bell had asked Melville to help him with his research and felt that the boy's work on statistical tables had been good practice in arithmetic. But Bert had asked that reading and literary composition not be slighted, and Bell had done his best in that line also. Melville, he reported triumphantly, "has developed his GREAT AMBITION IN LIFE, which is to be an editor like his father. . . . I have encouraged him in this and he has now produced three issues of the great new popular magazine known as 'Wild Acres Weekly.' " In the wartime summer of 1918, a Beinn Bhreagh picnic celebrated Melville's admission to the United States Naval Academy, from which in due course he graduated. But his "great ambition in life" remained what it had been in 1914, and he would eventually achieve it brilliantly as his father's successor in command of the *National Geographic Magazine*.

In teaching Melville, Bell had an educational philosophy ready at hand. Like his own grandfather, he was a thoughtful and experienced teacher. His boyhood experience had itself turned him against regimentation and toward the encouragement of natural curiosity. The successful methods of his little school for deaf children during the mid-eighties were expressed in what he told a Chicago reporter ten years later: "The system of giving out a certain amount of work which must be carried through in a given space of time, and putting the children into orderly rows of desks and compelling them to absorb just so much intellectual nourishment, whether they are ready for it or not, reminds me of the way they prepare pâté de foie gras in the living geese" — that is, forced feeding.

In 1911 *McClure's Magazine* heralded the news that in Italy Dr. Maria Montessori had developed a similar viewpoint more systematically and concretely, with remarkable success. There was no forced feeding. Instead, carefully prepared materials for learning were set before the children, and the teacher gave them such help as their natural impulses led them to seek. The idea of the child's inducing generalizations from observation appealed to that science-struck era. And whether Bell came to it through his frequent Wednesday Evening guest S. S. McClure or, as family tradition has it, through his daughter Marian Fairchild, the Montessori Method excited sympathetic vibrations in him especially.

At Washington in the spring of 1912, Roberta Fletcher introduced five-year-old Sandy Fairchild and his three-year-old sister Barbara to the Montessori Method. A half-dozen of the neighbors' children joined them in what the Dottoressa in Italy had called a "children's house" but the science-minded Fairchilds and Bells called the "Children's Laboratory." That summer, on the upper floor of the Beinn Bhreagh warehouse, Miss Fletcher opened the first Montessori class in Canada with five Grosvenor children, the two Fairchilds, and five others from the vicinity. Fascinated, Bell reviewed progress and discussed problems in regular afternoon conferences with Miss Fletcher. This in turn led Mrs. Bell to sponsor Miss Fletcher and Madame Montessori's foremost American disciple, Anne E. George, in a full-fledged Montessori school at the Bells' Washington home during the school year 1912–1913.

The Bell name had brought press coverage even to the modest program at Beinn Bhreagh. At Washington the Bells' enthusiasm aroused wider interest in the new method. In the spring of 1913 a group of Washington parents organized the Montessori Educational Association and unanimously elected Mrs. Bell president, whereupon she bought a house on Kalorama Road and rented it to the association for a still larger Montessori school in the fall of 1913. McClure brought Dr. Montessori herself to America that December for a lecture tour, in the course of which the Bells gave a reception for several hundred people in her honor. By the spring of 1915 the Montessori Educational Association was publishing a magazine and had elected Bell its president.

After 1916 the movement in America declined. The educational establishment launched a counterattack, charging (in the words of one leader) that the Montessori method had "the spirit of science, but not its content." The World War weakened its ties to its European base before enough trained leadership had developed; some thought it too Catholic-oriented; certain of its psychological assumptions seemed outmoded; new ideas, especially those of John Dewey, offered alternatives more congenial to American culture. By 1919 it had faded out in America, not to be revived for many years. Whatever benefits it may have brought Bell's grandchildren, Sandy Fairchild, for one, in later years found his career as an entomologist delayed by his lack of Latin and his belated entry into college.

But though the Montessori movement had to be introduced all over again to a later generation of Americans, its first brief vogue, in which the Bells were conspicuous leaders, at least stimulated thought and discussion about the nature of children and of their learning processes. In that respect it made a significant and lasting contribution to American life.

"When one gets to be Alec's age," Mabel wrote her mother in 1908, "the question will arise whether he is as fine and clear mentally as when at 27 he invented the telephone." She was sure he was; and in most affairs, large and small, the years that followed proved her right. But significant inventions required a level of creativity not likely to endure much beyond middle age, and Bell was then sixty-one. Five years later, in an interview with Frank Carpenter, he insisted bravely, "I have never felt stronger intellectually. My mind has a greater power of concentration than it ever had. It seems to be quicker, and it does not tire along the lines in which I am interested." Yet a few months earlier he had cleared out much of his Washington laboratory, disposing of Wheatstone bridges, X-ray apparatus, coils, rheotomes, induction balances, and other relics.

At Beinn Bhreagh the old kite house on the hillside now became a "museum," complete with meticulously dated models and apparatus in big glass cases and brooded over by big kites suspended from the roof. Despite this further symbolic resignation of inventive ambitions, Bell's habit of work was unbreakable; and so he partitioned off one end of the long building as an office, unpainted, bare-floored, and plainly furnished with a rough little wooden table, an old Morris chair, and a shabby sofa on which he read, thought, scribbled in his notebook, and napped. From time to time, he would resolve to observe a more conventional working day. Then after a few days, he would fall back comfortably into the old routine of rising at ten or eleven, giving over the afternoon and evening to the office and to family life, and finally withdrawing to his study to smoke, think, and write until three or four in the morning.

At Beinn Bhreagh, Bell's weekend retreats to the beached houseboat continued. Mabel sometimes spent a day or two alone with him there. Besides the undiluted company of her husband, she enjoyed cooking for him and puttering about in the comfort of a Japanese kimono. While at Washington, Bell now had an equivalent in the little house, called his "Retreat," that the Fairchilds built for him in the woods on their Maryland property, at a spot overlooking Rock Creek.

In 1905, after nearly fifteen years as private secretary and family friend, Arthur McCurdy left to promote his own invention of a small, portable photographic developing tank. Other secretaries came and went, until in 1914 Bell hired young Catherine MacKenzie, just out of a convent boarding school, who stayed with him until his death eight years later. She read the *New York Times* to him, took down his dictation of letters and notebook entries, and saw to it that his pencils, pens, and other office impedimenta were in their accustomed places. In later years she remembered especially his obsession with the keeping of precisely dated records, both written and photographic, an obvious mark of his telephone litigation.

In March 1910 the chief machinist and superintendent of the Beinn Bhreagh laboratory was fired for having falsified dates on photographs of

kite and "aerodrome" trials. After the world tour of 1910–1911, Casey
Baldwin became permanent·manager of both the estate and the laboratory
for $2500 a year and a rent-free house. Hindsight might suggest that in
settling for that snug berth, Baldwin showed less enterprise than, say,
Douglas McCurdy. But his work with Bell offered Baldwin scope for experiment, and he reciprocated the Bells' affection for him and his family.
Furthermore, Beinn Bhreagh was a lovely place to spend a lifetime. And
particularly seductive to a passionate yachtsman like Baldwin was its
sweep of salt water.

Casey Baldwin's bent for boating may have encouraged the sea change
in Bell's experimental interests during his last years. Casey had already been
on the scene for some months when Bell asked himself in October 1906,
"Why should we not have heavier than water machines as well as lighter
than water? I consider the invention of the hydroplane as the most significant of recent years." A few weeks later he sketched what is now called
a hydrofoil boat.

Bell's phrase "heavier than water machine" referred to the fact that such
a boat, while in motion, is sustained not by its displacement, like a balloon,
but by the lift of its hydrofoils — plates or blades acting in water as airplane wings do in air. As early as 1861 the idea had been successfully tested
by towing on an English canal. Bell had probably read the March 1906
Scientific American article by an American hydrofoil pioneer named William E. Meacham, explaining the basic principle. As the speed of a hydrofoil boat increases, the submerged hydrofoils raise the hull until it leaves
the water entirely, supported by the vertical stanchions to which the hydrofoils are attached. If speed is increased further, the hydrofoils themselves
emerge from the water to a point at which their lift is cut back again to
the weight of the boat. This automatic reduction of the acting surface area
was called "reefing" by Bell and Baldwin, by analogy with sails. Since the
ratio between horizontal and vertical forces is fixed by the angle at which
the hydrofoils are set, the horizontal force — the resistance to forward
motion — likewise tends to be kept constant by hydrofoil "reefing" as
speed increases; whereas the resistance of a conventional boat increases as
the square of the speed.

Bell and Baldwin did not begin hydrofoil experimentation until the
summer of 1908 and even then merely as a possible aid to airplane takeoff
from water. But as Baldwin studied the work of the Italian inventor Enrico
Forlanini and began testing models, he and Bell turned to hydrofoil water-
craft as well. Bell's interest was active and his suggestions astute. Towing
trials continued that fall, though plagued with troubles, including the failure of parts under stress. There was a strange, persistent blind spot in

Baldwin's thinking, like Bell's disregard of aerial drag; though he was a trained engineer, Baldwin was forever overestimating the strength of structural parts or underestimating the stress on them. Nevertheless, the hull of a towed hydrofoil-equipped boat lifted clear of the water with Baldwin aboard. And Baldwin designed a ladderlike set of hydrofoil blades slanted so that the lower end of one was on a level with the upper end of another, thus making the reefing action continuous rather than jumpy.

Tests of another hydrofoil-adapted boat ran through the late summer and fall of 1909. The world tour of 1910–1911 interrupted the new line of experiment, but during the journey Bell and Baldwin talked with Forlanini in Italy. "Both Father and Casey had rides in the [Forlanini hydrofoil] boat over Lake Maggiore at express train speed," wrote Mabel to her daughter Elsie in the spring of 1911. "They described the sensation as most wonderful and delightful. Casey said it was as smooth as flying through the air. The boat . . . at about 45 miles an hour glides above the water . . . being supported on slender hydro-planes which leave hardly any ripple."

When Bell and Baldwin returned to Beinn Bhreagh late in the summer of 1911, Baldwin designed and built a hydrofoil boat, which he and Bell referred to as a "hydrodrome" and designated the HD-1. Its aerial propeller and short biplane wings for extra lift made it look like a stunted seaplane skittering across the lake that fall in a series of unconsummated takeoffs. After being redesigned and rebuilt during the winter, it was tested again from July to October 1912; and before it cracked up from unknown causes, it had made fifty miles per hour. Rebuilt as the HD-2 and given the unpropitious name of *Jonah*, it failed to match that speed record before being disabled by a structural failure in December.

Meanwhile, in September 1912, Bell had an alluring vision of hydrofoil sailboats. Baldwin, the experienced sailor, pointed out that the swiftest sailing craft fell far short of the speed required to lift its hull above water by hydrofoils. Bell merely suggested that aerodynamic lift be added through horizontal sails, and he went on to dream of large passenger vessels skating smokeless across the Atlantic on the wings of the wind. So in the summer of 1913, Casey dutifully worked with Bell on test models.

Early in 1913, at Bell's urging, Baldwin had designed and built the HD-3, another chunky little quasi-airplane. Its most spectacular moment came when a structural failure led to its turning turtle during a demonstration for the prince of Monaco, whose yacht had put into Baddeck Bay. The HD-3 trials that fall were disappointing; and Casey felt that to put the boat into commission in 1914 would divert too much time and money from further testing of Bell's hydrofoil sailboat ideas.

Though the hydrofoil sailboat dream died in 1914 after a second summer

HD-1, July 16, 1912

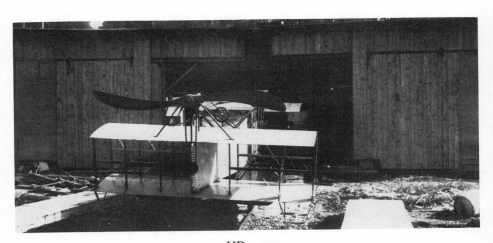

HD-3, 1913

of fruitless trials, the outbreak of the European war that August lengthened the hydrodrome moratorium by two years. Baldwin felt that a private purse could not finance so expensive a project as the hydrodromes were getting to be, and that a government in wartime would not. Bell himself decided that as a citizen of a neutral power he should not pursue experiments in Canada that might have naval uses. So in order to save the jobs of the workers, he got Casey to use the laboratory facilities in building small boats for local people and summer residents.

Hydrofoils or not, Bell kept busy. Sheep, eugenics, the National Geographic Society, and the Montessori Method jostled for his attention. So did the affairs of the AAPTSD and its *Association Review*. The *Review*'s circulation had been languishing at six or seven hundred when Mabel Bell asked Bert Grosvenor in 1909 what he thought would help it. She reported to her husband that their enterprising son-in-law recommended changing it "from cover to cover," widening its audience to friends and relatives of the deaf and hard of hearing, making it less technical and more popular. Bell and his other son-in-law, David Fairchild, as Association directors, urged those policies early in 1910. And so a new editor, Frederick K. Noyes, was hired, the name was changed to the *Volta Review*, its publication was stepped up to once a month instead of five times a year, and the "New Departure," explicitly emulating that of the *National Geographic*, was announced by Bell in the April 1910 number.

As an assistant and clerk at fifteen dollars a week, young Noyes brought along his gawky, red-haired Yale classmate Sinclair Lewis. "We are going to get right after people, and *try* to boom the circulation tremendously," Lewis wrote a friend upon arriving. In the Volta Bureau Library, Lewis read with fascination Bell's unpublished notebooks on his aeronautical experiments. After six months, Lewis quit his ill-paid drudgery for a job in New York. But out of his *Volta Review* interlude came Sinclair Lewis's first published book, *Hike and the Aeroplane*, written hastily for money in 1911 and published a year later. It was a boy's adventure story in the mold of *Tom Swift*, and its central gadget was a "tetrahedral aeroplane." Glenn Curtiss and Douglas McCurdy had read and liked it, Lewis told a friend.

The effort to make money for the Association by popularizing its *Review* proved to be too much for Noyes, as it would probably have been for anyone. The journal's regular readers objected to the new departure, and the general public ignored it. Noyes left for a better job early in 1912, by which time the circulation had edged up to only a thousand. Bell did not give up hope, however. As interim editor, he brought Fred DeLand to Washington.

When an article by DeLand, "Pioneer Telephone Exchanges," caught Bell's attention in 1905, DeLand, then in his early fifties, was treasurer of several telephone companies and an enthusiastic amateur student of telephone history. Bell wrote him to express surprise at the "accuracy and range of knowledge displayed . . . concerning my work." In 1908 DeLand published a book, *Dumb No Longer*, on Bell's work for the deaf as well as with the telephone. Early in 1912 Bell wrote wonderingly to a friend that DeLand (whom he had not yet met) "seems to know more about me and what I have done than I know myself." DeLand was an intelligent man but perhaps somewhat unstable; at least Bell mentioned hearing that he had suffered a nervous breakdown at some time. Bell's friendship evidently gave DeLand a rock to stand on. His subsequent feeling toward Bell can accurately be called devotion. Upon DeLand's arrival in Washington, he became to Bell's work with the deaf what Casey Baldwin was to Bell's experimental work at Beinn Bhreagh. After his stint as interim editor of the *Volta Review* and then for a time as Volta Bureau librarian, DeLand in 1914 became regular editor, serving in that capacity until 1920, and as superintendent of the bureau until 1921. With Bell's encouragement and advice over the first year or two, DeLand achieved a satisfactory balance between technical and general appeal, though circulation remained modest. And he also became as diligent and dedicated a chronicler of the deaf as he was of the telephone.

As an inveterate newspaper reader, Bell kept up with world affairs; and in his celebrity, his causes, and his Washington life, he sometimes touched them more directly. The twentieth century brought a rising cry for women's suffrage. From his first days at coeducational Boston University, Bell had supported equal rights for women, though old Mrs. Sanders took the other side. As late as 1901 Mabel also lagged behind him. To deny anyone the right to vote, he told her, was "to take away from the individual his right to protect himself." As for her suggestion of educational and property qualifications, he argued that "the ignorant have as much right to be represented as the educated. . . . I believe in universal suffrage, without qualification of education, sex, color, or property." His grandfather would have approved.

By 1910 he had converted Mabel. She and their daughters took part in the national convention that presented a huge petition to Congress that April. On his sixty-sixth birthday in March 1913, Bell and his wife cheered the Women's Suffrage Parade from a grandstand on Pennsylvania Avenue, while Elsie and the Grosvenor children rode in their automobile and Marian followed on a float. In March 1914 he introduced the main speaker at a public lecture for the cause given in the Bell home. "I spoke of myself as the first suffragist in our family," he noted afterward, "and of my wife as the second. And that of course we brought up our daughters in the way they should go, and that they had captured their husbands, so that we were a united family on the subject."

Bell's general political coloration seems to have become that of a progressive Republican. He displayed it in a 1909 article for the magazine *The World's Work*. Big business would wring excessive profits from the consumer, he believed, unless prevented by government. But breaking up big business would undo its real economies of scale and would lead to wasteful competition. He saw only two answers, socialism or federal regulation, and like Theodore Roosevelt he favored the latter.

Personal contacts, however, blurred Bell's ideological allegiances. In 1910 he and Mabel called on the Grosvenors' cousin, President William Howard Taft, as the Roosevelt-Taft break was developing. And Bell happened to attend the politically decisive dinner of the Association of Periodical Publishers at Philadelphia in February 1912, at which the near breakdown of the progressive Republican leader Robert LaFollette opened the way to Theodore Roosevelt's attempt at a comeback and thus to the Republican split that gave Woodrow Wilson the presidency. LaFollette, Bell noted afterward, "was a great disappointment. . . . I listened for about five minutes to a ranting stump speech of undignified character." But Wilson's speech that night "was the finest I have ever listened to." Bell became an ardent Wilsonian. In the fall of 1914 Bert Grosvenor joined him, registering as a Democrat for the first time in his life, because in the current world crisis, he explained, "everyone should uphold President Wilson and Secretary Bryan."

The European catastrophe that summer had shocked the Bells along with the rest of the world. "We were all pursuing the even tenor of our lives," wrote Mabel a few months later, "when suddenly the news was flashed across the ocean. . . . It came like a knife cutting sharply and forever our present world from that of yesterday, and left us stunned, bewildered, utterly unable to conceive how this dreadful, this impossible event had happened."

In 1911, on their round-the-world trip, Mabel had written Elsie with gratified wonder about "Papa's whole-souled devotion to America, the completeness with which he identifies himself with it." She herself felt that she would always be an American however long she lived elsewhere, yet for all her husband's love of his native Scotland, "his interests are American." It was no mere capitulation to the mood of the times that led Bell himself to note privately in 1915, "I am not one of those hyphenated

Americans who claim allegiance to two countries." And so President Wilson's call in 1914 for American neutrality in thought and deed weighed heavily on Bell.

Still, the Bells spent much of their time in Canada, and Canada was visibly and audibly at war. Bell stopped his hydrofoil work, but he could not shut off his natural sympathies. Catherine MacKenzie's three brothers died in action. At Beinn Bhreagh parties, everyone sang "Tipperary"; and "God Save the King" carried new overtones. The Bells ran a "Garden Fete" for the benefit of the Red Cross in the summer of 1915, with games, exhibits, refreshments, and produce for sale. The sea about Beinn Bhreagh was a reminder of German submarines in the North Atlantic. On the morning of December 6, 1917, windows shook and the ground rumbled from the great munition ship explosion at Halifax 177 miles away. The Bells packed up spare blankets and clothing for the survivors, and Casey Baldwin led a party of men to help build emergency shelters.

In his sympathy for the Allied cause, Bell's optimism never deserted him. "Paris is doomed," Charles Thompson remarked in August 1914; "the French aren't prepared, so how can they hold?" "I believe, sir," said Bell, "they have something up their sleeve; they have reserves somewhere, where are they?" He exulted when Joffre turned back the Germans at the Marne. ("The real battle of the Marne," said General Gallieni later, "was fought on the telephone" — by which he meant his persuading Joffre to order the timely counterattack.) Through the gloomiest days of the war, Bell maintained his conviction that the Germans would ultimately fail. "I notice," remarked Mabel to Bert early in 1918, "that Father dilates triumphantly on all Allied successes, and lets me find out for myself their reverses."

By then the United States was in it, and Bell's dilemma as a neutral in a belligerent country was resolved. He was no war-lover. "In the present state of the world no nation in Europe can find its interests in protracted war," his grandfather had written many years before. Bell took the same view. In May 1916 he had donated a thousand dollars to Taft's League to Enforce Peace, a harbinger of the postwar League of Nations. After the war, he would support Wilson's League to the bitter end, calling it, in a letter to the *New York Times*, "the noblest attempt yet made to bring about harmony and co-operation among the nations of the world," and even in the reaction of the Harding era still regarding the League Covenant as "*a great document*." Nevertheless, while the United States was at war, Bell was determined to do what he could for the cause.

The timing of American entry limited the direct participation of the Bell clan. Bell's sons-in-law were both in their forties, and his eldest grand-

472

son was only fifteen. Bell himself was seventy, and his health had perceptibly declined. Still, he hoped that his laboratory facilities and what remained of his inventive talent might produce something useful. A dozen years earlier he had commented that "nearly all my inventions and discoveries are now having a practical application in warfare" — in particular, the telephone, photophone, and telephonic probe. In 1913 he had made an odd reply to a request that he invest in a newly patented "Duplex Rifle Butt": "While I am always interested in new inventions, and especially those that make for the advancement of the race, I regret that I am not able to become financially interested in them." From a man of more subtle temperament this might have been taken as sarcasm. In Bell's case it was probably only the accidental irony of a form letter. In the cause of his adopted country, as shown by his reaction in 1898, Bell had no apparent qualms about applying technology to war.

For a few weeks in 1918, Bell returned to his old interest in underwater sound and experimented with techniques of detecting submarines. As with other aspects of telephony, however, his own expertise had long since been outdistanced by that of the specialists. The navy's hydrophones, as he readily admitted on trying them, were far more effective than anything he could come up with. With Mabel's help, he also dabbled with processes for dehydrating vegetables from the Beinn Bhreagh garden. Here, too, the experts were ahead of him. His oddest notion was one that had first occurred to him in 1897: a shell that would put out a small "rudder" near the end of its trajectory, thereby to be sent up again into another trajectory, and so on in a series of diminishing upward swoops, like the skipping of a stone on the surface of a pond. In 1918 he expounded the suggestion in a draft letter to the War Department. Fortunately he tried the draft first on his friend Charles Walcott, secretary of the Smithsonian, who pointed out tactfully that a more sensible approach to extending range would be a winged glide from the top of the trajectory, but that in any case, rifled guns introduced a difficulty by firing rotating shells. Bell destroyed the draft letter.

These were vagrant fancies. From the outset Bell had recognized that among possible applications of technology to warfare, only in the fields of aviation and hydrofoil boats did he have substantial recent experience. Bell had always appreciated the military potentialities of aircraft. In 1915 he asserted in a letter to the *New York Sun* that "the power that secures supremacy in the air will ultimately have all other methods of warfare at its mercy." He elaborated on that proposition in an address to the national convention of the Navy League in the spring of 1916, an address which the league circulated as a pamphlet and which was widely quoted in the press.

473

But in this matter, too, he had already been left behind by later developments. As early as 1915 he recorded his own private doubt that he could even be "of much assistance to a government board" of military aviation.

So it all came down to "hydrodromes." In February 1917, after the German announcement of unrestricted submarine warfare made American entry certain, Bell had begun experimenting with hydrofoil models designed to slide over obstacles. Two days after the formal declaration of war on April 6 he left Washington for Beinn Bhreagh.

In 1915 Bell had failed to interest the United States Navy Department in Baldwin's HD-3. His hopes revived after American entry into the war, when the department called for proposals to build submarine-chasers. After offering to build two hydrofoil craft for that purpose, Bell and Baldwin happily pored over old records of towing trials and debated new designs. Then came the government's rejection of the offer. It "fell like a bomb into the camp and shattered all our plans," Bell recorded. But Mabel offered to put up the money anyway, and Bell got the navy to promise the loan of two 400-horsepower Liberty engines. Design work resumed on what was to be the HD-4.

The HD-4 incorporated all that had been learned from the successes and mishaps of its predecessors. It was a sleek gray giant with a cigar-shaped hull sixty feet long, riding on two sets of reefing hydrofoils forward and one aft. On each side was a small hull attached to a solid, streamlined outrigger, for balance while floating at rest. Each outrigger also carried a motor and aerial propeller. Later, a spray shield on each side was tapered from the outrigger hulls to the nose, giving a winged look to the craft and in fact contributing some additional aerodynamic lift. The navy could not or would not furnish the promised Liberty engines, and even the two 250-horsepower Renault engines it sent did not arrive until July 1918. (By then someone had suggested that "HD" stood for "Hope Deferred.") So the first trials did not begin until October 1918.

With the Renault engines, the top speed was only 54 miles per hour. Nevertheless the HD-4 performed well, rising easily, accelerating rapidly, taking waves without difficulty, steering well, showing good stability. Bell's proud report to the navy early in 1919, along with the postwar availability of Liberty engines, moved the navy at last to send two 350-horsepower Liberties in July 1919. With these powering its two aerial propellers, the HD-4 on September 9, 1919, set a world's marine speed record of 70.86 miles per hour, a record that stood for ten years.

The sight alone was exhilarating, and the actual ride even more so. "At fifteen knots you feel the machine rising bodily out of the water," wrote

HD-4, 1919

one visitor, "and once up and clear of the drag she drives ahead with an acceleration that makes you grip your seat to keep from being left behind. The wind on your face is like the pressure of a giant hand and an occasional dash of fine spray stings like birdshot. . . . She doesn't seem to heel a degree as she makes the turn. It's unbelievable — it defies the laws of physics, but it's true." Bell himself would never ride in her. A newsreel photographer (one of several who showed up at Baddeck in that jubilant season) got Bell to sit in the cockpit of the moored craft, but Bell insisted on having Baldwin's small son beside him to negate any false imputation of daring. Mabel was furious with herself later for not having gone down to make her husband go for a spin while he was in the boat.

A year later Bell managed to get observers from both the British Admiralty and the United States Navy to watch demonstrations. Both delegations made enthusiastic reports. But after long deferred hope, neither navy saw fit to place an order. The United States Navy Department thought such craft too fragile for naval action at sea and preferred a combination of seaplanes and motorboats to any such hybrid of the two. The Naval Disarmament Conference of 1921 dampened British interest. In the fall of 1921, the HD-4 was dismantled. Its big gray hull lay unregarded for decades on the shore at Beinn Bhreagh.

Bell had for years been trying to make a businessman of Casey Baldwin by something like the Montessori Method, giving him more responsibility, edging him into situations that would engage his business sense. Whether it was Baldwin's fault or not, the results had been discouraging. The boatbuilding enterprise had lost money. In the summer of 1920, with HD-4 tests in progress and the American naval observers yet to arrive, Casey elected to sail his yacht to England and back. As in the case of the Canadian Aerodrome Company, he also left behind him the newly incorporated Bell-Baldwin Hydrodromes, Ltd., in the interests of which he and Bell had acquired the United States hydrofoil patents of Forlanini and Peter Cooper Hewitt and had made a working arrangement with William Meacham, as well as applying for patents of their own. (In the absence of any business, Bell-Baldwin Hydrodromes, Ltd., passed out of existence about the beginning of 1923.)

After some difficulty in persuading the Patent Office that the HD-4 had original features substantial enough to patent, a joint patent on certain of them was issued to Baldwin and Bell on March 28, 1922. On the same day they were also granted a patent for means to prevent fouling of hydrofoil sets, one for features of a "hydrodrome" that was designed but never built, and one (shared with a navy lieutenant who had worked with them) for a device to change hydrofoil blade angles. These turned out to be Bell's last patents. Forty-seven years had passed since his first patent, and he was now seventy-five years old.

34

Remembrance

The years had not diminished Bell's fame. On the contrary, it had spread with the telephone, as the Bells and Baldwins discovered on their world tour of 1910–1911. "One of the nice things," wrote Mabel to her daughters from New Zealand, "is the way so many perfect strangers have come to say how glad or honored they are that Papa should come to their country." A Sydney newspaper thought Bell "one of the two most interesting visitors Australia has ever had from the United States" (the other having been Mark Twain). Everywhere, in China and India as well as in Europe, officials received Bell as a very important person.

Official recognition in other forms had been coming to him since his Centennial medal, his honorary degree from Gallaudet College, and his Volta Prize. The twentieth century brought no slackening of formal honors. To the five honorary doctorates already bestowed on him, it added another seven, including those from St. Andrews, Edinburgh, and Oxford. He was often referred to as "Dr. Bell," though he did not use the title himself. The score or so of prize medals he received, mostly for the telephone, included those of leading American, French, British, Italian, and German technological societies. The Thomas Alva Edison Medal, awarded him in 1915 by the American Institute of Electrical Engineers, ought to have been especially gratifying in view of his friendly rivalry with Edison. On the other hand, Bell's numerous honorary memberships in American and foreign societies doubtless meant less to him than the Chicago and Cleveland public schools for the deaf which bore his name.

Occasionally he grumbled to Mabel that he would rather have been given all these honors when he was young and struggling. He probably meant that he wished he were still young when he was finally given them. Bell may have been a little jealous of the younger self from whom he now felt so detached and for whom most of these awards were, after all, ex-

477

pressly intended. Perhaps also the repetition had become cloying. In 1915 the movie executive Carl Laemmle asked permission to take some footage of him for a proposed newsreel special on "the ten most prominent men in the country," but Bell demurred because "publicity of this sort would be distasteful to me."

Nevertheless Bell uncomplainingly answered all the hundreds of autograph requests that came in the mail, though he never signed the blanks sent him, only a form reply to which he could close up his signature and thus forestall any documentary hanky-panky (another sign of the scar left by the telephone litigation). "I should think that every school child in the U.S. has an autograph by this time," his secretary Catherine MacKenzie wrote after his death. In his office at Beinn Bhreagh, she recalled, "he never failed to rise courteously and shake hands with a visitor no matter whether it was a cheeky tourist who had ignored the 'Private' on the door, or some embarrassed person who had blundered in while it stood open and was made at ease by Mr. Bell's friendliness."

The penalties of fame included an endless stream of letters from inventors seeking help. "Mr. Bell's patience in reading pages of description was infinite," testified Miss MacKenzie. "I think he was always a little touched by their confidence in his honesty. Every inventor thinks that someone wants to steal his device." His letterpress copybooks teem with his replies, generous with advice and encouragement, though politely declining financial involvement. With young inventors he was especially tender. In the midst of the hydrofoil work of 1918, for example, he took time to write a thoughtful essay in reply to one Joseph Strittwater's suggestion of a "vacuum balloon" made of metal. After a sketch of the notion's long history, he carefully explained the crushing force of atmospheric pressure and concluded: "Good luck and best wishes to you, and don't be discouraged by the fact that you don't know everything that has been done by others before you were born."

Bell occasionally used his fame and his access to the columns of periodicals to interest children in science and technology. As a boy he had invented the pen name "H. A. Largelamb," an anagram of A. Graham Bell; but not until the nineties did he actually begin to use it, both to write short popular science articles without inhibition and (as Mabel put it) to have "the fun of seeing if he can make another reputation for himself." Eventually, after several "Largelamb" articles had appeared in the *National Geographic*, the secret leaked out and Bell abandoned the game. Under his own name in 1912 and 1914, he wrote a series of articles for the *Volta Review*, "Simple Experiments," for children; and in 1914 and 1917, the *National Geographic* published two addresses he made to Washington schoolchildren on the wonders and opportunities of science and invention. The editor of the *Volta Review* commented perceptively that "few men possess the gift of seeing things from the viewpoint of a child so clearly as Alexander Graham Bell."

"What a glorious thing it is to be young and have a future before you," Bell told a school graduating class in February 1917, shortly before his seventieth birthday. But he added, "it is also a glorious thing to be old and look back upon the progress of the world during one's own lifetime. . . . I, myself, am not so very old yet, but I can remember the days when there were no telephones."

In 1904, urging Gilbert Grosvenor to make a career out of writing, Bell contemplated his own remarkable career. "Circumstances arise," he wrote, "that bend our lives hither or thither against our will. But our lives are bent, not broken — there is always continuity of growth. In looking back over my own life I realize how different has been the result from anything that I aimed at, and yet I can recognize continuity in the whole; one occupation has fitted me for the next, and that again for the next."

This rumination stands almost alone among Bell's recorded thoughts. He was not averse to discussing specific episodes of his life. But he shrank as if with pain from venturing or abetting any appraisal of the whole. "So far as I am concerned," he wrote a would-be biographer on the eve of his fortieth birthday, "I prefer a post mortem examination to vivisection without anesthetics." Two years later he held to his rule against furnishing material for a biographical overview even in the case of his old friend Fritz de Sumichrast, who wanted to contribute a sketch of Bell to a book to be called "Living Leaders of the World." Many years later still, Bell may have exposed the roots of his reticence. To a suggestion from Little, Brown and Company of Boston in 1916 that he write his reminiscences for publication, he replied, "I am as yet too much interested in the future and the development of new ideas to give the time to a book of this character." A formal biography or autobiography, even a brief one, to him evidently signified finality, a surrender to time, an acknowledgment of obsolescence. At forty he had feared that. At seventy he still resisted it.

In 1903 Bert Grosvenor got his father-in-law's permission to write an article or series of articles carrying the story only to the invention of the telephone. For three years, while editing the *National Geographic* and writing for other magazines, Bert assembled copies of family letters, abstracts of testimony, and notes. By 1906 Bell himself had been actively enlisted to the extent of writing down anecdotes of his boyhood. But in 1907 Bert decided that for the proper support of his growing family he had to postpone the project and concentrate on building up the *National Geographic Magazine*. That endeavor preoccupied him for the next forty-seven years.

Near the end of Bell's life, Grosvenor did manage to wangle from him the only autobiographical article Bell ever consented to publish. Even that was limited in coverage. It grew out of an address describing his youth and early struggles, which Grosvenor had a stenographer take down. Knowing that Bell lacked ready cash for a desired donation to the Clarke School for the Deaf, Grosvenor offered to send the National Geographic Society's check to the school in return for Bell's revision of the talk as a *Geographic* article. Once again Grosvenor had found a soft spot in Bell's defenses. The article appeared in the March 1922 issue as "Prehistoric Telephone Days."

Perhaps all the more because he fended off a biography in his own lifetime, Bell took pains to insure that his eventual biographer had plenty of material. "Mr. Bell never destroys letters," wrote his wife in 1899. The discovery of crucial telephone documents in a trash basket had given Bell a lifelong reverence for wastepaper; and in his later years, the historical motive reinforced his compulsion to save every scrap with writing on it — circulars, telegrams, invitations, greeting cards, even his grandchildren's scribbled drawings. "All the years I was with him," wrote Catherine MacKenzie, "I felt Mr. Bell's constant mindfulness of posterity and his biographer, in everything he wrote, and in a good deal he did. The sense of an audience is very evident to me in all his Home Note-books."

In July 1909, inspired by the recently discontinued *Aerial Experiment Association Bulletin*, Bell began issuing the *Beinn Bhreagh Recorder*, part research bulletin, part house organ, part news magazine, part court circular. As the years passed, the *Recorder* also incorporated more and more biographical material, Bell family history, and reprints of Bell's earlier writings. First issued in seven typed copies for the information of Baldwin and McCurdy during their absence on airplane trials, the *Recorder* came to be mimeographed in twenty copies for friends, family, the Volta Bureau, and the Smithsonian. Eventually the file lengthened to twenty-five bound volumes of four or five hundred pages each.

In March 1922 Bert Grosvenor persuaded Bell to let the National Geographic Society set aside a "Bell Room" in Hubbard Hall as a depository for Bell's massive accumulation of records and memorabilia. There, fifty years later, the collection remains.

During Bell's later life, occasional events and encounters reawakened memories of the telephone years. Shūji Isawa, the Boston University student who had been the first to speak a foreign language over the telephone, in

1901 sent Bell a book Isawa had written on Visible Speech. In 1905 Bell was an eyewitness at Portsmouth, New Hampshire, to the making of peace between Russia and Japan, in which Isawa's Harvard friends Komura and Kaneko played key roles. Three years later Bell reread his 1892 Government case deposition, which the American Telephone and Telegraph Company had just published in book form "because of its historical value and scientific interest." "Have enjoyed enormously fighting my old telephone battles all over again," he wrote Mabel from his houseboat.

Bell also reestablished communication with Tom Watson. After quitting the telephone business in 1881 and making a grand tour of Europe, Watson had married, farmed awhile near the Fore River in Braintree, Massachusetts, and then returned to his natural element, the machine shop. His success in making marine engines led him to expand the shop he had built on his farm. Watson's surplus energies during the nineties went into three years of geological study at MIT, leadership in the "Nationalist" movement of Edward Bellamy, the inception and management of Braintree's municipal electric company, and the chairmanship of the town's school board. During the depression of the nineties, Watson considered turning his shop to the making of steam automobiles, but the navy's expansion program tempted him in 1897 to make a successful bid for a destroyer contract. As more contracts came his way, he built one of the biggest shipyards in the nation and incorporated it in 1901 as the Fore River Ship and Engine Company. By 1903 the company was at work on twenty million dollars' worth of government contracts. During this frenzied expansion, Watson's two sons died, leaving him, like Bell, with two daughters.

A number of magazine articles were published on the spectacular new enterprise, and one of these may have stirred Bell in 1903 to write Watson a recommendation for a deaf boy who wanted to learn shipbuilding. "P.S.," Bell added, "Are you my old friend Thomas A. Watson of Salem, Mass.?" Watson replied cordially that he was that man, would gladly talk with the boy, had been following Bell's kite experiments with interest, and would like to show Bell around the Fore River plant. That fall, before Bell could take him up on the offer, Watson was squeezed out of his overextended company by financiers and in the process lost most of his fortune. He had the satisfaction of seeing the shipyard continue to grow, pass into the possession of Bethlehem Steel, and by 1919 give work to twenty thousand of his fellow Bay Staters. Meanwhile he lived comfortably on the income of a small trust fund he had prudently set up before his shipbuilding plunge. He studied art and music appreciation, traveled in Europe, and supplemented his income with public readings of poetry and plays. Watson especially enjoyed touring England, 1910–1911, as an actor with a company of

Shakespearean players. On his arrival there he had been told to await the call to rehearsals; and so we may speculate that on that occasion he heard still another classic telephonic line: "Don't call us, Mr. Watson, we'll call you."

Bell wrote Watson for historical material in 1905 when Bert Grosvenor was still working on the projected telephone story, but Watson could offer him little more than memories. Still, those memories were sharp and vivid. From 1913 onward Watson lectured hundreds of times on the "Birth and Babyhood of the Telephone." Bell congratulated him on the pamphlet version published by the AT&T Company in 1913.

One more historic telephone conversation remained for Bell and Watson. On the afternoon of January 25, 1915, by arrangement of the telephone company, Bell waited in New York and Watson in San Francisco for the formal opening of the transcontinental telephone line. The first official utterance over that line was Bell's: "Mr. Watson, are you there?" Watson was there and heard perfectly. President Wilson, Bell's political hero, spoke from the White House, congratulating Bell on "this notable consummation of your long labors." Theodore Vail (who had left the presidency of AT&T in 1887 but resumed it in 1907) spoke from Georgia. Presently a duplicate of the Centennial telephone was connected to the line at New York, and into it Bell repeated the nostalgic words, "Mr. Watson, come here, I want you." Watson protested that it would take him a week now.

At Boston in March 1916, on the occasion of the fortieth anniversary of Bell's first telephonic remark to Watson, Bell unveiled a plaque at Exeter Place commemorating "the First Complete and Intelligible Sentence by Telephone, March 10, 1876" and another at 109 Court Street, site of the Williams shop, proclaiming that "Here the Telephone Was Born June 2, 1875." At a Boston University luncheon and a City Club dinner, Bell once again told the story of those great days, and Clarence Blake recalled his own part in them.

The Boston ceremonies touched the civic pride of Brantford, Ontario. But in the matter of memorials, Brantford had the last word. As early as 1904, the president of the Brantford Board of Trade had proposed an Alexander Graham Bell monument there. Bell thought any monument to him should be posthumous, and he was in no hurry. But in 1917 he consented to speak in Brantford at the dedication of a memorial to the invention of the telephone. On October 24 of that year the governor-general of Canada unveiled the imposing memorial in a driving rain. Bell's speech brought sunshine into the hearts of municipal boosters, whose ambition was to advertise Brantford as the "Telephone City." "Brantford is right," said Bell, "in claiming the invention of the telephone here." He added, however, that the telephone "was conceived in Brantford in 1874 and born in Boston in 1875."

That afternoon the governor-general also dedicated the former Bell home at Tutelo Heights as a memorial to be administered by the Brantford park department. Erosion had claimed some of the bluff where the telephone had, by the Brantford definition, been invented, but the Bells posed jovially for a photograph as near the "dreaming place" as possible. Bell's retracing of the past had brought him back to "the days when there were no telephones."

Now and then over the years, reminders came to Bell of still remoter days in Scotland and England. On his European trip of 1881–1882 he revisited his boyhood friend and earliest collaborator in invention, Ben Herdman; and a few years later Ben tried unsuccessfully to enlist Alec in one more joint research, this one to find a remedy for the "coffee blight." At Melbourne, Australia, in 1910, the Bells paid a visit to Mrs. McBurney, the former Marie Eccleston, who had rejected Bell's proposal of marriage forty years before. After losing all his money in a bank failure, her husband had recently died; and his widow, now in her seventies, was trying to live by giving music and singing lessons. "She doesn't look as if there were much money in her profession," wrote Mabel, "but she is a bright, plucky, energetic woman."

In April 1906 the Bells spent a couple of weeks in Edinburgh, where the university gave Bell an honorary degree. Sentiment, family feeling, and his interest in heredity kept Bell busy searching parish records for Bell births, marriages, and deaths. But he found time to drive with Mabel to Trinity and hunt up Milton Cottage, the family retreat of his boyhood. Once it had been set amid woods and fields; now the suburbs had engulfed it. Its lot had been joined to another, its cottage enlarged, and so Bell did not recognize the place. Like many another in search of a time long gone, he looked for it where he had last seen it, only to discover that it was not there either.

Nevertheless something drew him back again. In May 1920, three days before the fiftieth anniversary of his brother Melly's death, Bell noted that he had dreamed about a visit from his parents and brother Edward, "and that they were very anxious for me to do something." Early in October 1920 the Bells, their fifteen-year-old granddaughter Mabel Grosvenor, and Miss MacKenzie sailed for England and Scotland on what Bell called a "farewell visit."

At London they saw old friends, including Chichester Bell and Adam Scott. Then they motored to Bell's former haunts elsewhere in England. But Edinburgh was the prime goal. There Bell dug deeper into family history, searching old directories, triumphantly discovering that his grand-father's brother had been a brewer. (Though Chichester Bell disapproved of Alec's dredging up "facts that are not so creditable to the family," Alec held to his view that the lowlier the start, the prouder was the rise.) Bell took his granddaughter to see his birthplace on Charlotte Street and drove out to Corstorphine Hill.

Even more strongly than in 1906 it was borne upon Bell that the past lives only in the mind. When he got to Corstorphine, he realized that he was too old now to climb to "Rest-and-be-thankful," where he had watched the birds so long ago. He grew more and more depressed. Many of the friends he had seen in 1906 were dead by 1920. Ben Herdman had died of a heart attack in 1908. Alec and Mabel drove north to Elgin and thence to Covesea, where they had spent their honeymoon. A cold November rain swept in from the North Sea, and they could not be sure which of the two or three deserted, tumbledown cottages had been theirs. At Pluscarden Abbey, near Elgin, the grassy floor had been paved with cement, and a wooden superstructure had been built around the moss-grown tower. The transformation may have been more useful for parish activities. But Bell remarked plaintively, "This is not Pluscarden."

One of his old Royal High School friends, however, proposed that the city councillors bestow on him the official "freedom of the city." Money running low and spirits lower, Bell was in London on the way home to America when word came that the honor had been voted. He canceled the sailing, and the Bells returned to Edinburgh.

Shaking off his uncharacteristic melancholy, Bell basked in the enthusiasm of a Royal High School student assembly. Then, at the City Chambers, Mabel Bell and her granddaughter proudly watched the "medieval pomp and pageantry" that culminated in the presentation of a silver casket symbolizing the honor. Bell's voice, still melodious and strong, carried to all of the large audience. "I have received many honors in the course of my life," he said, "but none that has so touched my heart as this gift of the freedom of my native city." To a true Scotsman, a still more signal tribute was the assertion of the city's leading newspaper next day that not even Sir Walter Scott had brought more honor to Edinburgh.

But the glow faded. In London a week later the weary sojourner told a reporter, "I visited Edinburgh, but I was a stranger in my own land. . . . My advice to those who have remained away from home and contemplate returning for a farewell visit is — don't."

484

35

Of Striving and Silence

A month after they were engaged in 1875 Alec Bell wrote Mabel Hubbard, "All my ancestors . . . have lived to extreme old age — so you need
not expect to get rid of me very soon." A quarter century later, the example of others had begun to shadow his own assurance. Early in 1901, the
death of Elisha Gray led Bell to reflect that "there is no one, I think, now
left alive of all the counsel who so ably defended me before the courts . . .
and as I look around to see who else remains who had personal knowledge
of the facts of the past, I recognize that I am left almost alone." "The
older one grows," he wrote Arthur McCurdy a few months later, "the more
rapidly does he find his friends dropping off all around him. . . . I am
sometimes afraid — since my return from Japan and Europe — to ask after
old friends for fear of finding a vacant place, or touching a wounded heart
by my ignorance." He was then fifty-four.

In 1908, the year Ben Herdman and John Hitz died, pneumonia claimed
Charles Williams, Jr., who had retired in 1886 with a fortune in the stock
of the Western Electric Company (his successor as manufacturing affiliate
of AT&T). Three years later, Thomas Sanders died of a heart attack at
seventy-one while driving up to his Vermont farm. "I begin to feel very
old," wrote Bell to George Sanders on hearing the news. "You are about the
last remaining link connecting me with those happy Salem days." Having
swept away his elders, death was taking Bell's contemporaries. A Royal
High School classmate wrote from London in 1915 that he could trace only
25 left alive of their class of 135. That year Bell heard without disbelief a
report that Sumner Tainter was dead. The report was mistaken. But Bell
was correct in noting early in 1919 that Clarence Blake and Edward Pickering had just died. And a year later Theodore Vail followed them.

By 1915 Bell himself was in the grip of that incurable and (in pre-insulin
days) perilous condition, diabetes. He had occasion for somber thought now

in his sessions of nocturnal solitude. He could not have been deaf to the meaning of what was upon him. Diabetes had been the death of his Uncle David fourteen years before. But Bell's motto was "keep on fighting."

Always a hearty eater, Bell broke loose now and then from the coils of medical caution and, to the distress of his family, defied restrictions on starch and sugar. "Melville," he would say to his grandson as they walked by a redolent bakeshop on Wisconsin Avenue, "would you like some apple pie?" Bell himself would then join in the snack. "Don't you say a word to your grandmother," he would caution the boy. But when he toyed with his dinner, Mabel would notice. "Alec, you stopped in that bakeshop, I know." Ignoring the smoke screen of an exciting story, she would keep after him until he confessed like a small boy caught out. Charles Thompson kept an eye on the state of the refrigerator, but one night Bell made a raid, washed the china, and brushed up every crumb. Called to treat his acute indigestion, the doctor extracted a confession. "To go downstairs at three in the morning, load up on Smithfield ham, cold potatoes, macaroni and cheese, and then go right to bed is the most ridiculous thing imaginable," said the doctor severely; "that meal might have put an end to you, sir." "Well, as it is," said Bell, "the game was worth the candle. It was the best meal I've enjoyed in an age."

Bell's defiance of death went on in other ways. "Never felt better in my life," he assured Fred DeLand in 1916, though he had begun at last to grow perceptibly thinner. From Beinn Bhreagh in April 1917, hard at work on hydrofoils for the war effort, he wrote Mabel, "Weather here just suits me — cold and bracing — and John McDermid makes me walk every day." When he was in Washington Bell would stop in at the Volta Bureau every week or so to encourage Fred DeLand and to draw encouragement himself from the evidence that the Association and the Volta Bureau now had a sturdy, independent life of their own. In 1917, reluctantly but dutifully, he accepted the presidency of the Clarke School's board of corporators and thereafter visited the school for a day or two whenever he traveled to or from Beinn Bhreagh. In his last months of life, he gave two lectures there on Visible Speech and carried on an extensive correspondence with educators, scientists, and others about establishing a department for research on the problems of hearing loss.

Bell's experimental work went on too — sheep, hydrofoils, and occasional revivals of interests from the distant past, such as selenium experiments. He talked of those and other notions in a magazine interview published in December 1921: the photophone, air conditioning, conservation of waste heat from stoves, the storing of solar heat in some fluid circulated on rooftops and passed into insulated tanks. "There cannot be mental atrophy," he said, "in any person who continues to observe, to remember what he observes, and to seek answers for his unceasing hows and whys about things."

His last serious line of experiment was not new to him. Many years earlier his ocean voyaging had made him conscious of "undrinkable water upon which so many people die from thirst." From shipboard in 1900 he wrote Mabel that he meant to look into the problem of condensing drinking water from the sea, and for several years thereafter he mingled distillation experiments with sheep, kites, and aeronautics. Wave-powered bellows pumped fog through sea-cooled bottles; an assistant patiently exhaled through them. The condensed moisture seemed scarcely worth the trouble. But after the war, Bell returned to the pursuit. From 1920 to 1922 his notebooks recorded experiments with various simple contrivances. Toward the end, his favorite was a shallow box about three feet square, with a sloping glass lid on which moisture condensed from sun-warmed seawater beneath it, and down which the distilled drops trickled into a container. (Many years after Bell's death, Melville Grosvenor saw a whole field of such devices on the Greek island of Patmos.) On the eve of his seventy-fifth birthday in 1922, weak and haggard as he had become, Bell wrote Bert Grosvenor, "I am just starting on water distillation experiments at low temperatures, and as I think of trying to patent the apparatus, am avoiding publication at the present time."

So Bell kept on fighting. Only once, to Mabel in 1879, had he ever blurted out the fear that his greatest achievement, coming early in life, might condemn him to a long anticlimax. Yet in his reluctance to submit to a comprehensive biography and in other subtle ways, there were hints that the fear never died. Now, even at the edge of ultimate silence, his striving went on; and in that very striving, he defeated anticlimax after all.

For the life of Alexander Graham Bell, whether or not he realized it himself, exemplified an uncommon sort of heroism. The conventional hero tale is one of long adversity culminating in a final crescendo of triumph. Bell holds up to us a different pattern. The longest struggle of his life was not against obscurity but against the deadening drug of early fame. And however we measure the practical results, in never forsaking that struggle Bell showed valor of as high an order as any other open to the will of man.

"Smile and look cheerful," Bell often said, "and pretty soon you'll feel so." He practiced the precept now, for Mabel's sake as well as his own. Indeed, after a Washington dinner in June 1921, he plumed himself on having got the dour Vice-President Calvin Coolidge to smile at least once. At Beinn Bhreagh, Mabel would go as usual to Alec's office in the late afternoon, and they would walk home contentedly hand in hand along the road to the point of the promontory.

With the Fairchilds in 1916, the Bells paid their first visit to southern Florida, where Bell heard William Jennings Bryan speak and as a former teacher of the art gave him high marks for elocution. David Fairchild had described the Experimental Garden at Miami with a botanist's enthusiasm; and so, in order to be awake in the early morning when the garden was at its best, Bell sat up all night. Fairchild found him by his hotel window. "David," said Bell excitedly, "I have just seen something which I have heard about but never quite believed. I have seen buzzards, there on the sand bar, face the breeze and rise into the air without flapping their wings."

Perhaps Bell's loss of weight had made him more tolerant of warm climates. He and Mabel toured the Caribbean in the winter of 1921–1922, and Mabel reported that "we are having the time of our lives." The acting governor of the Canal Zone put his car and yacht at their disposal. Bell called at consulates, studied government publications, and in Jamaica drove around for most of one day in a futile search for the grave of a seafaring relative who had died there more than a century before. At Coconut Grove, Florida, where the Fairchilds had a place, he worked awhile on his water distillation experiments. He was there on his seventy-fifth birthday when greetings came in from family and friends, including Emile Berliner, who had a comfortable fortune and like Bell had been experimenting in aeronautics in recent years.

But there had been signs of weakening. In Venezuela Bell had uncharacteristically admitted to being too tired to visit the Caracas School for the Deaf. In the Bahamas he professed to feel well, but would not go walking. "I guess our travelling days are pretty well over now," wrote Mabel from Florida, "and I am glad we had this trip to remember."

With rest and the coming of spring, Bell seemed to revive. He had the newest household marvel, a radio, installed in his Washington study; and remembering the musical tone he had heard upon sending an intermittent current directly through his ears almost half a century before, he wondered if somehow one might similarly "hear the radio concert without any special receiver at all." Perhaps he was dreaming of music for Mabel. At Beinn Bhreagh in June, he took pleasure in the fresh air and cool nights. The sheep had to be looked after and tabulated. What turned out to be the last issue of the *Beinn Bhreagh Recorder* appeared on July 11, closing with an account of lambs born that year to "Mrs. Bell's Multi-Nippled Twin-Bearing Stock." Bell had new ideas about connection pieces for tetrahedral construction. Casey Baldwin had become interested in combining that construction with hydrofoils to make a light, fast, strong, easily repaired naval towing target. Later that month Bell watched trials of Casey's hydrofoil target model and entered the details in his notebook.

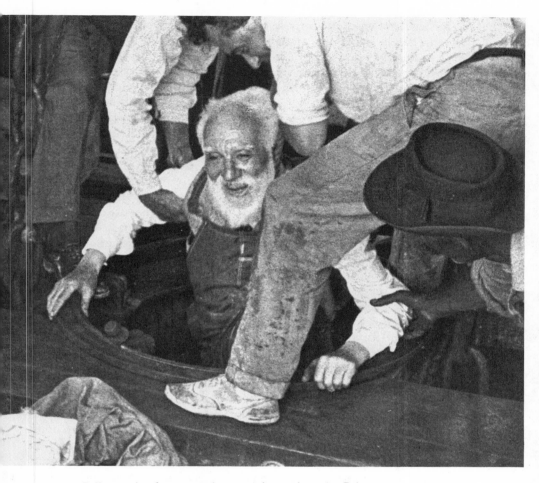

Bell emerging from an underwater observation tube, Bahamas, 1922

The end, so long foreseen, came as a surprise. Bert and Elsie Grosvenor were in Brazil. David and Marian Fairchild happened by chance to arrive for a summer visit on July 30, a few days earlier than planned, and found sudden concern at Beinn Bhreagh. Only a day or two before, Bell had been out looking over his sheep and then had sat up reading late into the night as usual. Now he was too weak to get up from his lounge. His appetite had failed utterly, and Mabel counted it "his last hard service to me" that he tried to force himself to eat because she wanted him to. The doctor thought that Bell's diabetes had affected his liver and pancreas.

On the evening of July 30, Bell sat up on his lounge, read a novel, and talked awhile. David Fairchild asked him if he thought that his weakness might be electrical in character and if life might not have an electrical basis. Bell shrugged and said, "Je ne sais pas, Monsieur. Je ne sais pas." No one seems to have raised the question of an afterlife. If they had, Bell's answer would presumably have been the same. He had remained steadfast in agnosticism and therefore, as Mabel took comfort in remarking, "he never denied God." Neither did he affirm God. He and Mabel occasionally attended Presbyterian services and sometimes Episcopalian, at which Mabel could follow the prayer book. Since otherwise she depended on Alec's interpreting, their church goings were rare; but their children attended Presbyterian services regularly. In 1901 Bell came across a Unitarian pamphlet and found its theology congenially undogmatic. "I have always considered myself as an Agnostic," he wrote Mabel, "but I have now discovered that I am a Unitarian Agnostic." He came no closer than that to formal identification with any organized sect.

Through July 31, Bell remained semiconscious, now and then murmuring confusedly, as he usually did when Charles Thompson tried to rouse him. He did not seem to be in pain. During the afternoon of August 1, he revived and with much effort and many pauses dictated a statement to Catherine MacKenzie. "Don't hurry," someone said. "I have to," said he. "I want to say that . . . Mrs. Bell and I have both had a very happy life together, and we couldn't have had better daughters than Elsie and Daisy or better sons-in-law than Bert and David, and we couldn't have had finer grandchildren." He struggled with the problem of how to provide financially for his children in the situation of Mabel's having legal title to nearly all his property, but he was finally content with an assurance that provision would somehow be made. And lastly, as his little reserve of strength gave out, he said, "We want to stand by Casey as he has stood by us . . . want to look upon Casey and his wife Kathleen as sort of children."

In his bed on the sleeping porch, facing Beinn Bhreagh, the "Beautiful Mountain," Bell sank back again into a half-sleep. Darkness gathered, and a lamp was lit on the table. Stars came out. A bright moon rose. There was stillness everywhere. Alec held Mabel's hand, and now and then his eyes opened and he smiled up at her. Midnight passed, and it was August 2, 1922. At about two in the morning, Mabel was resting on a sofa when David felt the dying man's pulse suddenly fade and called her to her husband's side. Alec's breathing grew slower and more labored. She spoke his name, and he opened his eyes for the last time and smiled at her again. "Don't leave me," she begged him. His fingers clasped hers with the old sign for "No." Even after his pulse could no longer be felt, she could feel his fingers move in the last feeble effort to comfort and communicate. Her face was the last he saw, her hand the last he touched, her voice the last he heard. Then silence closed about him forever.

Postlude:
The Sounding Name

On the day that followed her husband's death, Mabel Bell drugged herself with activity, composing telegrams and planning his funeral. Once, many years before, standing with her at the top of Beinn Bhreagh, Alec had remarked that he would like to be buried there when the time came. Never since then had the subject come up again, but she remembered. Under the tetrahedral tower, workmen now blasted a grave out of solid rock and lowered a steel vault into it. At the house, Bell's body, dressed in his familiar gray corduroy suit and knickerbockers, was laid in a coffin of Beinn Bhreagh pine, lined with airplane cloth. The day of the funeral, August 4, 1922, was gray and misty, the sort of day Bell had loved best. There were few other signs of gloom, which Mabel considered out of keeping with her husband's spirit. Children played around the house and ran up and down the stairs. No one wore mourning. Mabel felt that if black (which Alec had not cared for) were once worn, she could never bear to put it aside nor even to see others do so. The women wore white and the men their summer clothes.

John McDermid brought the coffin up the hill in a buckboard. The children had made a pall of fir boughs, and on it rested a floral tribute of roses, wheat sheaves, and laurel from the AT&T Company in New York. Many people had assembled in the panoramic setting of the grave site. They joined in the hymn "Bringing In the Sheaves." A Cape Breton girl sang the first verse of Stevenson's "Requiem," the local Presbyterian minister read Longfellow's "Psalm of Life," and all said the Lord's Prayer. The coffin was lowered. Sandy Fairchild and Casey Baldwin's son Bobby raised United States and British flags on temporary poles. And it was over.

Later, Mrs. Bell had the tower taken down lest it fall into disrepair. She intended to have a more permanent demonstration of her husband's tetrahedral system built at the site, but it was not. The grave is marked only by a rock in which is set a tablet bearing Bell's name, dates, vocation of "inventor," and the statement he himself had specified: "Died a Citizen of the United States."

David Fairchild had notified the Associated Press of Bell's death soon after it occurred, and telegrams of condolance began flooding into the Baddeck office. At the time of the burial, 6:25 P.M., August 4, 1922, all telephone service was stopped for one minute throughout the United States, and flags stood at half-mast on all buildings of the Bell System. Public tribute was paid to Bell's memory throughout the nation and the world. As a member of the advisory council of the American Association to Promote the Teaching of Speech to the Deaf, Thomas Edison (himself partially deaf) spoke of "my late friend, Alexander Graham Bell, whose world-famed invention annihilated time and space, and brought the human family in closer touch."

On August 2, Thomas Watson and his wife had been hiking through the Colorado mountains when a dense cloud swept over them and they lost the trail. After they had wandered for hours, the cloud suddenly lifted, and they found themselves near their camp. Waiting for them was a telegram from Mabel Bell announcing her husband's death. "I have often wondered," wrote Watson in his diary that night, "whether Bell or I would be the last surviving member of the four original associates. The fates have assigned to me that pathetic honor, but I feel a strange loneliness."

To all who knew Bell, his death brought sadness. To Mabel it brought desolation. "I am not reconciled," she wrote on the day Alec died. "I never will be." Two weeks later she knew it was "going to be terrible trying to live without him." And two months after that, it was still "unbelievable to me that one so strong and full of life should have gone. I thought I would always have him. . . . We lived so closely to one another that he was hardly ever out of my thoughts."

Her great object now was to provide for the continuation and completion of his work, especially the sheep and the hydrofoil projects, and the maintaining of Beinn Bhreagh for her children and grandchildren. Inseparable from these things was provision for Casey Baldwin and his family. For seventeen years Casey had given the Bells the love of "a third son-in-law," and his work had been much closer to Bell's than that of David or Bert. Mabel meant to "stand by Casey" as Alec had wanted. So she drew up and signed a contract binding herself and her estate to provide ten thousand dollars a year for ten years of Beinn Bhreagh laboratory work under Casey Baldwin.

Then there was the matter of the biography. Fred DeLand was to arrange and catalogue the material in the National Geographic Society's Bell Room. Mabel had written Bert several months before Alec died that she did not want a member of the family to write the biography. "I am a horrid old thing," she told Bert,

but my husband is so much to me that I know the very best account of him that could be written will seem to me wrong in some way. It would praise him perhaps, but in ways that did not seem to me true, and I would hate to have things attributed to him that were not so. He is big enough to stand as he is, a man, very imperfect, lacking in things that are lovely in other men, but a big good man all the same, broad-minded and generous and tolerant in some things beyond the comprehension of most, and then curiously the opposite in others. I do not know a person he has not fought some time or other, and sometimes I, his wife, have thought him badly in the wrong, and at others entirely in the right. But I could never say this publicly, it would seem disloyalty, and none of us would either, I think, which would mean that the book would be inadequate.

Mabel survived Alec by only five months. Her last words to Bert, two days before she died on January 3, 1923, were on the same subject. "You must see that the biography does not picture Father as a perfect man. He was a very clever man and a good man, but he had his faults, just like every other human being. And I loved him for his faults. Mr. Kennan has put Mr. [Edward H.] Harriman on a pedestal. I don't want you to do that for Father."

She was buried beside Alec in the grave on Beinn Bhreagh.

The earthly world of Alexander Graham Bell gave way gradually to the passing years. Fred DeLand died in 1928, Charles Bell and Emile Berliner in 1929. For eight years the house on Connecticut Avenue stood vacant while residences all around it yielded to business establishments, until in 1930 it was sold and demolished. A few weeks later, Glenn Curtiss died. At midnight on August 11, 1938, a taxi struck George Sanders on Park Avenue in New York City, and he died of shock the next day at the age of seventy-one. In April 1940 Sumner Tainter died in San Diego, where uncertain health had kept him in retirement for many years; he was eighty-five.

Tom Watson lived to publish his high-spirited autobiography, *Exploring Life*, in 1926. He also survived the Wall Street crash of 1929 in easy circumstances, dying in 1934 at eighty in his Florida winter home with a fortune of about half a million, only a small fraction of it in telephone stock.

Casey Baldwin stayed on at Beinn Bhreagh, died of a heart attack in August 1948, and was buried on the mountain not far from the Bells' grave. He had worked on hydrofoil naval targets during the early twenties and again during the Second World War with little success, in part because of the structural weaknesses he seemed compulsively to overlook. At intervals during the late twenties, mid-thirties, and war years, he designed several hydrofoil craft, of which three were built; but nothing significant grew out of these efforts, and the postwar revival of hydrofoil craft had other sources. Bell's descendants still maintain and use the house and estate at Beinn Bhreagh.

In 1928 Bell's last secretary, Catherine MacKenzie, published an account of Bell's life, for which (not having access to the Bell Collection) she disclaimed the title of biography. To the extent of her direct knowledge of Bell's later years, her book has permanent value. She married a magazine editor; and though childless herself, she earned a national reputation as writer of the *New York Times* regular Sunday feature on child development and mental health during the ten years before her death in 1949.

The longevity Bell had expected for himself appeared in his daughters — and for that matter, his sons-in-law. Elsie lived to be eighty-five, Marian to be eighty-two. David Fairchild died in 1954 at eighty-five, and Gilbert Grosvenor, the last of that generation, in 1966 at ninety. From Bell's eight married grandchildren have sprung great-grandchildren enough to make it probable that Bell descendants will walk the earth until doomsday, longevity or not.

None of them have inherited the Bell name. But the name lives. Its survival does not depend on the Arctic island to which the National Geographic Society assigned it in 1932 and which the Soviet Union, having jurisdiction, accepted as Graeme Bell Ostrov (Island). Neither does it rest on the popular movie biography premiered at the society in 1939 and at the time of this writing still occasionally rerun on television; nor on the Alexander Graham Bell Museum at Baddeck, which in a building with an architectural motif of triangles and tetrahedrons displays the models, photographs, notes, and other memorabilia of Bell's Beinn Bhreagh years; nor even on the fact that Bell has been enrolled in the Hall of Fame at New York University and, along with Curtiss and Selfridge, in the Aviation Hall of Fame at Dayton, Ohio.

In 1956, the AAPTSD became the Alexander Graham Bell Association for the Deaf. As Bell intended, it serves as a worldwide center of information on deafness, maintaining its library, publishing books and the *Volta Review*, administering scholarships, organizing parents, answering thousands of letters every year. With a staff of twenty-five and an annual budget presently exceeding half a million dollars, it seems likely to keep

Bell's name and his work alive for a long time among the deaf and hard of hearing.

When Bell died, the Bell System, one of history's greatest business enterprises, served less than ten million telephones; fifty years after his death, the system serves a hundred million telephones and employs a million people. This namesake of Bell has the look of permanence. (No doubt it was awareness of the Bell System that led four hundred corporation executives polled in 1967 to rank Bell seventh on the list of great American businessmen of the past, between Bernard Baruch and E. I. Du Pont; Bell, who considered himself no businessman at all, would have smiled at that.)

But Bell's name will live longest by virtue of history, which outlasts map names, scrolls, and charters. For history irrevocably attaches the name of Alexander Graham Bell to a technological creation too fundamental to be superseded and too useful for the world to do without. And in more ways than Thomas Edison suggested, history will also count Alexander Graham Bell among those who helped to bring the human family into closer touch.

Bibliography

References

Bibliography

The principal source for this biography was the immense Alexander Graham Bell Collection (of which Melville Bell Grosvenor is literary executor) deposited with the National Geographic Society in Washington, D.C. It includes all of Bell's voluminous incoming and outgoing correspondence (family and otherwise), correspondence and other material of members of his family and the family of Gardiner G. Hubbard, books, photographs, published and unpublished articles and reminiscences, laboratory records, clippings, journals and diaries, transcripts of legal testimony, and much else. Significant manuscript material is included in the collections of the Volta Bureau Library of the Alexander Graham Bell Association for the Deaf in Washington, D.C.; the Charles Sumner Tainter MSS, Division of Mechanical and Civil Engineering, Smithsonian Institution; and the Telephone Historical Collection of the American Telephone and Telegraph Company, New York. Among smaller collections, the Alexander Graham Bell MSS in the Special Collections of the Boston University Library, and the Thomas Borthwick MSS in the Scottish National Library, Edinburgh, furnished useful items. In the course of concurrent research for another book, I visited scores of manuscript depositories throughout the country and in each looked for Bell material, but turned up only a few items not duplicated or bettered in the principal Bell Collection.

Most of the 149 volumes of printed testimony in litigation unsuccessfully challenging the Bell telephone patents deal with alleged inventions of the telephone independent of and prior to Bell and so has no bearing on his story. From these volumes, the Bell Company's chief counsel, James J. Storrow, had all printed testimony by Bell or bearing on Bell's work culled for Storrow's private reference and bound together in two volumes, totaling 1674 printed pages, with the title "Proofs By and About Alexander G. Bell." These, with the volume entitled *The Bell Telephone*, published by the company in 1908 and containing Bell's 445 pages of testimony in 1892, were my chief reliance, beyond the important unpublished material in the Bell Collection, for the telephone story. Both the Bell Collection and the AT&T Historical Collection contain copies of the Storrow "Proofs." Since the AT&T Company has recently deposited microfilms of them in the Library of Congress and in the Special Collections of the Boston University Library, I have cited them directly as "Proofs," rather than citing the original volumes, usually less accessible or harder to locate, from which they were drawn. However, to indicate some of the principal sources briefly, I have cited four major cases by name: the Dowd Case, the Harmonic Interferences, the People's Telephone Case, and the Telephone Interferences. Material thus cited may be found in the "Proofs" volumes by means of the table of contents in each volume.

Published sources cited by short title in more than one chapter:

Beinn Bhreagh Recorder: issued periodically by mimeograph under this title by Alexander Graham Bell, 1909–1923.

Bell Telephone: American Telephone and Telegraph Company, *The Bell Telephone* (New York, 1908).

Bell, *Visible Speech:* Alexander Melville Bell, *Visible Speech: The Science of Universal Alphabetics* (London, 1867).

Best, *Deafness and the Deaf:* Harry Best, *Deafness and the Deaf in the United States* (New York, 1943).

DAB: Allen Johnson, et al., eds., *Dictionary of American Biography* (22 vols., New York, 1928–1958).

DeLand, *Dumb No Longer:* Fred DeLand, *Dumb No Longer* (Washington, D.C., 1908).

Dowd Case: see preceding paragraph.

Fairchild, "Alexander Graham Bell": David Fairchild, "Alexander Graham Bell: Some Characters of His Greatness," *Journal of Heredity,* XIII (1923).

Fairchild, *The World Was My Garden:* David Fairchild, *The World Was My Garden* (New York, 1939).

Finn, "Bell's Experiments": Bernard S. Finn, "Alexander Graham Bell's Experiments with the Variable-Resistance Transmitter," *Smithsonian Journal of History,* I (1966).

Grosvenor, "My Grandfather Bell": Lilian Grosvenor, "My Grandfather Bell," *The New Yorker,* November 11, 1950.

Grosvenor, *National Geographic Society:* Gilbert H. Grosvenor, *The National Geographic Society and Its Magazine* (Washington, D.C., 1957).

Haberman, "Elocutionary Movement": Frederick W. Haberman, "The Elocutionary Movement in England, 1750–1850" (Ph.D. dissertation, Cornell University, 1947).

Harmonic Interferences: see preceding paragraph.

Improvement of the Wisconsin System: Wisconsin Phonological Institute, *Improvement of the Wisconsin System of Education for Deaf Mutes* (Milwaukee, 1894).

Josephson, *Edison:* Matthew Josephson, *Edison* (New York, 1959).

Kennan, "Recollections": George Kennan, "A Few Recollections of Alexander Graham Bell," *The Outlook,* September 27, 1922.

MacKenzie, *Bell:* Catherine MacKenzie, *Alexander Graham Bell* (Boston, 1928).

Bibliography

Parkin, *Bell and Baldwin:* J. H. Parkin, *Bell and Baldwin* (Toronto, 1964).

Pier, *Forbes:* Arthur S. Pier, *Forbes: Telephone Pioneer* (New York, 1953).

Prescott, *Bell Telephone:* George B. Prescott, *Bell's Electric Speaking Telephone* (New York, 1884).

Proofs: see preceding paragraph.

Rhodes, *Beginnings of Telephony:* Frederick L. Rhodes, *Beginnings of Telephony* (New York, 1929).

Tosiello, "Bell System": Rosario J. Tosiello, "The Birth and Early Years of the Bell Telephone System, 1876–1880" (Ph.D. dissertation, Boston University, 1971).

Waite, *Make a Joyful Sound:* Helen E. Waite, *Make a Joyful Sound* (Philadelphia, 1961).

Watson, *Exploring Life:* Thomas A. Watson, *Exploring Life* (New York, 1926).

A full bibliography of Bell's published writings and addresses, as well as a list of his United States patents, may be found on pages 20–29 of Harold S. Osborne, "Biographical Memoir of Alexander Graham Bell, 1847–1922," National Academy of Sciences *Biographical Memoirs,* XXIII (Washington, D.C., 1943).

References

The following notes are confined exclusively to sources and occasional brief comments on them. Full bibliographical information on each published source is given with its first citation. All unpublished sources are in the Bell Collection at the National Geographic Society, Washington, D.C., unless otherwise stated. In citations of correspondence, AGB stands for Alexander Graham Bell, AMB for Alexander Melville Bell, and GGH for Gardiner Greene Hubbard. "Mabel" refers to Mabel Hubbard (later Mrs. AGB).

Sources for each paragraph are indicated by the last word of the paragraph and the number of the page on which it ends.

PRELUDE

5. (applications): J. Henry to H. Alexander, May 6, 1846, Joseph Henry MSS, Smithsonian Institution.

CHAPTER 1

9. (paper): *The Scotsman* (Edinburgh), Mar. 6, 1847.
10. (grandfather): *Edinburgh Advertiser*, Mar. 5, 1847.
10. (stage): *Beinn Bhreagh Recorder*, XXIV, 185; Catherine MacKenzie, *Alexander Graham Bell: The Man Who Contracted Space* (Boston, 1928), p. 14; genealogical data in the hand of AGB, folder, "Bell, Alexander — Biographical."
10. (workshop): MS memoir of Mrs. AMB by AMB, 1897.
10. (Prompter): *Beinn Bhreagh Recorder*, XXIV, 185; James C. Dibdin, *The Annals of the Edinburgh Stage* (Edinburgh, 1888), pp. 332, 337–339, 468, 498; *Edinburgh Evening Courant*, Dec. 9, 13, 1819; Theatre-Royal Playbills, Edinburgh Room, Edinburgh Public Library, Jan. 17, Aug. 16, 27, Sept. 1, 5, 14, 1821, Aug. 27, Nov. 1, 13, 1823, Feb. 27, 1824; "Re-Revised Condescendence, Poor Bell vs. Murray," Apr. 4, 1833, Jury Office Process No. 15, Bell vs. Murray, 1833, Scottish Record Office, Edinburgh.

References

11. (Andrews): *Beinn Bhreagh Recorder*, XXIV, 187; Edinburgh directories, 1821, 1822, Edinburgh Public Library.

11. (interest): Testimonials, July 13, 1825, July 5, 7, 1828, folder, "Bell, Alexander — Thesis on Public Reading & Poetry"; *Fife Herald*, Mar. 18, 1824; "Re-Revised Condescendence," Apr. 4, 1833; *Report of Trial in the Action of Damages for Crim. Con. Poor Alexander Bell, Teacher, Dundee, versus William Murray . . .* (Edinburgh, 1834), p. 19; Letter to "Editor," Mar. 6, 1841, folder, "Lecture on Stammering . . . by Alex. Bell."

11. (husband): MacKenzie, *Bell*, p. 15; "Revised Condescendence for Alexander Bell," July 1830, Bell vs. Bell, Court of Session Records, Extracted Processes, No. 58, Feb. 1833, Scottish Record Office; "Revised Case, Poor Alexander Bell," July 6, 1832, p. 29, in ibid.

11. (pounds): "Revised Condescendence," July 1830; "Re-Revised Condescendence," Apr. 4, 1833.

12. (irreproachable): "Defenders Proof in Divorce Alexr Bell agt Eliz Colville," Court of Session Records, Extracted Process, No. 58, Feb. 1833, Scottish Record Office; Testimonial of W. Lindsay, July 6, 1833, folder, "Bell, Alexander — Thesis on Public Reading & Poetry."

12. (in London): *Bell vs. Murray*, pp. 20, 62–63; "Re-Revised Condescendence," Apr. 4, 1833; "Certificate By Ministers & Elders of the Poverty of A. Bell," Dec. 28, 1831, Court of Session Records, Unextracted Processes, Scottish Record Office.

12. (to London): *Bell vs. Murray*, p. 13.

12. (confinement): D. Bell to E. Bell, Sept. 13, 1837.

13. (Elocution): Handbill, Dec. 3, 1834, folder, "Bell, Alexander — Thesis on Public Reading & Poetry"; Alexander Bell, *The Practical Elocutionist* (London, 1834), title page; Alexander Bell, "Lecture on Stammering," Apr. 27, 1841; Alexander Bell, *Stammering and Other Impediments of Speech* (London, 1836); Alexander Bell, *The New Testament . . .* (London, 1837); *London Morning Advertiser*, Apr. 5, 1838.

13. (later): *Beinn Bhreagh Recorder*, XXIV, 211.

13. (family): MacKenzie, *Bell*, p. 16.

13. (traits): AMB to mother and sister, Sept. 22, 1838; *St. Johns* (Newfoundland) *Public Ledger*, Feb. 8, 1842; MacKenzie, *Bell*, p. 16; N. Cousins to AMB, Feb. 7, 1842; J. Moore to AMB, Feb. 5, 1842.

14. (organs): AMB to G. Davis, July 5, 1842; Samuel S. Curry, *Alexander Melville Bell* (Boston, 1906), pp. 13–14.

14. (1842): *Beinn Bhreagh Recorder*, XXV, 86; AMB to G. Davis, July 5, 1842.

14. (admiration): MS memoir of Mrs. AMB by AMB.

14. (spinster): Notes by Mrs. AMB in a small notebook, "The Christian Remembrancer."

14. (Bell): Mrs. AMB to AGB, Aug. 18, 1868; MS memoir of Mrs. AMB by AMB; information in "Original Document File," Canadian Bell Telephone Co., Montreal, Canada.

15. (1844): MS memoir of Mrs. AMB by AMB.

15–16. (face, year): Alexander Melville Bell, *The Art of Reading* (Edinburgh, 1845).

CHAPTER 2

17. (Bell): *Edinburgh Advertiser*, Mar. 9, 1847; *The Scotsman*, Feb. 5, 1937; birth certificate of Edward C. Bell; C. MacKenzie to A. Donald, Dec. 27, 1915.

18. (arms): Original miniature in Bell Collection; *Beinn Bhreagh Recorder*, VII, 180.

18. (help): Ibid., 180–181.

18. (anticlimax): Ibid., 181–184.

19. (crowds): A. Ellis to S. Haldeman, July 16, 1866, Haldeman MSS, Edgar F.

Smith Collection, University of Pennsylvania; Benn Pitman, *Sir Isaac Pitman: His Life and Labors* (Cincinnati, 1902), pp. 184–185; scrapbook of clippings on AMB's lectures.

19. (field): Frederick W. Haberman, "The Elocutionary Movement in England, 1750–1850" (Ph.D. diss., Cornell University, 1947), chapter vii.

19. (terms): Alexander Melville Bell, *Visible Speech: The Science of Universal Alphabetics* (London, 1867), p. 14.

19. (him): AMB: "Memoranda written for the use of General Eaton," May 19, 1892.

20. (it): Alexander Melville Bell, *Steno-Phonography* (Edinburgh, 1852); Alfred Baker, *The Life of Sir Isaac Pitman* (London, 1913), pp. 115–116; Alexander Melville Bell, *The Standard Elocutionist* (Edinburgh, 1860), passim; fragment of MS biography of AGB by Gilbert H. Grosvenor.

20. (home): *Edinburgh Witness*, Nov. 19, 1851; "Book of Accounts" of AMB, passim; MacKenzie, *Bell*, p. 31; MS memoir of Mrs. AMB by AMB.

20. (north): *The Scotsman*, Dec. 24, 1936; plan of upper floors before alteration, in archives of Scottish Life Assurance Co., Edinburgh.

21. (aloud): AGB to G. Grosvenor, Oct. 14, 1906; Mrs. G. Fortesque to G. Grosvenor, May 14, 1923.

21. (upon): AGB to G. Grosvenor, Oct. 14, 1906; MS memoir of Mrs. AMB by AMB.

22. (fever): Mrs. AMB to S. Fuller, Sept. 26, 1873; AGB to Mabel, Aug. 1, 1876.

22. (also): AGB address before Telephone Society of Washington, Feb. 3, 1910; biographical sketch of AGB by AMB, folder, "Bell, A. Melville — Lecture on Telephone."

22. (Aleck): Cumberland Hill, *Historic Memorials & Reminiscences of Stockbridge* (Edinburgh, 1887), p. 55.

22. (Bell): "Book of Accounts" of AMB, pp. 60, 83; *Beinn Bhreagh Recorder*, XV, 140.

23. (father): James Hay, *The Sound of the Mill* (Edinburgh, 1924), pp. 9–10, 28; *Milling* (Liverpool), Feb. 26, 1938, p. 230.

23. (stream, caldron): John Geddie: *The Water of Leith from Source to Sea* (Edinburgh, 1896), pp. 132–133.

23. (Cottage, oratory, age): 1853 map of Edinburgh, Edinburgh Room, Edinburgh Public Library; Mrs. Mary Ross, "Notes Regarding the Family of A. Graham Bell"; Mary E. Symonds, "Early Recollections of Dr. A. Graham Bell"; *Beinn Bhreagh Recorder*, II, 75.

24–25. (deafness, games, boy, character): MS reminiscences by Annie Herdman, 1906; Symonds, "Early Recollections"; Ross, "Notes."

25. (wonder): AGB to Mabel, Dec. 6, 1876; M. Symonds to F. DeLand, July 13, 1908.

25. (teacher): *Beinn Bhreagh Recorder*, II, 59–61, 69.

26. (invention): AGB to G. Grosvenor, Oct. 14, 1906.

26. (help, off, fruit): *Beinn Bhreagh Recorder*, II, 58–59.

26. (notion): Herdman reminiscences.

26. (itself): James B. Gillies, *Edinburgh Past and Present* (Edinburgh, 1886), p. 144.

27. (space): Robert Louis Stevenson, *Edinburgh: Picturesque Notes* (London, 1889), pp. 128, 140.

27. (establishment): William C. A. Ross, *The Royal High School* (Edinburgh, 1949), p. 139.

27. (behalf): AGB to G. Grosvenor, Oct. 14, 1906.

27. (school): Ross, *Royal High School*, p. 63; James J. Trotter, *The Royal High School, Edinburgh* (London, 1911), p. 66.

28. (them): Ross, *Royal High School*, pp. 63, 67, 76; MacKenzie, *Bell*, p. 23; Walter S. Dalgliesh, *Memorials of the High School of Edinburgh* (Edinburgh, 1857), p. 37; *Beinn Bhreagh Recorder*, II, 73–74.

28. (lists): Dalgliesh, *Memorials*, p. 37; *Beinn Bhreagh Recorder*, II, 73–75; *Edinburgh Evening Courant*, July 21, 1858, July 26, 1860, July 25, 1862.
28. (Land): MacKenzie, *Bell*, p. 23; *Beinn Bhreagh Recorder*, IX, 206.
28. (life): Ibid., II, 75.

CHAPTER 3

29. (instruction): Handbills, 1842, in folder, "Bell, Alex.— Announcements of Lectures."
29. (art): Alexander Bell, "On the Character and Tragedy of Macbeth"; Alexander Bell, *The Tongue, A Poem, in Two Parts* (London, 1846), p. 57.
30. (*Annals*): Alexander Bell, *The Bride* (London, 1847), p. 1.
30. (Park): Handbills and cards in folder, "Bell, Alex.— Announcements of Lectures"; A. Bell to AMB, Nov. 26, 1849, May 26, 1850; AMB to A. Bell, Nov. 29, 1849.
30. (comfortably): A. Bell to AMB, May 26, 1850; photograph of 18 Harrington Square.
30. (years): *Beinn Bhreagh Recorder*, XXIV, 211; M. Eccleston to AGB, July 2, 1870.
31. (metropolis): AMB to AGB, Oct. 10, 1862.
31–32. (cane, passersby, boy, education): *Beinn Bhreagh Recorder*, II, 73–77.
32. (nonsense): Handbill, May 1843, in folder, "Bell, Alex.— Announcements of Lectures."
33. (assailants): Undated MS in folder, "Bell, Alexander — Parliament & the Social Order"; Alexander Bell, "On Humbug," pp. 12–13, 89–90, 111, 121–122; Alexander Bell, "The Commonwealth: Lecture First."
33. (speech): Bell, "Public Reading," in folder, "Bell, Alexander — Thesis on Public Reading," p. 1.
33. (life): In Bell Collection.
33. (independence): *Beinn Bhreagh Recorder*, II, 77.
34. (you): AMB to AGB, Mar. 2, 1863.
34. (man): *Beinn Bhreagh Recorder*, II, 73, 76–77.

CHAPTER 4

35. (him, speech): William T. Jeans, *Lives of the Electricians: Professors Tyndall, Wheatstone, and Morse* (London, 1887), pp. 111–116, 118, 123–205.
35. (apparatus): MacKenzie, *Bell*, p. 26; Jeans, *Lives of the Electricians*, pp. 117–118; *Beinn Bhreagh Recorder*, II, 61–62.
36. (own): Ibid., II, 77; *The Scotsman*, Dec. 3, 1862.
36. (butcher): *Beinn Bhreagh Recorder*, II, 68–70.
37. (again): Ibid., II, 62–72; AGB, "Making a Talking Machine."
37. (Scotland, Melly, himself): *Beinn Bhreagh Recorder*, II, 77–78.
38–39. (base, handily): John Shanks, *Elgin* (London, 1866), pp. 81–83, 92–94.
39. (house): *Elgin Courant*, Nov. 6, 13, 1863.
39. (Edinburgh): Ibid., June 24, 1864.
39. (circular): George Bernard Shaw, *Sixteen Self Sketches* (New York, 1949), pp. 58–59; G. Shaw to A. Henderson, Jan. 17, 1905, in Archibald Henderson, *Bernard Shaw: Playboy and Prophet* (New York, 1932), p. 119; *The Irish Times* (Dublin), Sept. 25, 1865.
40. (best): D. Bell to AMB, Oct. 1, 1865.
40. (first): Alexander Melville Bell, *Visible Speech: A New Fact Demonstrated* (London, 1865), pp. 51–53; diary of Joseph Henry, Jan. 9, 1858, Henry MSS.
40. (details): *Werner's Magazine*, May 1900, p. 213; Bell, *Visible Speech* (1867), pp. 1, 15–19; *Dundee Advertiser*, Feb. 6, 1897; *Washington Post*, Sept. 19, 1905.

40–42. (Speech, necessary): Bell, *Visible Speech* (1867), pp. vii–x, 26–27, 45–55, 65, 69, 71–72.

42. (press): Ibid., p. 22; Alexander Melville Bell, *Visible Speech: Every Language Universally Legible Exactly As Spoken* (Edinburgh, 1864), passim; testimonials, July 1846, folder, "Miscellaneous Writings of AMB"; clippings in folder, "Visible Speech — AMB Charts & Newspaper Articles."

42. (base): A. Bell to AMB, Apr. 13, June 19, 1864.

43. (produced): J. Hipkins scrapbook, "Science and Art. A Couple of Wise Men," MSS Room, Scottish National Library, Edinburgh; Bell, *Visible Speech: A New Fact Demonstrated*, p. 30.

43. (language): *London Morning Star*, Aug. 13, 1864.

43. (unprepared): A. Ellis to AMB, Aug. 29, 1864; *London Morning Star*, Aug. 31, 1864.

43. (Aleck): *Edinburgh Review*, Sept. 14, 1864; C. Barrington to AMB, Sept. 8, 1864; AMB to Queen Victoria, Sept. 13, 1864.

43. (optics): Pp. 62–63.

44. (1864): AMB to Queen Victoria, Sept. 13, 1864; L. Hammond to AMB, Sept. 24, 1864.

44. (Glasgow): MS notebook, "Alexr. Bell . . . Notes on Professor Blackie's Lectures"; Bell, *Visible Speech: A New Fact Demonstrated*, pp. 32–33; *Glasgow Daily Herald*, Dec. 17, 1864.

44. (possible): *London Morning Post*, July 6, 1865; *Phonetic Journal*, July 8, 15, Aug. 26, 1865.

44. (father): Testimonial by J. Skinner, July 14, 1865; Mrs. AMB to AGB, Sept. 15, 1865; AMB to AGB, Sept. 22, 1865.

45. (falling): AGB address before Telephone Society, Feb. 3, 1910; AGB dictation, "Telegraphic Transmission of Speech," Feb. 6, 1879.

45. (again): W. E. Watson, *Elgin Schools and Schoolmasters* (Elgin, n.d.), pp. 30–31.

45. (sunny): *Elgin Courant*, Dec. 1, 1865.

46. (remarked): AGB notebook, "The Result of some Experiments in connection with 'Visible Speech' made in Elgin in November 1865."

46. (school): Mrs. AMB to AGB, Jan. 13, 1866.

46. (mother): Mrs. AMB to AGB, Feb. 8, 1866; AMB to AGB, Feb. 14, 1866.

46. (necessary): Mrs. AMB to AGB, Mar. 17, 20, 1866, and undated.

47. (her): AGB to Mrs. A. Acklone, Sept. 7, 1906.

47. (House): MacKenzie, *Bell*, pp. 35–36.

47. (works): E. C. Bell to AGB, no date.

47–48. (indeed, priority, London): A. Ellis to AGB, Mar. 31, 1865.

48. (it): AGB to Mrs. A. Acklone, Sept. 7, 1906; Mrs. A. Acklone to AGB, Oct. 22, 1906.

CHAPTER 5

49. (well): Mrs. AMB to AGB, Jan. 13, Feb. 2, May 24, June 10, 1866; A. Scott to F. DeLand, Sept. 7, 1926.

49. (along): R. Browning to AMB, Oct. 12, 20, 1865, Mar. 12, 26, 1866; E. C. Bell to AGB, Sept. 23, 1865, and undated.

49. (alphabet): *Educational Times*, December 1865; E. C. Bell to AGB, Feb. 17, 1866; *Journal of the Society of Arts*, Mar. 16, 1866; AMB to editor of the London *Times*, July 2, 1866.

50. (alphabet): *Wakefield* (England) *Journal*, Nov. 16, 1866; A. Ellis to S. Haldeman, July 16, 1866, Haldeman MSS.

50. (future): A. Ellis to S. Haldeman, Jan. 13, 1868, Haldeman MSS; AMB to Editor, London *Times*, July 2, 1866; C. Josselyn to AMB, July 24, 1866; "Speech Made

Visible," reprinted from *Macmillan's Magazine*, in *Every Saturday*, II, 25–28, July 14, 1866.

51. (son): Dictated statement by AGB, Feb. 6, 1879.

52. (students): *The Post-Office Bath Directory* (Bath, 1865), p. 541; Robert E. Peach, *Rambles About Bath* (Bath, 1876), pp. 303–304.

53. (Kean): AGB, "Scribbling Diary," 1867, entries for Jan. 4, 5, 15, 29.

53. (telegraphically): Ibid., entries for Jan. 9, 11, Feb. 7, 8, 12, 18; AGB address before the Telephone Society, Feb. 3, 1910.

53. (time): E. C. Bell to AGB, Feb. 17, 1866; Mrs. AMB to AGB, June 10, 1866, Jan. 26, 1867.

53. (away): Mrs. AMB to AGB, Mar. 1, 22, May 4, 1867.

53. (AGB): AGB, "Scribbling Diary," 1867.

54. (cricket): *Bath Chronicle*, May 30, June 6, 1867.

54. (year): Mrs. AMB to AGB, June 1, 1867.

54. (breakfast): AGB to parents, June 9, 11, 1867; Mrs. AMB to AGB, June 13, 14, 1867.

54. (Square): *Bath Chronicle*, July 4, 1867; AGB to parents, June 11, 1867.

54. (book): AMB to AGB, July 25, 1867; Bell, *Visible Speech*, p. 126.

55. (characters): Wilfred G. R. Murray, *Murray the Dictionary-Maker* (Wynberg, Cape Colony, 1943), pp. 37–40; AGB to Lady Murray, Aug. 26, 1915.

55. (walks): A. Scott to AMB, Oct. 12, 1903, and to Mrs. AGB, Mar. 26, 1918.

56. (thought): AGB, "Prehistoric Telephone Days," *National Geographic Magazine*, March 1922, XLI, 229–230, 233.

56. (growl): AGB to G. Grosvenor, Oct. 14, 1906.

56. (eight): *Wakefield Journal*, Nov. 16, 1866; Frederick DeLand, *Dumb No Longer* (Washington, D.C., 1908), p. 194.

56–57. (sounds, Mama): Typed copy of AGB, "Extracts from Journal . . . May 21, 1868."

57. (etc.): M. Eccleston to AGB, July 2, 1870; AMB to T. Henderson, Nov. 18, 1869.

57. (independence): J. Symonds to AGB, July 26, 1868: Certificate of Alexander Graham Bell, signed by the University of London Registrar, July 30, 1868.

CHAPTER 6

58. (too): A. Ellis to S. Haldeman, Jan. 13, 1868, Haldeman MSS; Bell, *Visible Speech*, p. 24.

58. (buy): Haberman, "Elocutionary Movement," chapter vii.

58. (London): Mrs. AMB to AMB, Sept. 9, 1868.

59. (elegance): AMB to Mrs. AMB, Aug. 2, 11, 1868.

59. (gas): David Macrae, *The Americans at Home* (Edinburgh, 1870), pp. 34–37.

59. (Romic): Ibid., p. 19; *North American Review*, CVII (1868), 349–357; Haberman, "Elocutionary Movement," chapter vii.

60. (school): *Werner's Magazine*, May 1900, pp. 220–221; AMB to Mrs. AMB, Aug. 23, 1868.

60. (fall): T. Hill to AMB, Aug. 23, 1868; J. Lowell to AMB, Sept. 5, 15, 1868.

60. (country): AMB to Mrs. AMB, Sept. 1, 1868; Mrs. AMB to AMB, Sept. 16, 1868.

60. (course): AMB to Mrs. AMB, Sept. 6, 16, 1868.

60. (hesitate): AMB to Mrs. AMB, Sept. 16, Oct. 4, 1868; Mrs. AMB to AMB, Oct. 7, 14, 1868.

61. (mission): *Boston Transcript*, Oct. 24, 1868; AMB to Mrs. AMB, Nov. 1, 8, 1868.

61. (Aleck): DeLand, *Dumb No Longer*, p. 197.

61. (him): AGB to AMB, Aug. 1, Sept. 29, 1868; Mrs. AMB to AMB, Aug. 18, 30, Sept. 16, 21, 30, Oct. 4, 14, 21, 24, 1868; AMB to Mrs. AMB, Nov. 8, 1868.

61. (all): AGB to AMB, Sept. 29, 1868; Mrs. AMB to AMB, Aug. 18, 30, Oct. 24, 1868.
62. (again): AGB to AMB, Sept. 29, Oct. 5, 1868; AGB to G. Grosvenor, Nov. 2, 1906; Mrs. AMB to AMB, Aug. 23, Sept. 16, 23, 1868.
62. (wife): Mrs. AMB to AMB, Aug. 18, Sept. 30, Nov. 11, 1868.
62. (school): Mrs. AMB to AMB, Nov. 14, 1868.
62. (died): Mrs. AMB to AMB, Aug. 18, Sept. 6, 16, 23, 30, Oct. 14, 30, 1868; M. Bell to parents, July 4, Sept. 17, 1869; J. Symonds to AMB, Feb. 6, 1870.
63. (glottis): Mrs. AMB to AMB, Sept. 30, 1868; M. Bell to AMB, Oct. 30, 1868.
63. (so): Idem; A. Scott to F. DeLand, Sept. 7, 1926; M. Bell to AMB, July 2, 1869; AGB to Mabel, Sept. 26, 1875.
63. (country): M. Bell to Mr. and Mrs. AMB, July 2, Sept. 17, 1869, and undated; AMB to T. Henderson, Nov. 18, 1869; AMB to AGB, May 24, 1870; J. Symonds to AMB, June 15, Oct. 6, Dec. 5, 1869; AGB to G. Grosvenor, Nov. 2, 1906.
63. (strong): AGB to AMB, June 5, 1870; Mrs. AMB to AMB, July 1, 28, 1869; AMB to T. Henderson, Nov. 18, 1869; M. Eccleston to Mrs. AMB, undated; M. Bell to Mr. and Mrs. AMB, undated; J. Symonds to AMB, June 15, 1869.
64. (lyre): A. Scott to F. DeLand, Sept. 7, 1926.
65. (sound): Mrs. AMB to AGB, May 31, 1869; M. Bell to Mr. and Mrs. AMB, July 4, 1869.
65. (decline): AGB to parents, Sept. 2, 5, 7, 1869.
65. (headaches): M. Eccleston to AGB, July 2, 1870.
65. (marriage): M. Eccleston to Mrs. AMB, undated.
66. (herself): Alexander Melville Bell, *Universal Line-Writing and Steno-Phonography* (London, 1869); *Shorthand Writers' Journal*, March 1869, p. 37; M. Bell to Mr. and Mrs. AMB, undated; Mrs. AMB to AGB, Sept. 7, 1869.
66. (Melly): L. Monroe to AMB, Apr. 16, 1869; Mrs. AMB to AMB, July 1, 1869; M. Bell to AMB, July 2, 1869.
66. (emigration): M. Bell to Mr. and Mrs. AMB, July 2, 1869, and two undated; Mrs. AMB to AGB, May 31, 1869; AMB to T. Henderson, Nov. 18, 1869; J. Lowell to AMB, Dec. 20, 1869, Mar. 22, 1870; T. Coats to AMB, May 10, 1870.
66. (chap): M. Bell to AGB, Apr. 1, 1870.
67. (lucid): AGB to G. Grosvenor, Nov. 2, 1906.
67. (Bell): AMB to AGB, May 24, 28, 1870.
67. (spiritualism): J. Symonds to AMB, May 28, 1870; Mrs. J. Jones to M. B. Grosvenor, Apr. 4, 1972, copy in author's possession; AGB to Mabel, Sept. 26, 1875.
67. (land, consent): Fragment of MS biography of AGB by Gilbert H. Grosvenor, ca. 1906.
68. (word): AGB to parents, June 5, 1870; AGB to AMB, June 5, 1870.
68. (shorthand): M. Eccleston to AGB, July 2, 1870.
68. (Canada): Copy of W. Inglis to parents, June 16, 1870; AGB to parents, June 5, 19, 1870.
68. (there): M. Eccleston to AGB, July 2, 1870.
69. (elsewhere): AGB to parents, June 30, 1875.
69. (vacancy): Mrs. AMB to AGB, June 19, 1870; A. Scott to F. DeLand, Sept. 7, 1926.

CHAPTER 7

73. (purchase): T. Henderson to AMB, July 5, 1870; William Patten, *Pioneering the Telephone in Canada* (Montreal, 1926), p. 7.
73. (technology): Ibid., pp. 7-8; poster announcing sale of AMB's house and grounds on Apr. 27, 1881; AGB to AMB, Oct. 3, 1870.
74. (Parliament): Sale poster, Apr. 27, 1881; Mrs. AMB to AGB, Apr. 13, 19, 1871;

Alexander Mackenzie, *The Life and Speeches of Hon. George Brown* (Toronto, 1882), pp. 9, 118.
74. (can): AGB to Mabel, Sept. 26, 1875; AGB to AMB, Oct. 3, 10, 1870.
74. (series): AGB to AMB, Oct. 28, 1870; DeLand, *Dumb No Longer*, p. 197; S. Fuller to grandnephew, Apr. 18, 1925.
74. (convenient): Idem; AGB to AMB, Oct. 28, 1870; AMB to AGB, Nov. 14, 1870; AMB to Mrs. AMB, Nov. 22, 1870; D. King to AMB, Mar. 15, 1871.
75. (gift): P. Tait to S. Newcomb, Apr. 1, 1871, Simon Newcomb MSS, Library of Congress.
75. (friend): AGB to parents, Apr. 5, 1871; *Boston Transcript*, Apr. 5, 7, 1871.
75. (theory): AGB to parents, Apr. 9, 1871; *Boston Transcript*, Apr. 5, 7, 12, 1871; *King's Handbook of Boston* (Cambridge, Mass., 1878), pp. 75, 83, 91, 115.
75–76. (tapping, dies, oranges): AGB to parents, Apr. 9, 1871.
76. (thought, come, anticlimax): AGB to parents, Apr. 16, 1871.
77. (time): AGB to parents, ca. May 1871.
77. (wheels): AGB to parents, May 23, 1871; S. Fuller to grandnephew, Apr. 18, 1925.
77. (Teachers): E. Jordan to F. DeLand, July 14, 1923; AGB to parents, May 23, 1871.
78. (Square): L. Dudley to AGB, May 24, 29, 1871; AGB to parents, ca. June 1, 1871; J. Philbrick to I. Allen, June 19, 1871.
78. (Brantford): AGB to S. Fuller, June 22, 1871.
79. (again): AGB to Mabel, Sept. 26, 1875; E. Jordan to AGB, July 5, 1871; AGB to E. Jordan, Aug. 7, 1871.
79. (School): E. Stone to AGB, July 11, 1871; L. Dudley to AGB, Aug. 10, 1871.
79. (fellow): AGB to S. Fuller, July 16, 25, 1871.
79. (Boston): *Boston Transcript*, July 29, 1871; AGB to S. Fuller, Aug. 19, 1871; the advertisement appeared in the *Transcript*.

CHAPTER 8

80. (understood): AGB to S. Fuller, Aug. 23, 1871; AGB to parents, Aug. 31, 1871.
81. (hear it): AGB's notebook on work with Theresa Dudley, 1871–1872, pp. 4, 24–25, 37, 46, 67.
81. (Dumb): AGB to parents, Dec. 1, 1871; AGB to S. Fuller, Nov. 28, 1871; *Boston Transcript*, Nov. 29, 1871; *Boston Advertiser*, Dec. 6, 1871.
81. (magazine): L. Dudley to AGB, Jan. 18, 1872; E. Stone to AGB, Jan. 18, 1872; AGB to S. Fuller, Mar. 6, 1872; *Old and New*, VI (1872), 68–69.
81. (up): AGB to parents, Dec. 1, 1871; AGB to E. Stone, Jan. 15, 1872; AGB to S. Fuller, Jan. 16, Feb. 8, 1872.
82. (name): *Old and New*, VI, 69; *Boston Transcript*, Dec. 4, 8, 1871; J. Gordon to AGB, Dec. 14, 1885.
82. (world): Diary of J. Henry, Dec. 5, 1872, Henry MSS, Smithsonian Institution; F. Barnard to A. Mayer, Jan. 22, 1873, Hyatt-Mayer Collection, Princeton University.
82. (her): AGB to S. Fuller, Jan. 5, 1872; AGB notebook on Theresa Dudley, p. 185.
82. (home): AGB to parents, Mar. 5, Apr. 17, 1872; *Silent World*, October 1871, p. 4.
83. (offprint): AGB to parents, Apr. 9, 17, 28, 1872.
83. (deaf): AGB to parents, Apr. 17, 1872.
83. (face): AGB to parents, Apr. 9, 1872.
83. (Cambridge): Allen Johnson, et al. (eds.), *Dictionary of American Biography* (22 vols., New York, 1928–1958), IX, 324–325; Justin Winsor (ed.), *The Memorial History of Boston* (4 vols., Boston, 1886), II, xliv, IV, 29, 66–67.

83. (Company): American Antiquarian Society *Proceedings*, XII (New Series), 222–223; John A. Miller, *Fares Please!* (New York, 1941), pp. 20–21.
84. (America): G. Hubbard to J. Henry, Oct. 26, 1850, Henry MSS, Smithsonian Institution.
84. (deaf): Amer. Ant. Soc. *Proceedings*, XII, 221; Helen E. Waite, *Make a Joyful Sound* (Philadelphia, 1961), pp. 20, 27.
85. (tutoring): Harry Best, *Deafness and the Deaf in the United States* (New York, 1943), pp. 374–381, 389–393, 531–534.
86. (ally): Waite, *Make a Joyful Sound*, pp. 21–25; Harold Schwartz, *Samuel Gridley Howe* (Cambridge, Mass., 1956), pp. 276–278.
86. (argument): Ibid., pp. 277–279.
86. (time, difficulty): Waite, *Make a Joyful Sound*, pp. 26–35; Gardiner G. Hubbard, *The Education of Deaf Mutes* (Boston, 1867), p. 28.
86. (trustee): Schwartz, *Howe*, pp. 279–280; Waite, *Make a Joyful Sound*, pp. 36–44; *Report of Massachusetts Joint Special Committee on the Education of Deaf Mutes, May 27, 1867* (Mass. Senate Document No. 265), pp. 2–8, 42–44.
87. (articulation): AGB to AMB, Jan. 7, 1872.
87. (once): AGB to parents, May 2, 1872.
87. (investigate): *Silent World*, June 15, 1872, p. 9.
88. (earnestness): AGB to parents, ca. May 15, 1872; *Silent World*, Aug. 1, 1872, pp. 7–8; American Asylum, *57th Annual Report* (Hartford, 1873), pp. 23, 27–28; AGB to parents, June 24, 1872.
88. (Bell): AGB to parents, ca. May 15, 1872, June 17, 1872.
88. (excellent): AGB to AMB, June 4, 1872; AGB to parents, June 27, 1872; AGB to S. Fuller, July 4, 10, 1872.
88. (timbre): *Silent World*, Sept. 1, 1872, pp. 3–4.
89. (fall): AGB to S. Fuller, July 20, Aug. 2, 23, 1872.

CHAPTER 9

90. (master): AGB to parents, Sept. 22, Oct. 2, 1872.
90. (all): AGB to parents, Oct. 6, Nov. 19, 1872, Jan. 19, 27, 1873; *Silent World*, Sept. 1, 1872, p. 5.
91. (years): *DAB*, XVII, 336 ("Thomas Sanders"); S. Fuller to grandnephew, Apr. 18, 1925; AGB to parents, Oct. 2, 6, Nov. 19, 1872.
91. (windstorm): AGB to parents, Oct. 28, 1872; Russell H. Conwell, *History of the Great Fire in Boston* (Boston, 1873), pp. 53–65; *DAB*, VI, 280 ("Moses Farmer").
92. (Christmas): AGB to parents, Nov. 11, 12, 19, 1872; Mrs. AMB to AGB, Nov. 18, 1872.
92. (nation): Conwell, *The Great Fire*, p. 291.
92. (City): Matthew Josephson, *Edison* (New York, 1959), pp. 60–73.
93. (triumph): AGB to parents, Oct. 14, 25, 1872; *Boston Transcript*, Oct. 26, 1872.
93. (tuning fork): AGB to AMB, Oct. 28, 1872; *Boston Transcript*, Oct. 28, 1872; Telephone Interferences, A–L, Gray, pp. 150–154; American Telephone and Telegraph Company, *The Bell Telephone* (New York, 1908), pp. 221–223.
94. (year): Harmonic Interferences, Bell, pp. 5–6, 38.
94. (few): Harmonic Interferences, Bell, pp. 6, 30–31; People's Telephone, Exhibits, p. 980.
95. (warranted): Harmonic Interferences, Bell, pp. 88–89, 119–120.
95. (now): Harmonic Interferences, Bell, pp. 7–8; Mrs. AMB to AGB, Dec. 12, 1872.
95. (experiment): Harmonic Interferences, Bell, pp. 8–9; People's Telephone, Exhibits, pp. 979–980.

95. (tuning forks): AGB to parents, Jan. 19, 27, Apr., May 15, 17, 1873; AMB to AGB, May 19, 1873; AGB to parents, June 1, 4, 22, 1873.
95. (Brantford): AGB to parents, June 22, July 1, 1873; AMB to AGB, July 5, 1873.

CHAPTER 10

97. (strength): AGB to S. Fuller, July 22, Aug. 5, 1873; AGB to parents, July 1873.
97. (professor): AGB to S. Fuller, Aug. 5, Sept. 4, 15, 1873; AGB to AMB, Oct. 1, 1873.
98. (boy): AGB to parents, July 1, 1873.
98. (Coromandel): MacKenzie, *Bell*, pp. 59–60; AGB to AMB, Oct. 1, 1873.
98. (side): AGB to parents, Oct. 2, 1873; AGB to Mabel, Oct. 18, 1875.
100. (fact): Bell's manuscript storybook for Georgie Sanders is in the Bell Collection; MacKenzie, *Bell*, pp. 60–61.
100. (organs): AGB to parents, Nov. 1, 1873.
100. (wreckage): Ibid.; Mrs. AMB to AGB, Nov. 30, 1873.
100. (home): AGB to parents, Nov. 10, 1873.
100. (birthday): Entry dated Jan. 6, 1879, in journal of Mrs. AGB; unfinished reminiscences by Mrs. AGB, ca. 1922; notes by Mrs. AGB, ca. 1890–1895.
101. (Locke): Notes by Mrs. AGB, ca. 1890–1895.
101. (Cambridge): Mabel to Mrs. GGH, Nov. 19, 1873.
101. (back): Mabel to Mrs. GGH, ca. Nov. 25, 1873.
101. (again): Mabel to Mrs. GGH, Feb. 3, 1874.
102. (quality): Notebook in Bell Collection.
102. (name): *American Annals of the Deaf and Dumb*, XIX (1874), 179–180; AGB to GGH, Nov. 23, 1875.
102. (June): AGB to S. Fuller, Dec. 22, 1873; AGB to parents, Jan. 15, 1874; AGB to AMB, Jan. 26, 1874.
102. (lectures): AGB to parents, Feb. 1, 1874; AGB to AMB, Feb. 5, 22, Mar. 2, 1874; first issue of "Visible Speech Pioneer," Mar. 2, 1874.
102. (Boston): AGB to parents, Mar. 8, 1874, ca. Apr. 9, 1874.
103. (Apparatus): Ibid.

CHAPTER 11

104. (himself): Mrs. AMB to AGB, Apr. 8, 1875; AGB to parents, Dec. 6, 1873.
104. (fork): AGB to parents, Dec. 15, 1873; Harmonic Interferences, Bell, p. 11; People's Telephone, Exhibits, pp. 980–981, 993.
105. (mind): People's Telephone, Evidence before Swan, p. 1609; People's Telephone, Exhibits, p. 981; J. Baile, *The Wonders of Electricity* (New York, 1872), pp. 140–143.
105. (silently): Harmonic Interferences, Bell, p. 52; People's Telephone, Exhibits, p. 981.
106. (it): People's Telephone, Complainants in Reply, p. 112.
106. (unthinking): People's Telephone, Complainants in Reply, p. 113.
107. (paper): Mrs. GGH to GGH, undated; AGB to parents, Nov. 10, Dec. 6, 1873; Harmonic Interferences, Bell, p. 16; People's Telephone, Exhibits, p. 982.
107. (existence): Baile, *Wonders*, p. 131.
107. (line): People's Telephone, Exhibits, pp. 981–982.
108. (receiver): Harmonic Interferences, Bell, pp. 16–18; People's Telephone, Exhibits, p. 982; *Bell Telephone*, pp. 18–19.
108. (transmitter): People's Telephone, Complainants in Reply, p. 116; *Bell Telephone*, pp. 19–20.
108. (it): Ibid., p. 20.
109. (doing): Ibid., pp. 21–22.

109. (system): Ibid., pp. 22–23.
110. (fees): Harmonic Interferences, Bell, p. 32; AGB to parents, Dec. 6, 1873.
110. (proposition): Harmonic Interferences, Bell, p. 33; People's Telephone, Exhibits, p. 983.
110. (frequencies): AGB to parents, Mar. 8, 1874.
110. (later): Harmonic Interferences, Bell, p. 106; AGB to parents, Wednesday, April 1874.
111. (tracing): AGB to AMB, Saturday, April 1874; *American Journal of Science,* CVIII (1874), 130–131.
111. (intervals): AGB to parents, Wednesday, April 1874.
111. (eternity): AGB to parents, May 6, 1874.
111. (sound): People's Telephone, Complainants in Reply, pp. 115–116.
112. (flame): AGB to parents, May 6, 1874.
112. (waves): Pickering's standing as a specialist in acoustics is attested in A. Mayer to E. Pickering, Oct. 26, 1876, Pickering MSS; E. Harris to E. Pickering, May 15, 1874, ibid.; Edward C. Pickering, "Early Experiments in Telegraphing Sound," *American Academy of Arts and Sciences Proceedings,* XXI (1885–1886), part ii, p. 264; *Bell Telephone,* pp. 225–226.
112. (you): C. Cross to AGB, Apr. 23, 1874.
112. (auriculation): Address by Clarence J. Blake, Mar. 13, 1916, in *Telephone Topics* (New England Telephone Company), IX (Apr. 1916), 350.
112. (did): AGB to parents, Apr. 4, May 6, 1874; Mrs. AMB to AGB, Apr. 12, 1874.
113. (Bell): Mabel to Mrs. GGH, May 6, 1874; *American Annals of the Deaf and Dumb,* XIX (1874), 144, 179–180.
113. (MIT): Ibid., XIX (1874), 217–218.
113. (indeed): Diary of Frank W. Clarke, June 13, 1874, Frank W. Clarke MSS, Library of Congress.
113. (inventor): *DAB,* VII, 514 ("Elisha Gray"); Dowd Case, pp. 111–112.
113. (rheotome): Ibid., p. 112.
114. (research): *DAB,* VII, 514 ("Elisha Gray"); Dowd Case, pp. 112–113.
114. (suit): Speaking Telephone Interferences, Evidence for E. Gray, p. 41.
116. (receivers): Ibid., pp. 41–42.
117. (*Times*): Diary entry, Aug. 23, 1870, Frank W. Clarke MSS, Library of Congress; People's Telephone, Exhibits, pp. 153–154.
118. (fame): Bell's father recorded in his diary that his son arrived at Tutelo Heights on July 11, 1874, which implied that Bell left Boston on July 10, if not earlier; People's Telephone, Evidence before Swan, p. 1703.
118. (September): E. Gray to S. White, July 9, 1874, Telephone Historical Collection, American Telephone and Telegraph Company, New York; A. Hayes to E. Pickering, Aug. 6, Sept. 23, 1874, Pickering MSS.
118. (done): B. Osgood Peirce, "Biographical Memoir of Joseph Lovering," National Academy of Sciences *Biographical Memoirs,* VI (Washington, D. C., 1909), pp. 329–344; American Association for the Advancement of Science *Proceedings,* XVI (1867), 25–27, XVII (1868), 104–105; *Hartford Daily Courant,* Aug. 18, 20, 1874.

CHAPTER 12

120. (himself): AGB to parents, June 30, 1875.
120. (throat): *Report of the Eighth Annual Convention of American Instructors of the Deaf and Dumb in Belleville, Ontario, July 15–20, 1874.*
120. (speech): People's Telephone, Evidence before Swan, p. 1703.
120. (speech): Notebook, "Electrical Experiments by A.G.B., Vol. III," entry for Nov. 25, 1878.

121. (bones): People's Telephone, Complainants in Reply, pp. 120–121; AGB, "Early Telephony" (address before the Telephone Society of Washington, Feb. 3, 1910).
122. (plate): People's Telephone, Complainants in Reply, p. 117.
122. (electromagnet): *Bell Telephone,* pp. 34–35.
123. (current): People's Telephone, Complainants in Reply, p. 121; People's Telephone, Exhibits, p. 986; People's Telephone, Evidence before Swan, pp. 1576, 1600; People's Telephone, Exhibits, p. 1002.
123. (electromagnet): People's Telephone, Complainants in Reply, p. 121; AGB to parents, Nov. 23, 1874.

CHAPTER 13

125. (experiments): AGB to parents, Oct. 12, 1894; Mrs. AMB to AGB, Oct. 18, 1874.
125. (reglycerined): C. Blake to E. Pickering, Oct. 3, 1874, Pickering MSS; AGB to parents, Oct. 12, 1874; C. Cross to AGB, Oct. 14, 1874.
125. (library): Mrs. AGB to Mrs. J. Penman, Oct. 30, 1922.
126. (music): Idem; Mrs. AGB to Mrs. J. Penman, Oct. 15, 1922; autobiographical notes by Mrs. AGB, 1922.
126. (wire): People's Telephone, Exhibits, p. 955.
126. (*wire*): AGB to parents, Oct. 23, 1874.
126. (before): AGB to parents, Mar. 8, 1874; Gardiner G. Hubbard, *Union of the Post-Office and Telegraph* (Boston, 1868), pp. 25–27.
127. (industry): Idem; 43d Congress, 1st Session, Senate Report 242, p. 12.
127. (it): Mrs. GGH to GGH, Feb. 14, 1873.
127. (proceedings): Mrs. GGH to GGH, 1874 (otherwise undated).
127. (science): 43d Congress, 1st Session, Senate Report 242, pp. 101–102 and passim; *DAB,* XIV, 65 ("William Orton"); *N.Y. Times,* July 10, 1874.
128. (patent): AGB to parents, Oct. 23, 1874.
128. (funds): Idem.
128. (Gray): People's Telephone, Complainants in Reply, p. 342.
129. (conception): People's Telephone, Complainants in Reply, pp. 342–343, 337.
129. (partnership): People's Telephone, Complainants in Reply, pp. 345–346, 123, 349.
129. (cash): AGB to parents, ca. Oct. 30, 1874.
129. (so): Idem.
130. (night): AMB to AGB, Nov. 5, 1874; Harmonic Interferences, Bell, pp. 63, 65; People's Telephone, Evidence before Swan, pp. 1601–1602.
130. (papers): Harmonic Interferences, Bell, p. 64; People's Telephone, Exhibits, pp. 986–987; AGB to parents, Nov. 16, 1874.
130. (August): *Boston Commonwealth,* Nov. 14, 1874; Harmonic Interferences, Bell, p. 53; AGB to parents, Nov. 24, 1874.
130. (conception): People's Telephone, Exhibits, pp. 956, 978; Harmonic Interferences, Bell, p. 63; AGB to parents, Nov. 23, 1874.
130. (them): AGB to parents, Nov. 16, 1874.
131. (wished): AGB to parents, Nov. 23, 1874; People's Telephone, Evidence before Swan, pp. 1727–1728.
131. (well): AGB to parents, Nov. 23, 24, 1874.
131. (transmitter): AGB to AMB, Nov. 26, 1874; AGB to GGH, Nov. 26, 27, 1874.
131. (memory): Harmonic Interferences, Bell, p. 107.
132. (currents): AGB to AMB, Dec. 1, 1874; People's Telephone, Exhibits, p. 988; Harmonic Interferences, Bell, pp. 26–27.
132. (him): AGB to GGH, Nov. 30, 1874; AGB to parents, Dec. 1, 15, 1874.
132. (reeds): Idem; Harmonic Interferences, Bell, pp. 93–96, 27, 39; People's Telephone, Exhibits, p. 989.
132. (progress): People's Telephone, Evidence before Swan, p. 1704; AGB to parents, Thursday, January 1875.

132. (listener): C. Stowell to G. Barker, Mar. 3, 1875, George F. Barker MSS, University of Pennsylvania Archives.
132. (day): Thomas A. Watson, *Exploring Life* (New York, 1926), pp. 1–35.
134. (corner): Ibid., pp. 32–33; Harmonic Interferences, Bell, p. 117.
134. (inventors): Watson, *Exploring Life*, pp. 34–35, 47–53.
134. (University): Ibid., pp. 52–55.
135. (others): Ibid., pp. 57–59.
135. (interruptions): Harmonic Interferences, Bell, pp. 97–98, 27–28.
135. (myself): AGB to parents, Feb. 5, 1875.
135. (espionage): AGB to GGH, Jan. 26, 1875; AGB to parents, Feb. 5, 1875.
136. (yesterday): AGB to parents, Feb. 12, 13, 1875.
136. (receiver): Harmonic Interferences, Bell, p. 29.
137. (cylinder): AGB to parents, Feb. 12, 1875.
137. (far): AGB to parents, Feb. 21, 1875.
137. (lines): AGB to parents, Mar. 5, 1875; Harmonic Interferences, Bell, p. 63.
138. (concentration): Specifications of Patents No. 166,094, 166,095, 166,096.
138. (25): Dowd Case, p. 100; People's Telephone, Exhibits, pp. 990–991.
138. (thing): AGB to parents, Mar. 5, 1875.
139. (prevailing): Idem; People's Telephone, Exhibits, pp. 990–991.
139. (mankind): *Bell Telephone*, pp. 302–303.
139. (elocution): AGB to parents, Mar. 18, 1875; diary of Mary Henry, Dec. 12, 16, 22, 1866, Joseph Henry MSS, Smithsonian Institution.
139. (scheme): J. Henry to A. Bache, Dec. 5, 1865, Joseph Henry MSS, Smithsonian Institution Archives.
140. (it): AGB to parents, Mar. 18, 1875.
140. (problem): Idem.
140. (priorities): Idem.
141. (wire): AGB to parents, Mar. 18, 1875.
141. (interested): AGB to parents, Mar. 22, 1875.
142. (one): Idem.

CHAPTER 14

143. (telegraphy): AGB to parents, Mar. 18, 22, 1875.
143. (help): *Bell Telephone*, pp. 88–89; small notebook, "T. A. W. Cash a/c 1875 — Notes and Ideas 1874-5," Thomas A. Watson MSS, Boston Public Library.
143. (all): People's Telephone, Evidence before Swan, p. 1606.
143. (Gray): Note dated Mar. 11, 1875, in account book kept for AGB by Sarah Fuller; AGB to Pollok & Bailey, Apr. 6, 1875; *N.Y. Times*, Dec. 27, 1873; AGB to GGH, May 24, 1875.
144. (coincide): Watson, *Exploring Life*, pp. 61, 66.
144. (easy): People's Telephone, Complainants in Reply, p. 189.
144. (completed): *Bell Telephone*, pp. 55–56.
144. (it): Entry for Mar. 28, 1875, in notebook, "T. A. W. Cash a/c 1875," Watson MSS, Boston Public Library; AGB to J. Henry, Apr. 2, 1875, Official Incoming Correspondence, Smithsonian Institution; J. Henry to AGB, Apr. 29, 1875; People's Telephone, Evidence before Swan, p. 1604.
144. (thinking): Watson, *Exploring Life*, p. 62.
145. (vibrations): *Bell Telephone*, p. 53.
145. (transmitted): Ernest F. Fenollosa, *Epochs of Chinese and Japanese Art* (2 vols., New York, 1912), I, viii–xi; People's Telephone, Complainants in Reply, pp. 134–135; People's Telephone, Evidence before Swan, pp. 1610–1611; People's Telephone, Complainants in Reply, p. 202.
145. (hoped): People's Telephone, Complainants in Reply, pp. 131–132.
145. (headaches): Watson, *Exploring Life*, p. 66.

146. (happened): *Bell Telephone*, pp. 57–58.
146. (expedient): Ibid., p. 58.
146. (circuit): People's Telephone, Evidence before Swan, p. 1590.
147. (repeated): Watson, *Exploring Life*, p. 67.
147. (heard): People's Telephone, Exhibits, p. 1007.
147. (it): People's Telephone, Evidence before Swan, pp. 1590–1591.
147. (speech): Ibid., p. 1591.
147. (receiver): Ibid., p. 1590; People's Telephone, Complainants in Reply, pp. 202–203.
148. (diaphragm): Watson, *Exploring Life*, p. 69; *Bell Telephone*, p. 61.
148. (it): People's Telephone, Complainants in Reply, pp. 203, 139–140; Proofs, I, 499.
148. (frequency): GGH to AGB, June 19, 1875; *Bell Telephone*, pp. 62–64.
149. (me): AGB to parents, June 30, 1875.
149. (*accompli*): *Bell Telephone*, pp. 62, 66–67; AGB to S. Fuller, July 1, 1875.
149. (transmitter): GGH to AGB, July 2, 1875.
149. (speech): People's Telephone, Evidence before Swan, pp. 1575–1576; *Bell Telephone*, p. 71.
150. (patent): People's Telephone, Evidence before Swan, pp. 1662, 1674.

CHAPTER 15

151. (discussions): Journal of Mrs. AGB, Jan. 6, 1879.
151. (trouble): AGB to parents, June 30, 1875; AGB to Mrs. GGH, June 24, 1875.
151. (decision): Idem.
152. (much): AGB to parents, June 30, 1875; AGB journal, June 25, 27, 1875.
152. (relieved): AGB to Mabel, Aug. 8, 1875.
152. (sound): AGB to parents, June 30, 1875.
153. (return): AGB journal, July 23, 1875.
153. (please): Mabel to Mrs. GGH, ca. Aug. 2, 1875.
153. (go): AGB journal, Aug. 4, 1875; AGB to Mr. and Mrs. GGH, Aug. 5, 1875.
153. (it): AGB journal, Aug. 6, 1875.
154. (ill): AGB to Mabel, Aug. 8, 1875.
154. (you): Idem.
154. (encounter): AGB journal, Aug. 9, 1875.
154. (Hubbard): AGB to Mabel, Aug. 10, 1875; Mabel to AGB, Aug. 15, 1875.
154. (her): AGB journal, Aug. 17, 1875.
155. (Aleck): AGB to Mrs. AMB, Aug. 18, 1875.
155. (work): Mrs. AMB to AGB, Aug. 23, 30, 1875.
155. (*wait*): AGB to Mr. and Mrs. GGH, Aug. 24, 1875.
155. (call): Mrs. GGH to AGB, Aug. 25, 1875.
155. (FINIS): AGB to Mabel, Aug. 26, 1875; AGB journal, Aug. 26, 1875.
156. (Brantford): AGB to AMB, Aug. 31, 1875.
156. (Heights): AGB to Mabel, Sept. 1, 1875, quoting AMB to AGB.

CHAPTER 16

157. (me): AGB to Mabel, Sept. 12, 1875.
157. (sensation): Idem; People's Telephone, Evidence before Swan, p. 1704.
157. (others): People's Telephone, Evidence before Swan, p. 1676; AGB to GGH, Sept. 28, 1875; *Brantford Expositor*, Sept. 23, 1875; *Toronto Globe*, Sept. 27, 1875.
158. (customers): People's Telephone, Evidence before Swan, p. 1676.
158. (promised): Idem; AGB to AMB, Sept. 27(?), 1875; AGB to GGH, Sept. 28, 1875.
158. (advice): Draft of AGB to George Brown, Oct. 4, 1875.

158. (controversy): AGB to GGH, Aug. 14, 1875; GGH to AGB, Aug. 15, 1875.
159. (telegraphically): *Bell Telephone*, pp. 76–78.
159. (way): People's Telephone, Complainants in Reply, p. 316.
159. (meanwhile): *Bell Telephone*, pp. 89–90.
160. (out): AGB to AMB, Oct. 16, 1875, Oct. 25, 1875; AGB to S. Fuller, Nov. 3, 1875; AGB to parents, Nov. 11, 1875.
160. (trial): GGH to AGB, Oct. 29, 1875.
160. (trifling): AGB to Mabel, Nov. 5, 7, 10, 15, 1875.
160. (anyway): AGB to GGH, Nov. 23, 1875.
160. (concluded): Idem.
161. (spot): Mabel to M. True, ca. December 1875.
161. (Mabel): Idem; AGB to Mabel, Nov. 25, 1875.
161. (Alec): AGB to parents, Nov. 25, 1875.
161. (fame): AMB to Mabel, Dec. 6, 1875.
162. (all): People's Telephone, Evidence before Swan, pp. 1678–1679; GGH to AGB, Dec. 4, 1875.
162. (stand-still): AGB to Mabel, Jan. 7, 1876.
162. (Roberta): AGB to Mabel, Dec. 26, 1875.
162. (apparatus): AGB to Mabel, Dec. 27, 28, 1875; People's Telephone, Evidence before Swan, pp. 1679–1681.
163. (damages): AGB to parents, ca. Jan. 15, 1876; AGB to Mabel, Jan. 17, 1876.
164. (owl): AGB to GGH, Jan. 13, 1876; Mrs. AGB to Mrs. J. Penman, Oct. 30, 1922.
164. (broken): Spark-arrester specification sworn to by AGB, Jan. 31, 1876.
165. (current): *Bell Telephone*, pp. 83–88; Bernard S. Finn, "Alexander Graham Bell's Experiments with the Variable-Resistance Transmitter," *Smithsonian Journal of History*, I (1966), 2–3.
165. (patents): GGH to AGB, Jan. 15, 1876; Dowd Case, p. 127.
165. (day): AGB to GGH, Jan. 13, 1876; AGB to Mabel, Jan. 16, 19, 1876; AGB to AMB, Jan. 18, 1876.
165. (revised): People's Telephone, Exhibits, pp. 936–947; AGB to parents, Feb. 12, 1876.
165. (patents): People's Telephone, Exhibits, p. 951.
166. (mentioning): J. Storrow to E. Brown, June 28, 1887; J. Brown to AGB, Feb. 27, 1876; AGB to AMB, Mar. 8, 1876.
166. (added): AGB to parents, Feb. 12, 1876.
166. (telegraph): Idem.
166. (undertaking): Proofs, I, 57–59.
168. (clipping): Dowd Case, p. 127.
168. (it): AGB to parents, Feb. 29, 1876.
169. (farther): Dowd Case, p. 107; Telephone Interferences, A–L, Gray, p. 29.
169. (can): Dowd Case, pp. 122, 125; *Bell Telephone*, pp. 211–212.
169. (lead): Dowd Case, p. 125; Telephone Interferences, A–L, Gray, p. 49.
169. (instrument): Proofs, II, 331–333, 341.
170. (voice): Proofs, II, 333.
171. (current): E. Pickering to AGB, May 6, 1877.
171. (hour): Speaking Telephone Interferences, Evidence for E. Gray, p. 17; Proofs, II, 626, 629, 717, 729.
172. (dissolved): Proofs, I, 17–19; Proofs, II, 336, 725–726.
172. (February 26): AGB to AMB, Feb. 22, 25, 29, 1876.
172. (favor): AGB to AMB, Feb. 29, 1876.
172. (apparatus): Idem.
173. (examples): *Bell Telephone*, pp. 263–264; Telephone Interferences, A–L, Gray, p. 1609.
173. (gross): *Bell Telephone*, p. 434.
173. (myself): AGB to AMB, Feb. 29, 1876.

174. (mind): Josephson, *Edison*, pp. 138–140; George B. Prescott, *Bell's Electric Speaking Telephone* (New York, 1884), pp. 110–113.
174. (now): AGB to AMB, Feb. 29, 1876; Mabel to AGB, ca. Mar. 1, 1876.
174. (time): AGB to AMB, Feb. 29, 1876.
174. (to): Dowd Case, p. 127; Proofs, II, 863; W. Baldwin to S. White, Mar. 3, 1876, Telephone Historical Collection, American Telephone and Telegraph Company, New York.
174. (well-disposed): Proofs, II, 718–719.
175. (Washington): Mabel to AGB, Mar. 5, 1876.
176. (issued): Idem.

CHAPTER 17

177. (before): AGB to AMB, Feb. 15, 1876; People's Telephone, Evidence before Swan, p. 1720.
177. (experiments): Notebook, "Experiments made by A. Graham Bell. (Vol. I)," pp. 12–34.
179. (all): Entry for Mar. 8, 1876, ibid.
179. (all): Finn, "Bell's Experiments," pp. 6–7.
180. (effect): Entry for Mar. 20, 1876, "Experiments . . . (Vol. I)."
180. (sense): Entry for Mar. 9, 1876, ibid.
181. (transmitter): Entry for Mar. 10, 1876, ibid.
181. (removed): Idem.
181. (home): AGB to AMB, Mar. 10, 1876.
182. (letter): Entry for Mar. 10, 1876, in memo book of T. Watson, Box 1069, AT&T Historical Collection, N.Y.
182. (dramatic): Watson, *Exploring Life*, p. 78.
182. (1882): People's Telephone, Complainants in Reply, p. 310.
184. (membrane): Entries for Mar. 10, 11, 12, 1876, "Experiments . . . (Vol. I)."
184. (them): Entry for Mar. 13, 1876, ibid.
184. (March 23): Entry for Mar. 14, 1876, ibid.
184. (resistances): Entries for Mar. 14, 15, 1876, ibid.
185. (father): AGB to AMB, Mar. 18, 1876; AGB to parents, Apr. 12, 1876; AGB to Mabel, Apr. 2, 13, 1876.
185. (experimentally): Finn, "Bell's Experiments," pp. 10–11.
185. (foil): Entries for Mar. and Apr., "Experiments . . . (Vol. I)."
185. (go): Entries for Apr. 1, 2, 1876, ibid.; AGB to Mabel, Apr. 2, 1876.
186. (anything): Entry for Apr. 5, 1876, "Experiments . . . (Vol. I)"; AMB to AGB, Apr. 16, 1876; GGH to AGB, Apr. 26, 1876.
186. (telegraphy): Idem; AGB to Mabel, May 6, 1876.
186. (slightest): "Experiments . . . (Vol. I)," Apr. and May 1876, passim.
186. (home): AGB to Mrs. AMB, Apr. 30, 1876.
187. (son-in-law): GGH to AGB, Apr. 26, 1876.
187. (Papa): Entry for Mar. 27, 1876, "Experiments . . . (Vol. I)."
187. (Centennial): Entries for May 5, 22, 1876, ibid.

CHAPTER 18

189. (savants): E. Horsford to E. Pickering, Apr. 25, 1876, Pickering MSS; AGB to Mr. and Mrs. AMB, May 3, 1876.
189. (tide): AGB to parents, May 12, 1876; Alexander Graham Bell, "Researches in Telephony," *Proceedings of the American Academy of Arts and Sciences*, XII (1876), 1–10; *Bell Telephone*, p. 256. Years later, Bell thought he had demonstrated telephonic speech in the lecture of May 10. But he said nothing of that when he described the occasion in a letter to his parents two days afterward. His later recollection probably confused that lecture with its repetition on May 25.

190. (Telephony): *Boston Transcript,* May 31, 1876; AGB to Mrs. AMB, May 26, 1876.
190. (while): AMB to Mrs. AMB, Apr. 5, 1876; AGB to Mabel, Wednesday evening, 1876 (probably June).
190. (Brazil): AGB to Mrs. AMB, June 18, 1876.
190. (showing): Leverett Saltonstall, *Report of the Massachusetts Commissioner to the Centennial Exhibition at Philadelphia* (Boston, 1877), pp. 7–8; MacKenzie, *Bell,* pp. 118–119.
191. (well-prepared): AGB to Mrs. AMB, June 18, 1876.
191. (tell): AGB to Mrs. AMB, June 18, 1876.
191. (Apparatus): J. Watson to Editor, *Nature,* Dec. 5, 1878; AGB to Mabel, June 19, 20, 1876.
191. (English): AGB to Mabel, June 21, 1876.
193. (time): AGB to Mabel, June 21, 1876.
193. (reminiscences): AGB to Mabel, June 21, 22, 1876; AGB to parents, June 22, 1876; Proofs, II, 95; J. Storrow to AGB, May 14, 1896; G. Barker to H. Draper, June 21, 1876, Henry Draper MSS, New York Public Library.
193. (organs): Saltonstall, *Report of Mass. Commissioner,* pp. 9, 18.
194. (Survey): Unsigned, undated, unfinished MS, "The First Exhibition of Bell's Telephone at the Centennial Exposition," by the father of Dr. George W. Outerbridge, Society Collections, Pennsylvania Historical Society (the elder Outerbridge had assisted Professor Barker in preparing and conducting the judges' tour of June 25, 1876); G. Barker to H. Draper, June 21, 1876, Draper MSS; *Bell Telephone,* pp. 100–101; J. Watson to editor, *Nature,* Dec. 5, 1878; AGB to parents, June 27, 1876.
194. (books): AGB to parents, June 27, 1876.
195. (absence): AGB to parents, June 27, 1876.
195. (current): Outerbridge MS; *Bell Telephone,* pp. 96–100, 322–323.
196. (touch): Outerbridge MS; *Bell Telephone,* pp. 96–100.
196. (end): AGB to parents, June 27, 1876.
197. (marvel): AGB to parents, June 27, 1876.
197. (cheered): The independent testimony of three witnesses agrees to Dom Pedro's words; the expression "My God, it talks!," later ascribed to Dom Pedro, is mentioned in no contemporary accounts by witnesses and is presumably apocryphal. AGB to parents, June 27, 1876, quoting Willie Hubbard's report; P. Richards, Feb. 17, 1879, to Editor, *Boston Transcript,* Feb. 19, 1879; H. Oliver (one of the judges) to Editor, *Boston Transcript,* Feb. 26, 1879; testimony of Elisha Gray, Apr. 5, 1879, Proofs, II, 446; Proofs, I, 616; address by AGB, "The Pre-Commercial Period of the Telephone," Nov. 2, 1911, at the first meeting of the Telephone Pioneers' Association, Boston.
197. (day): AGB to parents, June 27, 1876.
198. (unfounded): AGB to parents, June 27, 1876; Proofs, I, 288, II, 449; AGB to A. Hayes, Oct. 20, 1876.
198. (compatriots): *Bell Telephone,* p. 101.
198. (understood): J. Watson to Editor, *Nature,* Dec. 5, 1878; telegram, W. Hubbard to AGB, July 26, 1876.
198. (year): Proofs, II, 436, 447, 542; J. Watson to Editor, *Nature,* Dec. 5, 1878; J. Watson to AGB, Dec. 20, 1878; Proofs, II, 449–451.

CHAPTER 19

199. (triumph): Proofs, I, 365; AGB to J. Storrow, Feb. 26, 1880.
199. (following): William D. Armes (ed.), *The Autobiography of Joseph LeConte* (New York, 1903), pp. 254–255.
200. (remained): AGB to J. Watson, June 28, 1876, James C. Watson MSS, Univer-

sity of Michigan Library; Notebook, "Electrical Experiments by AGB Vol. II," pp. 20–27; AGB to J. Watson, Jan. 14, 1879 (written as 1878), J. Watson MSS.

200. (diaphragm): "Electrical Experiments," II, 28–39; AGB to parents, July 17, 1876; Mrs. GGH to GGH, July 14, 1876.

201. (while): Mrs. GGH to GGH, July 14, 1876; AGB to Mabel, July 16, 1876.

201. (Heights): *Boston Globe*, July 19, 1876; AGB to Mabel, July 20, 24, 1876.

201. (working): AGB to Mabel, July 27, 1876; Mabel to AGB, July 26, 1876.

201. (distances): AGB to Mabel, Aug. 5, 1876.

202. (lines): AGB to Mabel, Aug. 4, 1876; Thomas B. Costain, *The Chord of Steel* (New York, 1960), pp. 186–189.

202. (exclusively): AGB to Mabel, Aug. 5, 1876.

202. (present): Costain, *Chord of Steel*, pp. 191–202.

203. (passed): AGB to Mabel, Aug. 7, 11, 1876; Costain, *Chord of Steel*, pp. 217–231.

203. (him): AGB to parents, Aug. 27, 1876, Sept. 11, 1876; AGB to AMB, Dec. 1, 1876; Watson, *Exploring Life*, pp. 88–89; Rosario J. Tosiello, "The Birth and Early Years of the Bell Telephone System, 1876–1880" (Ph.D. diss., Boston University, 1971), p. 16; Mabel to Mrs. AMB, Sept. 11, 1876; AGB notebook, "Electrical Experiments," II, 45 (Sept. 11, 1876).

204. (hand): AGB notebooks, "Electrical Experiments," II, 45–91, III, 1–4; AGB to parents, Oct. 8, 1876.

204. (apparatus): AGB notebook, "Electrical Experiments," III, 5–14; Watson, *Exploring Life*, pp. 91–96; AGB to parents, Oct. 10, 1876.

205. (listen): AGB to parents, Oct. 12, 1876; AGB to AMB, Oct. 14, 1876; Mabel to Mrs. AMB, Oct. 17, 1876; AGB notebook, "Electrical Experiments," III, 14–15; Proofs, I, 341; *Bell Telephone*, p. 131.

205. (patent): AGB notebook, "Electrical Experiments," III, 15, 17.

205. (uses): AGB to parents, Nov. 7, 1876; AGB to Mabel, Nov. 12, 14, 21, 1876; AGB to AMB, Dec. 1, 1876; AGB to Mrs. GGH, Nov. 13, 14, 1876; Watson: *Exploring Life*, p. 90; *Bell Telephone*, pp. 131–132.

206. (use): AGB to Mabel, Nov. 19, 1876.

206. (coil): *Bell Telephone*, p. 131; AGB to GGH, Nov. 12, 1876; AGB notebook, "Electrical Experiments," III, 18, 37, 43.

206. (telephony): AGB to Mabel, Nov. 12, 1876; Watson, *Exploring Life*, pp. 99–102; AGB notebook, "Electrical Experiments," III, 18, 37; Proofs, I, 341, 423. Bell was aware that if a field more powerful than that of a permanent magnet were desired, an electromagnet excited by a separate, local battery current could be used; the point was that even in that case, the current would not have to be sent over the high resistance of the main line. (*Bell Telephone*, pp. 134–135.)

207. (resistance): AGB notebook, "Electrical Experiments," III, 32.

207. (transmitter): AGB to Mabel, Dec. 19, 1876 (two letters of the same date).

208. (phones): *Bell Telephone*, pp. 136–137, 145–146; Watson, *Exploring Life*, pp. 103–105; AGB to Mabel, Dec. 6, 1876.

208. (that): *Bell Telephone*, p. 136.

209. (miles): Ibid., pp. 126–127, 139; Proofs, I, 365; AGB to parents, Sept. 28, 1876; A. Ellis to AMB, Oct. 16, 1876.

209. (name): AGB to Mabel, December 1876.

209. (Michigan): AGB to Mabel, Nov. 14, 22, 1876.

210. (available): Proofs, II, 737; AGB to Mabel, Nov. 10, 1876.

210. (advice): AGB to Mrs. AMB, Nov. 29, 1876.

210. (millionaire): Tosiello, "Bell System," pp. 83–84; AGB to Mabel, Dec. 8, 1876; Thomas Watson in two letters to Fred DeLand, Nov. 17, 28, 1907, confirms Ponton's early enthusiasm for the telephone exchange system, but adds that it "was a perfectly familiar idea in Prof. Bell's mind long before Ponton came forward."

211. (nations): AGB to parents, Oct. 10, 1876; AGB to AMB, Oct. 20, 22, 1876; AGB

to Mrs. AMB, Oct. 25, 1876; Letters Patent, Great Britain, No. 4765 (Bell Collection); AGB to Mabel, Oct. 29, 1876, Nov. 21, 23, 1876, Dec. 8, 1876; AGB to S. Fuller, Nov. 10, 1876; Draft of proposed agreement between Dion Boucicault and AGB, November 1876.

211. (magnet, magnets): AGB to AMB, Dec. 1, 1876; AGB to Mabel, Dec. 1, 8, 15, 24, 28, 1876; AGB to J. Hubbard, Dec. 5, 1876, Boston Public Library MSS.
211. (Gray): AGB to Mabel, Dec. 31, 1876, Jan. 3, 1877; AGB to parents, Jan. 6, 1877, Alexander Graham Bell MSS, Boston University.
212. (telephone): AGB to T. Sanders, Jan. 7, 1877, AT&T Historical Collection.
212. (telegraph): The two preceding paragraphs are based on a study of the subject made by my graduate research assistant at Boston University, Mr. George Wise.
214. (patent): AGB to Mabel, Jan. 3, 7, 1877; AGB to parents, Jan. 6, 1877, Bell MSS, Boston University; Proofs, I, 1639.
214. (experimentally): Proofs, II, 96.
214. (patents): AGB to Mabel, Jan. 12, 13, 1877; *Bell Telephone*, pp. 384–385, 461–469.
214. (invention): AGB to parents, Jan. 6, 1877, Bell MSS, Boston University; AGB to Mabel, Jan. 12, 13, 1877; AGB to parents, Jan. 13, 1877; *Washington Star*, Jan. 20, 1877; *Bulletin of the Philosophical Society of Washington*, II (1874–1878), 103.
214. (Boston): AGB to Mrs. GGH, Jan. 17, 1877.

CHAPTER 20

215. (instruments): AGB to GGH, Jan. 21, 1877; Mabel to C. McCurdy, Jan. 24, 1877; AGB to Mabel, Jan. 19, 1877.
215. (early): AGB to GGH, Jan. 21, 1877; AGB, "The Pre-Commercial Period of the Telephone" (address at the first meeting of the Telephone Pioneers' Association, Boston, Nov. 2, 1911).
215. (sufficiently): AGB to GGH, Feb. 15, 1877, Jan. 21, 1877.
216. (Bell): Mabel to C. McCurdy, Jan. 24, 1877; AGB to Mabel, Apr. 18, 22, 1877.
216. (circuit): Mabel to C. McCurdy, Mar. 16, 1877; AGB to AMB, Feb. 3, 1877; AGB to Mabel, Apr. 4, 1877; AGB to parents, Mar. 2, 1877; AGB to Mrs. AMB, Apr. 19, 1877.
216. (chose): AGB to AMB, Feb. 3, 1877; AMB to AGB, Feb. 8, 1877; AGB to Mabel, Jan. 21, 1877, Feb. 13, 1877.
217. (room): *Boston Transcript*, Feb. 1, 1877.
217. (table): Mabel to C. McCurdy, Feb. 16, 1877.
219. (Hello, success, Wires): *Boston Globe*, Feb. 13, 1877; *Salem Register*, Feb. 15, 1877; AGB to GGH, Feb. 13, 1877; Mabel to C. McCurdy, Feb. 16, 1877.
219. (famous): *Bell Telephone*, pp. 149–150; *Springfield* (Mass.) *Republican*, Feb. 15, 1877; *N.Y. Daily Graphic*, Mar. 6, 1877; *Leslie's Illustrated Weekly*, Mar. 31, 1877; *Scientific American*, Mar. 31, 1877; *The Athenaeum* (London), Mar. 3, 1877; *La Nature* (Paris), Apr. 21, 1877.
219. (each): AGB to Mabel, Feb. 13, 1877; AGB to GGH, Feb. 15, 1877.
219. (have): *Providence Morning Star*, Feb. 26, 1877; *Boston Herald*, Feb. 24, 1877; *Boston Globe*, Feb. 24, 1877; AGB to parents, Mar. 2, 1877; typed statement by Mrs. AGB, June 1922, suggesting that the silver model be deposited for safekeeping at the Smithsonian Institution.
220. (direction): *Bell Telephone*, pp. 165–166.
221. (patent): Ibid., pp. 167–168.
221. (invention): Ibid., p. 168.
223. (worked): The two preceding paragraphs are summarized from *Providence Star*, Feb. 26, 1877, May 8, 1877; *Providence Journal*, May 6, 8, 1877; *Lowell Citizen*, Apr. 26, 1877; *Boston Advertiser*, May 5, 8, 1877; *Boston Globe*, May 5, 8, 9, 1877; *Boston Herald*, May 5, 1877; *Boston Post*, May 6, 1877; *Manchester* (N.H.)

Union, May 9, 1877; *N.Y. Herald*, May 12, 18, 1877; *N.Y. Sun*, May 12, 1877; *N.Y. Times*, May 19, 20, 1877; *N.Y. Tribune*, May 12, 18, 1877; *N.Y. World*, May 12, 18, 1877; *Springfield* (Mass.) *Republican*, May 14, 1877; AGB to Mabel, Apr. 7, 22, 1877; Mabel to C. McCurdy, May 7, 1877; Watson, *Exploring Life*, pp. 113–119.

224. (supernatural): *Providence Press*, Mar. 12, 1877; *Providence Star*, Feb. 26, 1877; *Manchester* (N.H.) *Union*, May 9, 1877; *Boston Advertiser*, May 5, 1877; *N.Y. Herald*, May 12, 1877.

224. (this): AGB to Mabel, Apr. 7, 1877; *Bell Telephone*, p. 154; *Providence Journal*, May 8, 1877; *Boston Advertiser*, May 5, 1877; *N.Y. Sun*, May 12, 1877; *N.Y. Tribune*, May 12, 1877.

225. (Brooklyn): *Boston Times*, May 6, 1877; *Boston Globe*, May 5, 1877; *N.Y. Tribune*, May 12, 18, 1877; *N.Y. Herald*, May 12, 18, 1877.

225. (system): E. Blake to AGB, Feb. 10, 1877; W. Channing to AGB, July 16, 1877; *The Alumni Register* (University of Pennsylvania), V, 242 (April 1901); *DAB*, IV, 8 ("W. F. Channing").

225. (advance): E. Blake to AGB, Mar. 19, 23, 1877, July 3, 10, 1877; J. Peirce to AGB, Apr. 8, 13, 1877, July 7, 1877; W. Channing to AGB, July 16, 1877; W. Channing to GGH, Nov. 8, 1877; AGB to GGH, Mar. 18, 1878; AGB to Sanders, Hubbard, and Watson, Mar. 2, 1880; see Prescott, *Bell Telephone*, pp. 76–80, for Bell's public acknowledgment to the Providence group, and pp. 102–107, for a further illustrated commentary on their work.

226. (claimant): C. Blake to AGB, Feb. 26, 1878; W. Channing to AGB, Feb. 26, 1878; AGB to J. Peirce, Mar. 18, 1878; *The Alumni Register* (University of Pennsylvania), V, 242; see Tosiello, "Bell System," pp. 149–154, for a more extended analysis of the Providence group and its claims.

226. (besides): AGB to parents, Mar. 25, 1877; AMB to AGB, Mar. 27, 1877; Mabel to AGB, Apr. 6, 12, 1877; AGB to Mabel, Apr. 8, 12, 1877.

226. (it): GGH to AGB, Mar. 29, 1877; Mrs. GGH to GGH, Apr. 6, 1877; AGB to Mabel, Apr. 8, 1877.

227. (tickets): AGB to Mabel, Apr. 12, 1877; AGB to Mrs. AMB, Apr. 19, 1877; G. Coe to AGB, Mar. 18, 1913; AGB to parents, May 5, 1877; *Providence Morning Star*, May 8, 1877; *Springfield Republican*, May 14, 1877; GGH to Mrs. GGH, May 18, 1877; *N.Y. Mail*, May 19, 1877.

227. (silence): Mabel to Mrs. AMB, May 12, 1877; *N.Y. Evening Post*, May 16, 1877; AGB to Mabel, May 11, 1877; AGB to parents, May 11, 1877; facsimile of flyer in Bell Collection, advertising Bell lecture for May 28; Watson, *Exploring Life*, pp. 120–121; *Lawrence* (Mass.) *Daily American*, May 29, 1877.

227. (that): Watson, *Exploring Life*, p. 110.

228. (machinery): Tosiello, "Bell System," pp. 78–81.

228. (practice): Ibid., pp. 71–72; AGB to Mabel, Apr. 4, 1877.

228. ($20.00): Tosiello: "Bell System," pp. 71–75; Proofs, II, 277.

229. (property): GGH to A. Pollok, July 4, 1877.

229. (suits): Tosiello, "Bell System," pp. 81–82; Watson, *Exploring Life*, p. 107.

230. (offer): Josephson, *Edison*, pp. 142–146.

230. (February): E. Converse to GGH, Feb. 7, 1877; AGB to GGH, Feb. 13, 1877; GGH to AGB, Feb. 15, 1877; H. Howard to AGB, Feb. 16, 1877.

231. (locality): Tosiello, "Bell System," pp. 86–90; Mabel to Mrs. AMB, May 12, 1877.

231. (affairs): Tosiello, "Bell System," pp. 92–95.

231. (Fuller): Mabel to C. McCurdy, May 7, 13, 1877; AGB to S. Fuller, June 25, 1877; W. Reynolds to AGB, July 7, 1877.

232. (machine): AGB to Mabel, Jan. 17, 19, 1876, Dec. 6, 1876.

232. (telephone): AGB to Mrs. AMB, Oct. 27, 1876; AGB to Mabel, Nov. 9, 1876.

233. (ahead): AGB to Mabel, Nov. 10, 1876; Mabel to AGB, Jan. 11, 1877.

233. (Washington): Mabel to AGB, April 1877; AGB to Mabel, Apr. 8, 10, 1877; E. Gallaudet to AGB, Apr. 12, 1877.
233. (her): Mabel to C. McCurdy, Feb. 16, 1877; Mabel to Mrs. AMB, May 12, 1877.
233. (feet): Mabel to C. McCurdy, July 3, 1877.
234. (in): Watson, *Exploring Life*, p. 123.
234. (occasion): Mabel to Mrs. J. Penman, Oct. 15, 1922.
234. (Brantford): Mary E. Symonds, recollections of AGB, Apr. 3, 1923; Mrs. AGB to Mrs. GGH, July 20, 23, 1877.
234. (off): Mrs. AGB to Mrs. AMB, July 30, 1877.
235. (day): GGH to A. Pollok, July 4, 1877; GGH, Report to Board of Managers, Aug. 1, 1877; Proofs, I, 418; Tosiello, "Bell System," pp. 97-98.
235. (development): Tosiello, "Bell System," pp. 95-96.
235. (England): Mrs. AGB to Mrs. AMB, Aug. 4, 1877.

CHAPTER 21

236. (possible): Mrs. AGB to Mrs. GGH, Aug. 11, 1877; AGB to Mrs. GGH, Aug. 14, 1877.
236. (Towns): Mrs. AGB to Mrs. GGH, Aug. 15, 1877, Sept. 24, 1877.
237. (position): Mrs. AGB to Mrs. GGH, Sept. 24, 27, 1877, Oct. 1, 8, 1877.
237. (her): Mrs. AGB to Mrs. GGH, Sept. 4, 10, 1877; Mrs. AGB to Mrs. AMB, December 1877, Oct. 25, 1877.
237. (her): Mrs. AGB to Mrs. GGH, Aug. 23, 1877, Sept. 11, 1877; AGB to parents, Oct. 28, 1877; Mrs. AGB to Mrs. AMB, May 10, 22, 1878, Aug. 14, 1878; AGB to Mrs. AGB, Sept. 5, 1878.
238. (meadow): Mrs. AGB to Mrs. GGH, Sept. 10, 1877, Nov. 10, 27, 1877, Dec. 5, 1877; Mrs. AGB to Mrs. AMB, Nov. 12, 1877, Dec. 2, 1877, July 7, 1878.
238. (do): Mrs. AGB to Mrs. GGH, Sept. 10, 1877, Nov. 21, 27, 1877, Dec. 11, 1877; Mrs. AGB to Mrs. AMB, Oct. 19, 1877; AGB to parents, Oct. 28, 1877; AGB to GGH, Oct. 28, 1877; Mrs. AGB to AGB, Aug. 17, 1878.
238. (campaign): AGB to Mrs. AMB, Oct. 11, 1878; Mrs. AGB to Mrs. GGH, Dec. 5, 1877; Mrs. AGB to Mrs. AMB, December 1877.
239-240. (thin, planet): Mrs. AGB to Mrs. GGH, Sept. 11, 27, 1877, Dec. 26, 1877, Feb. 20, 1878; Mrs. AGB to Mrs. AMB, Sept. 23, 1877, Aug. 14, 1878; AGB to GGH, Oct. 28, 1877.
240. (do): Mrs. AGB to Mrs. GGH, Dec. 26, 1877; AGB to Mrs. GGH, Feb. 21, 1878.
240. (Alec): *Exeter and Plymouth* (England) *Gazette*, Aug. 20, 1877; *Western Morning News* (Plymouth, England), Aug. 22, 1877; Mrs. AGB to Mrs. GGH, Aug. 20, 1877.
241. (minutes): Mrs. AGB to Mrs. GGH, Aug. 15, 16, 17, 1877, Oct. 22, 1877, Nov. 1, 9, 1877; Mrs. AGB to Mrs. AMB, Sept. 23, 1877, Oct. 19, 1877; *Western Morning News* (Plymouth, England), Aug. 22, 1877; *Aberdeen Journal*, Sept. 26, 1877; *Leeds Mercury*, Oct. 26, 1877; *Birmingham Daily Post*, Oct. 30, 1877; *Glasgow News*, Nov. 9, 1877; *The Observer* (London), Dec. 2, 1877; *London Daily News*, Dec. 21, 1877; P. Foster to AGB, Dec. 1, 1877.
241. (Victoria): Mrs. AGB to Mrs. GGH, Sept. 21, 1877, late December 1877; AGB to S. Baird, Apr. 5, 1879, Bell MSS, Boston University.
241. (arm): Mrs. AGB to Mrs. GGH, Dec. 11, 1877, Jan. 9, 17, 1878; MacKenzie, *Bell*, pp. 191-193.
242. (telephone): Ibid., pp. 190-191; Mrs. AGB to Mrs. GGH, Jan. 12, 1878.
242. (himself): MacKenzie, *Bell*, pp. 194-195; Kate Field, *The History of Bell's Telephone* (London, 1878), p. 16 and passim.
242. (him): Mrs. AGB to Mrs. AMB, Feb. 27, 1878.
242. (prejudice): AGB to GGH, Oct. 28, 1877.
243. (so): AGB to GGH, July 28, 1880; AGB to Mrs. AGB, Jan. 6, 1880.

243. (infringements): F. Gower to AGB, Aug. 22, 1877; Mrs. AGB to Mrs. GGH, Aug. 29, 1877, Sept. 11, 1877, Nov. 27, 1877; Mrs. AGB to Mrs. AMB, Sept. 23, 1877.

243. (contest): Mrs. AGB to Mrs. GGH, Jan. 9, 17, 1878; Mrs. AGB to Mrs. AMB, Feb. 27, 1878; F. Warner to AMB, Mar. 23, 1878.

244. (backers): MacKenzie, *Bell*, pp. 200–201; Mrs. AGB to Mrs. GGH, Feb. 17, 1878; Mrs. AGB to Mrs. AMB, Mar. 22, 1878.

244. (£3000): F. Warner to AMB, Mar. 23, 1878.

244. (value): Depositions by AGB and Thomas A. Watson, Apr. 23, 24, 1879, "In the Matter of Interference between the applications respectively of A. G. Bell and David Brooks, for patents for Telephone Circuit," in Bell Collection; W. Preece to AGB, Dec. 3, 1877; AGB to C. Wollaston, Oct. 2, 1878; Mrs. AGB to Mrs. GGH, Oct. 7, 1877, Dec. 26, 1877; Mrs. AGB to Mrs. AMB, Oct. 19, 1877.

244. (response): MacKenzie, *Bell*, pp. 202–206.

245. (letters, Manager): Mrs. AGB to Mrs. GGH, Dec. 26, 1877; Mrs. AGB to Mrs. AMB, Apr. 30, 1878, July 19, 1878, Aug. 14, 1878; GGH to AGB, Aug. 11, 1879, Oct. 28, 1879; AGB to Mrs. AGB, July 17, 1878; printed copy of AGB to the Directors of the Telephone Company, Ltd., July 25, 1878; draft or copy of AGB to Directors of the Telephone Company, Ltd., Oct. 15, 1878.

245. (anticipated, summer): GGH to AGB, Apr. 18, 1879, Aug. 11, 1879, Oct. 28, 1879, July 1880; Josephson, *Edison*, pp. 149–155.

246. (expense): Mrs. AGB to Mrs. GGH, Nov. 27, 1877, Jan. 22, 1878; AGB to GGH, July 28, 1880.

246. (patronage, trust): AGB to GGH, Oct. 28, 1877, Nov. 1, 1877; Mrs. AGB to Mrs. GGH, Nov. 1, 21, 27, 1877, Dec. 11, 1877, Jan. 22, 25, 1878; GGH to AGB, Nov. 30, 1877, Apr. 18, 1879, July 1880; Mrs. AGB to Mrs. AMB, Jan. 25, 1878; C. Roosevelt to AGB, Sept. 7, 1878, Oct. 9, 1878; Prescott, *Bell Telephone*, pp. 192–198; Watson, *Exploring Life*, p. 125; *Boston Herald*, Feb. 20, 1881; Gower to AGB, Aug. 22, 1877, Oct. 15, 1877.

246. (business): Mrs. AGB to Mrs. AMB, Oct. 25, 1877; Mrs. AGB to Mrs. GGH, Oct. 12, 1877, Nov. 1, 2, 21, 1877, Dec. 11, 1877; F. Warner to AMB, Mar. 23, 1878; draft of AGB to GGH, Mar. 27, 1879.

247. (boons): AGB to GGH, July 28, 1880; G. Grosvenor to W. Langdon, Dec. 16, 1925, Box 1103, AT&T Collection; AGB to H. Peabody & Co., Jan. 29, 1880; AGB to W. Ker, Feb. 3, 1880; Mrs. AGB to Mrs. GGH, Oct. 26, 1877, Nov. 27, 1877, Jan. 22, 1878; AGB to C. Hubbard, July 3, 1880.

247. ($260,000): AGB to GGH, July 28, 1880; GGH to AGB, July 1880, July 25, 1887; *Washington Post*, Feb. 26, 1898.

248. (this): AGB to F. DeLand, Nov. 27, 1914; Preece quoted in MacKenzie, *Bell*, p. 200.

248. (life): AGB to Mrs. AGB, Sept. 9, 1878.

248. (strife): AGB to Mrs. AGB, Aug. 21, 1878.

248. (science): AGB to Mrs. AGB, Sept. 9, 1878.

249. (about): Mabel to AGB, ca. Aug. 2, 1876.

249. (something): AGB to Mrs. AGB, Apr. 5, 1879, Sept. 9, 1878.

250. (invent): AGB to Mabel, July 23, 1876.

251. (jelly): Mrs. AGB to Mrs. GGH, Aug. 19, 22, 1877, Oct. 1, 1877, Nov. 8, 9, 12, 1877; AGB, "Experimental Note-Book Vol. IV," Aug. 30, 1877, Sept. 9, 1877; AGB to GGH, Nov. 15, 1877.

251. (universities): *Glasgow Herald*, Nov. 15, 1877; Mrs. AGB to Mrs. GGH, Mar. 13, 1878; AGB to Mrs. AMB, Oct. 4, 1878; M. Müller to AGB, Oct. 12, 1878; Mrs. M. Müller to Mrs. AGB, October 1878; William F. Warren, *Annual Report of the President of Boston University* (Boston, 1878), pp. 35–36.

252. (scientist): AGB, "Experimental Note-Book Vol. IV," May 10, 12, 31, 1878, June 5, 12, 1878; W. Thomson to J. Clerk Maxwell, Aug. 30, 1877; James Clerk

Maxwell, *The Scientific Papers of James Clerk Maxwell*, ed. W. D. Niven (2 vols., Cambridge, 1890), II, 742–743, 751–753.

252. (it): Watson, *Exploring Life*, p. 97.
252. (sound): Mrs. AGB to Mrs. GGH, Jan. 9, 1878; AGB to GGH, Mar. 18, 1878.
253. (idea): AGB to GGH, ca. March 1878, Box 1103, AT&T Historical Collection; GGH to AGB, Jan. 22, 1878; C. Cheever to GGH, May 21, 1878, Box 1205, AT&T Historical Collection.
253. (follow): AGB, transcript of lecture, "Speech," before Royal Institution, London, May 17, 1878; A. Ellis to A. Mayer, Apr. 9, 1878, Hyatt-Mayer Collection, Princeton University.
253. (offered): Josephson, *Edison*, pp. 171–174; Tosiello, "Bell System," p. 258.
254. (hers): AGB to GGH, ca. March 1878, Box 1103, AT&T Historical Collection.
254. (ended): AGB to T. Watson, Aug. 12, 1878, photostat of catalogue of American Autograph Shop, Merion, Pa., 1937, in MSS Division, New York Public Library; T. Watson to AGB, Aug. 30, 1878.
254. (circuit): AGB, transcript of lecture, "Speech," before Royal Institution, London, May 17, 1878.
255. (experimentation): Watson, *Exploring Life*, pp. 153–154.
255. (bubbles): Mrs. AGB to Mrs. AMB, Sept. 23, 1877; Mrs. AGB to Mrs. GGH, Oct. 1, 1877.
255. (March): AGB, "Experimental Note-Book Vol. IV," Oct. 6, 10, 1877, Mar. 7, 8, 1878.
256. (deaf): Mrs. AGB to Mrs. GGH, Aug. 29, 1877, Sept. 27, 1877; AGB, "Experimental Note-Book Vol. IV," Sept. 3, 1877; *Aberdeen Daily Free Press*, Sept. 27, 1877; AGB to Mrs. AGB, Aug. 18, 1878; *Eastern Morning News* (Hull, England), Oct. 10, 1878.
256. (it): T. Borthwick to AGB, Nov. 17, 1877.
256. (services): Mrs. AGB to T. Borthwick, ca. Dec. 1, 1877; AGB to T. Borthwick, Feb. 18, 20, 25, 1878, Aug. 13, 24, 31, 1878; M. True to AGB, Aug. 27, 1878, all in Thomas Borthwick MSS, Scottish National Library, Edinburgh.
257. (over): AGB to Mrs. AGB, Sept. 5, 9, 1878; Mrs. AGB to Mrs. AMB, Sept. 30, 1878.
257. (himself): AGB to Mrs. AGB, Sept. 9, 1878.
257. (Greenock): S. Stevenson to G. Fellendorf, Feb. 22, 1963, copy in Bell Collection.
257. (Boston): AGB to Mrs. AGB, Sept. 5, 1878; Watson, *Exploring Life*, pp. 151–152.
257. (you): Ibid., p. 152.

CHAPTER 22

258. (greatest): Tosiello, "Bell System," p. 182.
259. (treasurer): T. Sanders to AGB, Mar. 20, 1878.
259. (Bell): Tosiello, "Bell System," pp. 158, 163, 185.
259. (it): T. Sanders to AGB, Mar. 20, 1878.
260. (business): Tosiello, "Bell System," p. 184; T. Sanders to AGB, Sept. 27, 1878.
260. (four-in-hand): Watson, *Exploring Life*, pp. 158–160; Tosiello, "Bell System," pp. 325, 329, 347.
260. (race): Tosiello, "Bell System," pp. 224, 228.
260. (harden, Hubbard, off): Ibid., pp. 225–235, 239, 247, 249–250, 256.
262. (transmitter): Ibid., pp. 262–264, 266–268, 273, 276, 334, 416–417.
262. (rights): Ibid., pp. 339–346; Josephson, *Edison*, pp. 147–148.
262. (competition): Tosiello, "Bell System," pp. 272, 354.
263. (all): Ibid., pp. 274, 355.
264. (College): Inga S. Dolbear, "Amos Emerson Dolbear: A Biography" (privately mimeographed, 1963), p. 4; clippings from *Newport* (R.I.) *Mercury*, July 14,

1888, and *Practical Electricity*, November 1888, in Dolbear Collection, Tufts University.
264. (twenty-nine): A. Dolbear to E. Pickering, Feb. 15, 1874, Pickering MSS; A. Dolbear to A. Winchell, Aug. 5, 1867, Sept. 10, 1867, Oct. 14, 1873, Winchell MSS, University of Michigan Library.
264. (Christmas): Proofs, II, 484–487.
264. (phone): Proofs, II, 486–488, 687; Dolbear, *Dolbear*, p. 132; *Bell Telephone*, pp. 170–171; A. Dolbear to AGB, July 8, 1877.
265. (invention): *Bell Telephone*, p. 172; A. Dolbear to A. Winchell, Mar. 20, 1877, Winchell MSS.
265. (friends): *Bell Telephone*, pp. 172–173.
266. (England): Dolbear, *Dolbear*, p. 140.
266. (recognition): Proofs, II, 97, 502–506; A. Dolbear to AGB, July 8, 1877.
266. (made): Proofs, II, 505–509.
266. (patent): Amos E. Dolbear, *The Telephone* (Boston, 1877), pp. vi, 116–117.
267. (Case): Proofs, II, 488–489, 510–514.
267. (alive): Watson, *Exploring Life*, pp. 152–153; *Bell Telephone*, pp. 158–159; AGB to Mrs. AGB, Nov. 14, 1878.
267. (battle): AGB to Mrs. AGB, Nov. 14, 1878, Jan. 22, 1879; C. Blake to AGB, Jan. 5, 1879.
268. (Wakefield): *DAB*, XVIII, 99 ("James J. Storrow"), XVII, 253 ("Chauncey Smith"); Herbert N. Casson, *The History of the Telephone* (Chicago, 1911), p. 102; *N.Y. Evening Post*, Nov. 10, 1885.
268. (quarter): AGB to Mrs. AGB, Jan. 26, 1879.
269. (experimenters): Charles H. Swan, "Narrative History of the Litigation on the Bell Patents, 1878–1896," AT&T Historical Collection, Box 1098, p. 111; AGB to Mrs. AGB, Apr. 18, 1879; Mrs. AGB to AGB, May 25, 1879.
269. (it): AGB to Mrs. AGB, Apr. 8, 1879.
269. (court): E. Gray to G. Barker, Apr. 30, 1879, Barker MSS; Swan, "Narrative History," pp. 112–114, 118.
270. (telephone): S. T. Cameron's reminiscences of AGB, January 1924; J. Storrow to AGB, May 18, 1883.
270. (value): Swan, "Narrative History," pp. 127–129; Mrs. AGB to Mrs. AMB, June 16, 1879; Watson, *Exploring Life*, p. 170.
271. (struggles, evident): Tosiello, "Bell System," pp. 489–491; Swan, "Narrative History," pp. 131–132.
271. (price): Tosiello, "Bell System," p. 490.
271. (one): Frederick L. Rhodes: *Beginnings of Telephony* (New York, 1929), pp. 53–54, 75; William C. Langdon, "Myths of Telephone History," in *Bell Telephone Quarterly*, April 1933, pp. 10–11.
272. (original, it): Langdon, "Myths," pp. 11–13; J. Storrow to AGB, Jan. 6, 1886, Dec. 28, 1886; Giovanni E. Schiavo, *Antonio Meucci, Inventor of the Telephone* (New York, 1958), passim.
272. (affords): Swan, "Narrative History," pp. 130–132; *Bell Telephone*, pp. 1–3.
273. ($6000): Prescott, *Bell Telephone*, pp. 500–506.
274. (1880): Prescott, *Bell Telephone*, p. 487; G. Roberts to AGB, Feb. 16, 1894; Rhodes, *Beginnings of Telephony*, p. 67.
274. (reticence): MacKenzie, *Bell*, pp. 215–216; Prescott, *Bell Telephone*, p. 505.
275. (Storrow): Arthur S. Pier, *Forbes: Telephone Pioneer* (New York, 1953), p. 151; *Bell Telephone*, pp. 186–187; C. Swan to AGB, Oct. 7, 1914.
275. (Marconi): Rhodes, *Beginnings of Telephony*, p. 65; *Bell Telephone*, pp. 186–187; Casson, *History of the Telephone*, p. 98.
275. (patents): Pier, *Forbes*, pp. 165–166; MacKenzie, *Bell*, p. 258; Rhodes, *Beginnings of Telephony*, pp. 71–72.
276. (Age): Rhodes, *Beginnings of Telephony*, p. 72.

276. (Company): MacKenzie, *Bell*, pp. 258–259, 262; Pier, *Forbes*, pp. 167–169.
276. (settled): *Bell Telephone*, p. 2; Rhodes, *Beginnings of Telephony*, pp. 217–218; MacKenzie, *Bell*, pp. 261–264, 268–275; Allan Nevins, *Grover Cleveland: A Study in Courage* (New York, 1932), pp. 294–295.
277. (elsewhere): AGB to A. Garland, Oct. 26, 1885, printed copy in Bell Collection; Journal of Mrs. AGB, Oct. 20, 23, 1885.
277. (pages): *Bell Telephone*, pp. 2–3 and passim.
277–278. (Court, claim): Dolbear, "Dolbear," pp. 147–149, 161–162, 164–165, 170–171, 177; Albert Stetson, "Lest We Forget: The Story of Professor Dolbear's Experiments in Telephony," *The Tufts College Graduate*, XXI, No. 3, March–May 1923, pp. 146–148, 150; *Boston Globe*, May 12, 1895; Frank W. Lovering, "The First Telephone," *Tufts Alumni Review*, Fall 1949.
278. (negotiations): Proofs, II, 690–695, 733–734; Tosiello, "Bell System," pp. 481–482 and chapter xiv, passim.
278. (significance): Proofs, II, 690–695.
279. (research): *DAB*, VII, 514 ("E. Gray").
279. (tighter): AGB to Mrs. AGB, Mar. 12, 1901; Lloyd W. Taylor, "The Untold Story of the Telephone," *American Physics Teacher*, December 1937, pp. 249, 251; *DAB*, VII, 514.
280. (loss): AGB to G. Maynard, Mar. 11, 1901.

CHAPTER 23

281. (out): AGB to Mrs. AGB, Jan. 24, 1879, Feb. 19, 21, 1879; Journal of Mrs. AGB, Mar. 12, 1879; Tosiello, "Bell System," pp. 397–402.
282. (Hubbard): W. Forbes to AGB, Mar. 5, 1879; AGB to GGH, Feb. 25, 1879; Watson, *Exploring Life*, pp. 173–179; AGB to A. Marble, Feb. 25, 1880.
282. (welcome): AMB to AGB, Nov. 14, 1877, Feb. 25, 1878, Dec. 3, 1879; *Hamilton (Ontario) Evening Times*, Aug. 30, 1877; *Brantford (Ontario) Expositor*, Sept. 28, 1877, Dec. 28, 1877; AMB to GGH, Feb. 5, 1878; AMB to T. Henderson, Oct. 7, 1881, Original Documents File, Canadian Bell Telephone Company, Montreal; AGB to AMB, Sept. 9, 1879; AGB to H. Bishop, Sept. 10, 1879.
282. (1880): AGB to W. Forbes, Mar. 29, 1879; Pier, *Forbes*, pp. 123–126.
283. (demand): United States Letters Patent, Nos. 213,090, 220,791, 228,507, 230,168, 238,833, 250,704; Mrs. AGB to AGB, Mar. 20, 1879; Proofs, I, 428.
283. (others): United States Letters Patent, Nos. 244,426, 241,184; AGB to R. Fenn, Mar. 30, 1888; AGB to Mrs. AGB, Dec. 23, 1891; Mrs. AGB to GGH, Dec. 3, 1895; *Washington Evening Star*, June 11, 1904.
283. (years): Watson, *Exploring Life*, p. 126; T. Watson to AGB, May 12, 1878.
284. (notes): AGB to W. Forbes, Mar. 29, 1879; AGB to parents, Apr. 20, 1879; Journal of Mrs. AGB, Mar. 6, 1879.
285. (station): AGB to T. Edison, May 25, 1879; AGB to F. Maguire, Oct. 20, 1886; AGB to Mrs. AGB, June 6, 1894; Mrs. AMB to AMB, Apr. 5, 1880, Volta Bureau Library.
285. (Sir): S. Clemens to GGH, Dec. 27, 1890.
286. (posterity): *Electrical Review*, Oct. 29, 1892.
286. (Gray): H. Pope to AGB, Mar. 20, 1911.
286. (law): AGB to T. Vail, Sept. 30, 1911; AGB to N. Kingsbury, Feb. 12, 1912.
287. (about): *Sydney (Australia) Morning Herald*, July 30, 1910.

CHAPTER 24

291. (it): Journal of Mrs. AGB, Mar. 8, 1879.
291. (total): Tosiello, "Bell System," pp. 95, 195, 312, 314, 400; "Bell, A. G. — Issue of Orig. Stock," Box 1104, AT&T Historical Collection.

References

292. (then): GGH to AGB, Dec. 1, 11, 1878; Mrs. AGB to AGB, Mar. 9, 1879.
292. (in): Mrs. AGB to Mrs. AMB, June 16, 1879; AGB to AMB, Sept. 9, 1879; GGH to Mrs. GGH, Aug. 26, 1879, Sept. 12, 1879.
293. (holder): AGB to Mrs. AMB, Oct. 10, 1879; C. Hubbard to AGB, Nov. 3, 11, 1879; AGB to AMB, Dec. 20, 1879.
293. (that): C. Hubbard to AGB, Mar. 27, 1880, Jan. 7, 1881; AGB to AMB, Jan. 18, 1881.
293. (millionaire): AGB to C. Hubbard, Apr. 4, 1881; C. Hubbard to AGB, Dec. 13, 1881, Jan. 13, 1882; Mrs. AGB to C. Hubbard, Dec. 12, 1881; AGB to GGH, Dec. 14, 1881.
293. ($31,000): Mrs. AGB to AGB, undated (ca. 1886), Nov. 7, 1904.
293. ($45,415.53): GGH to AGB, June 1, 1893; Mrs. AGB to GGH, Jan. 5, 1895; Mrs. AGB to AGB, Oct. 29, 1896, undated (ca. 1901); AGB to GGH, Feb. 1, 1897.
294. (Carnegies): Affidavit of Gilbert H. Grosvenor, Feb. 16, 1926; F. DeLand to J. Bleecker, Feb. 16, 1923; GGH to AGB, June 12, 1886; P. Schulze to GGH, Dec. 30, 1892; D. Lesh to GGH, Feb. 24, 1894; Mrs. AGB to AGB, Dec. 9, 1889, Oct. 29, 1892; AGB to Mrs. AGB, Feb. 18, 1901.
294. (exaggerated): 49th Cong., 1st session, H. of R., Misc. Doc. 355, p. 762; AGB to Mrs. AGB, Oct. 14, 1889.
294. (sale): Mrs. AGB to AGB, undated (ca. 1885); Melville Bell Grosvenor, "Life with Grandfather" (talk before Literary Society, Feb. 10, 1968).
295. (them): Mrs. AGB to AGB, July 14, 1883.
295. (response): Mrs. AGB to AGB, May 13, 1898; H. Tilcomb to AGB, Feb. 26, 1905, with AGB's endorsement; W. Mitchell to H. Tilcomb, Mar. 15, 1905; Longfellow Memorial Association to AGB, Mar. 2, 1900, with AGB's endorsement.
296. (deafness): AGB to Mrs. AGB, Sept. 24, 1881; folder, "Gifts"; J. Faber to AGB, Apr. 15, 25, 1885; AGB to J. Faber, Apr. 25, 1885; J. Saville to AGB, May 4, 1885; F. Pearson to AGB, Nov. 22, 1887; S. Tamura to AGB, June 12, 1905; P. Richards to AGB, Jan. 24, 1883, Feb. 23, 1883, May 1, 1883, Apr. 12, 1886, May 25, 1888, Aug. 23, 1888, Sept. 4, 20, 1888, Nov. 5, 1888; C. Hubbard to Mrs. AGB, May 14, 25, 1886; AGB to G. Roberts, Dec. 3, 1891; T. Sanders to AGB, Mar. 14, 1891, Apr. 8, 1891; L. Sanders to Mrs. T. Sanders, undated (probably 1890); Mrs. AGB to Elsie Bell, May 18, 1891; Mrs. AGB to AGB, May 13, 1898; AGB to G. Sanders, Oct. 25, 1899; AGB to R. Sparrow, Jan. 24, 1902.
296. (penniless): AGB to R. Moulton, Jan. 19, 1900; J. Wright to AGB, Mar. 9, 1901; AGB to Mrs. AGB, Mar. 15, 1901.
296. (on): J. Foster to AGB, Feb. 8, 1888; S. Burnett to AGB, Dec. 4, 1888; W. McCain to AGB, Apr. 30, 1909; G. Kennan to AGB, Mar. 15, 1891; F. Branagan to AGB, Jan. 5, 1898.
296. (it): AGB to Mrs. AGB, Mar. 9, 1879; Mrs. AGB to Mrs. AMB, June 16, 1879, Aug. 8, 1879; Mrs. AGB to Mrs. GGH, July 11, 1880.
297. (it): AGB to parents, Sept. 5, 1879; Mrs. GGH to AGB, Nov. 1878; GGH to AGB, Dec. 11, 1878; Mrs. AGB to Mrs. AMB, Oct. 9, 1879, Dec. 14, 1879; AGB to AMB, Feb. 12, 1880; AGB to C. A. Bell, Mar. 9, 1880.
297. (dozen): Notes in folder, "Washington Addresses of Alexander Graham Bell"; undated and unidentified clipping in folder, "Scott Circle House"; *Washington Post*, Jan. 11, 1887.
297. (insurance): *Washington Post*, Jan. 11, 1887; *Washington Star*, Jan. 11, 1887.
298. (life): Reminiscences of Charles F. Thompson, Feb. 23, 1923.
299. (hair): *Aurora* (Ill.) *Herald*, Mar. 12, 1889.
299. (1882): Notes in folder, "Washington Addresses of Alexander Graham Bell."
299. (Washington): *Washington Star*, June 4, 1891, June 11, 1930.
300. (Lakes): AGB to Mrs. AGB, May 6, 1902; chronology of the Bells' travels in the 1880s, in folder, "Itinerary of AGB"; MacKenzie, *Bell*, pp. 256–257.

References

300. (Baddeck): Recollections of Mr. and Mrs. Bell by Maude MacKenzie, 1923; MacKenzie, *Bell*, pp. 257, 276.
301. (refuge): Journal of Mrs. AGB, Sept. 17, 1885; Recollections by Maude MacKenzie; AGB to D. McAulay, Aug. 19, 1886; MacKenzie, *Bell*, p. 276.
301. (point): MacKenzie, *Bell*, p. 277; K. McKenzie to AGB, Apr. 24, 1889.
301. (Canada, rambling): Lilian Grosvenor, "My Grandfather Bell," *The New Yorker*, Nov. 11, 1950, p. 44; *Halifax* (Nova Scotia) *Morning Chronicle*, Dec. 1, 1893.
301. (view): Idem; *Halifax Chronicle*, Dec. 1, 1893.
303. (world): Idem; Mrs. AGB to Mrs. GGH, Sept. 15 (1889?).
303. (loon): George Kennan, "A Few Recollections of Alexander Graham Bell," *The Outlook*, Sept. 27, 1922, pp. 146–147.
304. (cold): Mrs. AGB to Mrs. GGH, undated.
304. (iris): AGB to Mrs. AGB, Apr. 19, 1889; Mrs. AGB to Mrs. GGH, Oct. 15, 1906, June 25, 1897, Aug. 4, 1907.
304. (world): Mrs. AGB to Mrs. GGH, Sept. 16, 1907.
305. (roads): MacKenzie, *Bell*, pp. 278, 283, 286; M. True to J. Hitz, Dec. 12, 1892, Volta Bureau Library; AGB to Mrs. AGB, May 25, 1899; Mrs. AGB to Mrs. GGH, "Friday" (1898?), June 1907.
305. (it): Mrs. AGB to GGH, Jan. 5, 1895.
305. (time): Mrs. AGB to AGB, Jan. 10, 1892.
305. (other): Mrs. AGB to AGB, Dec. 7, 1889; Mrs. AGB to Mrs. GGH, Feb. 13, 1892.
306. (indeed): Mrs. AGB to Mrs. AMB, Oct. 24, 1880; AGB to Mrs. AGB, Oct. 23, 1881; Mrs. AGB to Mrs. GGH, Nov. 15, 1898.
306. (awake): Thompson, Reminiscences, Feb. 23, 1923.
306. (junkets): AGB to Mrs. AGB, Mar. 26, 1901.

CHAPTER 25

307. (alone): David Fairchild, "Alexander Graham Bell: Some Characters of His Greatness," *Journal of Heredity*, XIII (1922), 195.
308. (part): AGB to Mrs. AGB, June 2, 1894, and note by G. Grosvenor, Sept. 23, 1923; Mrs. AGB to AGB, July 9, 1895.
308. (one): Mrs. AGB to G. Grosvenor, October 1906; AGB to Mrs. AGB, July 25, 1883.
309. (outside): AGB to Mabel, Aug. 1, 1876; AGB to Mrs. AGB, Nov. 20, 1881, May 25, 1899; AGB to Mrs. AGB, May 5, 1890.
310. (need): Mrs. AGB to AGB, July 9, 1895, Mar. 10, 1907.
310. (summer): Mrs. AGB to AGB, June 18, 1888.
310. (Alec): Mrs. AGB to Mrs. GGH, Apr. 24, 1882, May 1, 1883.
310. (people): Thompson Reminiscences, Feb. 23, 1923; Mrs. AGB to Elsie Bell, May 12, 1891.
311. (announced): Mrs. AGB to AGB, May 18, 1893.
311. (him): AGB to Secretary of Cosmos Club, Mar. 6, 1880; Mrs. AGB to Mrs. G. Grosvenor, August 1906.
311. (at): Journal of Mrs. AGB, Mar. 6, 1879; Ethel F. Fisk (ed.), *The Letters of John Fiske* (New York, 1940), p. 516.
312. (other): Kennan, "Recollections," p. 148.
312. (roof): Mrs. AGB to Elsie Bell, Apr. 24, 1896.
313. (him): C. Walcott to G. Grosvenor, Mar. 27, 1923; M. Benjamin to G. Grosvenor, Mar. 23, 1923.
313. (Bell): Mrs. AGB to AGB, Mar. 3, 1891; G. Kennan to Mrs. AGB, Oct. 9, 1922; author's interview with Gilbert H. Grosvenor, June 21, 1965.
313. (Halifax): Later note by Mrs. AGB, on Mrs. AGB to Elsie, Aug. 19, 1891;

M. Bell to Mrs. AMB, no date, ca. 1869; *A Public Document . . . [re] Roussy v. Roussy* (broadside, Windsor, Nova Scotia, 1886); "Summary of Diaries," pp. 11, 17, 19–25, Fritz de Sumichrast MSS, Harvard University Archives.

314. (time): Telegram, AGB to Bishop Binney, Dec. 10, 1886; F. de Sumichrast to AGB, Sept. 8, 1887; "Summary of Diaries," p. 66, Sumichrast MSS; F. de Sumichrast to AGB, May 15, 1889; AGB to F. de Sumichrast, Nov. 20, 1897.

314. (now): Journal of Mrs. AGB, Oct. 25, 1884.

314. (everything): Mrs. AGB to Mrs. GGH, Sept. 1, 1890.

315. (banker): Journal of Mrs. AGB, Mar. 14, 19, 1880.

315. (interest, arms, usual): Mrs. AGB to G. Grosvenor, October 1906; *Brantford* (Ontario) *Expositor,* May 6, 1881; AMB to AGB, Jan. 26, Apr. 6, 1881; Mrs. G. Grosvenor to Mrs. W. Peter, Mar. 27, 1950; Mrs. AGB to AGB, May 3, 1891.

316. (Daisy): Journal of Mrs. AGB, Mar. 14, 1880; Mrs. AGB to Mrs. AMB, February 1880.

316. (me): Mrs. AGB to G. Grosvenor, Oct. 21, 1906; Mrs. AGB to AGB, Aug. 26, 1882, July 22, 1883.

316. (die): AGB to Mrs. AGB, Sept. 19, 1881; Mrs. AGB to G. Grosvenor, Oct. 21, 1906; Irving W. Rolfe, "Alexander Graham Bell at Pigeon Cove," Box 1107, AT&T Collection; note by Mrs. AGB accompanying painting by Lobrichon of her dead baby; AGB to Mrs. AGB, Nov. 20, 1881; AGB to T. Borthwick, Dec. 23, 1881, Borthwick MSS; AGB to Mrs. AGB, Sept. 4, 1883.

317. (Sciences): Mrs. AGB to Mrs. GGH, Apr. 24, 1882; Journal of Mrs. AGB, Feb. 3, 1884.

317. (children): AGB to Mrs. AGB, Dec. 19, 1887, May 6, 1892; Mrs. AGB to AGB, Dec. 12, 1893, Jan. 25, July 22, 1895.

317. (furthest): Mrs. AGB to Mrs. GGH, July 26, 1898, Mar. 6, 1901.

318. (pneumonia): AGB to Mrs. AGB, Sept. 24, 1881; Mrs. AGB to Mrs. GGH, May 14, 1883, Sept. 17, Dec. 11, 1890; Mrs. AGB to AGB, Mar. 19, 1892, June 26, 1895, and three undated letters, ca. January 1891.

319. (difficulty): Mrs. AGB to AGB, May 8, 19, 24, 1895.

319. (deafness): Mrs. AGB to AGB, ca. 1885.

319. (it): Mrs. AGB to AGB, June 25, 1895; Mrs. AGB to Mrs. GGH, two undated letters, ca. summer of 1895.

320. (this): AGB to Mrs. AGB, May 25, 1899; AGB to Elsie Bell, Oct. 15, 1897.

320. (glory): Mrs. AGB to AGB, Nov. 24, 1904.

320. (knew): AGB to Mrs. AGB, Nov. 10, 1896; *Evening Wisconsin* (Milwaukee), Mar. 13, 1896.

321. (be): Mrs. AGB to E. Nitchie, Mar. 15, 1906.

321. (reply): Mrs. AGB to Miss M. Faircloth, Nov. 11, 1922; Mrs. G. Grosvenor, notes for speech at New York University, May 24, 1951; AGB, "Reminiscences of Early Days of Speech-Teaching," *Volta Review,* Dec. 1912, p. 581; Mabel H. Bell, "The Subtle Art of Speech-Reading," *Atlantic Monthly,* LXXV (1895), 165–167; AGB, address at Milwaukee, Mar. 12, 1896; G. Gower to AGB, Jan. 27, 1915.

322. (easily): Mrs. AGB to Mr. Gregory, Oct. 17, 1922; Mrs. AGB to F. DeLand, Sept. 21, 1906; AGB, remarks in transcript of "Discussion Following Mrs. Bell's Paper," possibly at Lake George in 1892.

322. (much): AGB to Mrs. AGB, Mar. 14, 1879; Mrs. AGB to AGB, undated, but probably 1882; Mrs. AGB to AGB, July 9, 1884.

323. (lives): Mrs. AGB to AGB, Dec. 3, 1889, May 28, 1894.

323. (time): Mrs. AGB to AGB, Oct. 4, 1893, May 1, 1895.

324. (you): AGB to Mrs. AGB, May 20, 1895, June 24, 1889, May 6, 1900.

324. (regularly): Journal of Mrs. AGB, Mar. 6, 1879; Mrs. AGB to AGB, undated, but possibly ca. 1885.

324. (course): AGB to Mrs. AGB, Dec. 20, 1891; dictation by AGB, June 19, 1904.

325. (also): Memorandum by G. Grosvenor, Dec. 2, 1924.
325. (laughing): Remarks by Mrs. G. Grosvenor at the opening of the Alexander Graham Bell Museum, Baddeck, Nova Scotia, Aug. 18, 1956; Thompson Reminiscences, Feb. 23, 1923.
325. (again): Ibid.; petition to Commissioners of the District, Mar. 21, 1903.
326. (day): Remarks by Mrs. G. Grosvenor, Aug. 18, 1956; Thompson Reminiscences, Feb. 23, 1923; Mrs. AGB to AGB, May 20, 1899.
326. (Notes): Thompson Reminiscences, Feb. 23, 1923; Fairchild, "Alexander Graham Bell," p. 196.
326. (now): Home Notes, October 1881–March 1882, passim; AGB to Mrs. AGB, Aug. 21, 1883; Journal of Mrs. AGB, Oct. 8, 1883.
327. (again): Mrs. AGB to Mrs. GGH, Aug. 20, 1886; AGB to J. Hitz, Oct. 15, 1888; Mrs. AGB, paper before Young Ladies Club of Baddeck, fall or winter of 1905–1906.
327. (correspondence): AGB to A. McCurdy, Dec. 13, 1896; A. McCurdy to AGB, Jan. 28, 1897; AGB to F. DeLand, Nov. 27, 1914.
328. (it): Mrs. AGB to J. Carty, Aug. 24, 31, 1922; Lilian Grosvenor, "My Grandfather Bell," p. 48; AGB to A. McInnis, May 6, 1905.
328. (moved): AGB to Mrs. AGB, Dec. 12, 1885; Mrs. AGB to Mrs. GGH, Oct. 5, no year; Mrs. AGB to C. Yale, Sept. 30, 1922.
329. (sir): Fairchild, "Alexander Graham Bell," p. 197; Thompson Reminiscences, Feb. 23, 1923.
329. (along): Ibid.
329. (world): AGB to W. Powell, Feb. 12, 14, 1899; W. Powell to AGB, Feb. 13, 1899.
330. (sentiments): AGB to Mrs. AGB, Feb. 26, 1892.
330. (word): AGB to Mrs. AGB, Nov. 28, 1904.
330. (two): Citizenship papers of AGB, granted in the District of Columbia, Nov. 10, 1882; AGB to S. Newcomb, Jan. 29, 1883; AGB to C. Tupper, June 12, 1896; AGB to Mrs. AGB, Nov. 9, 1894.
331. (agreed): Mrs. AGB to Elsie Bell, May 1, 1894.
331. (night): Thompson, Reminiscences of AGB, Feb. 23, 1923; H. Bell to J. Hitz, Oct. 3, 1902, Volta Bureau Library.
331. (whiskers): Thompson Reminiscences, Feb. 23, 1923; Mrs. AGB to AGB, June 26, 1895.
331. (woods): AGB, reply to a questionnaire, ca. 1894.
331. (paper): Mrs. Gilbert H. Grosvenor, "Memories of my Father," *The Transmitter*, Dec. 1936, p. 10.
332. (music): Grosvenor, "Life with Grandfather"; notes in folder, "Songs (Favorite)."
332. (sleep): Mrs. G. Grosvenor, talk before the Eistophos Club.

CHAPTER 26

333. (passion): Journal of Mrs. AGB, Feb. 11, 1879; AGB to Mrs. AGB, Jan. 3, 1892; Fairchild, "Alexander Graham Bell," pp. 196, 199.
333. (motives): AGB, Speech before a congress of inventors in Washington, April 10, 1891, celebrating the centennial of the United States Patent Office (transcript in Bell Collection).
334. (itself): Idem; AGB to Mabel, Sept. 26, 1876.
334. (me): AGB to Mrs. AGB, Apr. 5, 1879.
334. (known): Idem.
334. (experiment): AGB to Mrs. AGB, Nov. 29, 1880; Alexander Graham Bell, "Discovery and Invention," *National Geographic Magazine*, June 1914; Alexander

References

Graham Bell, "Observation: Twin Brother to Invention," *The Youth's Companion*, Feb. 7, 1898.

335. (death): Thomas H. Whitcroft, "Sonic Sounding," *United States Naval Institute Proceedings*, LXIX (1943), 216-219.

335. (photophone): AGB to Mrs. AGB, Feb. 19, 1879.

335. (shadow): Alexander Graham Bell, "On the Production and Reproduction of Sound by Light," *American Journal of Science*, October 1880, pp. 308-313.

335. (vu): AGB to Mrs. AGB, Feb. 19, 20, 1879.

336. (over): Agreement between AGB and Sumner Tainter, Oct. 15, 1879, Bell Collection; Charles Sumner Tainter: "Early History of Charles Sumner Tainter," Tainter MSS, pp. 1-5, 8, 11, 15, 19, 21, 25, 27-85, 87.

336. (search): Lab Notes, December 1879-January 1880, passim.

336. (1880): Ibid., Feb. 7, 24, 1880.

336. (room): Bell, "Production of Sound by Light," pp. 314-320.

337. (screen): AGB, Lab Notes, Mar. 6, 1880; *Boston Transcript*, Feb. 20, 1880; *N.Y. Times*, Feb. 25, Mar. 2, 1880; *Nature*, Apr. 15, 22, 1880; *The Times* (London), Apr. 22, 24, 1880; Mrs. AGB to Mrs. GGH, May 12, 1880.

338. (hand): AGB to S. Thompson, Mar. 8, 1880; AGB to C. A. Bell, Mar. 9, 1880.

338. (for): Lab Notes, Mar. 26, Apr. 1, 1880; Bell, "Production of Sound by Light," pp. 319-320; *The Crank* (published by students of Sibley College, Cornell University), I (March 1887), 7.

339. (candy): Bell, "Production of Sound by Light," pp. 322-323; Lab Notes, July 18, 1880.

339. (we): Bell, "Production of Sound by Light," passim; Lab Notes, Aug. 29, 1880.

339. (possess): AGB to W. Forbes, May 11, 17, 1880; W. Forbes to AGB, May 14, 1880.

339. (transmitter): AGB to AMB, June 3, 1880; AGB to Mrs. AGB, June 8, 23, July 18, 1880.

340. (civilization): MacKenzie, *Bell*, p. 228; AGB to Mrs. AGB, Nov. 29, 1880; *Journal of the Society of Arts*, Dec. 3, 1880.

340. (America, December): Mrs. AGB to Mrs. GGH, Dec. 26, 1877; Mrs. AGB to Mrs. AMB, Mar. 24, 1879; AGB to C. A. Bell, Mar. 9, 1880; *The Cosmos Club Bulletin*, May 1966, p. 2; Archibald Henderson, *Bernard Shaw: Playboy and Prophet* (New York, 1932), pp. 119, 297; AGB to Mrs. AGB, Nov. 25, 1880.

341. (Laboratory): AGB to D. Gilman, Feb. 5, 1881; AGB to W. Forbes, Feb. 2, 1881.

342. (invaluable): AGB to AMB, Apr. 11, 1881; *Bulletin of the Philosophical Society of Washington*, IV (1881), 161.

342. (feet): *American Catholic Quarterly Review*, IX (1884), 750-751; *Scientific American*, May 10, 1890; American Telephone and Telegraph Company, *The Radiophone* (St. Louis, 1904), p. 3.

342. (receiver): *Jacksonville* (Ill.) *Daily Journal*, Feb. 29, 1896; *Washington Evening Star*, Sept. 5, 1896; *The Radiophone*, pp. 3-5.

343. (perceptible): *N.Y. Sun*, Dec. 26, 1899; A. O. Rankine, "The Transmission of Speech by Light," *Nature*, June 2, 1923.

343. (physics): Interpolation by AGB in a biographical sketch of him by Mrs. AGB, 1898; Mary B. Mullett, "How to Keep Young Mentally," *American Magazine*, December 1921.

343. (recording): Lab Notes, May 31, 1881.

343. (it): Lab Notes, March-May 1881, passim; AGB to Dr. J. Kerr, June 4, 1881.

344. (recover): Charles E. Rosenberg, *The Trial of the Assassin Guiteau* (Chicago, 1968), pp. 2-4, 9; Stewart A. Fish, "The Death of President Garfield," *Bulletin of the History of Medicine*, XXIV (1950), 378-381.

344. (made): *The Medical Gazette*, Oct. 15, 1881; Simon Newcomb, *The Reminiscences of an Astronomer* (Boston, 1903), p. 356; Mrs. AGB to Mrs. GGH, June

14, 1881; Alexander Graham Bell, *Upon the Electrical Experiments to Determine the Location of the Bullet in the Body of the Late President Garfield; and upon a Successful Form of Induction Balance for the Painless Detection of Metallic Masses in the Human Body* (Washington, D.C., 1882), p. 6.

344. (metal): Ibid., pp. 47–48.

345. (Science): Transcript of paper by AGB before the American Academy of Arts and Sciences, Dec. 11, 1879; Bell, *Experiments*, pp. 2–3; AGB to Mrs. AGB, Mar. 9, 1879.

345. (Laboratory): Bell, *Experiments*, pp. 4–6; *N.Y. Tribune*, July 11, 1881.

345. (likewise): Bell, *Experiments*, pp. 7–8.

345. (President): Rosenberg, *Trial of Guiteau*, p. 9; Notebook, "Experiments with Induction Balance," entries for July 17–26, 1881; Bell, *Experiments*, pp. 10–20.

346. (this): AGB to Mrs. AGB, July 26, 1881.

346. (tomorrow): Idem.

346. (morning): Notebook, "Experiments with Induction Balance," entries July 27–31, 1881; Bell, *Experiments*, pp. 22–31.

347. (monotone): *N.Y. Tribune*, Aug. 1, 1881; *Boston Herald*, Aug. 1, 1881.

347. (mystery): Bell, *Experiments*, p. 33; Notebook, "Experiments on Induction Balance," entries for Aug. 1, 2, 1881.

347. (pounds): Notebook, "Experiments with Induction Balance," entries for August 1881; Bell, *Experiments*, pp. 41–42; S. Tainter to AGB, Aug. 18, 1881; Rosenberg, *Trial of Guiteau*, p. 9.

348. (probes): AGB to Mrs. AGB, Sept. 19, 1881; Bell, *Experiments*, p. 33; Fish, "Death of Garfield," p. 388; Rosenberg, *Trial of Guiteau*, p. 10.

348. (X-ray): *Medical Gazette*, Oct. 15, 1881; J. Girdner to AGB, Dec. 16, 1886, Mar. 3, 30, 1887; *Medical Record*, Feb. 12, 1887; *N.Y. Medical Journal*, Sept. 17, 1887; A. S. Aloe Company, *Surgical Catalogue* (New York, 1891), p. 102; *N.Y. Herald Tribune*, Oct. 27, 1933.

348. (glory): MacKenzie, *Bell*, p. 240; Sir James Mackenzie Davidson, "An Address on the Telephone Attachment in Surgery," *The Lancet*, Jan. 30, 1915; *N.Y. Herald*, Jan. 17, 1915, citing the *British Medical Journal*.

349. (poliomyelitis): Memorandum dictated by AGB to Mrs. AGB, Sept. 1, 1881.

349. (treatment): G. Minchin to AGB, Feb. 5, 1882; AGB to T. Gleason, August 1882; *Montreal Gazette*, Aug. 26, 1882.

350. (Laboratory): Memorandum by Tainter beginning "Some Facts relating to Patent #341,214," Tainter MSS; Oliver Read, *From Tinfoil to Stereo* (Indianapolis, 1960), pp. 31, 36.

350. (later): Testimony of Sumner Tainter, May 1896, in the case of American Graphophone Company vs. Edison Phonograph Works (transcript in Bell Collection), Questions 4, 13, 59; Leslie J. Newville, "Development of the Phonograph at Alexander Graham Bell's Volta Laboratory," United States National Museum Bulletin 218 (Washington, D.C., 1959), p. 77.

351. (Institution): Lab Notes, Oct. 23, 1881.

351. (us): S. Tainter to AGB, Dec. 14, 1881, Feb. 7, 1882.

352. (adjourned): Tainter testimony in American Graphophone vs. Edison Phonograph, May 1896, Question 17; for further descriptions and illustrations of their work, see Newville, "Development," pp. 72–77.

352. (primacy): Tainter testimony in American Graphophone vs. Edison Phonograph, May 1896, Questions 18, 52.

352. (development): Testimony of AGB, May 1896, in American Graphophone vs. Edison Phonograph, Question 24; Lab Notes, 1882–1885, passim.

352–353. (form, only): AGB to T. Vail, Nov. 11, 1884; C. A. Bell to AGB, Dec. 29, 1884; AGB to C. Williams, Dec. 30, 1884; AGB to C. A. Bell and S. Tainter, June 14, 1885; AGB to D. Fairchild, Nov. 6, 1907; Newville, "Development," p. 78.

353. (them, condition): Memorandum by Tainter beginning "Some facts relating to Patent #341,214," Tainter MSS; AGB to S. Tainter, Mar. 30, 1888.
353. (own): Josephson, *Edison*, pp. 317–321; AGB to E. Johnson, Sept. 29, 1885.
354. (backer): AGB to F. Noyes, Apr. 27, 1910; Agreement of AGB, C. A. Bell, and C. S. Tainter with Charles J. Bell and James H. Saville, Jan. 6, 1886; E. D. Easton to F. DeLand, Nov. 13, 1912.
354. (contributions): Newville, "Development," pp. 78–79; Read, *Tinfoil to Stereo*, pp. 36, 120; *San Diego* (Cal.) *Evening Tribune*, July 31, 1937; *N.Y. Times*, Apr. 22, 1940.
354. (1924): *Cosmos Club Bulletin*, May 1966; C. A. Bell to Mrs. AGB, Aug. 9, 1889.
354. (financially): AGB to F. Noyes, Apr. 27, 1910.

CHAPTER 27

355. (development): AGB to T. Edison, May 25, 1879.
356. (deaf): *Science*, Sept. 19, 1884; Lab Notes, Nov. 18, 1885; AGB to Mrs. AGB, Feb. 8, 1885.
356. (model): AGB to Mrs. AGB, June 14, 1887.
356. (ideas): Home Notes, Dec. 4, 1889; Mrs. AGB to AGB, Dec. 16, 1889; AGB to D. Fairchild, Nov. 6, 1907.
357. (things): *Detroit Free Press*, Oct. 20, 1895.
357. (abandoned): Home and Lab Notes, December 1889–March 1890 and June 1890, passim; AGB to Mrs. AGB, June 25, 1890.
357. (them): Lab and Home Notes, 1891–1897, passim.
358. (temperature): *DAB*, X, 594–595 ("Samuel P. Langley").
358. (madmen): Langley interview of December 1896, transcribed from an unidentified clipping in *Beinn Bhreagh Recorder*, July 9, 1913.
359. (someday): Mrs. AGB to AGB, Apr. 21, 1891; Home Notes, May 26, 1891; AGB to Mrs. AGB, June 14, 15, 1891.
359. (second): Home Notes and Lab Notes, 1891–1897, passim; Mrs. AGB to Mrs. GGH, Apr. 16, 1894; loose sheet of penciled notes by AGB, Apr. 16, 1895.
359. (practical): Home Notes, Nov. 14, 1893, and 1891–1894, passim.
359. (notion): AGB to Mrs. AGB, May 30, 1893; Home Notes, June 1, 1893, Mar. 29, Apr. 11, 1894, Jan. 10, 1895.
361. (on): Home and Lab Notes, 1894–1897, passim; Mrs. AGB to AGB, May 28, June 26, 1895.
361. (life): AGB to Mrs. AGB, May 9, 1896; address by AGB on presenting the Langley Medal to Gustave Eiffel and Glenn Curtiss, Feb. 13, 1913; AGB to the Editor of *Science*, May 12, 1896.
361. (switchboard): Home and Lab Notes, 1895–1897, passim; AGB to Mrs. AGB, Oct. 8, 1896.
362. (visits): Thompson Reminiscences, Mar. 20, 1924.
362. (innings): *McClure's Magazine*, I (1893), 39; *Cincinnati Enquirer*, Mar. 1, 1896; *N.Y. World*, May 17, 1896.
363. (inventor): *McClure's Magazine*, I, 40.
363. (subject): Mrs. AGB to Mrs. GGH, Sept. 25, 1897; Lord Kelvin to Mrs. AGB, Apr. 20, 1898.
363. (experiments): AGB to Mrs. AGB, May 12, June 9, 1898, May 9, 1899.
364. (men): Lab Notes, Sept. 5, 28, Oct. 3, 9, 21, 28, 1896, Aug. 31, 1897; AGB to Mrs. AGB, June 12, 1898.
364. (am): AGB to Mrs. AGB, May 9, 1899.
364. (kite-flying): AGB to Mrs. AGB, May 14, 17899.
366. (strategy): Home Notes, May 28–June 1, 1890.
366. (away): Quotation from Home Notes, July 17, 1880.
366–367. (study, life): AGB to M. King, Dec. 15, 1882; Home Notes, Oct. 5, Dec. 26,

1889, Feb. 7, 1890, Oct. 9, 1892, Sept. 11, 1893; Lab Notes, Feb. 21, 1896; AGB to Mrs. AGB, Oct. 8, 1896. For several of my generalizations about Bell's experimental work, I am indebted to my graduate assistant Mr. George Wise, who scanned every page of Bell's notebooks with care, insight, and the advantage of a master's degree in electrical physics. His excellent analytical essay on the notebooks is deposited in the Special Collections of the Mugar Library at Boston University.

367. (life): AGB to Mrs. AGB, Oct. 6, 1896.

367. (last): Mrs. AGB to AGB, Oct. 21, 1901.

368. (too): *Washington Times*, Jan. 19, 1902.

CHAPTER 28

369. (through): Journal of Mrs. AGB, Mar. 8, 1879; Mrs. AGB to Mrs. AMB, Mar. 24, 1879; AGB to parents, Apr. 2, 1879; AGB to S. Baird, Apr. 17, 1879, Spencer Baird MSS, Smithsonian Institution; GGH to AGB, Apr. 18, 1879; Mrs. AGB to AGB, Apr. 25, 1879.

369. (mid-eighties): Alexander Graham Bell, "Experiments Relating to Binaural Audition," *American Journal of Otology*, July 1880; bibliography of Bell's writings and addresses in Harold S. Osborne, "Biographical Memoir of Alexander Graham Bell, 1847–1922," National Academy of Sciences *Biographical Memoirs*, XXIII

370. (Washington, D.C., 1943).

370. (America): Mrs. AGB to AGB, Oct. 2, 1893; Mrs. AGB to GGH, Dec. 3, 1895; *Beinn Bhreagh Recorder*, Mar. 31, 1910; *Science*, July 31, 1903; *Washington Star*, Oct. 13, 1903; R. Weir to AGB, May 30, 1910.

370. (reputation): AGB to D. Gilman, Jan. 23, 1889; Home Notes, May 14, 1890; Home and Lab Notes, 1890–1899, passim; J. Powell to AGB, Sept. 14, 1892; Mrs. AGB to Mrs. GGH, Nov. 30, 1892; AGB to J. Powell, Feb. 28, 1898; AGB, MSS essay on motion, Apr. 24, 1896, with note by A. McCurdy, May 22, 1897, on its return by S. Langley a year earlier; William Culp Darrah, *Powell of the Colorado* (Princeton, N.J., 1951), pp. 354–360, 375–384.

371. (up): Folder, "Associations and Clubs"; biographical resumes by AGB for *American Men of Science*, in folder, "Autobiographical"; note by AGB with bill for dues from American Forestry Association, Jan. 1, 1899.

371. (1897): Folder, "Science, National Academy."

371. (time): Mrs. AGB to Mrs. GGH, Oct. 19, 1902; Thompson Reminiscences, Feb. 23, 1923.

371. (Bell): *DAB*, VII, 14 ("Charles L. Freer").

372. (Detroit): Geoffrey T. Hellman, *The Smithsonian* (Philadelphia, 1967), pp. 159–161; undated paper by Mrs. Gilbert H. Grosvenor.

372. (million): Hellman, *The Smithsonian*, pp. 161–166.

372. (transportation): D. Gilman to AGB, Feb. 21, 1906; Mrs. AGB to AGB, Oct. 2, 1893; *N.Y. Tribune*, Oct. 6, 1893; C. Crane to AGB, May 6, no year, introducing H. G. Wells.

373. (force): Darrah, *Powell*, pp. 263–264, 386–387.

373. (anecdote): Thompson Reminiscences, Mar. 20, 1924; *Boston Transcript*, Sept. 28, 1901.

373. (work): Mrs. AGB to Mrs. GGH, May 1, 1883; AGB to W. James, Mar. 18, 1906; W. James to AGB, Mar. 20, 1907, Apr. 14, 1910; A. James to AGB, May 5, 1912.

373. (experiments): AGB to S. Newcomb, Feb. 27, 1901; S. Newcomb to AGB, Sept. 30, 1904; R. Chittenden to AGB, Dec. 20, 1912; Paul H. Oehser, *Sons of Science* (New York, 1949), pp. 119–121; S. Langley to AGB, June 18, 1898.

374. (physicists): Nathan Reingold, *Science in Nineteenth Century America* (New

York, 1964), pp. 275–287; Bernard Jaffe, *Michelson and the Speed of Light* (New York, 1960), pp. 49–58.

374. (undulate): Reingold, *Science*, pp. 287–288.
375. (anyway): Ibid., pp. 288–289; A. Michelson to C. A. Bell, Dec. 22, 1880.
375. (make): Reingold, *Science*, pp. 288–291.
375. (later): Jaffe, *Michelson*, pp. 76–77, 83–89.
375. (ether): Ibid., pp. 100–102, 146–147.
376. (topics): AGB to Mrs. AGB, July 25, 1883.
376. (abroad): J. Michels to AGB, May 1, 27, July 18, 1881, June 9, 1882.
377. (offer): J. Michels to AGB, June 9, 14, 1882; AGB to J. Michels, June 19, 1882; AGB to S. Kneeland, July 18, 1882.
377. (publisher): AGB to J. Michels, June 19, 1882; GGH to J. Michels, Sept. 10, 1882; receipt for $5000 signed by J. Michels, Mar. 24, 1883; agreement between S. Scudder and AGB, Dec. 21, 1882; *DAB*, XVI, 526 ("Samuel H. Scudder").
377. (backers): AGB to S. Scudder, Dec. 15, 1882, Feb. 8, 1885; AGB to M. King, Dec. 15, 1882.
377. (articles): S. Scudder to J. L. LeConte, Nov. 19, 1882, John L. LeConte MSS, Academy of Natural Sciences, Philadelphia.
378. (Bell): R. Lacey to *Science*, Dec. 15, 1883; AGB to M. King, ca. July 25, 1883; AGB to S. Scudder, Jan. 18, 1883, Feb. 8, 1885; C. Condit to AGB, Dec. 20, 1883; AGB to GGH, Dec. 20, 1891; GGH to AGB, Apr. 6, 1892.
378. (community): Copy of AAAS report, Aug. 22, 1894; GGH to AGB, Sept. 2, 1894; J. Cattell to AGB, Nov. 6, 1894, Feb. 8, 1899; *Science*, Oct. 8, 1926, July 3, 1964.

CHAPTER 29

379. (telephone): AGB to parents, Sept. 5, 1879; AGB to H. Wood, May 5, 1916.
379. (force): AGB to Mabel, July 31, 1876.
380. (countries): AGB, Address at the Gallaudet Centennial Commemoration, Philadelphia, Dec. 12, 1887 (transcript in the Bell Collection).
380. (people): Mrs. AGB to G. Grosvenor, Oct. 11, 1921.
381. (man, Bell): Fred DeLand, *The Story of Lip-Reading* (Washington, D.C., 1968), pp. 146–149; Mrs. AGB to AGB, July 3, 1894; P. Gillett to GGH, Sept. 1, 1894.
381. (later): *Report of the Committee on the School for Deaf Mutes* (Boston, 1873), pp. 7–9; *Silent World*, Dec. 1, 1874.
382. (closed): AGB, President's Address, in 1884 Convention of Articulation Teachers of the Deaf, *Official Report* (Albany, 1884), pp. 27–28.
382. (language): *Volta Review*, XXV, 37, 39 (Jan. 1923), 94 (Feb. 1923); AGB to J. Wright, Aug. 25, 1915.
383. (teachers): *Proceedings of the National Educational Association Convention of 1884* (Washington, D.C., 1885), p. 59; *Volta Review*, XIV, 579 (Dec. 1912); *The Silent Worker*, October 1915; AGB to F. DeLand, Oct. 15, 1915.
383. (other side): AGB to C. M. Kendall, Aug. 27, 1913.
383. (Gallaudet): Maxine T. Boatner, *Voice of the Deaf* (Washington, D.C., 1959), pp. 7–10, 18–20, 37, 166.
384. (English, speechreading): Boatner, *Voice of the Deaf*, pp. 49–67; Joseph C. Gordon (ed.), *Education of Deaf Children* (Washington, D.C., 1892), pp. 7–10, 13–14, 23–25.
385. (indispensable): *Deaf-Mutes' Register* (Rome, N.Y.), Feb. 15, 1894.
385. (degrees): Boatner, *Voice of the Deaf*, pp. 103–106, 123, 127–129.
385. (naturally): AGB to E. Gallaudet, Dec. 2, 1884, Bell Correspondence, Volta Bureau Collection.
385. (year): Gordon, *Education of Deaf Children*, p. 7; AGB, transcript of address at Chautauqua, 1894.

386. (feel): AGB to E. Gallaudet, February 21, 1887.
386. (out): AGB to T. Kiesel, Nov. 15, 1888; AGB to Mrs. AGB, Mar. 18, 1891.
386. (ultimately): Boatner, *Voice of the Deaf*, pp. 129–130; *Volta Review*, XXV (1923), 37; J. Hitz to Mrs. AGB, June 14, 1888; AGB to E. Gallaudet, Nov. 8, 28, 1889; E. Gallaudet to AGB, Dec. 3, 1889.
387. (deaf): Boatner, *Voice of the Deaf*, pp. 131–133; AGB to J. R. Dobyns, Feb. 21, 1891.
387. (students): *The Companion* (Faribault, Minn.), Mar. 7, 1891.
387. (teachers): Thompson Reminiscences, Feb. 23, 1923, pp. 2–3; AGB to W. Cogswell, Feb. 28, 1891; Boatner, *Voice of the Deaf*, pp. 133–135; *N.Y. Evening Post*, Dec. 15, 1897.
388. (knife, later): E. Gallaudet to AGB, Mar. 18, Apr. 7, 1891, Apr. 22, 1893; AGB to E. Gallaudet, Mar. 30, 1891; *The National Exponent* (Chicago), July 11, 1895; AGB to Mrs. AGB, July 22, 1895; *Detroit Tribune*, July 6, 1895; S. Fuller to Mrs. AGB, Aug. 31, 1895; Boatner, *Voice of the Deaf*, pp. 138–139.
388–389. (was, away): E. Gallaudet to Mrs. AGB, Feb. 10, 1896; Boatner, *Voice of the Deaf*, p. 139; AGB to F. Booth, Aug. 14, 1900; E. Gallaudet to AGB, July 4, 1903.
389. (school): AGB to T. Borthwick, Borthwick MSS; T. Jones to AGB, June 17, 1881; *Greenock* (Scotland) *Advertiser*, May 29, 1882.
390. (Washingtonians): AGB to M. True, Jan. 11, 1883; AGB to "My Dear Sir," undated, presumably 1883; John D. Hitz, "Dr. A. Graham Bell's Private Experimental School," in Edward A. Fay (ed.), *Histories of American Schools for the Deaf, 1817–1893* (3 vols., Washington, D.C., 1893), III, chapter xc, *Washington Evening Star*, Oct. 31, 1883.
390. (learned): Hitz, "Bell's Private School," pp. 13–16; AGB to Mrs. J. R. Dobell, Apr. 25, 1885.
390. (up): AGB to Miss Littlefield, Oct. 11, 1884; AGB to T. Borthwick, Dec. 29, 1884, Borthwick MSS; AGB to Mrs. Bingham, Nov. 8, 1885; Journal of Mrs. AGB, Nov. 19, 1885; Hitz, "Bell's Private School," p. 23.
390. (circumstances): Journal of Mrs. AGB, Nov. 19, 1885; Mrs. AGB to F. DeLand, Sept. 14, 1922.
391. (soured): AGB to E. Garrett, Oct. 11, 1883; AGB to R. Spencer, Jan. 16, 1884; Convention of Articulation Teachers of the Deaf . . . 1884, *Official Report* (Albany, 1884), p. 29.
391. ($25,000): Fay, *History of Schools for the Deaf*, III, chapter xciii, p. 12.
391. (patched): Idem; AGB to S. Fuller, Dec. 24, 1904; AGB to F. DeLand, May 22, 1912; AGB to J. Spencer, Feb. 23, 1917; *Deaf-Mutes' Journal*, May 26, 1892; *Volta Review*, XXV (1923), 149.
392. (affair): AGB to Mrs. AGB, July 2, 3, 1891.
392. (classes): AGB, "Opening Address of the President," June 29, 1892; *The Educator*, September 1893; *Volta Review*, XXI (1919), 525, 527, 528, 581–584, 664, 701; AGB to S. Fuller, Dec. 24, 1904.
392. (failure): Tributes to AGB at Horace Mann School graduation exercises, June 22, 1923; Mrs. AGB to Mrs. GGH, Nov. 30, 1892.
392. (up, years): Minutes of the Board of Directors of the AAPTSD, I, 148, 159, 190, Volta Bureau Collection; AGB to P. Gillett, Oct. 28, 1893, Mar. 2, 1894, in ibid.; *Deaf-Mutes' Register*, Feb. 15, 1894; A. Crouter to GGH, June 5, 1894, Volta Bureau Collection; GGH to AGB, Jan. 25, 1895; *Volta Review*, XIV (1912), 245; DeLand, unpublished portion of "An Ever-Continuing Memorial," pp. 120–122; J. Wright to AGB, Oct. 10, 1916.
392. (per cent): Harry Best, *Deafness and the Deaf in the United States* (New York, 1943), p. 537.
393. (respond): Mrs. M. Jones to AGB, Feb. 2, 1914; AGB to Mr. Strawn, Apr. 10, 1897; J. Gordon to AGB, Mar. 20, 1901; AGB to E. Nevis, July 27, 1898; AGB to F. Booth, Jan. 7, 1901; AGB to R. Spencer, Jan. 11, 1901.

393. (schools, institutions, experimentation): The best brief summary of Bell's views on day schools for the deaf is a fourteen-page typescript by Fred DeLand, "Hard of Hearing Pupils in Public Schools," in the Bell Collection; AGB, "Deaf-Mute Instruction in Relation to the Work of the Public Schools," July 16, 1884, an address at a meeting of the National Education Association (published in Washington, D.C., 1885); AGB, "Deaf Classes in Connection with the Public Schools," *American Annals of the Deaf and Dumb*, XXIX (1884), 313–318; *N.Y. Tribune*, Oct. 25, 1884; AGB, "Education of the Deaf," *National Educational Association Proceedings*, 1897, pp. 96–104; AGB, "Open Letter to Wisconsin Legislative Committees, Feb. 18, 1885," in Wisconsin Phonological Institute, *Improvement of the Wisconsin System of Education for Deaf Mutes* (Milwaukee, Wisc., 1894); AGB to J. Spencer, Feb. 23, 1917.

394. (decibel): Lab Notes, Dec. 20, 1879.

394. (campaign): *Deaf-Mute Journal*, Oct. 9, 1884; *Science*, May 1, 1885; *N.Y. Tribune*, Dec. 21, 1886.

394. (schools): *Volta Review*, XVIII (1916), 155; AGB to E. Lyon, Dec. 24, 1901; Wisconsin Phonological Institute, *Improvement of the Wisconsin System*, pp. 16–17; R. Spencer to AGB, November 30, 1884.

394. (job): AGB to Mrs. AGB, Feb. 14, 1885.

395. (deaf): *Delavan* (Wisc.) *Republican*, Mar. 7, 1895; R. Spencer to AGB, Jan. 27, 1896; *Oshkosh* (Wisc.) *Times*, Mar. 17, 1896; AGB to Mrs. AGB, Oct. 19, 1896; Lavilla A. Ward, "Handicapped Children in Our Public Schools," *Wisconsin Journal of Education*, March 1936, p. 329.

395. (proponents): AGB correspondence files, 1884–1914, passim; AGB to C. Brown, Dec. 20, 1897.

395. (go): S. Wesselius to AGB, Mar. 1, 1899; AGB to D. French, Nov. 26, 1901; W. Martindale to AGB, June 14, 1905; *Daily Eastern Argus* (Portland, Me.), Jan. 19, 1894; AGB to and from Mrs. C. McGuigan, 1897–1899, passim; J. Holden to AGB, September 1898, passim; W. Payne to AGB, Apr. 27, 1898; J. Tate to AGB, Feb. 14, Mar. 13, 1914.

395. (battles): AGB to E. Garrett, Oct. 11, 1883; DeLand, "Ever-Continuing Memorial," p. 114; *Report of the President of the Boston Parents' Education Association . . . April 29, 1896* (printed sheet in Bell Collection); letters from heads of oral day schools in reply to inquiries by AGB, Feb. 1896; AGB to F. Booth, Aug. 12, 1899; AGB to M. Rusch, Oct. 8, 1913.

396. (most): AGB to Mrs. C. Crane, Feb. 3, Dec. 14, 1896, June 26, 1897; AGB to Mrs. AGB, Apr. 1, 1897; AGB to F. Booth, Aug. 12, 1899; DeLand, "Hard of Hearing Pupils in Public Schools."

CHAPTER 30

397. (hand): M. Stark to AGB, Mar. 30, 1886; E. Ingham to AGB, Apr. 9, 1886.

397. (reply): Letters in folder, "Lip-Reading for Adults," Bell Collection; F. Radcliffe to M. Stark, May 6, 1886; AGB to Mrs. E. Turner, July 1891; J. Carty to F. De-Land, Aug. 7, 1923; AGB to Mrs. AGB, Jan. 29, 1901.

398. (family): Notes by Gilbert H. Grosvenor of a conversation with Charles R. Crane, Mar. 26, 1923.

398. (arms): B. Munger to AGB, Apr. 11, 1892; Helen Keller, *The Story of My Life* (New York, 1905), p. 137.

398. (costume!): AGB to Mrs. AGB, March 1893.

399. (them): *Annual Report of the Committee on the Horace Mann School for the Deaf*, 1877 (School Document No. 21), pp. 7–9; 1888 (School Document No. 14), p. 4; 1891 (School Document No. 24), p. 6.

399. (deaf): AGB to S. Fuller, Feb. 14, 1880; *Improvement of the Wisconsin System*,

p. 8; AGB to W. Forbes, Dec. 10, 1883; C. Whitcomb to W. Forbes, Dec. 17, 1883, Box 1097, AT&T Historical Collection.

399. (be): J. Denison to AGB, Nov. 22, 1882; AGB to Mrs. S. Howe, Jan. 8, 1884, Box 1105, AT&T Historical Collection; Mrs. B. Yaffey to M. Payne, Jan. 8, 1963.

399. (too): AGB to Mrs. AGB, May 5, 1890; L. Swett to Mrs. Sanders, undated, probably ca. 1890.

400. (ceased): Mrs. AGB to Elsie Bell, May 18, 1891; S. Sanders to AGB, Mar. 31, 1892; J. Hitz to AGB, Jan. 17, 1893; Mrs. AGB to AGB, May 13, 1898; AGB to G. Sanders, Oct. 25, 1899; G. Sanders to AGB, Oct. 30, 1899.

400. (time): AGB to Mrs. AGB, Feb. 12, 1876.

400. (light): A. Keller to AGB, Feb. 3, 1887; E. Keller to AGB, Apr. 20, 1887, Keller Correspondence, Volta Bureau Collection; Keller, *Story*, pp. 18–19.

401. (Sullivan, Bell): Nella Braddy, *Anne Sullivan Macy* (Garden City, N.Y., 1933), pp. 16–49, 105–106, 310; Helen Keller, *Teacher: Anne Sullivan Macy* (New York, 1955), p. 11; E. Keller to AGB, Apr. 20, 1887, Keller Correspondence.

401. (Deaf): Keller, *Story*, pp. 145, 311, 315–316; AGB to H. Keller, May 2, 1888; AGB to J. Hitz, Dec. 7, 1891, Bell Correspondence, Volta Bureau Collection.

402. (Helen Keller): Braddy, *Anne Sullivan Macy*, pp. 164–165; transcript of conversation between AGB and Helen Keller, May 7, 1890; AGB, "The Method of Instruction Pursued with Helen Keller," *The Silent Educator*, June 1892.

402. (soar): DeLand, "Ever-Continuing Memorial," p. 117; *The Educator*, September 1893; Braddy, *Anne Sullivan Macy*, pp. 169–171; Keller, *Story*, pp. 75–77, 392–393.

404. (me): Braddy, *Anne Sullivan Macy*, pp. 166–167.

404. (right): Keller, *Story*, pp. 148–149; H. Keller to AGB, July 5, 1918.

404. (years): Keller, *Story*, pp. 63–72, 396–418; Braddy, *Anne Sullivan Macy*, pp. 162–163.

405. (good): Keller, *Story*, pp. 402–403; Braddy, *Anne Sullivan Macy*, pp. 162–163; AGB to Mrs. AGB, Mar. 10, 1892.

405. (hear): AGB to A. Keller, July 20, 1894, Keller Correspondence.

405. (parted): Braddy, *Anne Sullivan Macy*, pp. 182–186; A. Gilman to AGB, Nov. 19, Dec. 4, 23, 24, 28, 1897, Jan. 3, 1898; AGB to A. Gilman, Dec. 31, 1897; copy of statement by Annie M. Sullivan, Dec. 30, 1897, Keller Correspondence.

405. (fiasco): AGB to Mrs. Chamberlin, Dec. 8, 10, 1900; Braddy, *Anne Sullivan Macy*, pp. 193–194; Keller, *Story*, pp. 268–269.

405. (known): AGB to C. Warner, May 8, 1896; AGB to E. King, Mar. 6, 1897; H. Keller to AGB, Sept. 3, 1896, Apr. 30, 1905; AGB to H. Keller, May 28, 1899, Apr. 14, 1905; A. Sullivan to J. Hitz, Feb. 1, 1898, Sullivan Correspondence, Volta Bureau Collection.

406. (world): Van Wyck Brooks, *Helen Keller: Sketch for a Portrait* (New York, 1956), pp. 31–32.

406. (blossoms, benediction): Ibid., pp. 31–32, 58–59; Keller, *Story*, pp. 137, 278; H. Keller to Mrs. AGB, Sept. 17, 1901, Keller Correspondence.

407. (on): AGB to Mrs. Pratt, Oct. 13, 1900; AGB to H. Keller, Nov. 20, 1908.

407. (detail): H. Keller to AGB, Mar. 9, 1900, AGB to H. Keller, Mar. 23, 1907, June 5, 1908; H. Keller to J. Hitz, Oct. 11, 1902, June 2, 1907, Keller Correspondence.

407. (occasion): H. Keller to AGB, Jan. 12, 1907; W. Holt to AGB, Jan. 17, 1907.

408. (history): H. Keller to AGB, July 5, 1918; AGB to H. Keller, July 18, 1918.

CHAPTER 31

409. (deaf): AGB to parents, Jan. 27, 1873; Mrs. AMB to AGB, Feb. 2, 1873.

409. (findings): AGB to A. Hyatt, Dec. 28, 1878, Jan. 13, 1879; P. Gillett to AGB, Mar. 4, 1879; *American Annals of the Deaf*, April 1879, p. 126.

409. (Race): Alexander Graham Bell, "Memoir upon the Formation of a Deaf Variety of the Human Race," *National Academy of Sciences Memoirs*, II (1884),

179–262; A. Hunter Dupree, *Science in the Federal Government* (Cambridge, Mass., 1957), pp. 277–278.

410. (recognized): Mark H. Haller, *Eugenics: Hereditarian Attitudes in American Thought* (New Brunswick, N.J., 1963), pp. 8–10, 21–22.

410. (training): Bell, "A Deaf Variety," pp. 179–180, 189, 192–195, 203, 222–224.

410. (breeding): Ibid., p. 179; Haller, *Eugenics*, p. 10.

411. (traits): Bell, "A Deaf Variety," pp. 190, 201.

411. (years): *Deaf-Mutes' Journal*, Nov. 3, 1887.

411. (America): Edward A. Fay, *An Inquiry Concerning the Results of Marriages of the Deaf in America* (Washington, D.C., 1895), pp. 1–2, 7–11.

412. (schools): Fay, *Marriages of the Deaf*, pp. 13, 122–133.

412. (invention): Haller, *Eugenics*, p. 33.

412. (Squibnocket): *Volta Review*, XXIV (1922), 362–363; AGB to A. Gilman, Jan. 27, 1885, Bell Collection, Boston University; *Boston Sunday Herald*, Jan. 20, 1895 (an article on Squibnocket called to my attention by my colleague Robert E. Moody).

412. (1900–4): AGB to F. DeLand, July 16, 1913, Bell Correspondence, Volta Bureau Collection; AGB to Mrs. AGB, May 22, 1887; Mrs. AGB to AGB, Apr. 30, 1900; AGB to Mrs. AGB, Nov. 14, 1900.

413. (1890): AGB to F. Noyes, Apr. 27, 1910; AGB to J. Hitz, Oct. 15, 1888; correspondence between Helen Keller and John Hitz in Keller Correspondence, Volta Bureau Collection, e.g., H. Keller to J. Hitz, Apr. 11, 1899; *Volta Review*, XXV (1923), 98.

413. (columns): *Volta Review*, LXXII (1970), 148–149; AGB to F. Noyes, Apr. 27, 1910; AGB to J. Hitz, Dec. 16, 1891; Mrs. AGB to AGB, June 6, 1893.

413. (library): *Volta Review*, LXXII (1970), 152.

414. (research): *Rochester* (N.Y.) *Union and Advertiser*, Mar. 18, 1887; AGB to S. Green, Jan. 13, 1889; S. Green to AGB, Jan. 18, 1889.

414. (debt): AGB to C. Wright, May 16, 1895; undated copy of a formal protest by AGB, in folder, "Census."

414. (training): AGB to E. Fay, Oct. 28, 1889; *Volta Review*, XXV (1923), 190–192.

414. (library): Ibid., 193–195; AGB to F. Wines, Jan. 26, 1889; AGB to J. Billings, Dec. 21, 1889.

415. (census): Folder, "Statistics (Deaf)"; AGB to Senator Carter, Jan. 3, 1900; *Volta Review*, XXV (1923), 195–196; AGB to Mrs. AGB, Nov. 13, 1904; Mrs. AGB to G. Grosvenor, Oct. 19, 1904; AGB to F. DeLand, Aug. 11, 1915.

415. (fog): AGB to Mrs. AGB, Dec. 16, 1889; Horne notes, Dec. 27, 1900.

415. (fruitful): AGB to L. Knorr, Jan. 15, 1886, Dr. Bessels, Mar. 9, 1886, E. Scholtz, June 15, 1887, F. Wines, Aug. 20, 1887, W. German, Aug. 30, 1906, C. Davenport, Sept. 14, 1906; *Science*, III (1884), 171, 243.

416. (cats): *Science*, XIX (1904), 767; T. Sanders to AGB, Oct. 31, 1889, Oct. 4, 1891.

416. (Sheepville): *Science*, XIX (1904), 767; Mrs. AGB to Mrs. GGH, "Sunday," early 1890s; *Journal of Heredity*, XIV (1923), 100; Lab Notes, July–October 1890, passim.

416. (dogs): *Journal of Heredity*, XIV (1923), 100.

417. (hereditary): *Science*, XIX (1904), 767–768; *Journal of Heredity*, XIV (1923), 109–111; Melville B. Grosvenor, "Life with Grandfather," transcript of address to Literary Society, Feb. 10, 1968; Catherine MacKenzie, "Some Notes about Alexander Graham Bell," p. 13.

417. (production): *Journal of Heredity*, XV (1924), 75, 77, 81; U.S. Department of Agriculture Technical Bulletin 909, December 1945, pp. 2–3, 15.

417. (thought): Haller, *Eugenics*, p. 62.

417. (blood): Ibid., pp. 17–20, 62.

418. (heredity): Ibid., pp. 62–65; AGB to J. Smith, Jan. 20, 1908; *Washington Post*, Oct. 28, 1909; David Fairchild, *The World Was My Garden* (New York, 1939), pp. 403–404, 423–424.

418. (genocide): Haller, *Eugenics*, pp. 144–159.

419. (out): *Beinn Bhreagh Recorder*, Dec. 11, 1909; *Journal of Heredity*, November 1920, p. 341.

419. (race): AGB to F. DeLand, Oct. 15, 1915; Fairchild, "Alexander Graham Bell," p. 198; *Beinn Bhreagh Recorder*, Dec. 11, 1909.

420. (disease): Ibid., Sept. 2, 1915, Jan. 28, 1916; Alexander Graham Bell, *The Duration of Life and Conditions Associated with Longevity: A Study of the Hyde Genealogy* (Washington, D.C., 1918), passim; *National Geographic Magazine*, XXXV (1919), 505–514.

420. (milestone): AMB, Memoir of Mrs. AMB, 1897; *Boston Transcript*, Aug. 9, 1894.

420. (time): Mrs. AGB to Mrs. GGH, December 1895; Mrs. AGB to AGB, June 9, 1895.

420. (Georgetown): *Beinn Bhreagh Recorder*, Sept. 21, 1920.

421. (work): Journal of Mrs. AGB, Feb. 16, 1884; GGH to "Mr. Smith," May 16, 1892; GGH to AGB, Apr. 22, 1894.

421. (age): *N.Y. World*, Dec. 18, 1897; AGB to Mrs. AGB, Nov. 30, 1900.

421. (should): Mrs. AGB to AGB, June 12, 19, 1898; AGB to Mrs. AGB, June 8, 13, 1898; *N.Y. Sun*, Dec. 26, 1899.

422. (Europe): Mrs. AGB to Mrs. G. Grosvenor, Feb. 11, 1901.

422. (Gloucester): Home Notes, July 8, 1905; Mrs. AGB to Mrs. G. Grosvenor, Aug. 28, 1905.

422. (breath): Mrs. AGB to Mrs. GGH, ca. 1901; Mrs. AGB to Mrs. G. Grosvenor, Aug. 9, 1905.

422. (eighty-two): Mrs. AGB to G. Grosvenor, May 28, 1909; *Beinn Bhreagh Recorder*, Oct. 26, 1909.

423. (Gilbert): Gilbert H. Grosvenor, *The National Geographic Society and its Magazine* (Washington, D.C., 1957), pp. 8–13, 20.

423. (object): Mrs. AGB to Elsie, May 20, June 27, 1897; *Harper's Weekly*, July 17, 1897; Mrs. AGB to GGH, ca. Aug. 1, Aug. 29, Sept. 5, 1897; Mrs. AGB to Mrs. GGH, ca. July 1, ca. Oct. 1, 1897.

423. (employee): Home Notes, Nov. 4, 1915; galley proofs of an unpublished article by Gilbert H. Grosvenor, ca. 1923, "Alexander Graham Bell's Contribution to the National Geographic Society," in Bell Collection; AGB to G. Grosvenor, July 13, 1899; Grosvenor, *National Geographic Society*, p. 21.

424. (text): Conversation of Gilbert H. Grosvenor with author, June 21, 1965; AGB to G. Grosvenor, July 14, 1899; Grosvenor, "Bell's Contribution."

424. (greatness): Peter Lyon, *Success Story* (New York, 1963), pp. 130–134.

424–425. (reliable, press, theme): AGB to G. Grosvenor, July 14, Aug. 16, Sept. 24, 28, 1899; AGB to various publishers, Aug. 16, 1899; Grosvenor, "Bell's Contribution."

425. (effective): Ibid.

425. (insight): Ibid.; AGB to A. Greeley, Sept. 19, 1899.

427. (later): Mrs. AGB to AGB, May 3, 20, 1899, May 4, 1900; Grosvenor, *National Geographic Society*, pp. 32–33.

427. (mother): Grosvenor, "Life with Grandfather"; *Beinn Bhreagh Recorder*, Sept. 17, 1910.

427. (number): *National Geographic Magazine*, XI (1900), 403; AGB to Mrs. AGB, Sept. 22, 1900; AGB to G. Grosvenor, Jan. 18, 31, 1901, Feb. 15, 1902.

427. (thereafter): *National Geographic Magazine*, IX (1898), 289; Grosvenor, *National Geographic Society*, p. 39.

428. (1912): Ibid., pp. 42–44; AGB to G. Grosvenor, Dec. 7, 1905.

428. (duty): AGB to W. McGee, Sept. 19, 1899; AGB to Mrs. AGB, Sept. 22, 1900; *National Geographic Magazine*, CXI (1957), 420; Mrs. AGB to AGB, May 12, 1903.
428. (character): Fairchild, *The World Was My Garden*, pp. 288–290.
429. (Maryland): Ibid., pp. 311–313, 316–318.
429. (environment): Mrs. AGB to AGB, Aug. 21, 1906.
429. (kites): Thompson Reminiscences, Feb. 23, 1923.

CHAPTER 32

430. (problem): Alexander Graham Bell, "The Tetrahedral Principle in Kite Structure," *National Geographic Magazine*, XIV (1903), 219.
430. (him): Ibid., p. 220.
431. (currents): Alexander Graham Bell, "Aerial Locomotion," *National Geographic Magazine*, XVIII (1907), p. 10.
431. (Baddeck): J. H. Parkin, *Bell and Baldwin* (Toronto, Canada, 1964), pp. 6–7.
431. (kite): Bell, "Tetrahedral Principle," pp. 226–229.
432. (thereafter): Conversation of Gilbert H. Grosvenor with author, June 21, 1965.
432. (cell): Parkin, *Bell and Baldwin*, pp. 7–8; AGB to Mrs. AGB, Oct. 25, 1901.
432. (triangles): Home Notes, Aug. 1, 1899, Sept. 30, 1901, Mar. 15, 1902; Bell, "Tetrahedral Principle," pp. 221–224.
432. (suddenness): Home Notes, Mar. 29, Aug. 25, 1902.
433. (construction): Idem.
433. (stanzas): Mrs. AGB's typed copy is in the Bell Collection.
433. (Magazine): Mrs. AGB to J. Hitz, Nov. 18, 1902, Volta Bureau Collection; Bell, "Tetrahedral Principle," pp. 228–231; dictation by AGB, Nov. 13, 1903.
436. (Monthly): *Boston Herald*, Sept. 29, 1903; *N.Y. Herald*, Nov. 1, 1903; *Popular Science Monthly*, Nov. 1903; *N.Y. Tribune*, Nov. 26, 1903.
436. (successfully): *DAB*, X, 596 ("Langley"), XII, 239 ("Manly"); MacKenzie, *Bell*, pp. 298–300.
436. (it): Mark Sullivan, *Our Times* (6 vols., New York, 1927–1935), II, 560–566; MacKenzie, *Bell*, pp. 303–304.
436. (endure): Bell, "Aerial Locomotion," p. 7; *Washington Post*, Mar. 4, 1906.
436. (failure): Bell, "Aerial Locomotion," p. 7.
437. (made): *Washington Star*, June 11, 1904.
437. (stability): Parkin, *Bell and Baldwin*, pp. 12, 15–16.
438. (winds): Ibid., pp. 12–13.
438. (tail): Ibid., pp. 13–14.
438. (order): AGB to Mrs. AGB, Nov. 30, 1904; Home Notes, Nov. 11, 1905.
438. (business): Bell, "Aerial Locomotion," pp. 14, 16, 21.
439. ("space frame"): Bell, "Tetrahedral Principle," p. 225; Carl W. Condit, *American Building* (Chicago, 1968), pp. 52–53, 198–199.
440. (elements): Bell, "Tetrahedral Principle," pp. 231, 248.
440. (1881): Mrs. AGB to Mrs. GGH, ca. August 1904; Osborne, "Bell," p. 29.
440. (corners): U.S. Patent No. 770,626, p. 4.
440. (literal): Mrs. AGB to C. J. Bell, Oct. 21, 1904.
441. (him): W. Burr to C. J. Bell, Mar. 30, 1906; E. Snodgrass to Mrs. AGB, May 10, 1906; Mrs. AGB to P. Mauro, May 15, 1906.
441. (aeronautics): Parkin, *Bell and Baldwin*, pp. 26–28.
441. (later): Green, *Silver Dart*, pp. 11–13, 33–34.
442. (engineering): Mrs. AGB to D. McCurdy, ca. June 1, 1906; F. Baldwin to AGB, May 30, 1906.
443. (Bell): Parkin, *Bell and Baldwin*, pp. 29, 32–33.
444. (all): Ibid., pp. 32–33; MacKenzie, *Bell*, p. 310.

444. (October): Parkin, *Bell and Baldwin*, p. 33; *National Geographic Magazine*, XVIII (1907), 672–675; *Scientific American*, Oct. 5, 1907.
444. (parts): Condit, *American Building*, pp. 199–200.
444. (up): AGB to C. A. Bell, Feb. 18, 1901, Mar. 25, 1903; C. A. Bell to AGB, Apr. 10, 1901.
445. (ways): Mrs. AGB to G. Grosvenor, Oct. 19, 1904; Mrs. AGB to AGB, Mar. 10, 1907; Mrs. AGB to F. Baldwin, ca. 1907; entry in a notebook of Mrs. AGB, ca. fall, 1907.
445. (gentlemen): Elsbeth E. Freudenthal, *Flight into History* (Norman, Okla., 1949), pp. 88–91, 110–111; Fairchild, *The World Was My Garden*, pp. 333–334; O. H. Tittmann in folder, "Reminiscences by Friends."
445. (year): *N.Y. World*, Feb. 2, 1906; Freudenthal, *Flight into History*, p. 152.
446. (cleat): Parkin, *Bell and Baldwin*, p. 34.
447. (Mrs. Bell): Ibid., pp. 36–39; *DAB*, XXI, 213–214 ("Curtiss").
447. (salary): Summary of the history of the Aerial Experiment Association, written for Gilbert Grosvenor by Mrs. AGB in the summer of 1909; Parkin, *Bell and Baldwin*, pp. 40–41.
447. (coach): Lab Notes, Sept. 22, 1907; Parkin, *Bell and Baldwin*, p. v.
448. (unhurt): Ibid., pp. 46–48.
448. (man): Fairchild, *The World Was My Garden*, pp. 333–334; Parkin, *Bell and Baldwin*, pp. 48–49; AGB to Mrs. AGB, Dec. 1907; Mrs. AGB to Mrs. GGH, Dec. 12, 1907.
448. (airplane): Parkin, *Bell and Baldwin*, pp. 49–50.
448. (trial): Freudenthal, *Flight into History*, p. 182; Parkin, *Bell and Baldwin*, pp. 50–53.
449. (Baldwin): Ibid., pp. 53–54; Freudenthal, *Flight into History*, p. 174; Green, *Silver Dart*, pp. 41–43.
449. (maintained): Parkin, *Bell and Baldwin*, pp. 54–55.
449. (feet): Ibid., pp. 54–58.
449. (prize): Lab Notes, May 16, 1908; Parkin, *Bell and Baldwin*, pp. 58–61; Green, *Silver Dart*, pp. 49–51; Fairchild, *The World Was My Garden*, p. 343.
451. (contributions): Parkin, *Bell and Baldwin*, pp. 63–66, 158.
451. (*Silver Dart*): Ibid., pp. 67–69, *Beinn Bhreagh Recorder*, Apr. 4, July 8, 1913.
451. (agreed): Ibid., pp. 71–72; Mrs. AGB to AGB, Sept. 20, 1908.
451. (feet): Parkin, *Bell and Baldwin*, pp. 78–88; Green, *Silver Dart*, pp. 58–64.
452. (moment): Parkin, *Bell and Baldwin*, p. 40; AGB to Mrs. AGB, Apr. 2, 1909.
452. (June Bug): Parkin, *Bell and Baldwin*, pp. 160–165.
453. ($2,000,000): Mrs. AGB to G. Grosvenor, Dec. 26, 1915; F. Baldwin to AGB, Mar. 6, 1915; Home Notes, Dec. 28, 1914, Feb. 20, 1917; Parkin, *Bell and Baldwin*, p. 165.
453. (died): Ibid., pp. 273–284; Green, *Silver Dart*, pp. 106–108.
453. (history): Parkin, *Bell and Baldwin*, pp. 285–310; Grosvenor, "Life with Grand-father."
454. (March 1909): Parkin, *Bell and Baldwin*, pp. 30–31, 91–98.
454. (enough): Ibid., pp. 316–325.
454. (went): Ibid., pp. 329–332.
454. (hope): Ibid., p. 331.

CHAPTER 33

455. (be): Mrs. AGB to AGB, late Aug. 1904.
455. (century): Journal of Mrs. AGB, Mar. 3, 1879; *Arkansas Gazette* (Little Rock), Apr. 25, 1895.
455. (pounds): Mrs. AGB to AGB, Apr. 23, 1891, Oct. 27, 1901, AGB to Mrs. AGB, May 14, 1902.

456. (father): Mrs. AGB to AGB, May 24, 1898; Catherine D. MacKenzie, "Notes about Dr. Alexander Graham Bell."

456. (alternative): Home Notes, Sept. 25, 1914; Mrs. AGB to A. Crouter, June 6, 1898; AGB to Mrs. AGB, May 6, 1902; *Washington Post*, July 3, 1901; *N.Y. Times*, July 30, 1911; *Milwaukee Sentinel*, Apr. 16, 1918.

457. (young): Waite, *Make a Joyful Sound*, p. 208; Mary Blatchford to Grace Bell, Sept. 30, Oct. 2, 1911, *Beinn Bhreagh Recorder*, XXIII, Appendix, p. 81.

457. (junior): M. Blatchford to G. Bell, Nov. 1, 1911, ibid., p. 124; Home Notes, May 19, 1910.

457. (now): M. Blatchford to G. Bell, Sept. 24, 1911, *Beinn Bhreagh Recorder*, XXIII, Appendix, p. 69; AGB to Mrs. AGB, Feb. 27, 1909; Mrs. AGB to AGB, Feb. 11, 1903, June 22, 1906, Apr. 15, 1917.

459. (purported): Grosvenor, "Life with Grandfather."

459. (careers): *Beinn Bhreagh Recorder*, XXIII, Appendix, p. 69; Mrs. AGB to G. Grosvenor, Sept. 19, 1915.

459. (tea): Grosvenor, "Life with Grandfather"; conversation of Dr. Mabel H. Grosvenor with author, Jan. 15, 1972; AGB to G. Grosvenor, Oct. 1, 1906; Mrs. AGB to Mrs. G. Grosvenor, Nov. 22, 1911.

460. (suffer): Home Notes, Sept. 9, 10, 1911.

460. (beans): Grosvenor, "Life with Grandfather."

460. (lived): Mrs. AGB to G. Grosvenor, June 13, Oct. 17, 1914; G. Grosvenor to AGB, Sept. 14, 1914.

462. (Magazine): AGB to G. Grosvenor, Dec. 2, 1914; Home Notes, Aug. 6, 1918.

462. (feeding): *Chicago Evening Post*, Mar. 5, 1896.

462. (especially): Lyon, *Success Story*, p. 350; Mary L. K. Wills, "Conditions Associated with the Rise and Decline of the Montessori Method of Kindergarten-Nursery Education in the United States from 1911–1921" (Ph.D. diss., Southern Illinois University, 1967), p. 123; Melville B. Grosvenor, Recollections of AGB, as told to Helen Waite, Nov. 1961.

463. (1912–13): *Beinn Bhreagh Recorder*, July 20, Oct. 12, 14, 1912, July 25, 26, 1913; *Freedom for the Child*, Apr. 1915, pp. 5–7.

463. (president): *Beinn Bhreagh Recorder*, July 25, 1913, Jan. 9, 1914; *Freedom for the Child*, April 1915.

463. (college): Wills, "Conditions," pp. 124–125; Grosvenor, Recollections of AGB, Nov. 1961.

463. (life): Wills, "Conditions," p. 125.

464. (relics): Mrs. AGB to Mrs. GGH, Oct. 4, 1908; *Philadelphia Press*, Aug. 3, 1913; unsigned letter, probably from Volta Bureau librarian, to Williams, Brown & Earle, Feb. 27, 1913.

464. (morning): MacKenzie, *Bell*, pp. 346–347; *Philadelphia Press*, Aug. 3, 1913.

464. (Creek): Lab Notes, Nov. 10, 1907; Grosvenor, Recollections of AGB, Nov. 1961; Fairchild, *The World Was My Garden*, pp. 326–327.

464. (litigation): MacKenzie, "Some Notes about AGB"; MacKenzie, *Bell*, pp. 347–349.

466. (water): Home Notes, Mar. 22, 1910; Mrs. AGB "to my Children," Jan. 27, 1919.

466. (boat): Home Notes, Oct. 16, Nov. 11, 1906.

466. (speed): *Scientific American*, Mar. 3, 1906.

467. (jumpy): Parkin, *Bell and Baldwin*, pp. 103–119.

467. (ripple): Ibid., pp. 370–373; Mrs. AGB to Mrs. G. Grosvenor, Apr. 3, 1911.

467. (December): Parkin, *Bell and Baldwin*, pp. 376–396.

467. (models): Ibid., pp. 340–355.

467. (ideas): Ibid., pp. 396–402.

469. (residents): Ibid., pp. 356, 402.

469. (number): Mrs. AGB to AGB, Mar. 7, 1909; DeLand, "Ever-Continuing Memorial," pp. 125–126, 134–137; *Volta Review*, XII (1910), 7, XIV (1912), 245.

469. (friend): Mark Schorer, *Sinclair Lewis: An American Life* (New York, 1961), pp. 167–168, 170, 174, 203.
469. (Washington): *Volta Review*, XIV (1912), 245, 249.
470. (telephone): AGB to F. DeLand, Sept. 29, 1905, Feb. 25, 1913, Apr. 13, July 2, Sept. 5, 1914, Oct. 15, 1915; AGB to H. Rogers, Jan. 31, 1912; F. DeLand to AGB, Mar. 11, 1915; *Volta Review*, XVI (1914), 230.
470. (approved): AGB to Mrs. AGB, Mar. 28, 1901.
471. (subject): Parkin, *Bell and Baldwin*, p. 22; Home Notes, Mar. 5, 1913, Mar. 6, 1914.
471. (latter): Alexander Graham Bell, "When Does Profit Become Usury," *The World's Work*, March 1909, pp. 11314–11316.
471. (Secretary Bryan): *Beinn Bhreagh Recorder*, Apr. 30, 1910; Home Notes, Feb. 3, 1912; G. Grosvenor to Mrs. AGB, Sept. 28, 1914.
471. (happened): *Beinn Bhreagh Recorder*, Nov. 11, 1914.
472. (Bell): Mrs. AGB to Mrs. G. Grosvenor, Jan. 26, 1911; Home Notes, Oct. 16, 1915.
472. (shelters): Mrs. AGB to G. Grosvenor, Oct. 17, 1914; *The Victoria News* (Baddeck, N.S.), Aug. 4, 1915; *Beinn Bhreagh Recorder*, Jan. 29, 1918.
472. (reverses): Thompson, Reminiscences of AGB, Mar. 20, 1924; Mrs. AGB to G. Grosvenor, Jan. 3, 1918.
472. (cause): Alexander Bell, "On Humbug" (in Bell Collection), p. 148; Home Notes, May 25, 1916; *N.Y. Times*, Aug. 3, 1919; Home Notes, Apr. 14, 1921.
473. (war): AGB to A. Greely, Mar. 27, 1905; AGB to G. Sues, Aug. 28, 1913.
473. (letter): Home Notes, May 1917, passim, Nov. 10, 1917, Sept. 20, 1918, Mar. 14, 1919; Parkin, *Bell and Baldwin*, p. 20; statement by AGB, Sept. 24, 1897, in file "Rockets and Projectiles," Bell Collection; memorandum by AGB, Mar. 29, 1918, ibid.; AGB to C. Walcott, Apr. 1, 1918; C. Walcott to AGB, Apr. 15, 1918.
474. (aviation): Home Notes, Mar. 31, 1917; *Beinn Bhreagh Recorder*, July 28, 1915, Apr. 20, 1916; *Washington Star*, Apr. 13, 1916; Home Notes, July 15, 1915.
474. (Beinn Bhreagh): Parkin, *Bell and Baldwin*, pp. 357–361.
474. (HD-4): Parkin, *Bell and Baldwin*, pp. 404–406; AGB to Mrs. AGB, Apr. 22, 1917; Home Notes, June 29, 1917; AGB to E. Bayley, Apr. 28, 1921.
474. (1918): Parkin, *Bell and Baldwin*, pp. 405–411.
474. (years): Ibid., pp. 412–414.
475. (boat): MacKenzie, *Bell*, pp. 344–345; Mrs. AGB to G. Grosvenor, October 21, 1919.
475. (Beinn Bhreagh): Parkin, *Bell and Baldwin*, pp. 426–438.
476. (1923): Home Notes, Mar. 7, 1913, Sept. 8, 1917, July 17, Sept. 4, 1920; AGB to E. Bayley, Apr. 28, 1921; Parkin, *Bell and Baldwin*, pp. 426, 438.
476. (old): Ibid., p. 425.

CHAPTER 34

477. (person): Mrs. AGB to Mrs. G. Grosvenor, Sept. 4, 1910; *Sydney* (Australia) *Morning Herald*, July 30, 1910; MacKenzie, *Bell*, pp. 322–325.
477. (name): Summary of AGB's formal honors, prepared for William C. Langdon by Fred DeLand.
478. (me): Mrs. AGB to AGB, Mar. 26, 1907; C. Laemmle to AGB, Dec. 24, 1915; AGB to C. Laemmle, Dec. 28, 1915.
478. (friendliness): MacKenzie, Notes about AGB.
478. (born): Ibid.; AGB to J. Strittwater, Aug. 13, 1918, Bell MSS, Boston University.
479. (Bell): Dictation by AGB, Oct. 24, 1889; Mrs. AGB to Mrs. G. Grosvenor, Dec. 2, 1905; *Volta Review*, 1912, passim, July, September 1914; *National Geographic Magazine*, June 1914, February 1917.
479. (telephones): Ibid.

479. (next): AGB to G. Grosvenor, Apr. 4, 1904.
479. (it): AGB to Mrs. M. Lincoln, Feb. 18, 1887; F. Sumichrast to AGB, Feb. 16, 1889; AGB to F. de Sumichrast, Feb. 18, 1889; AGB to H. Jenkins, Mar. 22, 1916.
480. (years): Memorandum by Gilbert Grosvenor, Apr. 2, 1924; G. Grosvenor to AGB, Oct. 3, 1904; G. Grosvenor to Mrs. AGB, Nov. 2, 1922; G. Grosvenor to D. Fairchild, Oct. 27, 1926.
480. (Days): Grosvenor, *National Geographic Society*, pp. 103–104.
480. (Note-books): Mrs. AGB to J. Hitz, Oct. 4, 1899, Mabel Bell Collection, Volta Bureau Library; MacKenzie, Notes about AGB; MacKenzie, *Bell*, pp. 348–349.
480. (each): Ibid., pp. 239–231.
480. (remains): Grosvenor, *National Geographic Society*, p. 105.
481. (houseboat): AGB to S. Isawa, Oct. 14, 1901; copy of paper by Mrs. AGB on experiences at Portsmouth Conference, in journal of Mrs. AGB for 1905; *Bell Telephone*, p. 2; AGB to Mrs. AGB, Feb. 28, 1909.
481. (daughters): Watson, *Exploring Life*, pp. 180–194, 197–199, 203, 208–210, 213, 220, 223, 226, 235, 248.
482. (you): Ibid., pp. 240–242, 244, 248, 250–295.
482. (1913): AGB to T. Watson, Jan. 25, 1905; T. Watson to AGB, Jan. 31, 1905; Watson, *Exploring Life*, p. 305.
482. (now): *Electrical Review and Western Electrician*, Jan. 30, 1915; Watson, *Exploring Life*, pp. 307–309.
482. (them): *Boston Globe*, Mar. 14, 1916.
483. (1875): AGB to W. Cockshutt, Mar. 16, 1904; *Volta Review*, XX (1918), 233; *Beinn Bhreagh Recorder*, Jan. 11, 1918.
483. (telephones): *Volta Review*, XX (1918), 233.
483. (woman): B. Herdman to AGB, May 20, 1882, Feb. 27, 1887, Feb. 21, 1889; Mrs. AGB to Mr. and Mrs. G. Grosvenor, Aug. 21, 1910.
483. (either): MacKenzie, *Bell*, pp. 307–308.
483. (visit): Home Notes, May 25, Oct. 2, 1920.
484. (Hill): Ibid., Oct. 12, 1920; Mrs. AGB to G. Grosvenor, Oct. 25, 1920; MacKenzie, *Bell*, pp. 356–357.
484. (Pluscarden): Ibid., pp. 357–359.
484. (Edinburgh): Undated clipping, ca. 1940, quoting an account by R. S. Skinner, in Press Cuttings, "Alexander Graham Bell," Edinburgh Room, Edinburgh Public Library; MacKenzie, *Bell*, p. 359.
484. (Edinburgh): Mrs. AGB to Mrs. G. Grosvenor, Nov. 21, 28, 1920, Jan. 3, 1921; *Schola Regia* (Royal High School, Edinburgh), Christmas 1920, pp. 15–16; *The Royal Scotsman* (Edinburgh), Dec. 1, 1920.
484. (don't): *N.Y. World*, Dec. 19, 1920.

CHAPTER 35

485. (four): AGB to Mabel, Dec. 26, 1875; AGB to G. Maynard, Mar. 11, 1901; AGB to A. McCurdy, June 15, 1901.
485. (them): Unidentified obituary of Charles Williams in Box 1071, AT&T Collection; AGB to G. Sanders, Aug. 14, 1911; J. Jamieson to AGB, Mar. 12, 1915; AGB to C. Walcott, July 8, 1915; Home Notes, Jan. 21, 1919.
486. (fighting): W. Langsche to Mrs. AGB, Sept. 25, 1915; AGB to Mrs. AGB, Nov. 30, 1900.
486. (age): Grosvenor, "Life with Grandfather"; Thompson Reminiscences, Feb. 23, 1923.
486. (loss): AGB to F. DeLand, Aug. 22, 1916, Bell Collection, Volta Bureau Library; AGB to Mrs. AGB, Apr. 21, 1917; *Volta Review*, XXIV (1922), 365; in folder, "Tribute to Dr. Bell (Clarke School)," resolution of the Board of Corporators, on the death of AGB, and other undated typescripts.

487. (things): Mary B. Mullett, "How to Keep Young Mentally," *American Magazine,* December 1921.
487. (time): AGB to Mrs. AGB, Nov. 4, 1900; Mrs. AGB to Mrs. G. Grosvenor, July 15, 1920; AGB to G. Grosvenor, Feb. 23, 1922.
487. (promontory): MacKenzie, Notes about AGB; Mrs. AGB to G. Grosvenor, June 12, 1921; Mrs. AGB to Mrs. G. Grosvenor, Mar. 4, 1921.
488. (wings): Fairchild, *The World Was My Garden,* p. 451.
488. (years): Mrs. AGB to Mrs. G. Grosvenor, Jan. 13, 28, 1922; MacKenzie, *Bell,* pp. 361–362; Home Notes, Mar. 3, 1922.
488. (remember): MacKenzie, *Bell,* p. 361; Mrs. AGB to Mrs. G. Grosvenor, Feb. 14, 23, 1922.
488. (notebook): Home Notes, Apr. 29, 1922; MacKenzie, *Bell,* p. 362; Parkin, *Bell and Baldwin,* p. 440.
490. (pancreas): Mrs. D. Fairchild to Mrs. G. Grosvenor, Aug. 1, 1922; Mrs. AGB to Mrs. G. Grosvenor, Aug. 2, 1922; D. Fairchild to G. Grosvenor, Aug. 6, 1922.
490. (sect): Ibid.; Mrs. AGB to J. MacKinnon, Aug. 16, 1922; M. Grosvenor to Mrs. R. Rickard, Apr. 13, 1953; typed statement by Mrs. G. Grosvenor, in folder, "Grosvenor, Elsie Bell, Reminiscences"; AGB to Mrs. AGB, Mar. 12, 1901.
490. (children): D. Fairchild to G. Grosvenor, Aug. 6, 1922; MacKenzie, *Bell,* p. 363; Home Notes, Aug. 1, 1922.
491. (forever): Mrs. AGB to Mrs. G. Grosvenor, Aug. 2, 1922; Mrs. D. Fairchild to Mr. and Mrs. G. Grosvenor, Aug. 6, 1922; D. Fairchild to G. Grosvenor, Aug. 6, 1922; Mrs. AGB to J. MacKinnon, Aug. 16, 1922.

POSTLUDE

493. (clothes): Mrs. AGB to C. Yale, Sept. 30, 1922; Mrs. D. Fairchild to Mrs. G. Grosvenor, Aug. 2, 6, 1922; D. Fairchild to G. Grosvenor, Aug. 6, 1922.
493. (over): Mrs. D. Fairchild to Mr. and Mrs. G. Grosvenor, Aug. 6, 1922.
494. (United States): Ibid.
494. (touch): D. Fairchild to G. Grosvenor, Aug. 6, 1922; *The Transmitter* (Chesapeake & Potomac Telephone Co.), Supplement, August 1922; *Volta Review,* XXIV (1922), 346.
494. (loneliness): Watson, *Exploring Life,* pp. 313–314; diary entry for Aug. 2, 1922, Watson MSS, Boston Public Library.
494. (thoughts): Mrs. AGB to Mrs. G. Grosvenor, Aug. 2, 1922; Mrs. AGB to W. Warren, Aug. 20, 1922; Mrs. AGB to Miss J. Timberlake, Oct. 19, 1922.
494. (Baldwin): Mrs. AGB to Mrs. G. Grosvenor, Aug. 2, 1922; Mrs. D. Fairchild to Mrs. G. Grosvenor, Aug. 2, 1922; Affidavit of Gilbert H. Grosvenor, Feb. 12, 1926.
495. (inadequate): Mrs. AGB to G. Grosvenor, May 4, 1922.
495. (Father): Memorandum by G. Grosvenor, Jan. 1, 1923.
495. (eighty-five): *Volta Review,* December 1928; *N.Y. Times,* Aug. 4, Oct. 1, 1929, July 24, 1930, Aug. 15, 1938, Apr. 22, 1940; *Washington Star,* June 11, 1930.
495. (stock): Statement of securities in Watson estate, Box 8, Watson MSS, Boston Public Library.
496. (Bhreagh): Parkin, *Bell and Baldwin,* pp. 443–525.
496. (1949): *N.Y. Times,* Oct. 25, 1949.
497. (hearing): *Volta Review,* LXXII (1970), 149–152.
497. (that): AT&T Company, *1971 Annual Report,* pp. 4, 20; U.S. Bureau of the Census, *Historical Statistics of the United States* (Washington, D.C., 1960), p. 480; Herman E. Krooss and Charles Gilbert, *American Business History* (Englewood Cliffs, N.J., 1972), p. 9.

Picture Credits

The Bell Family © National Geographic Society: pages 15 (both), 21 (right), 32, 38, 84, 85, 91 (left), 95, 105, 106, 107, 183, 192, 220, 222–223, 239, 286, 298, 302 (top), 318, 322, 326, 349, 426 (bottom), 434–435, 442, 450 (top and middle), 468 (both) and 489

Copyright © National Geographic Society: pages 178 and 360 (bottom)

Gilbert H. Grosvenor © National Geographic Society: pages 302 (middle), 458, 461, 475 and 492

H. M. Benner, The Bell Family © National Geographic Society: pages 450 (bottom) and 452

I. D. Boyce, The Bell Family © National Geographic Society: page 465

E. A. Holton, The Bell Family © National Geographic Society: page 72

Charles Martin © National Geographic Society: page 302 (bottom)

David G. McCurdy, The Bell Family © National Geographic Society: page 426 (top)

John A. D. McCurdy, The Bell Family © National Geographic Society: pages 437 and 443

Courtesy of the Volta Bureau Library, Alexander Graham Bell Association for the Deaf: pages 99 (both) and 403

Courtesy of AT&T: pages 78, 91 (right), 133 (both), 146, 194, 218, 261 and 341

Courtesy of Gallaudet University, Washington, D.C.: page 384

Courtesy of Scottish Life Assurance Co., Ltd., Edinburgh: page 21 (left)

Sumichrast Papers, Harvard University Archives: page 55

Courtesy of the Library of Congress: page 128

Courtesy of the Smithsonian Institution: pages 114 and 360 (top)

Index

of concept, 105, 108–109, 121–124, 131–132, 140, 144; diaphragm, 106, 117, 123, 129, 140, 144, 147–149, 169–170, 200, 203–205, 214–215, 243, 264, 266, 439; naming of, 118; business organization, 128–129, 188–189, 203, 210, 227–231, 243–244, 247, 258–260, 281–282, 284, 291–294, 470; first vocal sounds, 145–149, 152, 206, 482; first membrane telephone, 147–148, ill., 148; permanent magnet armature, 147, 206, 213–214, 225–226, 264–266, ill., 207; press reports, 157, 199, 201–202, 204, 207–209, 216–224, 240, 242, 246; British patent, 158–159, 162–163, 165–166, 201, 205, 210–211, 231, 238, 243–244; first U.S. patent, 158–159, 161–168, 171–176, 220–221, 339, ill., 175; development of first effective transmitters, 177–185, 187, 283, 482, ills., 179–180, 183; first intelligible sentence, 181–183, 482, ill., 183; commercial application, 181, 185, 210, 216, 218–219, 221, 228, 232, 234–236, 241, 244, 285–286; public demonstrations, 189, 201–204, 214, 217–227, 240–242; Centennial exhibit, 190–191, 193, 195–199, 209, ill., 196; outdoor lines tested, 198–203, 205–208, 214; technical development, 203–208, 211–213, 215–216, 225–226, 244–246, 282–283; first conversation, 204; foreign rights, 210–211, 216, 230, 243, 245–247; second U.S. patent, 213–214, 248, 264–265, ill., 213; experimental applications, 225, 334, 355, 369, 394; significance, 234, 241, 406, 472–473, 494, 497; in Great Britain, 240–245; in Canada, 246–247, 282; litigation, 248, 252–253, 263, 267–272, 274–277, 284, 353, 356, 377; and the hard of hearing, 255–256, 397; first directory, ill., 261; automatic switchboard, 283, 357, 361, 366; N.Y.–Chicago opening, ill., 286. *See also* AGB and the telephone; Edison, Thomas A., and telephone

Telephone Pioneers of America, 286–287
telephonic probe, 347–348, 369, 473
television, 337–338, 357, 366
tetrahedral construction: ills., 434–435; 365, 432–435, 437–447, 469, 488
tetrahedral tower: ill., 443; 442–444, 446, 493
Thompson, Charles F.: ill., 326; 297–298, 305–306, 310, 325–326, 328, 330–331, 362, 371, 387, 429, 472, 486, 490
Thomson, Lady, 197–198, 251
Thomson, Sir William (later Lord Kel-

vin), 165, 191, 193–200, 208–209, 224, 240, 243, 250–251, 363, 367, 375
thought transference, 357, 367
Titusville, Pa., 210, 228, 230
Toronto Globe, 74, 92, 157–158, 168, 202, 208
Totten, George O., 423
Trinity (in Edinburgh). *See* Milton Cottage
Trouve (dog), 56, 69
True, Mary: ill., 78; 86, 100, 256, 392
Tufts College, 214, 263–264, 278
Tutelo Heights, 201–202, 211, 234, 315, 483. *See also* Brantford, Bell home in
Twain, Mark, 87, 285, 343, 404, 477
Tyndall, John, 75–76, 93, 119, 128, 135, 334, 340
type-setting machine for the deaf, 356–357

University of Edinburgh, 9, 20, 37, 44, 477
University of London, 46, 49, 57, 61, 255
University of New Hampshire, 417

vacuum jacket: ill., 349; 316, 348–349, 352, 357, 366, 369
Vail, Theodore N., 258, 260, 262, 287, 482, 485
Van der Weyde, Philip H., 117–118, 269
Vanderbilt, William H., 271
Varley, Cromwell, 200
vibratory circuit breaker, 136, 138, 167, 195–196
Victoria, Queen, 12, 28, 43–44, 241–242, 254, 421
Visible Speech: ill., 41; evolution of, 19, 40; use and publicizing of, 42–44, 49–50, 54–61, 65, 74, 76–77, 81–83, 87–88, 102, 110–113, 129, 132, 140, 150, 158–160, 166, 186, 190–191, 194, 200–201, 204, 210–211, 251, 257, 381, 390, 421, 481, 486
"Visible Speech Pioneer," 102, 112
Volta Associates, 343, 350–351, 353–354, 440, 444–445, 447
Volta Bureau, 413–414, 419, 469, 480, 486
Volta Graphophone Company, 354
Volta Laboratory, 341–343, 345–347, 350–352, 354–355, 377, 413
Volta Prize, 340, 351, 374–375, 477
Volta Review, 392, 412, 469–470, 478–479, 496. See also *Association Review*

Wachsmann, Konrad, 444
Walcott, Charles, 312, 473